RILEM State-of-the-Art Reports

RILEM STATE-OF-THE-ART REPORTS
Volume 19

RILEM, The International Union of Laboratories and Experts in Construction Materials, Systems and Structures, founded in 1947, is a non-governmental scientific association whose goal is to contribute to progress in the construction sciences, techniques and industries, essentially by means of the communication it fosters between research and practice. RILEM's focus is on construction materials and their use in building and civil engineering structures, covering all phases of the building process from manufacture to use and recycling of materials. More information on RILEM and its previous publications can be found on www.RILEM.net.

The RILEM State-of-the-Art Reports (STAR) are produced by the Technical Committees. They represent one of the most important outputs that RILEM generates—high level scientific and engineering reports that provide cutting edge knowledge in a given field. The work of the TCs is one of RILEM's key functions.

Members of a TC are experts in their field and give their time freely to share their expertise. As a result, the broader scientific community benefits greatly from RILEM's activities.

RILEM's stated objective is to disseminate this information as widely as possible to the scientific community. RILEM therefore considers the STAR reports of its TCs as of highest importance, and encourages their publication whenever possible.

The information in this and similar reports are mostly pre-normative in the sense that it provides the underlying scientific fundamentals on which standards and codes of practice are based. Without such a solid scientific basis, construction practice will be less than efficient or economical.

It is RILEM's hope that this information will be of wide use to the scientific community.

More information about this series at http://www.springer.com/series/8780

Carlo Pellegrino · José Sena-Cruz
Editors

Design Procedures for the Use of Composites in Strengthening of Reinforced Concrete Structures

State-of-the-Art Report of the RILEM Technical Committee 234-DUC

rilem

Springer

Editors
Carlo Pellegrino
Civil, Architectural and Environmental Engineering
University of Padua
Padua
Italy

José Sena-Cruz
Civil Engineering
University of Minho
Guimarães
Portugal

ISSN 2213-204X ISSN 2213-2031 (electronic)
RILEM State-of-the-Art Reports
ISBN 978-94-024-0413-5 ISBN 978-94-017-7336-2 (eBook)
DOI 10.1007/978-94-017-7336-2

Springer Dordrecht Heidelberg New York London

Softcover re-print of the Hardcover 1st edition 2016

Printed on acid-free paper

Springer Science+Business Media B.V. Dordrecht is part of Springer Science+Business Media (www.springer.com)

Contents

RILEM Publications

The following list is presenting the global offer of RILEM Publications, sorted by series. Each publication is available in printed version and/or in online version.

RILEM Proceedings (PRO)

PRO 1: Durability of High Performance Concrete (ISBN: 2-912143-03-9); *Ed. H. Sommer*
PRO 2: Chloride Penetration into Concrete (ISBN: 2-912143-00-04); *Eds. L.-O. Nilsson and J.-P. Ollivier*
PRO 3: Evaluation and Strengthening of Existing Masonry Structures (ISBN: 2-912143-02-0); *Eds. L. Binda and C. Modena*
PRO 4: Concrete: From Material to Structure (ISBN: 2-912143-04-7); *Eds. J.-P. Bournazel and Y. Malier*
PRO 5: The Role of Admixtures in High Performance Concrete (ISBN: 2-912143-05-5); *Eds. J. G. Cabrera and R. Rivera-Villarreal*
PRO 6: High Performance Fiber Reinforced Cement Composites—HPFRCC 3 (ISBN: 2-912143-06-3); *Eds. H. W. Reinhardt and A. E. Naaman*
PRO 7: 1st International RILEM Symposium on Self-Compacting Concrete (ISBN: 2-912143-09-8); *Eds. Å. Skarendahl and Ö. Petersson*
PRO 8: International RILEM Symposium on Timber Engineering (ISBN: 2-912143-10-1); *Ed. L. Boström*
PRO 9: 2nd International RILEM Symposium on Adhesion between Polymers and Concrete ISAP '99 (ISBN: 2-912143-11-X); *Eds. Y. Ohama and M. Puterman*
PRO 10: 3rd International RILEM Symposium on Durability of Building and Construction Sealants (ISBN: 2-912143-13-6); *Ed. A. T. Wolf*
PRO 11: 4th International RILEM Conference on Reflective Cracking in Pavements (ISBN: 2-912143-14-4); *Eds. A. O. Abd El Halim, D. A. Taylor and El H. H. Mohamed*

PRO 12: International RILEM Workshop on Historic Mortars: Characteristics and Tests (ISBN: 2-912143-15-2); *Eds. P. Bartos, C. Groot and J. J. Hughes*
PRO 13: 2nd International RILEM Symposium on Hydration and Setting (ISBN: 2-912143-16-0); *Ed. A. Nonat*
PRO 14: Integrated Life-Cycle Design of Materials and Structures—ILCDES 2000 (ISBN: 951-758-408-3); (ISSN: 0356-9403); *Ed. S. Sarja*
PRO 15: Fifth RILEM Symposium on Fibre-Reinforced Concretes (FRC)—BEFIB'2000 (ISBN: 2-912143-18-7); *Eds. P. Rossi and G. Chanvillard*
PRO 16: Life Prediction and Management of Concrete Structures (ISBN: 2-912143-19-5); *Ed. D. Naus*
PRO 17: Shrinkage of Concrete—Shrinkage 2000 (ISBN: 2-912143-20-9); *Eds. V. Baroghel-Bouny and P.-C. Aïtcin*
PRO 18: Measurement and Interpretation of the On-Site Corrosion Rate (ISBN: 2-912143-21-7); *Eds. C. Andrade, C. Alonso, J. Fullea, J. Polimon and J. Rodriguez*
PRO 19: Testing and Modelling the Chloride Ingress into Concrete (ISBN: 2-912143-22-5); *Eds. C. Andrade and J. Kropp*
PRO 20: 1st International RILEM Workshop on Microbial Impacts on Building Materials (CD 02) (e-ISBN 978-2-35158-013-4); *Ed. M. Ribas Silva*
PRO 21: International RILEM Symposium on Connections between Steel and Concrete (ISBN: 2-912143-25-X); *Ed. R. Eligehausen*
PRO 22: International RILEM Symposium on Joints in Timber Structures (ISBN: 2-912143-28-4); *Eds. S. Aicher and H.-W. Reinhardt*
PRO 23: International RILEM Conference on Early Age Cracking in Cementitious Systems (ISBN: 2-912143-29-2); *Eds. K. Kovler and A. Bentur*
PRO 24: 2nd International RILEM Workshop on Frost Resistance of Concrete (ISBN: 2-912143-30-6); *Eds. M. J. Setzer, R. Auberg and H.-J. Keck*
PRO 25: International RILEM Workshop on Frost Damage in Concrete (ISBN: 2-912143-31-4); *Eds. D. J. Janssen, M. J. Setzer and M. B. Snyder*
PRO 26: International RILEM Workshop on On-Site Control and Evaluation of Masonry Structures (ISBN: 2-912143-34-9); *Eds. L. Binda and R. C. de Vekey*
PRO 27: International RILEM Symposium on Building Joint Sealants (CD03); *Ed. A. T. Wolf*
PRO 28: 6th International RILEM Symposium on Performance Testing and Evaluation of Bituminous Materials—PTEBM'03 (ISBN: 2-912143-35-7; e-ISBN: 978-2-912143-77-8); *Ed. M. N. Partl*
PRO 29: 2nd International RILEM Workshop on Life Prediction and Ageing Management of Concrete Structures (ISBN: 2-912143-36-5); *Ed. D. J. Naus*
PRO 30: 4th International RILEM Workshop on High Performance Fiber Reinforced Cement Composites—HPFRCC 4 (ISBN: 2-912143-37-3); *Eds. A. E. Naaman and H. W. Reinhardt*
PRO 31: International RILEM Workshop on Test and Design Methods for Steel Fibre Reinforced Concrete: Background and Experiences (ISBN: 2-912143-38-1); *Eds. B. Schnütgen and L. Vandewalle*

PRO 32: International Conference on Advances in Concrete and Structures 2 vol. (ISBN (set): 2-912143-41-1); *Eds. Ying-shu Yuan, Surendra P. Shah and Heng-lin Lü*
PRO 33: 3rd International Symposium on Self-Compacting Concrete (ISBN: 2-912143-42-X); *Eds. Ó. Wallevik and I. Níelsson*
PRO 34: International RILEM Conference on Microbial Impact on Building Materials (ISBN: 2-912143-43-8); *Ed. M. Ribas Silva*
PRO 35: International RILEM TC 186-ISA on Internal Sulfate Attack and Delayed Ettringite Formation (ISBN: 2-912143-44-6); *Eds. K. Scrivener and J. Skalny*
PRO 36: International RILEM Symposium on Concrete Science and Engineering—A Tribute to Arnon Bentur (ISBN: 2-912143-46-2); *Eds. K. Kovler, J. Marchand, S. Mindess and J. Weiss*
PRO 37: 5th International RILEM Conference on Cracking in Pavements—Mitigation, Risk Assessment and Prevention (ISBN: 2-912143-47-0); *Eds. C. Petit, I. Al-Qadi and A. Millien*
PRO 38: 3rd International RILEM Workshop on Testing and Modelling the Chloride Ingress into Concrete (ISBN: 2-912143-48-9); *Eds. C. Andrade and J. Kropp*
PRO 39: 6th International RILEM Symposium on Fibre-Reinforced Concretes—BEFIB 2004 (ISBN: 2-912143-51-9); *Eds. M. Di Prisco, R. Felicetti and G. A. Plizzari*
PRO 40: International RILEM Conference on the Use of Recycled Materials in Buildings and Structures (ISBN: 2-912143-52-7); *Eds. E. Vázquez, Ch. F. Hendriks and G. M. T. Janssen*
PRO 41: RILEM International Symposium on Environment-Conscious Materials and Systems for Sustainable Development (ISBN: 2-912143-55-1); *Eds. N. Kashino and Y. Ohama*
PRO 42: SCC'2005—China: 1st International Symposium on Design, Performance and Use of Self-Consolidating Concrete (ISBN: 2-912143-61-6); *Eds. Zhiwu Yu, Caijun Shi, Kamal Henri Khayat and Youjun Xie*
PRO 43: International RILEM Workshop on Bonded Concrete Overlays (e-ISBN: 2-912143-83-7); *Eds. J. L. Granju and J. Silfwerbrand*
PRO 44: 2nd International RILEM Workshop on Microbial Impacts on Building Materials (CD11) (e-ISBN: 2-912143-84-5); *Ed. M. Ribas Silva*
PRO 45: 2nd International Symposium on Nanotechnology in Construction, Bilbao (ISBN: 2-912143-87-X); *Eds. Peter J. M. Bartos, Yolanda de Miguel and Antonio Porro*
PRO 46: ConcreteLife'06—International RILEM-JCI Seminar on Concrete Durability and Service Life Planning: Curing, Crack Control, Performance in Harsh Environments (ISBN: 2-912143-89-6); *Ed. K. Kovler*
PRO 47: International RILEM Workshop on Performance Based Evaluation and Indicators for Concrete Durability (ISBN: 978-2-912143-95-2); *Eds. V. Baroghel-Bouny, C. Andrade, R. Torrent and K. Scrivener*
PRO 48: 1st International RILEM Symposium on Advances in Concrete through Science and Engineering (e-ISBN: 2-912143-92-6); *Eds. J. Weiss, K. Kovler, J. Marchand, and S. Mindess*

PRO 49: International RILEM Workshop on High Performance Fiber Reinforced Cementitious Composites in Structural Applications (ISBN: 2-912143-93-4); *Eds. G. Fischer and V.C. Li*

PRO 50: 1st International RILEM Symposium on Textile Reinforced Concrete (ISBN: 2-912143-97-7); *Eds. Josef Hegger, Wolfgang Brameshuber and Norbert Will*

PRO 51: 2nd International Symposium on Advances in Concrete through Science and Engineering (ISBN: 2-35158-003-6; e-ISBN: 2-35158-002-8); *Eds. J. Marchand, B. Bissonnette, R. Gagné, M. Jolin and F. Paradis*

PRO 52: Volume Changes of Hardening Concrete: Testing and Mitigation (ISBN: 2-35158-004-4; e-ISBN: 2-35158-005-2); *Eds. O. M. Jensen, P. Lura and K. Kovler*

PRO 53: High Performance Fiber Reinforced Cement Composites—HPFRCC5 (ISBN: 978-2-35158-046-2); *Eds. H. W. Reinhardt and A. E. Naaman*

PRO 54: 5th International RILEM Symposium on Self-Compacting Concrete (ISBN: 978-2-35158-047-9); *Eds. G. De Schutter and V. Boel*

PRO 55: International RILEM Symposium Photocatalysis, Environment and Construction Materials (ISBN: 978-2-35158-056-1); *Eds. P. Baglioni and L. Cassar*

PRO 56: International RILEM Workshop on Integral Service Life Modelling of Concrete Structures (ISBN 978-2-35158-058-5); *Eds. R. M. Ferreira, J. Gulikers and C. Andrade*

PRO 57: RILEM Workshop on Performance of cement-based materials in aggressive aqueous environments (e-ISBN: 978-2-35158-059-2); *Ed. N. De Belie*

PRO 58: International RILEM Symposium on Concrete Modelling—CONMOD'08 (ISBN: 978-2-35158-060-8); *Eds. E. Schlangen and G. De Schutter*

PRO 59: International RILEM Conference on On Site Assessment of Concrete, Masonry and Timber Structures—SACoMaTiS 2008 (ISBN set: 978-2-35158-061-5); *Eds. L. Binda, M. di Prisco and R. Felicetti*

PRO 60: Seventh RILEM International Symposium on Fibre Reinforced Concrete: Design and Applications—BEFIB 2008 (ISBN: 978-2-35158-064-6); *Ed. R. Gettu*

PRO 61: 1st International Conference on Microstructure Related Durability of Cementitious Composites 2 vol., (ISBN: 978-2-35158-065-3); *Eds. W. Sun, K. van Breugel, C. Miao, G. Ye and H. Chen*

PRO 62: NSF/ RILEM Workshop: In-situ Evaluation of Historic Wood and Masonry Structures (e-ISBN: 978-2-35158-068-4); *Eds. B. Kasal, R. Anthony and M. Drdácký*

PRO 63: Concrete in Aggressive Aqueous Environments: Performance, Testing and Modelling, 2 vol., (ISBN: 978-2-35158-071-4); *Eds. M. G. Alexander and A. Bertron*

PRO 64: Long Term Performance of Cementitious Barriers and Reinforced Concrete in Nuclear Power Plants and Waste Management—NUCPERF 2009 (ISBN: 978-2-35158-072-1); *Eds. V. L'Hostis, R. Gens, C. Gallé*

PRO 65: Design Performance and Use of Self-consolidating Concrete—SCC'2009 (ISBN: 978-2-35158-073-8); *Eds. C. Shi, Z. Yu, K. H. Khayat and P. Yan*

PRO 66: 2nd International RILEM Workshop on Concrete Durability and Service Life Planning—ConcreteLife'09 (ISBN: 978-2-35158-074-5); *Ed. K. Kovler*
PRO 67: Repairs Mortars for Historic Masonry (e-ISBN: 978-2-35158-083-7); *Ed. C. Groot*
PRO 68: Proceedings of the 3rd International RILEM Symposium on 'Rheology of Cement Suspensions such as Fresh Concrete (ISBN 978-2-35158-091-2); *Eds. O. H. Wallevik, S. Kubens and S. Oesterheld*
PRO 69: 3rd International PhD Student Workshop on 'Modelling the Durability of Reinforced Concrete' (ISBN: 978-2-35158-095-0); *Eds. R. M. Ferreira, J. Gulikers and C. Andrade*
PRO 70: 2nd International Conference on 'Service Life Design for Infrastructure' (ISBN set: 978-2-35158-096-7, e-ISBN: 978-2-35158-097-4); *Ed. K. van Breugel, G. Ye and Y. Yuan*
PRO 71: Advances in Civil Engineering Materials—The 50-year Teaching Anniversary of Prof. Sun Wei' (ISBN: 978-2-35158-098-1; e-ISBN: 978-2-35158-099-8); *Eds. C. Miao, G. Ye, and H. Chen*
PRO 72: First International Conference on 'Advances in Chemically-Activated Materials—CAM'2010' (2010), 264 pp., ISBN: 978-2-35158-101-8; e-ISBN: 978-2-35158-115-5, *Eds. Caijun Shi and Xiaodong Shen*
PRO 73: 2nd International Conference on 'Waste Engineering and Management—ICWEM 2010' (2010), 894 pp., ISBN: 978-2-35158-102-5; e-ISBN: 978-2-35158-103-2, *Eds. J. Zh. Xiao, Y. Zhang, M. S. Cheung and R. Chu*
PRO 74: International RILEM Conference on 'Use of Superabsorsorbent Polymers and Other New Addditives in Concrete' (2010) 374 pp., ISBN: 978-2-35158-104-9; e-ISBN: 978-2-35158-105-6; *Eds. O.M. Jensen, M.T. Hasholt, and S. Laustsen*
PRO 75: International Conference on 'Material Science—2nd ICTRC—Textile Reinforced Concrete—Theme 1' (2010) 436 pp., ISBN: 978-2-35158-106-3; e-ISBN: 978-2-35158-107-0; *Ed. W. Brameshuber*
PRO 76: International Conference on 'Material Science—HetMat—Modelling of Heterogeneous Materials—Theme 2' (2010) 255 pp., ISBN: 978-2-35158-108-7; e-ISBN: 978-2-35158-109-4; *Ed. W. Brameshuber*
PRO 77: International Conference on 'Material Science—AdIPoC—Additions Improving Properties of Concrete—Theme 3' (2010) 459 pp., ISBN: 978-2-35158-110-0; e-ISBN: 978-2-35158-111-7; *Ed. W. Brameshuber*
PRO 78: 2nd Historic Mortars Conference and RILEM TC 203-RHM Final Workshop—HMC2010 (2010) 1416 pp., e-ISBN: 978-2-35158-112-4; *Eds J. Válek, C. Groot, and J. J. Hughes*
PRO 79: International RILEM Conference on Advances in Construction Materials Through Science and Engineering (2011) 213 pp., e-ISBN: 978-2-35158-117-9; *Eds Christopher Leung and K.T. Wan*
PRO 80: 2nd International RILEM Conference on Concrete Spalling due to Fire Exposure (2011) 453 pp., ISBN: 978-2-35158-118-6, e-ISBN: 978-2-35158-119-3; *Eds E.A.B. Koenders and F. Dehn*
PRO 81: 2nd International RILEM Conference on Strain Hardening Cementitious Composites (SHCC2-Rio) (2011) 451 pp., ISBN: 978-2-35158-120-9, e-ISBN:

978-2-35158-121-6; *Eds R.D. Toledo Filho, F.A. Silva, E.A.B. Koenders and E.M.R. Fairbairn*

PRO 82: 2nd International RILEM Conference on Progress of Recycling in the Built Environment (2011) 507 pp., e-ISBN: 978-2-35158-122-3; *Eds V.M. John, E. Vazquez, S.C. Angulo and C. Ulsen*

PRO 83: 2nd International Conference on Microstructural-related Durability of Cementitious Composites (2012) 250 pp., ISBN: 978-2-35158-129-2; e-ISBN: 978-2-35158-123-0; *Eds G. Ye, K. van Breugel, W. Sun and C. Miao*

PRO 85: RILEM-JCI International Workshop on Crack Control of Mass Concrete and Related issues concerning Early-Age of Concrete Structures—ConCrack 3—Control of Cracking in Concrete Structures 3 (2012) 237 pp., ISBN: 978-2-35158-125-4; e-ISBN: 978-2-35158-126-1; *Eds F. Toutlemonde and J.-M. Torrenti*

PRO 86: International Symposium on Life Cycle Assessment and Construction (2012) 414 pp., ISBN: 978-2-35158-127-8, e-ISBN: 978-2-35158-128-5; *Eds A. Ventura and C. de la Roche*

PRO 87: UHPFRC 2013—RILEM-fib-AFGC International Symposium on Ultra-High Performance Fibre-Reinforced Concrete (2013), ISBN: 978-2-35158-130-8, e-ISBN: 978-2-35158-131-5; *Eds F. Toutlemonde*

PRO 88: 8th RILEM International Symposium on Fibre Reinforced Concrete (2012) 344 pp., ISBN: 978-2-35158-132-2, e-ISBN: 978-2-35158-133-9; *Eds Joaquim A.O. Barros*

PRO 89: RILEM International workshop on performance-based specification and control of concrete durability (2014) 678 pp, ISBN: 978-2-35158-135-3, e-ISBN: 978-2-35158-136-0; *Eds. D. Bjegović, H. Beushausen and M. Serdar*

PRO 90: 7th RILEM International Conference on Self-Compacting Concrete and of the 1st RILEM International Conference on Rheology and Processing of Construction Materials (2013) 396 pp, ISBN: 978-2-35158-137-7, e-ISBN: 978-2-35158-138-4, *Eds. Nicolas Roussel and Hela Bessaies-Bey*

PRO 91 draft: CONMOD 2014—RILEM International Symposium on Concrete Modelling (2014), ISBN: 978-2-35158-139-1; e-ISBN: 978-2-35158-140-7

PRO 92: CAM 2014—2nd International Conference on advances in chemically-activated materials (2014) 392 pp., ISBN: 978-2-35158-141-4; e-ISBN: 978-2-35158-142-1, *Eds. Caijun Shi and Xiadong Shen*

PRO 93: SCC 2014—3rd International Symposium on Design, Performance and Use of Self-Consolidating Concrete (2014) 438 pp., ISBN: 978-2-35158-143-8; e-ISBN: 978-2-35158-144-5, *Eds. Caijun Shi, Zhihua Ou, Kamal H. Khayat*

PRO 94 (online version): HPFRCC-7—7th RILEM conference on High performance fiber reinforced cement composites, e-ISBN: 978-2-35158-146-9, *Eds. H.W. Reinhardt, G.J. Parra-Montesinos, H. Garrecht*

PRO 95: International RILEM Conference on Application of superabsorbent polymers and other new admixtures in concrete construction, ISBN: 978-2-35158-147-6; e-ISBN: 978-2-35158-148-3, *Eds. Viktor Mechtcherine, Christof Schroefl*

PRO 96 (online version): XIII DBMC: XIII International Conference on Durability of Building Materials and Components, e-ISBN: 978-2-35158-149-0, *Eds. M. Quattrone, V.M. John*
PRO 97: SHCC3—3rd International RILEM Conference on Strain Hardening Cementitious Composites, ISBN: 978-2-35158-150-6; e-ISBN: 978-2-35158-151-3, *Eds. E. Schlangen, M.G. Sierra Beltran, M. Lukovic, G. Ye*
PRO 98: FERRO-11—11th International Symposium on Ferrocement and 3rd ICTRC—International Conference on Textile Reinforced Concrete, ISBN: 978-2-35158-152-0; e-ISBN: 978-2-35158-153-7, *Ed. W. Brameshuber*
PRO 99 (online version): ICBBM 2015—1st International Conference on Bio-Based Building Materials, e-ISBN: 978-2-35158-154-4, *Eds. S. Amziane, M. Sonebi*
PRO 100: SCC16—RILEM Self-Consolidating Concrete Conference, ISBN: 978-2-35158-156-8; e-ISBN: 978-2-35158-157-5
PRO 101 (online version): III Progress of Recycling in the Built Environment, e-ISBN: 978-2-35158-158-2, *Eds. M. Quattrone, V.M. John*

RILEM Reports (REP)

Report 19: Considerations for Use in Managing the Aging of Nuclear Power Plant Concrete Structures (ISBN: 2-912143-07-1); *Ed. D. J. Naus*
Report 20: Engineering and Transport Properties of the Interfacial Transition Zone in Cementitious Composites (ISBN: 2-912143-08-X); *Eds. M. G. Alexander, G. Arliguie, G. Ballivy, A. Bentur and J. Marchand*
Report 21: Durability of Building Sealants (ISBN: 2-912143-12-8); *Ed. A. T. Wolf*
Report 22: Sustainable Raw Materials—Construction and Demolition Waste (ISBN: 2-912143-17-9); *Eds. C. F. Hendriks and H. S. Pietersen*
Report 23: Self-Compacting Concrete state-of-the-art report (ISBN: 2-912143-23-3); *Eds. Å. Skarendahl and Ö. Petersson*
Report 24: Workability and Rheology of Fresh Concrete: Compendium of Tests (ISBN: 2-912143-32-2); *Eds. P. J. M. Bartos, M. Sonebi and A. K. Tamimi*
Report 25: Early Age Cracking in Cementitious Systems (ISBN: 2-912143-33-0); *Ed. A. Bentur*
Report 26: Towards Sustainable Roofing (Joint Committee CIB/RILEM) (CD 07) (e-ISBN 978-2-912143-65-5); *Eds. Thomas W. Hutchinson and Keith Roberts*
Report 27: Condition Assessment of Roofs (Joint Committee CIB/RILEM) (CD 08)(e-ISBN 978-2-912143-66-2); *Ed. CIB W 83/RILEM TC166-RMS*
Report 28: Final report of RILEM TC 167-COM 'Characterisation of Old Mortars with Respect to Their Repair (ISBN: 978-2-912143-56-3); *Eds. C. Groot, G. Ashall and J. Hughes*
Report 29: Pavement Performance Prediction and Evaluation (PPPE): Interlaboratory Tests(e-ISBN: 2-912143-68-3); *Eds. M. Partl and H. Piber*

Report 30: Final Report of RILEM TC 198-URM 'Use of Recycled Materials' (ISBN: 2-912143-82-9; e-ISBN: 2-912143-69-1); *Eds. Ch. F. Hendriks, G. M. T. Janssen and E. Vázquez*

Report 31: Final Report of RILEM TC 185-ATC 'Advanced testing of cement-based materials during setting and hardening' (ISBN: 2-912143-81-0; e-ISBN: 2-912143-70-5); *Eds. H. W. Reinhardt and C. U. Grosse*

Report 32: Probabilistic Assessment of Existing Structures. A JCSS publication (ISBN 2-912143-24-1); *Ed. D. Diamantidis*

Report 33: State-of-the-Art Report of RILEM Technical Committee TC 184-IFE 'Industrial Floors' (ISBN 2-35158-006-0); *Ed. P. Seidler*

Report 34: Report of RILEM Technical Committee TC 147-FMB 'Fracture mechanics applications to anchorage and bond' Tension of Reinforced Concrete Prisms—Round Robin Analysis and Tests on Bond (e-ISBN 2-912143-91-8); *Eds. L. Elfgren and K. Noghabai*

Report 35: Final Report of RILEM Technical Committee TC 188-CSC 'Casting of Self Compacting Concrete' (ISBN 2-35158-001-X; e-ISBN: 2-912143-98-5); *Eds. Å. Skarendahl and P. Billberg*

Report 36: State-of-the-Art Report of RILEM Technical Committee TC 201-TRC 'Textile Rein-forced Concrete' (ISBN 2-912143-99-3); *Ed. W. Brameshuber*

Report 37: State-of-the-Art Report of RILEM Technical Committee TC 192-ECM 'Environment-conscious construction materials and systems' (ISBN: 978-2-35158-053-0); *Eds. N. Kashino, D. Van Gemert and K. Imamoto*

Report 38: State-of-the-Art Report of RILEM Technical Committee TC 205-DSC 'Durability of Self-Compacting Concrete' (ISBN: 978-2-35158-048-6); *Eds. G. De Schutter and K. Audenaert*

Report 39: Final Report of RILEM Technical Committee TC 187-SOC 'Experimental determination of the stress-crack opening curve for concrete in tension' (ISBN 978-2-35158-049-3); *Ed. J. Planas*

Report 40: State-of-the-Art Report of RILEM Technical Committee TC 189-NEC 'Non-Destructive Evaluation of the Penetrability and Thickness of the Concrete Cover' (ISBN 978-2-35158-054-7); *Eds. R. Torrent and L. Fernández Luco*

Report 41: State-of-the-Art Report of RILEM Technical Committee TC 196-ICC 'Internal Curing of Concrete' (ISBN 978-2-35158-009-7); *Eds. K. Kovler and O. M. Jensen*

Report 42: 'Acoustic Emission and Related Non-destructive Evaluation Techniques for Crack Detection and Damage Evaluation in Concrete'—Final Report of RILEM Technical Committee 212-ACD (e-ISBN: 978-2-35158-100-1); *Ed. M. Ohtsu*

Chapter 1
Introduction

Carlo Pellegrino and José Sena-Cruz

Abstract This book analyses the current knowledge on structural behaviour of RC elements and structures strengthened with composite materials (experimental, analytical and numerical approaches for EBR and NSM), and the comparison of the predictions of the current available codes/recommendations/guidelines with selected experimental results. The book shows possible critical issues (discrepancies, lacunae, relevant parameters, test procedures, etc.) related to current code predictions or to evaluate their reliability, in order to develop more uniform methods and basic rules for design and control of FRP-strengthened RC structures. General problems/critical issues are clarified on the basis of the actual experiences, detect discrepancies in existing codes, lacunae in knowledge and, concerning these identified subjects, provide proposals for improvements. The book will help to contribute in promoting and consolidating a more qualified and conscious approach towards rehabilitation and strengthening existing RC structures with composites and their possible monitoring.

Keywords FRP · FRCM · Composites · Strengthening · Reinforced concrete

Introduction

Strengthening and retrofitting of existing structures have been widely discussed topics for the last few decades. A great number of existing structures need rehabilitation or strengthening because of improper design or construction, change of the design loads, damage caused by environmental and/or human factors, seismic events, etc. Several different systems have been developed and used to strengthen

C. Pellegrino (✉)
University of Padua, Padua, Italy
e-mail: carlo.pellegrino@unipd.it

J. Sena-Cruz
University of Minho, Guimarães, Portugal

C. Pellegrino and J. Sena-Cruz (eds.), *Design Procedures for the Use of Composites in Strengthening of Reinforced Concrete Structures*, RILEM State-of-the-Art Reports 19, DOI 10.1007/978-94-017-7336-2_1

existing structures. They include replacing structural members, adding new material to improve their performance, modifying the restraint conditions, introducing post-tension, etc. These techniques have been proven to be effective, but in some cases they can be expensive and difficult to apply. The use of fibre reinforced composites applied to existing structural elements may represent a cost-effective alternative to such traditional strengthening techniques. Among fibre reinforced composites, strengthening by means of fibre reinforced polymers (FRP) has gained great popularity because of its high mechanical properties and relatively low cost. FRP composites are comprised of high strength fibres (e.g. carbon, glass, aramid) applied to the element surface through thermosetting organic matrices, usually epoxy resin. FRP composites can be externally bonded (ER) to the element surface or placed within groove carved into the element and filled with organic matrices (Near Surface Mounted technique, NSM). The fibres are meant to carry the tensile forces, whereas the matrix transfers the stress to the concrete support. They are easy to install, have a high strength-to-weight ratio, and have suitable mechanical properties. The structural behaviour of FRP composites applied to reinforced concrete (RC) elements and structures has been widely studied over the last decades and these studies have resulted in some design guidelines. American ACI 440.2R-08 (2008), European fib T.G. 9.3 (2001), and Italian CNR-DT 200 R1 (2013) are examples of such guidelines. Although a large number of experimental, analytical, and numerical studies regarding FRP composites are available in the literature, the predictions provided by guidelines and analytical models are sometimes contrasting and disagreeing with the experimental results. For this reason, the scientific community is still discussing about some important issues and new improved guidelines are under preparation all around the world. It has been observed that the use of organic resins, though effective, may represent an issue for the durability of the intervention. Indeed, organic matrices degrade when exposed to UV radiation and loose most of their mechanical properties when subjected to temperatures close (or higher) to their glass transition temperature. Promising newly-developed types of matrix that potentially represent a valid alternative to organic resins are the so-called inorganic matrices. Within the broad category of inorganic matrices, polymer-modified cement-based mortars have raised some interest in recent years. Composite materials that employ modified cement-based mortars are usually referred to as fiber-reinforced cementitious matrix (FRCM) composites. Although several works about FRP strengthening are available in the literature very few studies can be found regarding FRCM composites.

The international Rilem Technical Committee 234-DUC was created in 2009 with the aim of facing the issues connected with the use of fibre reinforced composites to strengthen RC elements and structures. This committee is composed by a team of experts representing most of the main international institutions working on the subject. Members become from international academic and research institutions, other Rilem technical committees working on reinforced concrete and composites, standardization groups, and national and international groups who have contributed to the development of the current codes/recommendations/guidelines. The committee includes members from Italy, Cyprus, United States, England, Portugal,

China, Germany, Switzerland, Sweden, Spain, and other countries. Through the fruitful collaboration between members, promoted by the annual meetings, the committee could analyse many different aspects of the FRP strengthening technique and give insights for improvement and development of the existing analytical models. Furthermore, newly-developed promising fibre reinforced inorganic composites (FRCM) were analysed as well.

This book collect the results of 4 years of work by the Technical Committee 234-DUC. Chapter 2 provides the outcomes of a statistical analysis based on the indication of EN1990 and extended to the case of EBR FRP system. This chapter shows a procedure to evaluate the statistical parameters of the capacity models and to evaluate its characteristic values, which is the aim for application in design. Furthermore, some applications are reported to prove the feasibility of the proposed procedure. Chapter 3 provides a deep analysis of the bond behavior of FRP composites externally applied onto RC structures. The bond is described through a fracture mechanics approach and the theoretical models are compared with the experimental results available in the literature. An assessment of the most important analytical models for the estimation of the bond strength of FRP-concrete joints is provided as well. Finally, the chapter faces the critical issues of models and experimental procedures employed to investigate the FRP bond behavior. In Chap. 4 shear strengthening by means of EBR FRP is analyzed. The main analytical formulation for the evaluation of the shear strength of shear strengthened RC elements are recalled and new improved formulations are provided. The models analyzed are assessed through a wide experimental database to evaluate their accuracy with respect to the experimental evidences. Chapter 5 describes the use of FRP jackets for confining RC members mainly subjected to axial loading. The current formulations for the evaluation of the effective ultimate strain in the FRP are provided and discussed. The influence of the internal steel on the EBR FRP jackets and particularly the effects of possible bar buckling are discussed as well. Finally, experimental results available in the literature are compared with analytical provision to assess the accuracy of the proposed models. Chapter 6 gives an overview on the state-of-the-art about verifications of reinforced concrete structures using Externally Bonded (EB) Fibre Reinforced Polymers (FRP) under serviceability loading conditions (long term behaviour, durability under severe conditions), fatigue load, and fire and high temperature. Furthermore, the use of anchoring systems and mechanically fastened system is described. Chapter 7 describes the use of prestressed EBR FRP system. The chapter provides information regarding commercially available prestressing systems and their anchorage procedures. The newly-developed technique of "gradient anchorage" and various current prototypes at the laboratory-scale level are shown as well. Chapter 8 gives an overview on the state-of-the-art of the NSM technique for structural retrofitting of reinforced concrete structures using FRP composites. The chapter firstly describes the technique and addresses the existing knowledge on the bond behaviour. Furthermore, two formulations for predicting the NSM shear carrying capacity are provided. Finally, the needs for future research on this topic are identified. Chapter 9 describes the use of FRCM composites for strengthening existing RC and masonry structures. After introducing the commercially available composites, the chapter gives a state-of-the-art about the test methods under development for characterizing this materials and provides a fracture mechanics approach that allows for describing FRCM-concrete joints bond behaviour. Finally, the effectiveness of the FRCM technique for flexural strengthening and confinement of RC elements is shown and some experimental campaigns are described.

Chapter 2
Design by Testing and Statistical Determination of Capacity Models

Giorgio Monti, Antonio Bilotta, Annalisa Napoli, Emidio Nigro, Floriana Petrone and Roberto Realfonzo

Abstract In this chapter, the procedure proposed in EN1990 is adopted and extended to the case of EBR FRP systems, with the aim of attaining a uniform reliability level among all equations developed in this technical report. This approach will allow comparing experimental results and theoretical predictions in a consistent manner, and also identifying possible sources of error in the formulations. Any capacity model should be developed on the basis of theoretical considerations and subsequently fine-tuned through a regression analysis based on tests results. The validity of the model should then be checked by means of a statistical interpretation of all available test data. The formulation should include in the theoretical model a new variable that represents the model error. This variable is assumed to be normally distributed whit unit mean and standard deviation to be evaluated from comparison with experimental results. Once the statistical parameters of the model error are known, it is possible to define the statistical parameters of the capacity model and to evaluate its characteristic value, which is the aim for application in design. Some applications are shown to prove the feasibility of the proposed procedure.

Introduction

When developing a design equation, the predictive ability of the analytical capacity model, regardless of how it has been obtained, whether through a mechanics-based approach or through a regression from test data, must be validated over a reasonably large(r) set of experimental data. Thus, the definition of a reliable capacity

G. Monti (✉) · F. Petrone
Sapienza University of Rome, Rome, Italy
e-mail: giorgio.monti@uniroma1.it

A. Bilotta · E. Nigro
University of Naples Federico II, Naples, Italy

A. Napoli · R. Realfonzo
University of Salerno, Salerno, Italy

C. Pellegrino and J. Sena-Cruz (eds.), *Design Procedures for the Use of Composites in Strengthening of Reinforced Concrete Structures*, RILEM State-of-the-Art Reports 19, DOI 10.1007/978-94-017-7336-2_2

model to be used in practical design applications requires to follow a rigorous procedure that eventually will aim at calibrating the safety factor to apply to the equation so that it meets an assigned reliability target. This procedure requires the model to be formulated in a probabilistic way, so that both inherent and epistemic uncertainties of the underlying basic variables (geometry and materials, essentially) can be dealt with, as well as the uncertainties associated to the capacity equation itself. All of these uncertainties can be easily incorporated into a model, through the adoption of a random variable that represents the difference between actual and predicted response.

Probabilistic Capacity Models: Analytical Definition

As widely illustrated by Monti et al. (2009) (and also by Monti and Petrone 2014 who extended the procedure to the case of additive uncertainties), a capacity model should be defined as:

$$C\{f; a\} = R\{f; a\} \cdot \delta\{X_i\} \tag{2.1}$$

where: R is a function that "explains" the resisting mechanism capacity, given certain mechanical properties f and geometrical properties a, and represents the "deterministic" part of the capacity model; $\delta\{X_i\}$ is a random variable containing information about the overall model error prediction with respect to experimental results, and represents the "random" part of the capacity model. It is characterized by mean $\mu_\delta = b$ and variance σ_δ^2 and is a function of all the X_i parameters characterized by uncertainties. In the following it will be assumed: $\{X\}_i = \{f; a; m\}$, where f, a and m are parameters related to materials, geometry and model, respectively.

In common practice, only the deterministic part of the capacity model is considered, usually given as:

$$C_{\text{det}}\{f; a\} = b \cdot R\{f; a\} \tag{2.2}$$

Many interpret b as a model fine-tuning coefficient (sometimes called "ignorance" coefficient), so that, when the "deterministic" (mean) part of the capacity model is used, see Eq. (2.3), it predicts the experimental capacity with zero error "in the average". The coefficient b is computed through a least-square approach, by minimizing "in the average" the difference between predicted and experimental values. Pictorially speaking, the coefficient b brings the "cloud" of "theoretical-experimental" points closer to the bisectrix of the Cartesian plan.

Therefore, once the functional form $R\{f; a\}$ is found, in order to completely describe the random variable δ, we perform n experimental tests T, by "appropriately" changing the values of f and a. So, by selecting n sets of values $\{f_{\text{exp},i}; a_{\text{exp},i}\}$, we obtain n experimental values of the capacity, expressed as:

$$C = T\{f_{\exp,i}; a_{\exp,i}\} \cdot \tau \quad (i = 1 \ldots n) \tag{2.3}$$

where τ represents a random error, with unit mean and assigned variance, the latter being due to test setup imprecisions, load application modality, and measurement errors that may affect the results. Usually, this term is disregarded (in the sense that the measured values $C_{\exp,i}$ already include it).

The corresponding values predicted by the "deterministic" model are:

$$C_{\det,i}\{f_{\exp,i}; a_{\exp,i}\} = b \cdot R\{f_{\exp,i}; a_{\exp,i}\} \tag{2.4}$$

By comparison of n experimental and theoretical values, b is found from the well-known minimization:

$$\min \sum_{i=1}^{n} \left(\frac{C_{\exp,i}}{C_{\det,i}} - 1 \right)^2 \rightarrow b \tag{2.5}$$

Equation (2.7) can be analytically developed to finally obtain:

$$b = \frac{\sum_{i=1}^{n} \left(\frac{C_{\exp,i}}{R\{f_{\exp,i}; a_{\exp,i}\}} \right)^2}{\sum_{i=1}^{n} \frac{C_{\exp,i}}{R\{f_{\exp,i}; a_{\exp,i}\}}} \tag{2.6}$$

The variance of δ is obtained as well, as:

$$\sigma_{\delta}^2 = Var\left[\frac{C_{\exp}}{C_{\det}} \right] \tag{2.7}$$

with its unbiased estimate being (after b has been fine-tuned):

$$s_{\delta}^2 = \frac{\sum_{i=1}^{n} \left(\frac{C_{\exp,i}}{b \cdot R\{f_{\exp,i}; a_{\exp,i}\}} - 1 \right)^2}{n - 2} \tag{2.8}$$

By replacing Eq. (2.2) into Eq. (2.1), the following relation holds:

$$C\{f; a\} = C_{\det}\{f; a\} \frac{\delta\{X_i\}}{b} \tag{2.9}$$

Letting:

$$\bar{\delta}\{X_i\} = \frac{\delta\{X_i\}}{b} \tag{2.10}$$

$\overline{\delta}\{X_i\}$ represents the model error to be applied to the fine-tuned deterministic capacity model. It is characterized by mean $\mu_{\delta} = 1$ and variance:

$$\sigma_{\overline{\delta}}^2 = \frac{1}{b^2} Var\left[\frac{C_{\exp}}{C_{\det}}\right] \tag{2.11}$$

with its estimate being (after b has been fine-tuned):

$$s_{\overline{\delta}}^2 = \frac{\sum_{i=1}^{n}\left(\frac{C_{\exp,i}\{f_{\exp,i};a_{\exp,i}\}}{b\cdot R\{f_{\exp,i};a_{\exp,i}\}} - 1\right)^2}{b^2(n-2)} \tag{2.12}$$

Calibration of Partial Safety Factors Based on Testing

Once the random part has been "adjusted" by means of experiments, a common, though wrong, further step is to use only the deterministic part to predict the capacity, with the assumption that its characteristic and design values can be obtained by plugging in the argument, respectively, characteristic and design values. Worse usual mistakes regard an arbitrary reduction of b, to obtain a "safer" estimate. In the following, a rigorous and effective procedure is instead developed, where the "modeling" part is clearly distinguished from the "safety" part.

The probabilistic capacity model so developed has first-order-approximation mean and variance given by, respectively:

$$\mu_C = b \cdot R\{\mu_f; \mu_a\} \cdot \mu_{\overline{\delta}} = b \cdot R\{\mu_f; \mu_a\} \tag{2.13}$$

$$\begin{aligned}\sigma_C^2 &= C_{,f}^2\{\mu_f; \mu_a\} \cdot \sigma_f^2 + C_{,a}^2\{\mu_f; \mu_a\} \cdot \sigma_a^2 + C_{,\overline{\delta}}^2\{\mu_f; \mu_a\} \cdot \sigma_{\overline{\delta}}^2 \\ &= b^2 \cdot R_{,f}^2\{\mu_f; \mu_a\} \cdot \sigma_f^2 + b^2 \cdot R_{,a}^2\{\mu_f; \mu_a\} \cdot \sigma_a^2 + b^2 \cdot R^2\{\mu_f; \mu_a\} \cdot \sigma_{\overline{\delta}}^2\end{aligned} \tag{2.14}$$

where $C_{,f} = \partial C/\partial f$, $C_{,a} = \partial C/\partial a$ and $C_{,\overline{\delta}} = \partial C/\partial\overline{\delta}$ are the partial derivatives of the function C with respect to f, a, and $\overline{\delta}$, respectively, and $R_{,f} = \partial R/\partial f$, $R_{,a} = \partial R/\partial a$ and $R_{,\overline{\delta}} = \partial R/\partial\overline{\delta}$ are the partial derivatives of the function R with respect to f, a, and $\overline{\delta}$, respectively. Also, note that all variables have been assumed as statistically independent, so that all covariance are zero.

In Eq. (2.14), the first term of the second member represents the intrinsic (material) uncertainty, the second term represents the parametric (geometry) variability, and the third term represents the epistemic (model) uncertainty.

Case of limited number of tests

When predicting a limited number of tests n, we can only obtain estimates of mean and variance of the capacity model, as follows:

$$C\{\bar{f};\bar{a}\} = b \cdot R\{\bar{f};\bar{a}\} \tag{2.15}$$

$$\begin{aligned} & s_C^2\{\bar{f};\bar{a}\} \\ & = b^2 \cdot R_{,f}^2\{\bar{f};\bar{a}\} \cdot s_f^2 + b^2 \cdot R_{,a}^2\{\bar{f};\bar{a}\} \cdot s_a^2 + b^2 \cdot R^2\{\bar{f};\bar{a}\} \cdot s_{\bar{\delta}}^2 \end{aligned} \tag{2.16}$$

where:

$$\bar{f} = \frac{\sum_{i=1}^{n} f_i}{n}; \quad \bar{a} = \frac{\sum_{i=1}^{n} a_i}{n} \tag{2.17}$$

$$s_f^2 = \frac{\sum_{i=1}^{n} (f_i - \bar{f})^2}{n-1}; \quad s_a^2 = \frac{\sum_{i=1}^{n} (a_i - \bar{a})^2}{n-1} \tag{2.18}$$

and where $R_{,f} = \partial R/\partial f$, $R_{,a} = \partial R/\partial a$ and $R_{,\bar{\delta}} = \partial R/\partial \bar{\delta}$ are the partial derivatives of the function R with respect to f, a, and $\bar{\delta}$, respectively.

Under the hypothesis that C is normally distributed, when we have a limited number of tests, the characteristic value of the capacity model has a non-central t-Student distribution with n–1 degrees of freedom and with non-centrality parameter equal to $u_\alpha \sqrt{n}$ (notice that, when looking for the characteristic values in a normal distribution, which is the 5 % fractile, we have: $u_\alpha = 1.645$). An unbiased estimate (mean) of the characteristic value is (see e.g., Madsen et al. 1986):

$$C_k\{\bar{f};\bar{a}\} = C\{\bar{f};\bar{a}\} - k_{\alpha,n} \cdot s_C\{\bar{f};\bar{a}\} \tag{2.19}$$

where $k_{\alpha,n}$ is:

$$k_{\alpha,n} = u_\alpha \varepsilon_n = u_\alpha \sqrt{\frac{n-1}{2} \frac{\Gamma\left(\frac{n-1}{2}\right)}{\Gamma\left(\frac{n}{2}\right)}} \tag{2.20}$$

with Γ the Gamma function. An excellent approximation to the above equation is here proposed as:

$$k_{\alpha,n} = u_\alpha \varepsilon_n = u_\alpha \frac{4n-5}{4n-6} = u_\alpha \frac{n-1.25}{n-1.50} \tag{2.21}$$

For the purpose of the following developments, it is expedient to rewrite Eq. (2.19) as:

$$C_k\{\bar{f};\bar{a}\} = C\{\bar{f};\bar{a}\} - u_\alpha \varepsilon_n \cdot s_C\{\bar{f};\bar{a}\} \tag{2.22}$$

where ε_n becomes a sort of “scaling” coefficient of the capacity axis, easily determined for practical purposes as function of the number n of tests performed as:

$$\varepsilon_n = \frac{n - 1.25}{n - 1.50} \tag{2.23}$$

Having determined the characteristic value of the capacity model, the next step is the determination of its design value. This is given as:

$$C_d\{\bar{f};\bar{a}\} = \frac{C_k\{\bar{f};\bar{a}\}}{\gamma_C} = \frac{C\{\bar{f};\bar{a}\} - u_\alpha \varepsilon_n \cdot s_C\{\bar{f};\bar{a}\}}{\gamma_C} \tag{2.24}$$

where the safety factor γ_C is to be calibrated by considering that the design value is found as:

$$C_d\{\bar{f};\bar{a}\} = C\{\bar{f};\bar{a}\} - \beta_{LS}\alpha_C \varepsilon_n \cdot s_C\{\bar{f};\bar{a}\} \tag{2.25}$$

where it should be noted that the axis has been scaled by means of the same coefficient ε_n to account for the limited amount of tests; in the above equation, β_{LS} is the safety index associated to the acceptable exceeding probability of the considered Limit State in a given time period, and α_C is the FORM sensitivity coefficient associated to capacity variables.

Thus, the safety factor is found as:

$$\gamma_C = \frac{C_k\{\bar{f};\bar{a}\}}{C_d\{\bar{f};\bar{a}\}} = \frac{C\{\bar{f};\bar{a}\} - u_\alpha \varepsilon_n \cdot s_C\{\bar{f};\bar{a}\}}{C\{\bar{f};\bar{a}\} - \beta_{LS}\alpha_C \varepsilon_n \cdot s_C\{\bar{f};\bar{a}\}} \tag{2.26}$$

This is the factor that, once applied to the characteristic value $C_k\{\bar{f};\bar{a}\}$, gives the design value of the capacity model, $C_d\{\bar{f};\bar{a}\}$. Notice that there is an explicit dependence of the safety factor on the number of tests performed.

The above procedure needs now to be applied to the format adopted in Eurocode 0 (EN 1990) for all capacity equations, which reads as follows:

$$C_{d,EC0}\{f_d;\bar{a}\} = \frac{1}{\gamma_{Rd}} b \cdot R\{f_d;\bar{a}\} = \frac{1}{\gamma_{Rd}} b \cdot R\left\{\frac{f_k}{\gamma_m};\bar{a}\right\} \tag{2.27}$$

According to the Eurocode philosophy, safety factors are divided into "internal" ones, such as the different γ_m's, applied to material properties, and an "external" one, such as γ_{Rd}, applied to the capacity model. The former are meant to cover the intrinsic uncertainties in the material properties, while the latter deals with epistemic uncertainties related to the model. This format allows calibrating γ_{Rd} separately from the γ_m's, which may as well be taken as those already given in the code.

Thus, we should impose:

$$C_{d,EC0}\{f_d;\bar{a}\} = C_d\{\bar{f};\bar{a}\} \tag{2.28}$$

$$\frac{1}{\gamma_{Rd}} b \cdot R\left\{\frac{f_k}{\gamma_m};\bar{a}\right\} = C\{\bar{f};\bar{a}\} - \beta_{LS}\alpha_C\varepsilon_n \cdot s_C\{\bar{f};\bar{a}\} \tag{2.29}$$

$$\begin{aligned}&\frac{1}{\gamma_{Rd}} R\left\{\frac{f_k}{\gamma_m};\bar{a}\right\}\\&= R\{\bar{f};\bar{a}\} - \beta_{LS}\alpha_C\varepsilon_n\sqrt{R^2_{,f}\{\bar{f};\bar{a}\}\cdot s_f^2 + R^2_{,a}\{\bar{f};\bar{a}\}\cdot s_a^2 + R^2\{\bar{f};\bar{a}\}\cdot s_\delta^2}\end{aligned} \tag{2.30}$$

Finally, the sought general expression for the "external" safety factor accounting for the number of tests performed is found as:

$$\gamma_{Rd} = \frac{R\left\{\frac{f_k}{\gamma_m};\bar{a}\right\}}{R\{\bar{f};\bar{a}\} - \beta_{LS}\alpha_C\varepsilon_n\sqrt{R^2_{,f}\{\bar{f};\bar{a}\}\cdot s_f^2 + R^2_{,a}\{\bar{f};\bar{a}\}\cdot s_a^2 + R^2\{\bar{f};\bar{a}\}\cdot s_\delta^2}} \tag{2.31}$$

Notice that the safety factor depends on the number of tests performed through: ε_n, $\bar{f}$, $\bar{a}$, s_f^2, s_a^2, and s_δ^2, where the latter also contains the "ignorance" coefficient b, which represents a measure of the prediction capability of the capacity equation.

The above equation can be used to calibrate the "external" safety factor of any capacity equation, with a known functional form $R\{\cdot\}$, and with the "internal" safety factors already provided by the relevant code.

Also notice that, when dealing with quality-controlled materials, such as steel, one may replace the sample estimates $\bar{f}$ and s_f^2 with the corresponding population parameters μ_f and σ_f^2, so that the characteristic value of the material property may also be found as: $f_k = \mu_f - 1.645\sigma_f$. In this case, by knowing the coefficient of variation V_f, mean and variance can be easily found from the characteristic value as:

$$\mu_f = \frac{f_k}{1 - 1.645V_f} \tag{2.32}$$

$$\sigma_f^2 = \left(\frac{V_f f_k}{1 - 1.645V_f}\right)^2 \tag{2.33}$$

The safety factor then becomes:

$$\gamma_{Rd} = \frac{R\left\{\frac{f_k}{\gamma_m};\bar{a}\right\}}{R\{\mu_f;\bar{a}\} - \beta_{LS}\alpha_C\varepsilon_n\sqrt{R^2_{,f}\{\mu_f;\bar{a}\}\cdot \sigma_f^2 + R^2_{,a}\{\mu_f;\bar{a}\}\cdot s_a^2 + R^2\{\mu_f;\bar{a}\}\cdot s_\delta^2}} \tag{2.34}$$

Therefore, the design capacity is:

$$R_d\left\{\frac{f_k}{\gamma_m};\bar{a}\right\} = \frac{1}{\gamma_{Rd}}b \cdot R\left\{\frac{f_k}{\gamma_m};\bar{a}\right\} \quad (2.35)$$

Application 1—End Debonding

In this section, an application of the design by testing procedure is shown for the assessment of a design formulation to predict the end debonding load in Reinforced Concrete (RC) members strengthened with FRP Externally Bonded Reinforcement (EBR). Indeed, the high performances of FRP materials often cannot be properly exploited, since a typical failure is the debonding of the external reinforcement, namely the loss of bond at the concrete/FRP interface. This makes the bond strength at the interface a key issue in the strengthening design procedure. Usually debonding occurs within a thin layer of concrete and is related to its very low strength.

Several theoretical formulations have been proposed by researchers and international codes to predict the maximum stress in the FRP reinforcement when the end (Chen and Teng 2001; fib bulletin 2001; CNR-DT 200 2004; Smith and Teng 2002) or the intermediate debonding (Teng et al. 2003) occurs. Most of these formulations, characterized by similar structures, are calibrated by numerical factors based on experimental results.

Even though the assessment of models for bond strength has been widely dealt with by various researchers, the definition of safety factors to calculate design values is still an open item. Thus, detailed statistical analyses have been performed using a wide experimental database of bond tests in order to calibrate a bond strength model based on the fracture energy approach. The final proposed strength model is similar to other well-known models suggested in the literature and codes, but it is based on a detailed and consistent statistical analysis according to the 'design by testing' procedure suggested in the Eurocode 0 (Monti et al. 2009; Bilotta et al. 2011a; Monti and Petrone 2014). Different corrective factors allow both mean and characteristic values of debonding load to be predicted in order to follow a limit state design approach and associate a structural safety to the chosen model.

The approach to calculate the bond strength based on the fracture energy at the FRP-to-concrete interface has been summarized.

In order to develop statistical analyses, the experimental debonding loads of several bond tests have been collected and compared with three well-known relationships providing the end-debonding load in order to assess their reliability. Then, the same data have been used to assess a new relationship for the end-debonding-load according to the 'design by testing' procedure. In particular, numerical factors for both mean and percentiles provisions have been calibrated in order to furnish design provisions.

Moreover, the preformed and cured in situ EBR FRP systems have been distinguished to better exploit the performance of the latter ones.

Theoretical Formulations of Debonding Load

The maximum tensile force, F_{max}, at debonding in an FRP external reinforcement characterized by an infinite bonded length can be calculated as:

$$F_{\max} = b_f \int_0^{\infty} \tau_b(x)dx \tag{2.36}$$

being $\tau_b(x)$ the bond shear stress distribution along the concrete-FRP interface and b_f the width of the FRP reinforcement.

Moreover, the fracture energy corresponding to a generic bond shear stress-slip law, $\tau_b(s)$, can be expressed as:

$$\Gamma_F = \int_0^{\infty} \tau_b(s)ds\,[\mathrm{F/L}] \tag{2.37}$$

This expression has the meaning of energy [FL] for unit surface [L^2].

Moreover, under the hypothesis that the concrete member has a stiffness much larger than the reinforcement, at the section in which the maximum stress, $\sigma_{f,max}$, is applied, the following relationship can be written:

$$\int_{A_f} \frac{1}{2} \cdot \sigma_f \cdot \varepsilon_f dA = b_f \cdot \int_0^{\infty} \tau_b(s)ds \tag{2.38}$$

This expression assumes the equality of the energy [F L] for unit length [L] associated to the tensile stress at the FRP section (area $A_f = b_f\, t_f$) with the fracture energy [F L] for unit length [L] developed at the FRP-concrete interface. Furthermore, assuming constant stresses along the FRP reinforcement section and a linear-elastic stress-strain relationship, Eq. (2.38) can be written as follows:

$$\frac{\sigma_f^2}{2 \cdot E_f} \cdot t_f \cdot b_f = b_f \cdot \Gamma_F \qquad \frac{\left(\sigma_f \cdot t_f \cdot b_f\right)^2}{2 \cdot E_f} \cdot = b_f^2 \cdot t_f \cdot \Gamma_F \tag{2.39}$$

that gives the expression:

$$F_{\max} = b_f \cdot \sqrt{2 \cdot E_f \cdot t_f \cdot \Gamma_F} \tag{2.40}$$

where t_f, b_f, E_f are the thickness, the width, and the Young's modulus of the FRP reinforcement.

The fracture energy, Γ_F, depends on both the strength properties of adherents, concrete, and adhesive, and the characteristics of the concrete surface. If the FRP reinforcement is correctly applied, the debonding occurs in the concrete and the specific fracture energy of the interface law can be written in a form similar to that used for the shear fracture (Mode I). Therefore, the fracture energy can be expressed as a function of the concrete shear strength: $\Gamma_F(\tau_{b,max})$, where $\tau_{b,max}$ depends on both tensile and compressive concrete strength.

In most formulations, the fracture energy depends directly on the concrete tensile strength and on a shape factor that is function of the FRP-to-concrete width ratio (b_f / b_c). The formulations proposed by Neubauer and Rostasy (1997) and Lu et al. (2005), e.g., are:

$$G_f = 0.204 \cdot k_b^2 \cdot f_{ctm} \tag{2.41}$$

$$G_f = 0.308 \cdot \beta_w^2 \cdot \sqrt{f_{ctm}} \tag{2.42}$$

being f_{ctm} the mean tensile strength of concrete, k_b and β_w the shape factors defined as:

$$k_b = 1.06\sqrt{\frac{2 - b_f/b_c}{1 + b_f/400}}; \quad \beta_w = \sqrt{\frac{2 - b_f/b_c}{1 + b_f/b_c}} \tag{2.43}$$

Based on formulations of fracture energy similar to Eqs. (2.41) and (2.42) and on experimental results of bond tests, several theoretical formulations to evaluate the bond strength have been proposed in the past (Taljsten 1994; Neubauer and Rostasy 1997; Brosens and Van Gemert 1997; fib bulletin 2001; Chen and Teng 2001; Smith and Teng 2002; CNR-DT 200 2004). These expressions allow for predicting the end debonding load. In some cases, the same expressions are suitably modified by changing some factors in order to predict the intermediate crack debonding load in RC beams (Teng et al. 2003; Chen et al. 2006; CNR-DT 200 2004). The lay-out of these formulations is often similar, while the numerical coefficients calibrated on experimental results are different. Moreover, the safety factors, which are needed in order to calculate design provisions as part of the Limit State approach, are not always considered. This last point is an important issue, if a safety level (mean, characteristic or design) has to be associated to the provision.

The theoretical approaches suggested by fib bulletin 14 (2001), Chen and Teng (2001), CNR-DT 200 (2004) are considered. In particular, the bond strength expressed in terms of maximum tensile load in the FRP reinforcement, $N_{f,max}$, and the effective length, L_e, which is the minimum length required to full transfer the load, are defined as follows by the three approaches:

(1) fib bulletin 14 (2001):

$$N_{f,\max} = \alpha \cdot c_1 \cdot k_c \cdot k_b \cdot b_f \cdot \beta_L \cdot \sqrt{E_f \cdot t_f \cdot f_{ctm}}; \quad L_e = \sqrt{\frac{E_f \cdot t_f}{2f_{ctm}}}$$
$$\beta_L = \frac{L_b}{L_e} \cdot \left(2 - \frac{L_b}{L_e}\right) \; if\; L_b \le L_e, \quad \beta_L = 1 \; otherwise; \; \frac{b_f}{b} \ge 0.33 \tag{2.44}$$

where b_f, t_f, E_f, L_b are width, thickness, Young's modulus and bonded length of the FRP reinforcement, b_c is the width of the concrete element, f_{ctm} is the mean tensile strength of concrete, $c_1 = 0.64$ and $c_2 = 2$ are coefficients related to an experimental calibration of the fracture energy (Neubauer and Rostasy 1997), $\alpha = 0.9$ is a reduction factor to account for the influence of inclined cracks on the bond strength, and k_c takes into account the state of compaction of concrete and usually is assumed equal to 1.00, or 0.67 for FRP bonded to concrete faces with low compaction. Finally, the shape factor k_b is given by Eq. (2.43).

(2) Chen and Teng (2001):

$$N_{f,\max} = \alpha \cdot \beta_w \cdot \beta_L \cdot b_f \cdot L_e \cdot \sqrt{f'_c}; \quad L_e = \sqrt{\frac{E_f \cdot t_f}{\sqrt{f'_c}}}$$
$$\beta_L = \sin\frac{\pi L_b}{2L_e} \; if\; L_b \le L_e, \quad \beta_L = 1 \; otherwise \tag{2.45}$$

f'_c being the mean cylindrical compressive strength of concrete and α a coefficient equal to 0.427 or 0.315 to calculate a mean or a design provision, respectively. The shape factor β_w is given by Eq. (2.43). Note that the debonding strain values of Eq. (2.45) should be divided by an appropriate safety factor $\gamma_b = 1.25$ for design purpose, according to suggestion in Sect. 3.4 of Teng et al. (2001).

(3) CNR-DT 200 (2004):

$$N_{f,\max} = \frac{1}{\gamma_{f,d}\sqrt{\gamma_c}} \cdot \beta_L \cdot b_f \cdot \sqrt{k_G \cdot k_b} \cdot \sqrt{2 \cdot E_f \cdot t_f \cdot \sqrt{f_{ck} \cdot f_{ctm}}};$$
$$L_e = \sqrt{\frac{E_f \cdot t_f}{2 \cdot f_{ctm}}}$$
$$k_b = \sqrt{\frac{2 - b_f/b_c}{1 + b_f/400}} \ge 1 \; \beta_L = \frac{L_b}{L_e} \cdot \left(2 - \frac{L_b}{L_e}\right) \; if\; L_b \le L_e, \; \beta_L = 1 \; otherwise \tag{2.46}$$

where f_{ck} is the characteristic value of the cylindrical compressive strength of concrete and k_G is an experimentally calibrated coefficient, which is 0.064 or 0.03 for mean or design provision, respectively. The shape factor k_b is the same given by Eq. (2.43), except for the coefficient 1.06.

The safety factor for debonding failure, $\gamma_{f,d}$, is usually assumed equal to 1.2 or 1.5 (non-controlled or controlled gluing application), while γ_c is the safety factor for concrete (equal to 1.5).

Experimental Database

A lot of results have been collected from the technical literature concerning bond tests on concrete elements externally strengthened with CFRP cured in situ (sheets) and preformed (plates) systems. Several set-ups (Yao et al. 2005) have been realized by the researchers and each of them can be considered more or less reliable for the right prediction of the actual loading conditions and, thus, of the end debonding load in existing elements. Moreover, constructive detailing of specimens can influence the reliability of these results.

In this application, the only results of push-pull bond shear tests have been considered to perform the statistical analyses; moreover, the results of cured in situ and preformed systems have been distinguished in two different groups. In the push-pull test set-up (see Fig. 2.1) the concrete block is loaded by a pushing force that is applied at a certain distance, *a*, from the FRP reinforcement that is loaded in tension by a pulling action. Several experimental programs have shown that the push-pull set-up can be simply realized, according to different set-ups (Bilotta et al. 2009b). Such a set-up is less sensitive to construction details and, thus, furnishes low scattered results in terms of debonding loads. This is the reason for which it is widely used to predict the bond strength for both shear and flexural strengthening in RC beams (Yao et al. 2005).

The realization of bond tests where both the FRP reinforcement and the concrete block are loaded by tension forces (pull-pull scheme) requires special attention in detailing, especially concerning the symmetry of the reinforcements on the sides of the concrete specimens (Leone et al. 2009). In these schemes the set-ups are more sensitive to the geometrical inaccuracies and thus, the repeatability or the variability of the results can be strongly affected by detailing.

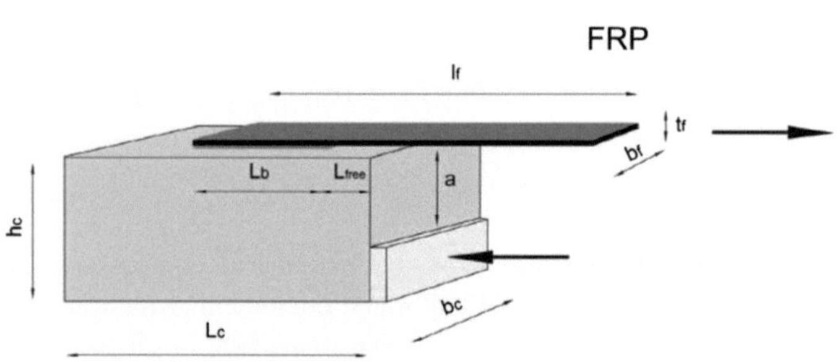

Fig. 2.1 General scheme of an asymmetrical push-pull bond test

In existing RC element (i.e. in RC beams) the FRP external reinforcement is usually applied on the tension side, so that the pull-pull scheme seems to replicate better the actual loading conditions. However, in the push-pull scheme, suitable values of the distance a (see Fig. 2.1) ensure the development of a bond failure at the concrete-FRP interface, similarly to what occurs in pull-pull scheme. On the contrary, for low values of a, the compressive stresses induced by the pushing force can limit the volume of concrete involved in the failure mechanism and, thus, furnish safe values of debonding load due to smaller values of fracture energy (see Eq. 2.40).

The results of specimens with bonded length, width and Young's modulus of the FRP reinforcement lower than 75 mm, 25 mm, and 80 GPa, respectively, were not considered.

For specimens strengthened with FRP sheets (Aiello and Leone 2005; Yao et al. 2005; Bilotta et al. 2009a; Ceroni and Pecce 2010; Lu et al. 2005; McSweeney and Lopez 2005; Takeo et al. 1997; Travassos et al. 2005; Ueda et al.1999; Wu et al. 2001; Zhao et al. 2000) the main parameters (see Fig. 2.1 for the geometrical parameters) vary in the following ranges: concrete width b_c = 100–500 mm, width of FRP reinforcement b_f = 25–100 mm, b_f/b_c = 0.17–1, thickness of FRP reinforcement t_f = 0.083–0.507 mm, bonded length of FRP reinforcement L_b = 75–500 mm, number of layers of FRP reinforcement n = 1–3, Young's modulus of FRP reinforcement E_f = 82–390 GPa, mean compressive strength of concrete f_{cm} = 17–62 MPa, and mean tensile strength of concrete f_{ctm} = 1.3–4.3 MPa.

Analogously, for specimens strengthened with FRP plates (Chajes et al. 1996; Faella et al. 2002; Mazzotti et al. 2009; Nigro et al. 2008; Bilotta et al. 2009a, b, 2011b) the main parameters vary in the following ranges: b_c = 150–230 mm, b_f = 25–100 mm, b_f/b_c = 0.11–0.63, t_f = 1.0–1.6 mm, n = 1, L_b = 150–400 mm, E_f = 108–400 GPa, f_{cm} = 15–53 MPa, and f_{ctm} = 1.10–3.8 MPa.

Totally, 216 data of bond tests for sheets and 68 for plates have been collected.

Calibration Procedure

According to the design assisted by testing approach, the random variable δ is defined as the ratio of the experimental debonding load, N_{exp}, to the theoretical one representing the strength model, N_{th}:

$$\delta_i = \frac{N_{\exp,i}}{N_{th,i}} \quad (2.47)$$

The mean value, the variance, the standard deviation and the *CoV* of this variable are defined as:

$$\overline{\delta} = \frac{1}{n}\sum_{i=1}^{n}\delta_i; \quad s_{\overline{\delta}}^2 = \frac{1}{n-1}\sum_{i=1}^{n}(\delta_i - \delta)^2; \quad \sigma_\delta = \sqrt{s_{\overline{\delta}}^2}; \quad CoV = \frac{\sigma_\delta}{\overline{\delta}} \tag{2.48}$$

The strength model expressed by N_{th} should be fine-tuned by a least-square coefficient, which minimizes the difference between each theoretical, $N_{th,i}$, and experimental, $N_{exp,i}$, value. Usually, this can be simply carried out considering the regression line of the graph $N_{th,i} - N_{exp,i}$. The slope of this line intercepting the origin furnishes the least-square coefficient, k_m.

Then, the random variable δ_m is defined as the ratio of the experimental debonding load N_{exp}, to the theoretical one, N_{th}, adjusted by means of the fine-tuning parameter, k_m:

$$\delta_{m,i} = \frac{N_{\exp,i}}{k_m \cdot N_{th,i}} \tag{2.49}$$

Mean value, variance, standard deviation and CoV of δ_m are defined as:

$$\begin{aligned} \overline{\delta_m} &= \frac{1}{n}\sum_{i=1}^{n}\delta_{m,i} = \frac{1}{n}\sum_{1=1}^{n}\frac{N_{\exp,i}}{k_m \cdot N_{th,i}} = \frac{\overline{\delta}}{k_m} \\ s_{\overline{\delta}_m}^2 &= \frac{1}{n-1}\sum_{i=1}^{n}(\delta_{m,i} - \overline{\delta_m})^2 = \frac{1}{n-1}\sum_{i=1}^{n}\left(\frac{N_{\exp,i}}{k_m \cdot N_{th,i}} - \frac{\overline{\delta}}{k_m}\right) = \frac{s_{\overline{\delta}}^2}{k_m} \end{aligned} \tag{2.50}$$

Thus, the mean provision for the debonding load can be assumed as:

$$N_{th,m} = k_m \cdot \overline{\delta_m} \cdot N_{th} \tag{2.51}$$

In Eq. (2.51) the model error is represented by the mean value of the variable δ_m, which is not 1, because the regression line was forced to intercept the origin.

In the Limit State approach, any strength is assumed as a random variable and, in general, the 0.05 percentile (named 'characteristic value') of its frequency distribution is used for design purposes. A very suitable distribution is the Gaussian one, but to use it the check of the normality hypothesis of the random variable is required. Several statistical tests (Shapiro-Wilk, Anderson-Darling, Martinez-Iglewicz, Kolmogorov-Smirnov, D'Agostino skewness, D'Agostino kurtosis, D'Agostino omnibus) can be performed (Mood et al. 1974; Shapiro and Wilk 1965) to verify the normality or log-normality hypothesis of the experimental distributions.

If the debonding load is assumed as a random variable and the Young's modulus of the FRP reinforcement, the tensile and compressive strength of concrete are assumed as the only parameters influencing the bond strength, the general expression of the strength model adjusted by the fine-tuning coefficient, k_m, is:

$$N_{th,m} = N_{th,m}\left(E_f, f_{cm} f_{ctm}, \overline{\delta_m}, k_m\right) \tag{2.52}$$

Moreover, under the hypothesis of Gaussian distribution, the 0.05 percentile of the variable debonding load can be calculated as:

$$N_{th,k,0.05} = N_{th,m} - 1.64 \cdot \left[Var(N_{th,m})\right]^{0.5} \tag{2.53}$$

where the variance of $N_{th,m}$ can be expressed as:

$$\begin{aligned} Var\left(N_{th,m}\right) = {} & C_{Efm}^2 \cdot Var(Ef) + C_{fcm}^2 \cdot Var(fcm) + C_{fctm}^2 \cdot Var(fctm) \\ & + C_{\delta m}^2 \cdot Var(\delta_m) \end{aligned} \tag{2.54}$$

$$C_{Efm} = \left|\frac{\partial N_{th,m}}{\partial E_f}\right|_{\overline{E_f}}, \quad C_{f_{cm}} = \left|\frac{\partial N_{th,m}}{\partial f_{cm}}\right|_{\overline{f_{cm}}}, \quad C_{f_{ctm}} = \left|\frac{\partial N_{th,m}}{\partial f_{ctm}}\right|_{\overline{f_{ctm}}}, \quad C_{f_{ctm}} = \left|\frac{\partial N_{th,m}}{\partial \delta_m}\right|_{\overline{\delta_m}} \tag{2.55}$$

If Eqs. (2.54) and (2.55) are substituted in the Eq. (2.53), the 0.05 percentile of the debonding load becomes:

$$\begin{aligned} & N_{th,k,0.05} \\ & = N_{th,m} - 1.64 \cdot N_{th,m} \cdot \left[a \cdot CoV_{E_f}^2 + b \cdot CoV_{f_{cm}}^2 + c \cdot CoV_{f_{ctm}}^2 + CoV_{\delta_m}^2\right]^{0.5} \end{aligned} \tag{2.56}$$

where the coefficients a, b, c depend on the functional relation of N_{th} from the parameters E_f, f_{cm} and f_{ctm}. The $CoVs$ are defined for each parameter as the ratio of its mean value to the standard deviation:

$$CoV_{E_f} = \frac{\overline{E_f}}{s_{E_f}}, \quad CoV_{f_{ctm}} = \frac{\overline{f_{ctm}}}{s_{f_{ctm}}}, \quad CoV_{f_{cm}} = \frac{\overline{f_{cm}}}{s_{f_{cm}}}, \quad CoV_{\delta_m} = \frac{\overline{\delta_m}}{s_{\delta_m}} \tag{2.57}$$

Note that the standard deviations of E_f, f_{cm}, and f_{ctm} have been assessed according to some literature information:

$$s_{E_f} = 0.05 \cdot \overline{E_f}; \quad s_{f_{ctm}} = 0.183 \cdot \overline{f_{ctm}}; \quad s_{f_{cm}} = 4.88 \tag{2.58}$$

Clearly, the coefficient of variation of the variable δ_m, $CoV_{\delta m}$, depends on the data distribution. Equation (2.56) can be written as:

$$N_{th,k,0.05} = k_k \cdot N_{th} \tag{2.59}$$

$$k_{k,0.05} = k_m \cdot \overline{\delta_m} \cdot \left(1 - 1.64 \cdot \left[a \cdot CoV_{E_f}^2 + b \cdot CoV_{f_{cm}}^2 + c \cdot CoV_{f_{ctm}}^2 + CoV_{\delta_m}^2\right]^{0.5}\right) \tag{2.60}$$

If the coefficients of variation of the materials are neglected, Eq. (2.60) becomes the well-known following one:

$$N_{th,k,0.05} = N_{th} \cdot \overline{\delta_m} \cdot k_m \cdot (1 - 1.64 \cdot CoV_{\delta_m}) = N_{th} \cdot k_m \cdot \left(\overline{\delta_m} - 1.64 \cdot s_{\delta_m}\right) \tag{2.61}$$

However, in this application all coefficients of variation have been taken into account. The percentiles 0.05 (characteristic values) are usually divided to safety factors, which take into account the model uncertainness (EN1990—Annex D). Furthermore, percentiles lower than 0.05 can be obtained by replacing in the Eq. (2.56) the coefficient 1.64 with the coefficients 2.58 and 3.08 corresponding to the 0.005 and 0.001 percentiles, respectively. These lower percentiles can be used as alternative to the characteristic values divided to the safety factors.

Application to the Experimental Database

The general Eq. (2.52) for debonding load can be particularized by introducing the dependence on the bond shear strength. Indeed, the bond shear strength depends on the concrete strength and can be related to the circle of Mohr representing the stress condition in the concrete at failure. Thus, different formulations for shear strength have been considered varying the dependence on the concrete strength. In particular, the following five expressions for the debonding load are examined:

$$\text{Case 1: } N_{th} = \beta_L \cdot b_f \cdot \sqrt{2 \cdot E_f \cdot t_f \cdot k_b \cdot \sqrt{f_{cm} \cdot f_{ctm}}} \tag{2.62}$$

$$\text{Case 2: } N_{th} = \beta_L \cdot b_f \cdot \sqrt{2 \cdot E_f \cdot t_f \cdot k_b \cdot f_{cm}^{2/3}} \tag{2.63}$$

$$\text{Case 3: } N_{th} = \beta_L \cdot b_f \cdot \sqrt{2 \cdot E_f \cdot t_f \cdot k_b \cdot f_{cm}^{0.6}} \tag{2.64}$$

$$\text{Case 4: } N_{th} = \beta_L \cdot b_f \cdot \sqrt{2 \cdot E_f \cdot t_f \cdot k_b \cdot \frac{f_{cm} \cdot f_{ctm}}{f_{cm} + f_{ctm}}} \tag{2.65}$$

$$\text{Case 5: } N_{th} = \beta_L \cdot b_f \cdot \sqrt{2 \cdot E_f \cdot t_f \cdot k_b \cdot 0.9 \cdot f_{ctm}} \tag{2.66}$$

In the Eq. (2.62), if a Coulomb failure criterion is adopted, the term $\sqrt{f_{cm} \cdot f_{ctm}}$ is 2 times the cohesion associated to the Mohr circle of an interface concrete element subjected to both shear and normal (peeling) stresses. The presence of peeling stresses has been often experimentally evidenced by the visual inspection of the debonded surface configuration (Mazzotti et al. 2008). Thus, Case 1 reproduces better the actual physical phenomenon, because it takes into account both the presence of shear and normal interfacial stresses.

Moreover, in Eq. (2.63) the term $f_{cm}^{2/3}$ is a simplification of $\sqrt{f_{cm} \cdot f_{ctm}}$, if the concrete tensile strength is calculated by the Eq. (2.47).

Analogously in Eq. (2.65), the term $\frac{f_{cm} \cdot f_{ctm}}{f_{cm}+f_{ctm}}$ is the maximum shear stress compatible with the strength f_{cm} and f_{ctm} in an interface concrete element subjected to only shear stresses (the Mohr circle has centre in the axis origin in this case).

In Eq. (2.66), the term $0.9 \cdot f_{ctm}$ is a simplification of the term $\frac{f_{cm} \cdot f_{ctm}}{f_{cm}+f_{ctm}}$ under the assumption that the compressive strength is about 10 times the tensile one.

Finally in Eq. (2.64), the term $f_{cm}^{0.6}$ is a further modification of $0.9 \cdot f_{ctm}$ according to the expression for the concrete tensile strength given by Model Code 90.

$$f_{ctm} = 0.32 \cdot f_{cm}^{0.6} \tag{2.67}$$

For all cases the mean and the characteristic provisions of debonding load can be calculated using the previously introduced Eqs. (2.51) and (2.59):

$$N_{th,m} = k_m \cdot \overline{\delta_m} \cdot N_{th} \tag{2.68}$$

$$N_{th,k,0.05} = k_{k,0.05} \cdot N_{th} \tag{2.69}$$

where k_m is the least square coefficient associated to the regression line intercepting the origin, $\overline{\delta_m}$ is given by Eq. (2.16), and $k_{k,0.05}$ is given by Eq. (2.60).

For each equation, the best fitting coefficient k_m has been calculated considering the experimental results distinguished in two series: sheets and plates.

In Table 2.1 the coefficient k_m and the R^2 value of the corresponding least-square line, which is a measure of the reliability of the regression, are reported for all the

Table 2.1 Statistical data for different bond strength models

Case	FRP type	k_m	R^2	$\overline{\delta_m}$	CoV_{δ_m} (%)	$k_m \cdot \overline{\delta_m}$	$k_{k,0.05}$	$k_{k,0.005}$	$k_{k,0.001}$
1	Sheet	0.270	0.855	1.027	17.7	0.278	0.192	0.143	0.117
	Laminate	0.236	0.349	1.064	23.2	0.251	0.152	0.095	0.064
2	Sheet	0.258	0.878	1.010	17.6	0.261	0.182	0.137	0.112
	Laminate	0.221	0.534	1.034	20.4	0.229	0.149	0.103	0.078
3	Sheet	0.291	0.881	1.006	17.7	0.293	0.204	0.154	0.126
	Laminate	0.248	0.565	1.030	20.0	0.255	0.169	0.119	0.092
4	Sheet	0.535	0.862	1.022	17.6	0.547	0.370	0.269	0.214
	Laminate	0.466	0.375	1.060	23.0	0.496	0.294	0.180	0.118
5	Sheet	0.544	0.863	1.021	17.5	0.555	0.375	0.270	0.213
	Laminate	0.473	0.379	1.059	22.9	0.502	0.297	0.180	0.117

equations. The mean value of the variable δ_m, defined by Eqs. (2.50) and (2.51), and its *CoV* are reported too. In all cases the *CoV*, which is a measure of the model significance, is lower than the threshold value of 40 % (Monti et al. 2009), so that all the models can be considered reliable. However, it can be observed that, while for the sheets the R^2 value is low sensitive to the model and is quite elevated (0.855–0.881), on the contrary for the plates the choice of the model can be significant considering that R^2 varies in the range 0.349–0.565. Despite their better quality control in factory, the preformed FRP systems present a higher *CoV* and a smaller R^2 value with respect to the in situ sheets. This is justifiable by the larger sensitivity of this system to the detailing of the experimental procedure. Indeed, increasing the stiffness of the FRP system results in more inaccuracies of the experimental set-up, which can influence the debonding load.

In particular, Case 3 results the best-fitting model because of the highest value of R^2; this relationship depends on the compressive strength of concrete with an exponent 0.6. Note that for design aim, the choice of the best-fitting model has the clear advantage to furnish characteristic values more close to the mean ones, because the theoretical loads show a smaller gap with the experimental results.

In Table 2.1 the coefficients k_k to calculate different percentiles (5, 0.5, and 0.1 %) are reported too. The coefficients $k_m \cdot \overline{\delta_m}$, which allow for calculating the mean provisions, differ of about 12–15 % for the sheets and the plates. The factors $k_{k,0.05}$, which allow for calculating the 0.05 percentiles, differ of about 17–21 %.

The general Eq. (2.52) for the mean provision for the best-fitting model (Case 3) is:

$$N_{th,m} = k_m \cdot \delta_m \cdot \beta_L \cdot b_f \cdot \sqrt{2 \cdot E_f \cdot t_f \cdot k_b \cdot f_{cm}^{0.6}} \tag{2.70}$$

The corresponding 0.05 percentile provision given by Eqs. (2.56) and (2.60) is:

$$\begin{aligned} N_{th,k,0.05} &= N_{th,m} - 1.64 \cdot N_{th,m} \cdot \left[0.5^2 \cdot CoV_{E_f}^2 + (3/10)^2 \cdot CoV_{f_{cm}}^2 + CoV_{\delta_m}^2\right]^{0.5} \\ &= k_k \cdot N_{th} \end{aligned} \tag{2.71}$$

$$k_{k,0.05} = k_m \cdot \delta_m \cdot \left[1 - 1.64 \cdot \left[0.5^2 \cdot CoV_{E_f}^2 + (3/10)^2 \cdot CoV_{f_{cm}}^2 + CoV_{\delta_m}^2\right]^{0.5}\right] \tag{2.72}$$

$$\begin{aligned} C_{Efm} &= \left|\frac{\partial N_{th,m}}{\partial E_f}\right|_{\overline{E_f}} = N_{th,m} \cdot \frac{0.5}{\overline{E_f}}, \\ C_{Efm}^2 \cdot Var(Ef) &= N_{th,m}^2 \cdot \frac{0.5^2}{\overline{E_f^2}} \cdot Var(E_f) = N_{th,m}^2 \cdot 0.5^2 \cdot CoV_{E_f}^2 \end{aligned} \tag{2.73}$$

$$\begin{aligned} C_{f_{cm}} &= \left|\frac{\partial N_{th,m}}{\partial f_{cm}}\right|_{\overline{f_{cm}}} = N_{th,m} \cdot \frac{3/10}{\overline{f_{cm}}}, \\ C_{fcm}^2 \cdot Var(f_{cm}) &= N_{th,m}^2 \cdot \frac{(3/10)^2}{\overline{f_{cm}^2}} \cdot Var(f_{cm}) = N_{th,m}^2 \cdot (3/10)^2 \cdot CoV_{f_{cm}}^2 \end{aligned} \tag{2.74}$$

$$\begin{aligned} C_{f_{ctm}} &= \left|\frac{\partial N_{th,m}}{\partial \delta_m}\right|_{\overline{\delta_m}} = \frac{N_{th,m}}{\overline{\delta_m}}, \\ C_{\delta m}^2 \cdot Var(\delta_m) &= N_{th,m}^2 \cdot \frac{Var(\delta_m)}{\overline{\delta_m}} =, \quad N_{th,m}^2 \cdot CoV^2 \delta_m \end{aligned} \tag{2.75}$$

$$C_{f_{ctm}} = \left|\frac{\partial N_{th,m}}{\partial f_{ctm}}\right|_{\overline{f_{ctm}}} = 0 \tag{2.76}$$

The last term related to the tensile strength of concrete, f_{ctm}, is clearly absent, because the debonding load in Eq. (2.64) depends only on the compressive strength.

However, as it occurs for all other cases, the variance of the materials is less significant compared with the variance of the model. Indeed, for the cured in situ systems the coefficient of variation of the variable δ_m is:

$$CoV_{\delta_m} = \frac{\overline{\delta_m}}{s_{\delta_m}} = \frac{1.006}{0.178} = 0.177 \rightarrow CoV_{\delta_m}^2 = 0.031 \tag{2.77}$$

By contrast, the contributes related to the CoVs of the materials are:

$$0.5^2 \cdot CoV_{E_f}^2 + (3/10)^2 \cdot CoV_{f_{cm}}^2 = 0.0026 \approx 0.003 \approx 0.1 \cdot CoV_{\delta_m}^2$$

In Fig. 2.2 the experimental debonding loads are compared with the theoretical ones given by Case 3—Eq. (2.64); the regression line intercepting the origin of axis is reported too. Figure 2.2a refers to the cured in situ systems and Fig. 2.2b to the preformed ones.

In Fig. 2.3 the experimental values of strain in the FRP reinforcement at debonding are plotted together with the mean and characteristic provisions given by Eq. (2.64) using the values of k_m and k_k, for the three percentiles 5, 0.5, and 0.1 %, listed in Table 2.1. The characteristic provision (5 % percentile) divided to the safety factor $\gamma_f = 1.2$ (related to good application conditions, according to CNR-DT 200 2004) is plotted too.

Both theoretical and experimental strains are plotted vs. the term $E_f \cdot t_f / 2 \cdot \Gamma_f$, assuming $\Gamma_f = k_b \cdot f_{cm}^{0.6}$ according to Eq. (2.64). This allows for graphing the theoretical curves normalized to the axial stiffness of the FRP reinforcement and the concrete strength.

The theoretical curves show that the 0.5 % percentile can be a good choice to warrant a reliable safety level to the aim of furnish design provisions. Note that the

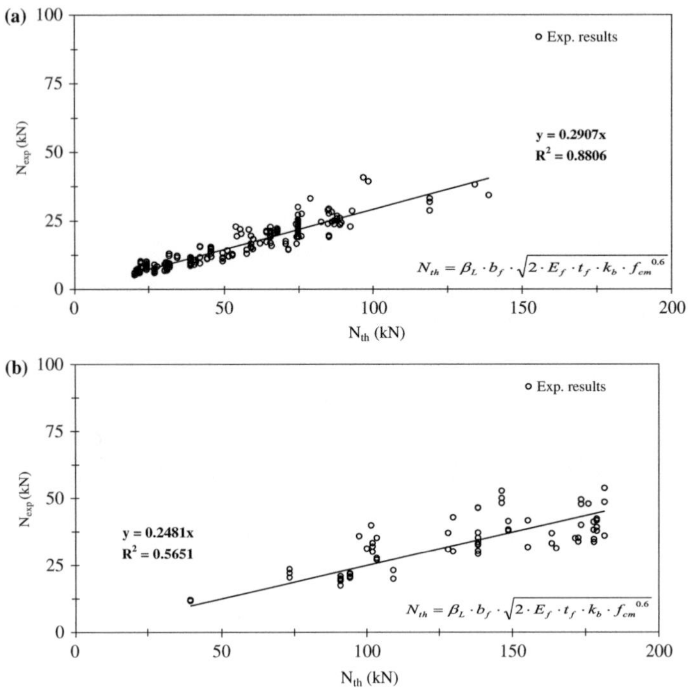

Fig. 2.2 Regression line: **a** cured in situ systems: 216 data; **b** preformed systems: 68 data

assessment of the percentiles has been carried out taking into account the variance of the materials.

Another possibility, which is adopted by the Italian guidelines (CNR–DT 200 2004) and, more in general, is included in the Eurocode approach, consists into divide the characteristic provision (5 % percentile) to a safety factor γ_f that depends on the quality of the application. Figure 2.3 shows that the provisions corresponding to the 5 % percentile divided to the factor $\gamma_f = 1.2$ are, however, less safe than the 0.5 % percentile.

It should be noted that the percentiles provisions were calculated under the hypothesis of Gaussian distribution. Some statistical tests (Shapiro-Wilk, Anderson-Darling, Martinez-Iglewicz, Kolmogorov-Smirnov, D'Agostino skewness, D'Agostino kurtosis, D'Agostino omnibus) were performed to verify this assumption but the comparison between the cumulate frequency curves of N_{exp} and the Gaussian distribution, having the same mean value and standard deviation, highlighted a bad agreement, especially for the cured in situ sheets. This was confirmed also by the responses of the statistical tests, which in most cases rejected the normality assumption for sheets and accepted it for plates. For the sheets, the experimental debonding loads seemed better represented by a log-normal distribution.

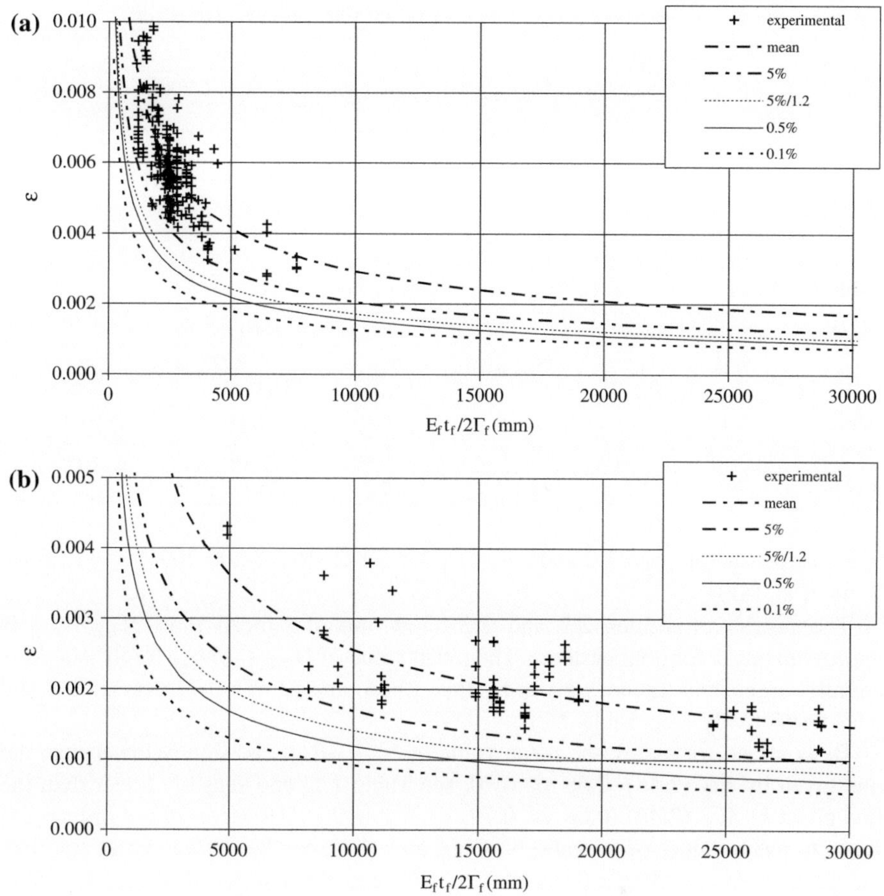

Fig. 2.3 Experimental versus theoretical failure strains: **a** cured in situ systems (216 data); **b** preformed systems (68 data)

However, the values of percentiles calculated under the hypothesis of log-normal distributions are larger than the ones reported in Table 2.1 (Gaussian distribution). Thus, the normal distribution can be considered safe to the aim of furnishing design provisions.

Finally, in Tables 2.2 and 2.3 the mean value, the standard deviation, and the CoV of the ratio N_{th}/N_{exp} are reported for both cured in situ (sheets) and preformed systems (plates).

The theoretical values N_{th} given by the new proposal refer to both mean (Eq. 2.34) and characteristic provisions (Eq. 2.35); in particular the percentiles 0.05 and 0.005 have been considered. The characteristic provision (5 % percentile) divided to $\gamma_f = 1.2$ is reported too. Finally, the design provisions of CNR-DT 200 (2004), Teng et al. (2001) are also listed. The characteristic provisions were divided

Table 2.2 Values of the ratio N_{th}/N_{exp} for cured in situ FRP systems (216 results)

N_{th}/N_{exp}	New calibration				Design	
	$N_{th,m}$	$N_{th,k,0.05}$	$N_{th,k,0.05}/1.2$	$Nt_{h,k,0.005}$	CNR DT 200 (2004)	Teng et al. (2001)
Mean	1.03	0.72	0.60	0.54	0.43	0.54
St. dev.	0.173	0.121	0.101	0.091	0.083	0.08
CoV (%)	16.8	16.8	16.8	16.8	19.4	15.7

Table 2.3 Values of the ratio N_{th}/N_{exp} for preformed FRP systems (68 results)

N_{th}/N_{exp}	New calibration				Design	
	$N_{th,m}$	$N_{th,k,0.05}$	$N_{th,k,0.05}/1.2$	$N_{th,k,0.005}$	CNR DT 200 (2004)	Teng et al. (2001)
Mean	1.03	0.68	0.57	0.48	0.47	0.63
St. dev.	0.20	0.13	0.11	0.09	0.12	0.120
CoV (%)	19.1	19.1	19.1	19.1	25.1	19.1

to the safety factors: $\gamma_f = 1.2$ and $\sqrt{\gamma_c} = \sqrt{1.5}$ for CNR-DT 200 2004; $\gamma_b = 1.25$ for Teng et al. 2001.

The results of Tables 2.2 and 2.3 show that the mean ratio $N_{th,m}/N_{exp}$ is approximately 1 for both systems. The mean value of $N_{th,k,0.05}/N_{exp}$ for the sheets is slightly larger than the ones given by Eqs. (2.46) and (2.45): 0.72 versus 0.63 and 0.67 respectively (see Table 2.1).

By contrast, for plates, the mean value of $N_{th,0.05}/N_{exp}$ is slightly lower than the one given by Eq. (2.46) (0.68 vs. 0.70, see Table 2.1) and sensibly lower than the one given by Eq. (2.45) (0.68 vs. 0.79, see Table 2.1). However, these differences relate to mean values of the ratio N_{th}/N_{exp} and, thus, can be misleading to compare the predictions of different models. In contrast, the curves of Fig. 2.4, where experimental and theoretical debonding strains of the only FRP sheets are plotted

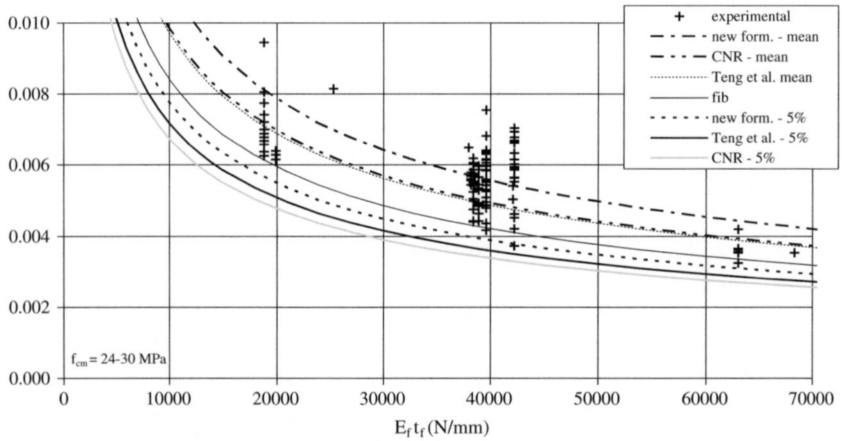

Fig. 2.4 Experimental maximum strain versus mean and 0.05 percentile provisions for sheets

vs. the parameter $E_f \cdot t_f$, show the actual variations between the different formulations examined. Since the theoretical strain depends on the concrete strength, a range of variability for this parameter has been fixed (24–30 MPa that corresponds to a mean value of about 27 MPa). The mean and characteristic (5 % percentile) provisions ($N_{th,k,0.05}$) given by the new formulation, by (CNR-DT 200 2004) and by (Chen and Teng 2001) are plotted in the graph. In particular, both mean and characteristic provisions (5 % percentile) given by the new formulation are larger than the predictions of (Chen and Teng 2001) and (CNR-DT 200 2004). Moreover, the formulation of (CNR-DT-200 2004) gives the most safe results in terms of 5 % percentile, while furnishes mean predictions very similar to (Chen and Teng 2001). Furthermore, it can be observed that the formulation of (fib bulletin 14 2001), lies between the mean and the characteristic curves given by the new formulation.

In Fig. 2.5 several design proposals coming from the new formulation (0.01, 0.5, and 5 % percentile divided to $\gamma_f = 1.2$), the model of fib bulletin 14 (2001), the design provisions of CNR-DT 200 (2004) and Teng et al. (2001) are plotted together with the experimental debonding strains of the same tests considered in Fig. 2.4.

For the sheets, the graph of Fig. 2.5a shows that two design proposals coming from the new formulation (5 % percentile divided to 1.2 and 0.5 % percentile) are less safe than the ones currently furnished by CNR-DT 200 (2004). Note that the latter introduces the additionally safety factor of concrete, $\sqrt{\gamma_c}$, for design. This coefficient has been omitted in the new formulation because the variance of the concrete has been taken into account in the calibration procedure by means of the *CoV* of its compressive strength (see Eq. 2.72). Moreover, it can be observed that, the current design values of CNR-DT 200 (2004) are comparable with the 0.1 % percentile provisions of the new formulation. By contrast, the design formulation of Teng et al. (2001) is comparable with the 0.5 % percentile of the new formulation. Finally for the cured in situ systems, if the 0.5 % percentile of the new formulation is chosen as design provision, the debonding load increases of about +25 % compared with the current CNR provisions and is comparable with the design values of Teng et al. (2001).

On the contrary, in case of plates, Fig. 2.5b shows that the 0.5 % percentile of the new formulation is comparable with those currently furnished by CNR-DT 200 (2004), while the 5 % percentile divided to 1.2 is less safe (about 15 %). Moreover, the design formulation of Teng et al. (2001) is less safe compared with both the 5 % percentile of the new formulation divided to 1.2 and the 0.5 % one.

In conclusion, the proposed formulation for the end debonding load has a clear statistical meaning and allows for separating the provisions for the cured in situ FRP systems and the preformed ones. This distinction is mainly due to the larger scatter of the experimental results collected for this strengthening system.

Both aspects let to better exploit the strength of the cured in situ systems; indeed the 0.05 % percentile values of the new formulation are larger than the design values furnished by the current Italian Guidelines and, however, allow for assessing the same safety level of model of Teng et al. (2001). Moreover, it is worth noticing that the formulation of fib Bulletin 14 (2001) results excessively unsafe compared to the experimental results.

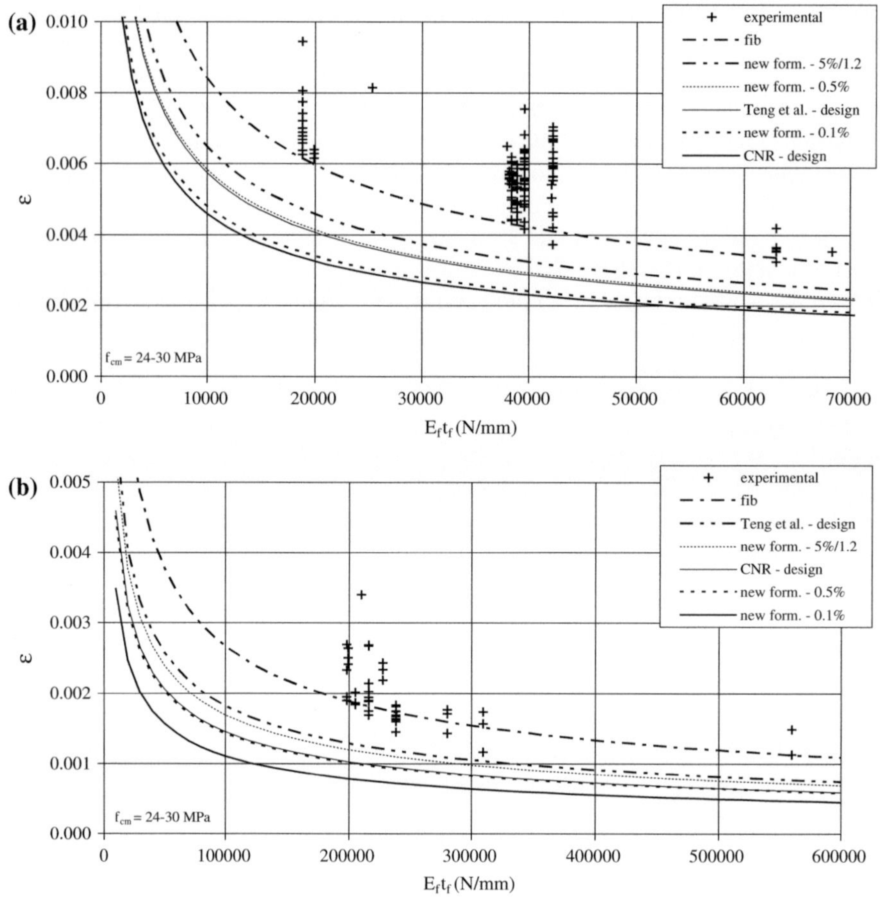

Fig. 2.5 Experimental maximum strain versus design provisions: **a** sheets; **b** plates

By contrast, the approach of Teng et al. (2001) and fib Bulletin 14 (2001) are found to be less safe when applied to preformed systems. Moreover, the 0.5 % percentile of the proposed design formulation provides a higher safety level compared with these models, while is similar to the current design provisions of CNR-DT 200 (2004). Thus, these results confirm that the distinction of the two strengthening systems seem to be reliable in predicting the debonding load.

Application 2—Intermediate Debonding

In this section the procedure is used to calibrate a design relationship for intermediate debonding on a wide database assembled by collecting data of experimental tests on FRP-strengthened RC beams.

Experimental Database

The database was obtained by merging the data considered by Ferracuti et al. (2007) with those collected by Wu and Niu (2007) and about thirty further experimental cases reported in the scientific literature (Beber 1999, 2003; Grace et al. 1999; Khomwan et al. 2004; Pham and Al-Mahaidi 2004; Sharif et al. 1991; Triantafillou and Plevris 1992). The resulting database collects the geometric and mechanical data describing the RC beams and their steel and composite reinforcement, the latter being made of externally bonded composite laminates based on carbon, glass or aramid fibers.

For specimens strengthened with FRP systems cured in situ (sheets), the relevant geometric and mechanical parameters range in the following intervals: concrete width b_c = 75–960 mm, FRP width b_f = 30–480 mm, b_f/b_c = 0.17–1, FRP thickness t_f = 0.11–2.55 mm, Young's modulus of FRP E_f = 21–390 GPa, mean compressive strength of concrete f_{cm} = 21–61 MPa, mean tensile strength of concrete f_{ctm} = 2.3–4.3 MPa.

For specimens strengthened with preformed FRP systems (laminates), the key parameters vary in the following ranges: concrete width b_c = 180–800 mm, FRP width b_f = 25–280 mm, b_f/b_c = 0.13–1, FRP thickness t_f = 1.0–6 mm, Young's modulus of FRP E_f = 190–220 GPa, mean compressive strength of concrete, f_{cm} = 12.6–53.4 MPa, mean tensile strength of concrete, f_{ctm} = 1.62–4.25 MPa.

A total number of 214 experimental results have been collected (164 FRP cured in situ systems and 50 FRP preformed system).

Intermediate debonding failure have been observed in all these tests. In principle, the maximum bending moment M_{db} observed in the experimental tests at debonding is smaller than the ultimate one M_u, corresponding to FRP rupture. The following parameter γ could be introduced for quantifying how premature is failure with respect to the ultimate flexural strength:

$$\gamma = \frac{M_{db} - M_y}{M_u - M_y} \tag{2.78}$$

M_y being the bending moment of the strengthened section at yielding of rebar: both M_u and M_y can be determined theoretically adopting the usual assumptions for RC sections at ULS. The parameter γ is closer to zero as debonding occurs for small values of the maximum axial strain in FRP; on the contrary, it is close to the unity as axial strain at debonding is close to the corresponding ultimate value $\varepsilon_{f,u}$. Figure 2.6 points out that the values of γ determined for the beams collected in the database generally range between zero and one; only in few cases (less than ten out of the total 214) it is slightly larger than the unity, mainly as a result of hardening of the materials.

The values of γ have been represented in Fig. 2.6 against the square root of the ratio between (twice) the fracture energy, G_F, and the specific axial stiffness of the FRP reinforcement, $E_f t_f$; the former parameter has been evaluated as a function of

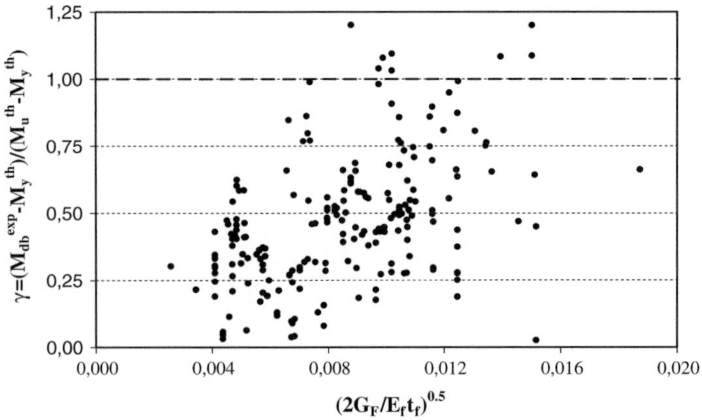

Fig. 2.6 Values of γ parameter against some mechanical parameters

both concrete tensile, f_{ct}, and compressive, f_c, strengths through the relation proposed in the Italian Code. The ratio $2G_F/E_f t_f$ is often considered in various proposals as the key parameter for determining the value of axial strain ε_{db} developed in FRP at debonding onset. However, Fig. 2.6 can only point out a general trend resulting in values of γ as large as the parameter represented on the x-axis, but it is quite hard to recognize a consistent correlation between γ (or, even, the maximum axial strain ε_{db} developed in FRP at debonding) and the quantity on the x-axis possibly depending on the two following reasons:

- fracture energy, G_F, basically depends on concrete (tensile) strength and, consequently, is widely scattered;
- besides the one reported on the x-axis, other parameters play an important role on the occurrence and extent of debonding failure.

For instance, the role of both the amount of steel rebar and their yielding stress/strain values have been emphasized in Faella et al. (2008a). Furthermore, load distribution also affects the possible premature failure of FRP strengthened beams as confirmed by Fig. 2.7 showing a strict correlation between the yielding moment M_y of the strengthened section and the maximum bending moment at debonding M_{db} at least in the case of three- or four-point-bending, while a completely different behavior results in the case of uniformly distributed load.

Figure 2.8 shows the distribution of the parameter γ for the experimental results considered within the database. It points out that such values are quite uniformly distributed since the cumulative frequency distribution is not so far from the ideally uniform straight curve, meaning that cases of very premature debonding are considered within the database as well as other cases whose failure is close to the complete development of the strength on the external reinforcement.

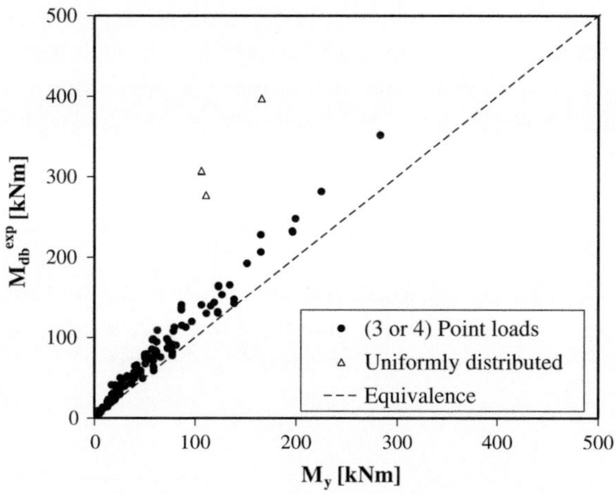

Fig. 2.7 Relation between bending moment at debonding and yielding

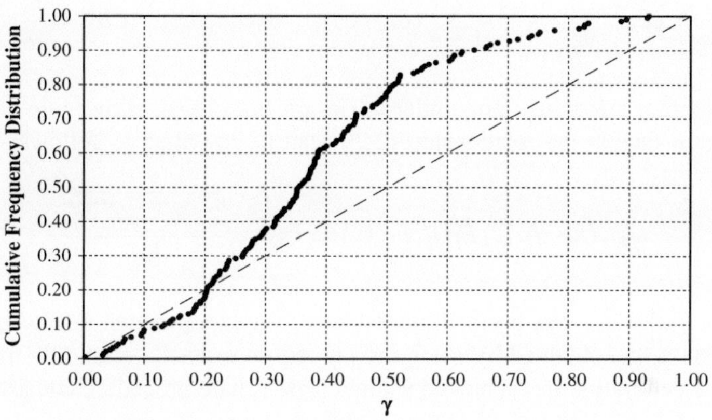

Fig. 2.8 Distribution of the parameter γ within the experimental database

Finally, it is worth noting that only the experimental results characterized by values of $\gamma \in (0,1)$ will be considered in the following, as that is a necessary condition for recognizing the cases of beams failure in intermediate debonding.

Procedure and Application

As already shown in the previous section for the end debonding, the mean value of the maximum axial strain in FRP corresponding to failure in intermediate crack-induced debonding (IC debonding strain) can be expressed by means of a relationship obtained by a deterministic model and fine-tuned on experimental data by a numerical coefficient k_{IC}. The assessment of k_{IC} through the design by testing procedure gives a clear probabilistic meaning to the provisions.

An error function δ can cover the uncertainties of the simplified model considered in the above mentioned calibration:

$$\varepsilon_{fd} = \varepsilon_{fd,m}\left(k_{IC}, f_c, f_{ct}, E_f, t_f, b_f, k_b\right) \cdot \delta \tag{2.79}$$

The random variable δ is defined, for each ith test, as the ratio of the experimental debonding strain, $\varepsilon_{exp,i}$, to the theoretical one, $\varepsilon_{fd,m}$ evaluated by considering the geometric and mechanical data characterizing that test:

$$\delta_i = \frac{\varepsilon_{\exp,i}}{\varepsilon_{th,i}} \tag{2.80}$$

Moreover, the mean value, the variance, the standard deviation and the CoV of this variable are defined as:

$$\overline{\delta} = \frac{1}{n}\sum_{i=1}^{n} \delta_i; \quad s_{\overline{\delta}}^2 = \frac{1}{n-1}\sum_{i=1}^{n} (\delta_i - \overline{\delta})^2; \quad \sigma_{\delta} = \sqrt{s_{\overline{\delta}}^2}; \quad CoV = \frac{\sigma_{\delta}}{\overline{\delta}} \tag{2.81}$$

By assuming a formulation similar to design Eq. (2.60), taking into account no safety partial factors the relationship (2.79) can be rewritten as follows:

$$\varepsilon_{fd,m}\left(k_{IC}, f_c, f_{ct}, E_f, t_f, b_f, k_b\right) = k_{IC} \cdot \sqrt{\frac{2 \cdot k_b \cdot \sqrt{f_c \cdot f_{ct}}}{E_f t_f}} \tag{2.82}$$

The coefficient k_{IC} has been calibrated based on experimental results in terms of deformation $\varepsilon_{fd,exp}$ obtained by using the procedure in Faella et al. (2010) as stated above. The calibration has achieved using a least-square procedure consisting in the resolution of the following minimum problem:

$$k_{IC,m} = \arg\min_{k_{IC}} \sum_{i=1}^{n} \left[\varepsilon_{fd,m}\left(k_{IC}, f^{(i)}, f_{ct}^{(i)}, E_f^{(i)}, t_f^{(i)}, b_f^{(i)}, k_b^{(i)}\right) - \varepsilon_{fd,\exp}^{(i)}\right]^2 \tag{2.83}$$

Moreover, the mean value of the intermediate debonding strain can be obtained by a coefficient $k_{IC,m}$ adjusted by means of the mean value of the error parameter $\overline{\delta}$, being in general $\overline{\delta} \neq 1$ because the regression line was imposed to intercept the origin.

$$k_{IC,m} = k_{IC,bf} \cdot \overline{\delta} \tag{2.84}$$

Thus, the mean provision for the intermediate debonding strain can be assumed as:

$$\varepsilon_{th,m} = k_{IC,m} \cdot \varepsilon_{th} \tag{2.85}$$

being ε_{th} the strain obtained by the deterministic model. Obviously, this strain is linearly proportional to the debonding strain being the FRP constitutive law linear elastic.

If the random variable represents strength, its characteristic value is often defined for design purposes as the 0.05 percentile of the frequency distribution associated to the examined variable. Gauss distribution is the most generally considered for describing the errors. The so-called "hypothesis of normal distribution" for the variable δ should be checked by comparing the experimental curve of the cumulative frequency to the theoretical one corresponding to a Gaussian distribution having the same mean value and standard deviation (see Fig. 2.9).

Assuming that the Young's modulus, E_f, of the FRP reinforcement, the concrete tensile and compressive strength, f_{ctm} and f_c, are the only mechanical parameters influencing the value of the maximum axial strain developed in FRP at debonding, the expressions for the general and calibrated models involving the coefficient $k_{IC,bf}$ as well as $\overline{\delta}$ are:

$$\varepsilon_{th} = \varepsilon_{th}\left(E_f, f_{cm} f_{ctm}\right) \tag{2.86}$$

$$\varepsilon_{th,m} = \varepsilon_{th,m}\left(E_f, f_{cm} f_{ctm}, \overline{\delta}, k_{IC,bf}\right) \tag{2.87}$$

In the following, the same assumptions already considered above in defining a characteristic value for plate end debonding strength are accepted (Bilotta et al. 2011). In particular, both E_f and f_c and f_{ct} have been assumed as normally and independently distributed random variables, with the following values of the coefficients of variation:

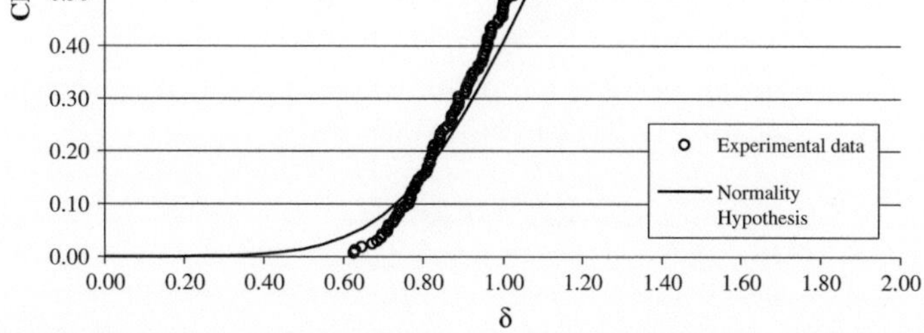

Fig. 2.9 Experimental data cumulative frequency against theoretical in normality hypothesis

$$s_{E_f} = 0.05 \cdot \overline{E_f} \quad s_{f_{ctm}} = 0.183 \cdot \overline{f_{ctm}} \quad s_{f_{cm}} = 4.88, \tag{2.88}$$

according to the design relationships provided by EN 1992-1-1 and literature information (Di Ludovico et al. 2009).

Hence, under the hypothesis of normal distribution for the variable δ, the provision corresponding to the 0.05 percentile of the Gaussian distribution is:

$$\varepsilon_{th,k} = \varepsilon_{th,m} - 1.64 \cdot \left[Var(\varepsilon_{th,m})\right]^{0.5} \tag{2.89}$$

where the variance of $\varepsilon_{th,m}$ can be expressed as:

$$\begin{aligned} Var(\varepsilon_{th,m}) = {} & C_{Efm}^2 \cdot Var(Ef) + C_{fcm}^2 \cdot Var(fcm) \\ & + C_{fctm}^2 \cdot Var(fctm) + C_{\delta m}^2 \cdot Var(\delta_m) \end{aligned} \tag{2.90}$$

$$\begin{aligned} C_{Efm} &= \left|\frac{\partial \varepsilon_{th,m}}{\partial E_f}\right|_{\overline{E_f}} \\ C_{f_{cm}} &= \left|\frac{\partial \varepsilon_{th,m}}{\partial f_{cm}}\right|_{\overline{f_{cm}}} \\ C_{f_{ctm}} &= \left|\frac{\partial \varepsilon_{th,m}}{\partial f_{ctm}}\right|_{\overline{f_{ctm}}} \\ C_{f_{ctm}} &= \left|\frac{\partial \varepsilon_{th,m}}{\partial \delta_m}\right|_{\overline{\delta_m}} \end{aligned} \tag{2.91}$$

If Eqs. (2.90) and (2.91) are substituted in Eq. (2.89), the following general expression is obtained for the characteristic provision of the debonding load:

$$\varepsilon_{th,k} = \varepsilon_{th,m} - 1.64 \cdot \varepsilon_{th,m} \cdot \left[a \cdot CoV_{E_f}^2 + b \cdot CoV_{f_{cm}}^2 + c \cdot CoV_{f_{ctm}}^2 + CoV_{\delta_m}^2\right]^{0.5} \tag{2.92}$$

where the coefficient a, b, c depend on the functional relationship of E_f, f_{cm} and f_{ctm} in the expression of ε_{th} and the coefficients of variation are defined for each parameter as the ratio of the mean value to its standard deviation:

$$CoV_{E_f} = \frac{\overline{E_f}}{s_{E_f}}, \quad CoV_{f_{ctm}} = \frac{\overline{f_{ctm}}}{s_{f_{ctm}}}, \quad CoV_{f_{cm}} = \frac{\overline{f_{cm}}}{s_{f_{cm}}} \quad CoV_{\delta_m} = \frac{\overline{\delta_m}}{s_{\delta_m}}, \tag{2.93}$$

Clearly the coefficient of variation of the variable δ_m, $CoV_{\delta m}$, depends on the data distribution. Equation (2.92) can be written as:

$$\varepsilon_{th,k} = k_{cr,k} \cdot \varepsilon_{th} \tag{2.94}$$

assuming:

$$k_{cr,k} = k_{cr,m} \cdot \left(1 - 1.64 \cdot \left[a \cdot CoV^2_{E_f} + b \cdot CoV^2_{f_{cm}} + c \cdot CoV^2_{f_{ctm}} + CoV^2_{\delta_m}\right]^{0.5}\right) \tag{2.95}$$

As already stated in the application related to the end-debonding phenomenon, lower percentiles can be obtained by substituting in Eq. (2.95) the coefficient 1.64, related to the 0.05 percentile of the frequency distribution, with the coefficients 2.58 and 3.08 corresponding to the 0.005 and 0.001 percentiles, respectively.

The use of percentiles lower than 0.05 can be alternative to the use of safety factors that usually have to be additionally applied to characteristic provision to take into account the model uncertainness (EN1990—Annex D).

The following values of the coefficients defined above have been derived by considering the experimental results in a least-square procedure:

$$k_{IC,bf} = 0.53, \quad k_{IC,,m} = 0.56, \quad k_{IC,5\%} = 0.32 \text{ and } k_{IC,0.5\%} = 0.18. \tag{2.96}$$

From the coefficients $k_{IC,m}$ and $k_{IC,5\%}$ the following coefficients can be easily defined:

$$\begin{aligned} k_{Gm,2} &= \left(k_{IC,m}\right)^2 = 0.32\,\text{mm} \\ k_{Gk,2} &= \left(k_{IC,k}\right)^2 = 0.10\,\text{mm} \end{aligned} \tag{2.97}$$

These coefficients provide a clear statistical meaning to the formulation proposed to assess the intermediate debonding strain.

References

ACI Committee 318. (2005). Building Code Requirements for Structural Concrete (ACI 318–05) and Commentary (318R-05) (p. 430). Farmington Hills: American Concrete Institute.

ACI Committee 440.2 R-08. (2008). Guide for the Design and Construction of Externally Bonded FRP Systems for Strengthening Concrete Structures.

Aiello, M. A., & Leone, M. (2005). Experimental bond analysis of concrete-FRP (fiber reinforced polymer) reinforced. *Proceedings of fib Symposium "Keep Concrete Attractive", Budapest, Hungary*.

Ang, A. H. S., & Tang, W. H. (1975). Probability concepts in engineering planning and decision. Basic principles (Vol. 1). New York: Wiley.

Arduini, M., & Nanni, A. (1997). Behavior of precracked RC beams strengthened with carbon FRP Sheets. *Journal of Composites for Construction, 1*(2), 63–70.

Beber, A. J. (2003). *Comportamento Estrutural de Vigas de Concreto Armado Reforçadas com Compósitos de Fibra de Carbono*. UFRGS, Porto Alegre: Tese de Doutorado. (in Portuguese).

Bilotta, A., Ceroni, F., Di Ludovico, M., Nigro, E., & Pecce ,M. (2011a). Bond tests on concrete elements strengthened with EBR and NSM FRP systems, *Proceedings Of FRP RCS 10*, April 2–4, 2011, Tampa, Florida, USA.

Bilotta, A., Ceroni, F., Di Ludovico, M., Nigro, E., & Pecce, M. (2011b). Design by testing of debonding load in RC element strengthened with EBR FRP materials, *Proceedings Of FRP RCS 10*, April 2–4, 2011, Tampa, Florida, USA.

Bilotta, A., Ceroni, F., Di Ludovico, M., Nigro, E., Pecce, M., & Manfredi, G. (2011). Bond efficiency of EBR and NSM FRP systems for strengthening concrete members, *Journal of Composites for Construction*, (Vol. 15, Issue 5, pp 757–772), DOI: 10.1061/(ASCE)CC.1943-5614.0000204.

Bilotta, A., Di Ludovico, M., & Nigro, E. (2009a). Influence of effective bond length on FRP-concrete debonding under monotonic and cyclic actions, *Proceedings of the 9th International Symposium on Fiber Reinforced Polymer Reinforcement for Concrete Structures Sydney*, Australia 13–15 July 2009.

Bilotta, A., Ceroni, F., Di Ludovico, M., Fava, G., Ferracuti, B., Mazzotti, C., Nigro, E., Olivito, R., Pecce, M., Poggi, C. & Savoia, M. (2009b). Experimental round robin test on FRP concrete bonding, *Proceedings of FRP RCS9*, 13–15 July, Sydney, Australia, CD ROM.

Brosens, K., & Van Gemert, D. (1997). Anchoring stresses between concrete and carbon fibre reinforced laminates, *Proceedings of the 3rd International Symposium on Non-metallic (FRP) Reinforced for Concrete Structures*, (Vol. 1, October 1997, pp. 271–278).

Burdekin, F. M. (2007). General principles of the use of safety factors in design and assessment. *Engineering Failure Analysis, 14*(3), 420–433.

CEB-FIP (1993). Model Code 1990, Final Draft, Bulletin d'Information n. 213/214.

Ceroni, F., & Pecce, M. (2010). Evaluation of bond strength and anchorage systems in concrete elements strengthened with CFRP sheets, *Journal of Composites in Construction*, ASCE. (In press)

Chajes, M. J., Finch, W. W., Januszka, T. F., & Thomson, T. A. (1996). Bond and force transfer of composite material plates bonded to concrete, *ACI Structural Journal*, *93*(2), 208–217.

Chen, J. F., & Teng, J. G. (2001). Anchorage strength models for FRP and Steel Plates bonded to concrete. *ASCE Journal of Structural Engineering, 127*(7), 784–791.

CNR-DT 200. (2004). Guide for the design and construction of externally bonded FRP systems for strengthening, Council of National Research, Rome, http://www.cnr.it/documenti/norme/IstruzioniCNR_DT203_2006_eng.pdf.

Di Ludovico, M., Piscitelli, F., Prota, A., Lavorgna, M., Manfredi, G., & Mensitieri, G. (2009). CFRP Laminates Behavior at Elevated Temperature. In: *Proceedings of the 7th International Conference on Composite Science and Technology*, January 20–22, 2009, Sharjah, United Arab Emirates.

EN-1992, EC2. (2005). Design of concrete structures.

EN-1993, EC3. (2005). Design of steel structures.

EN-1994, EC4. (2004). Design of composite steel and concrete structures.

EN-1990, EC0. (1990). Basis of structural design.

European Commitee for Standardization. (2002). EN 1990-Eurocode: Basic of Structural Design.

European Commitee for Standardization. (2004). EN 1992–1-1-Eurocode 2: Design of Concrete Structures. Part 1-1: General Rules and Rules for Buildings, ENV 1992-1-1.

European Commitee for Standardization. (2005). EN 1998–3-Eurocode 8: Design of structures for earthquake resistance. Part 3: Assessment and retrofitting of buildings.

Faella, C., Martinelli, E., & Nigro, E. (2010). A simplified design formula for intermediate debonding failure in RC beams externally strengthened by FRP, 3rd fib International Congress, May 29—June 2, 2010, Paper ID: 672, 8, Washington, USA.

Faella, C., Martinelli, E., & Nigro, E. (2004). Debonding in FRP-strengthened RC beams: comparison between code provisions, *Proceedings of the 2nd International Conference on FRP Composites in Civil Engineering*, Paper 074, Adelaide (Australia), December 8–10, 2004.

Faella, C., Martinelli, E., & Nigro, E. (2008a). Direct versus Indirect Method for Identifying FRP-to-Concrete Interface Relationships. *Journal of Composites for Construction, 13*(3), 226–233.

Faella, C., Martinelli, E., & Nigro, E. (2008). Formulation and Validation of a Theoretical Model for Intermediate Debonding in FRP Strengthened RC Beams, Composites Part B 2008, (39(4), pp. 645–655).

Faella, C., Martinelli, E., Nigro, E., Salerno, N., Sabatino, M., & Mantegazza G. (2002). Aderenza tra calcestruzzo e fogli di FRP utilizzati come placcaggio di elementi inflessi.:Parte prima: risultati sperimentali, *Proceedings of* the *XIV C.T.E. Conference*, Mantova, Italy, Nov. 2002, (in Italian).

Faella, C., Martinelli, E., & Nigro, E. (2008). Some remarks on the parameters affecting debonding in FRP strengthened RC beams, *Proceedings of CICE 2008*, Zurich (CH), 22–24 July, Paper E 145.

Ferracuti, B., Martinelli, E., Nigro, E., & Savoia, M. (2007). Fracture Energy and Design Rules against FRP-Concrete Debonding, *Proceedings of 8th FRPRCS Conference*, 16–18 July 2007, Patras (GR).

fib. (2001). Externally Bonded FRP Reinforcement for RC Structures. fib Bulletin 14, Technical Report, Task Group 9.3—FRP Reinforcement for Concrete Structures, International Federation for Structural Concrete.

Grace, N. F., Sayed, G. A., Soliman, A. K., & Saleh, K. R. (1999). Strengthening Reinforced Concrete Beams Using Fiber Reinforced Polymer (FRP) Laminates. *ACI Structural Journal, 96*(5), 865–875.

JSCE. (2001). Recommendations for upgrading of concrete structures with use of continuous fiber sheets, Concrete Engineering Series 41.

Khomwan, N., Foster, S. J., & Smith, S. T. (2004). Debonding failure in CFRP strengthened concrete beams, *Proceedings of the 2nd International Conference on FRP Composites in Civil Engineering*, (pp. 505–514), CICE 2004, Adelaide (Australia).

Leone, M., Aiello, M. A., & Matthys, S. (2009). Effect of elevated service temperature on bond between FRP EBR systems and concrete. *Composites Part B: Engineering, 40*(1), 85–93.

Lu, X. Z., Teng, J. G., Ye, L. P., & Jiang, J. J. (2005). Bond–slip models for FRP sheets/plates bonded to concrete, Engineering Structures, (27, pp. 920–937) Elsevier.

Madsen, H. O., Krenk, S., & Lind, N. C. (1986). *Methods of structural safety*. Ellingwood Cliffs: Prentice-Hall.

Mazzotti, C., & Savoia, M. (2009). Experimental Tests on Intermediate Crack Debonding Failure in FRP—Strengthened RC Beams Advances in Structural Engineering (Vol. 12 No. 5).

Mazzotti, C., Savoia, M., & Ferracuti, B. (2008). An experimental study on delamination of FRP plates bonded to concrete. *Construction and Building Materials, 22*(7), 1409–1421.

Mazzotti, C., Savoia, M., & Ferracuti, B. (2009). A new single-shear set-up for stable debonding of FRP–concrete joints. *Construction and Building Materials, 23*(4), 1529–1537.

McSweeney, B. M., & Lopez, M. M. (2005). FRP-concrete bond behavior: a parametric study through pull-off testing. In C. K. Shield & J. P. Busel (Eds.), *Proceedings of 7th International Symposium FRP Reinforcement for Concrete Structures* (pp. 441–460). Kansas City, Missouri.

Melchers, R. E. (1987). Structural Reliability. Analysis and Prediction, John Wiley & Sons, New York, USA.

Monti, G., Alessandri, S., & Santini, S. (2009). Design by Testing: A Procedure for the Statistical Determination of Capacity Models, *Journal of Construction and Building Materials*, Special Issue on FRP Composites (Vol. 23, pp. 1487–1494), Elsevier.

Monti, G., & Petrone, F. (2014). Calibration of Capacity Models. *ASCE, Journal of Structural Engineering*, (submitted).

Mood, A. M., Graybill, F. A., & Boes, D. C. (1974). *Introduction to the Theory of Statistics* (3rd ed.). New York: McGraw-Hill.

Napoli, A., Matta, F., Martinelli, E., Nanni, A., & Realfonzo, R. (2010). Modelling and verification of response of RC slabs strengthened in flexure with mechanically fastened FRP laminates. *Magazine of Concrete Research, 62*(8), 593–605.

Neubauer, U., & Rostásy, F. S. (1997). Design aspects of concrete structures strengthened with externally bonded CFRP-plates, *Proceedings of 7th International Conference on Structural Faults and Repair Concrete + Composites* (Vol. 2, pp. 109–118).

Nigro, E., Di Ludovico, M., & Bilotta, A. (2008). FRP-concrete debonding: experimental tests under cyclic actions, *Proceedings of the 14th World Conference on Earthquake Engineering* . October 12–17, 2008, Beijing, China.

Pan, J., Chung, T. C. F., & Leung, C. K. Y. (2009). FRP Debonding from Concrete Beams under Various Load Uniformities Advances in Structural Engineering (Vol. 12 No. 6).

Pham, H., & Al-Mahaidi, R. (2004) Bond characteristics of CFRP fabrics bonded to concrete members using wet lay-up method, *Proceedings of the 2nd International Conference on FRP Composites in Civil Engineering*, CICE 2004 (pp. 407–412) Adelaide (Australia).

Said, H., & Wu, Z. (2008). Evaluating and proposing models of predicting IC Debonding Failure. *ASCE Journal of Composites for Construction, 12*(3), 284–299.

Sedlacek, G., & Kraus, O. (2007). Use of safety factors for the design of steel structures according to the Eurocodes. *Engineering Failure Analysis, 14*(3), 434–441.

Shapiro, S. S., & Wilk, M. B. (1965). An analysis of variance test for normality (complete samples). *Biometrika, 52*, 591–611.

Sharif, A., Al-Sulaimani, G. J., Basunbul, I. A., Baluch, M. H., & Ghaleb, B. N. (1991). Strengthening of Initially Loaded Reinforced Concrete Beams using FRP Plates. *ACI Structural Journal, 91*(2), 160–168.

Smith, S. T., & Teng, J. G. (2002). FRP-strengthened RC beams-I: review of debonding strength models. *Engineering Structures, 24*(4), 385–395.

Takeo, K., Matsushita, H., Makizumi, T., & Nagashima, G. (1997). Bond characteristics of CFRP sheets in the CFRP bonding technique. *Proceedings of Japan Concrete Institute, 19*(2), 1599–1604.

Taljsten, B. (1994). Plate bonding: Strengthening of existing concrete structures with epoxy bonded plates of steel or fibre reinforced plastics, Doctoral thesis, Lulea, University of Technology, Sweden.

Task Group 9.3. (2001). Externally Bonded FRP Reinforcement for RC Structures, Technical Report Bulletin 14, fib-CEB-FIP.

Teng, J. G., Chen, J. F., Smith, S. T., & Lam, L. (2001). FRP-strengthened RC structures (p. 245.) John Wiley and Sons: Chichester, West Sussex, UK.

Teng, J. G.; Lu, X. Z.; Ye, L. P., & Jiang, J. J. (2004). Recent Research on Intermediate Crack Induced Debonding in FRP Strengthened Beams, *Proceedings of the 4th International Conference on Advanced Composite Materials for Bridges and Structures*, Calgary, AB, Canada.

Teng, J. G., Smith, S. T., Yao, J., & Chen, J. F. (2003). Intermediate Crack Induced Debonding in RC Beams and Slabs. *Construction and Building Materials, 17*(6–7), 447–462.

Travassos, N., Ripper, T., & Appleton, J. (2005). Bond stresses characterization on CFRP-RC interfaces, *Proceedings of 3rd International Conference Composites in Construction*, CCC2005, Lyon, France.

Triantafillou, T., & Plevris, N. (1992). Strengthening of RC beams with epoxy-bonded FRP composite materials. *Materials and Structures, 25*, 201–211.

Ueda, T., Sato, Y., & Asano, Y. (1999). Experimental study on bond strength of continuous carbon fiber sheet, *Proceedings of 4th International Symposium on Fiber Reinforced Polymer reinforcement for Reinforced Concrete Structure* (pp. 407–16).

Wu, Z., & Niu, H. (2007). Prediction of Crack-Induced Debonding Failure in R/C Structures Flexurally Strengthened with Externally Bonded FRP Composite. *JSCE Journal of Materials, Concrete Structures and Pavements, 63*(4), 620–639.

Wu, Z. S., Yuan, H., Yoshizawa, H., & Kanakubo, T. (2001). Experimental/analytical study on interfacial fracture energy and fracture propagation along FRP-concrete interface, ACI International SP-201–8 (pp. 133–52).

Yao, J., Teng, J.G., & Chen, J.F. (2005). Experimental study on FRP-to-concrete bonded joints, Composites: Part B Engineering (Vol.36, pp. 99–113) Elsevier.

Yua, Z., Dinga, F., & Caib, C. S. (2007). Experimental behavior of circular concrete-filled steel tube stub columns. *Journal of Constructional Steel Research, 63*, 165–174.

Zhao, H. D., Zhang, Y., & Zhao, M. (2000). Research on the bond performance between CFRP plate and concrete, *Proceedings of 1st Conference on FRP Concrete Structures of China* (pp. 247–53).

Chapter 3
Bond Between EBR FRP and Concrete

Claudio Mazzotti, Antonio Bilotta, Christian Carloni, Francesca Ceroni, Tommaso D'Antino, Emidio Nigro and Carlo Pellegrino

Abstract This chapter provides an overview of the debonding process between the FRP reinforcement and the concrete substrate. The main aspects of the debonding phenomenon are described and discussed, showing also mechanical interpretation of different processes. Experimental techniques to study the bond behavior between FRP and concrete are also described and corresponding available experimental results are shown to compare performances of different set-ups. Finally, an extensive description of the existing bond capacity predicting models is reported, together with the main international Codes provisions, allowing the designer for operating in common practice.

Keywords FRP · Strengthening · Bond · Concrete

General Aspects

Mechanism of Debonding Failure

Debonding of FRP reinforcement is one of most important failure modes to be considered in design of strengthening of reinforced concrete structures by means of composite materials (Teng et al. 2001). Following recent design Codes and Recommendations (*fib* 2001; ACI 2008; CNR 2013), anchorage force (bond

C. Mazzotti (✉) · C. Carloni
University of Bologna, Bologna, Italy
e-mail: claudio.mazzotti@unibo.it

A. Bilotta · E. Nigro
University of Naples Federico II, Naples, Italy

F. Ceroni
University of Sannio, Benevento, Italy

T. D'Antino · C. Pellegrino
University of Padua, Padua, Italy

C. Pellegrino and J. Sena-Cruz (eds.), *Design Procedures for the Use of Composites in Strengthening of Reinforced Concrete Structures*, RILEM State-of-the-Art Reports 19, DOI 10.1007/978-94-017-7336-2_3

capacity) of FRP sheets attached to the concrete surface is the basis of design rules for shear strengthening applications, since it determines maximum force that can be applied to external reinforcement. Moreover, intermediate crack-induced debonding (Teng et al. 2003) between concrete and external FRP plate/sheet reinforcement (debonding initiating at flexural cracks along the beam) is the failure mode governing maximum FRP strain which can be adopted for flexural design. For reinforcement with FRP plates, design maximum strain against debonding is usually significantly smaller than corresponding to composite failure.

When strengthening reinforced concrete members with FRP composites, the role of bond between concrete and FRP is of great relevance due to the brittleness of the failure mechanism by debonding (loss of adhesion). According to the capacity design criterion, such a failure mechanism shall occur prior to flexural or shear failure of the strengthened member. The loss of adhesion between FRP and concrete may concern both laminates or sheets applied to reinforced concrete beams as flexural and/or shear strengthening. As shown in Fig. 3.1, debonding may take place within the adhesive, between concrete and adhesive, in concrete itself, or within the FRP reinforcement (e.g. at the interface between two adjacent layers bonded each other) with different fibre inclination angles. When proper installation is performed, because the adhesive strength is typically much higher than the concrete tensile strength, debonding always takes place within the concrete itself with removal of a layer of material, whose thickness may range from few millimeters to the whole concrete cover.

Debonding failure modes for laminates or sheets used for flexural strengthening may be classified in the following four categories (Fig. 3.2):

1. Plate/sheet end debonding
2. Intermediate debonding caused by flexural cracks
3. Debonding caused by diagonal shear cracks
4. Debonding caused by irregularities and unevenness of concrete surface

The first two modes are the most frequent in ordinary situations.

The tension in the plate/sheet is transferred to the concrete mainly by shear stresses in the adhesive. When a limit shear stress (bond strength) is attained,

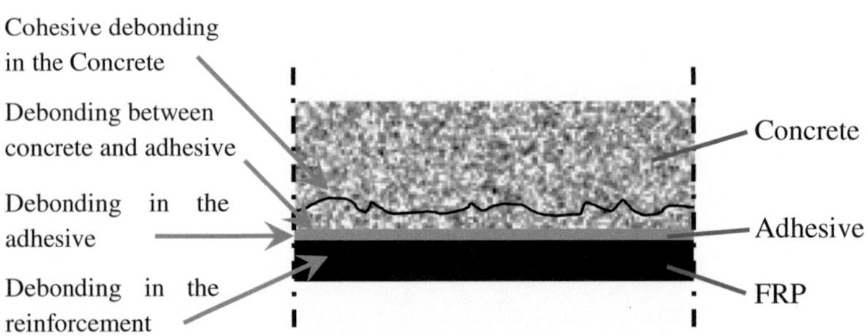

Fig. 3.1 Debonding mechanisms between FRP and concrete

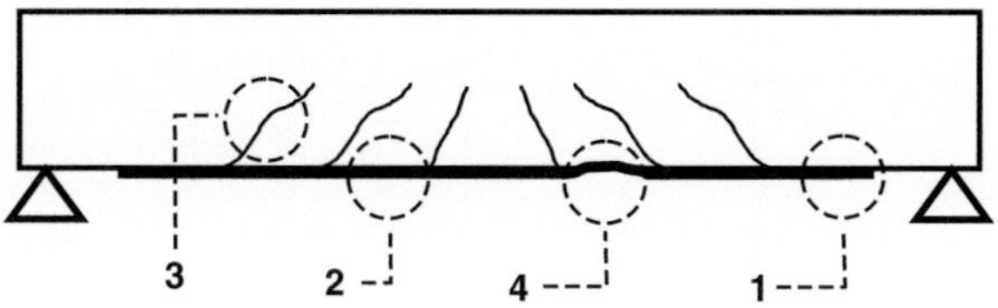

Fig. 3.2 Modes of debonding failure in a beam under flexure and shear

Fig. 3.3 Interface shear stress-slip local behavior

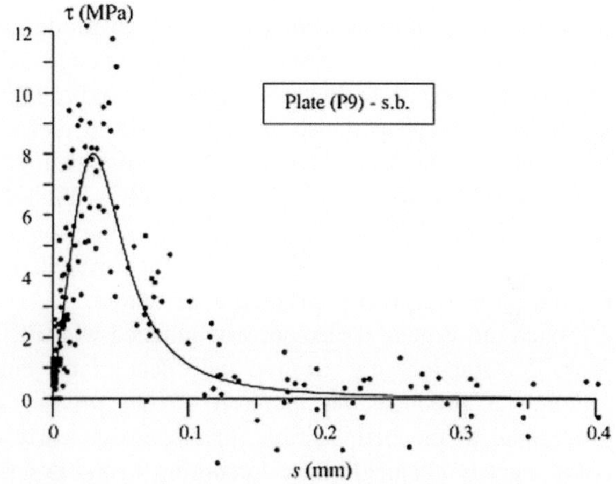

debonding starts according to one of previous modes. During the process, a shifting of the active zone (that able to transfer force) can be observed, which means that only part of the bonded area is effective. That is, as cracking in the concrete propagates, bond resistance is gradually lost in the zone near the load, but in the meantime it is activated farther away from the load. The implication is, then, that the anchorage force cannot increase with an increase in the bond length, and that the ultimate tensile strength of a plate may never be reached, no matter how long the bonded length is. This leads to the important concept of effective bond length, beyond which any increase in the bond length cannot increase the anchorage strength. From a mechanical point of view, it can be explained by considering a local non-linear brittle shear stress-slip relationship showing a post-peak softening behaviour, as confirmed by a number of experimental tests (Fig. 3.3).

Normal stresses in the FRP reinforcement are mainly transmitted through the substrate and the adhesive by means of shear stresses applied to its surface, usually producing Mode II shear condition (Buyukozturk et al. 2004). In fact, only a small layer of concrete close to the interface is subject to very high shear stresses and criterion of the maximum release rate requires that fracture propagates along it. During debonding, the portion of concrete where shear stresses are transmitted is in fact very small: 3–5 cm depth.

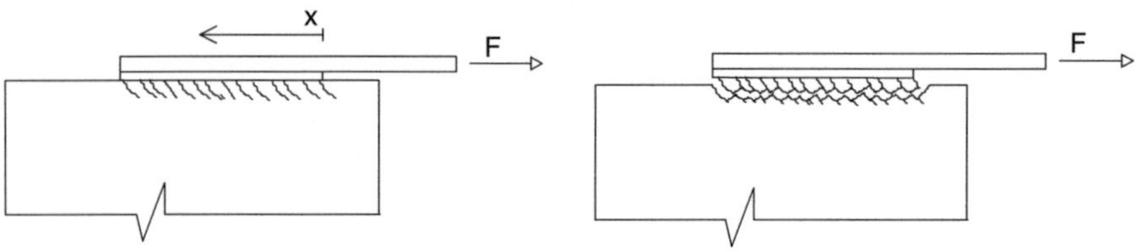

Fig. 3.4 Failure mechanism at the interface level

Failure mechanism, according to Mohr-Coulomb criterion can be described as follow: inclined microcracks start locally in Mode I condition in the small superficial layer of concrete because its tensile strength is much lower than that of the adhesive (see Fig. 3.4). Inclined cracks cannot propagate more than few millimetres inside the concrete specimen because stresses decrease very rapidly with depth from FRP-concrete interface. Then, a series of inclined struts clamped to concrete substrate are subject to compression and bending. Final failure can be due to concrete crushing in compression or transverse cracking on tensile side of concrete struts, depending on dimensions of struts, and a corrugated debonding surface parallel to the interface is typically detected after failure.

When the structural element strengthened with FRP is also subject to flexural or shear deformation (e.g. beams) the effect of the curvature has to be taken into account; it produces an increase of the peeling stresses along the direction orthogonal to the FRP surface, leading to an early detachment when this stress component is not negligible. According to the described mechanisms, the failure mode is sometimes addressed like a mix-mode failure and, for simplicity, it is usually governed by a Mode II fracture energy empirically defined.

Anchorage Length

The formulation suggested by *fib* (2001) and CNR (2004) for the effective bond length is:

$$L_e = \sqrt{\frac{E_f \cdot t_f}{2 \cdot f_{ctm}}} \tag{3.1}$$

whereas the formulation provided by Chen and Teng (2001) is:

$$l_{ed} = \sqrt{\frac{E_f \cdot t_f}{\sqrt{f_c'}}} \tag{3.2}$$

Both relationships show the obvious inverse dependence of l_{ed} on the substrate strength. If a rigid-softening bilinear law is assumed, the following theoretical relationship can be written (Faella et al. 2002):

$$l_e = \frac{\pi}{2}\sqrt{\frac{E_f \cdot t_f}{k_u}} = \frac{\pi}{2}\sqrt{\frac{E_f \cdot t_f}{\tau_{max}/s_u}} \tag{3.3}$$

in which the effective bond length is expressed as function of the stiffness of the substrate. Equation (3.3) can be rewritten as follows:

$$l_e = s_u\sqrt{\frac{\pi^2 \cdot E_f \cdot t_f}{8 \cdot \Gamma_f}} \tag{3.4}$$

This relationship provides mean values of the effective length, if the mean values of fracture energy is used. About the estimation of s_u, in Bilotta et al. (2011) the value 0.25 mm allowed the theoretical distribution of strain along the FRP reinforcement for fitting quite well the experimental ones by means of an inverse analysis procedure (Faella et al. 2009). Under the assumption of bilinear bond law, indeed, the procedure minimizes the scatter between the numerical and the experimental strains as the main parameters of the bond law are changed. However, the numerous experimental results investigated were referred to specimens with concrete compressive strength of about 20 MPa. A wider range of concrete strength should be investigated in order to have a more reliable estimation of s_u. Note that the following expression of s_u (Eq. 3.5) can be deduced by the relationships provided in Lu et al. (2005) under the bilinear bond law hypothesis:

$$s_u = \frac{2 \cdot G_f}{\tau_{max}} = \frac{0.41 \cdot \beta}{f_{ct}^{0.5}} \tag{3.5}$$

being G_f = 0.308 N/mm and τ_{max} = 1.5 MPa; β is a shape factor similar to k_b. For β = 1 and f_c = 20 MPa, the value $s_u \approx 0.25$ mm is confirmed.

Figure 3.5 shows the strain profiles recorded during shear tests carried out on two series of three sheets (V18a,V19a and V20a—see Fig. 3.5a) and three plates (laminates) (V1a,V2a and V3a—see Fig. 3.5b) glued on concrete specimens for a length l_b = 400 mm (Bilotta 2010). The strain profiles were recorded when the first debonding load, P_{fd}, that identified the beginning of debonding, was attained. Hence the distance from FRP end at which strains equal to zero were recorded represents the experimental effective bond length value, $l_{e,exp}$. The agreement between $l_{e,exp}$ and the value l_e calculated through the formulation (3.4) is good. However, in order to obtain a design value for the effective bond length, the Eq. (3.4) is modified as follows:

$$l_{ed} = \frac{1}{\gamma_{Rd} \cdot \tau_{max}}\sqrt{\frac{\pi^2 \cdot E_f \cdot t_f \cdot \Gamma_{Fd}}{2}} \text{ with } \tau_{max} = \frac{2 \cdot \Gamma_{Fd}}{s_u} \tag{3.6}$$

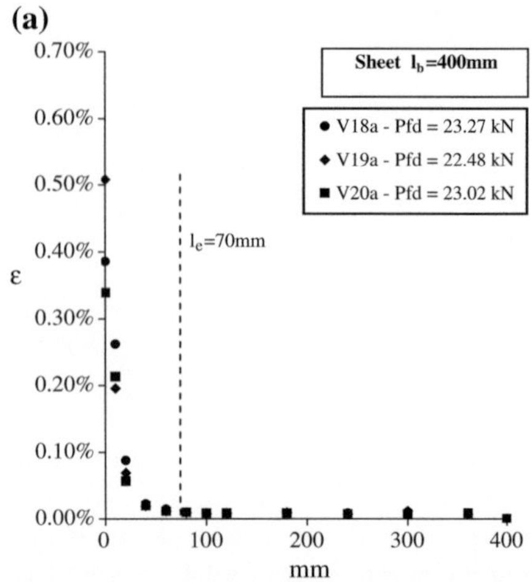

Fig. 3.5 Strain profiles along FRP: **a** sheets; **b** plates

In Eq. (3.6), which is included in CNR (2013), the design value of the fracture energy, Γ_{Fd}, is introduced and $\gamma_{Rd} = 1.25$ is a model factor introduced because it is not currently possible to calibrate a design value for s_u with the available experimental data.

The effective bond length calculated through Eqs. (3.1), (3.2), (3.4) and (3.6) are plotted against the concrete compressive strength, f_{cm}, for sheets ($E_f = 216$ GPa and $t_f = 0.166$ mm) and plates ($E_f = 171$ GPa and $t_f = 1.4$ mm) in Fig. 3.6a, b, respectively.

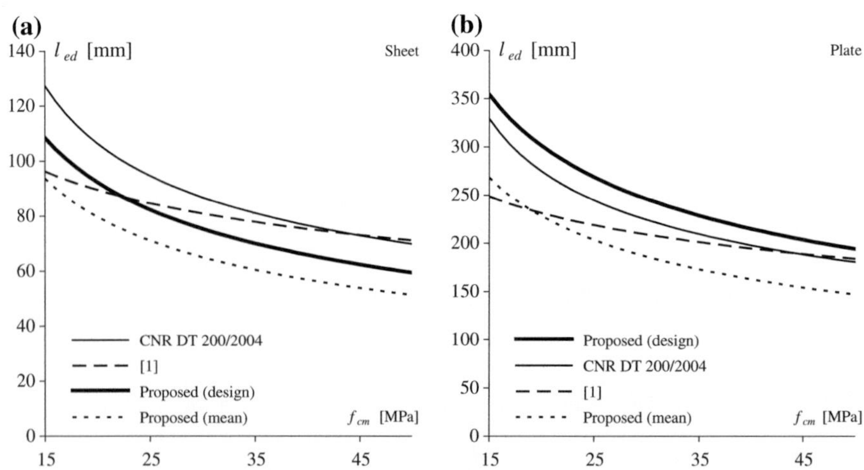

Fig. 3.6 Bond length versus concrete compressive strength: **a** Sheets **b** Plates

The mean values provided by the proposed formulation are lower than provisions given by Chen and Teng (2001) for both sheets and plates, whichever the concrete compressive strength is. On the contrary, the design values are slightly less safe for sheets and slightly more safe for plates.

Effect of Surface Preparation

As well recognized in literature, the bond capacity between the FRP system and the concrete depends on a number of factors, including the material properties of the epoxy matrix and of the fibers as well as the properties of the concrete substrate (Miller and Nanni 1999). Among these, particularly important are the concrete strength and surface roughness and cleanliness. For this reason, the surface preparation methods are a key issue governing the possible success of the strengthening intervention. In more details, surface preparation is the process by which the concrete substrate must be sound, clean, and suitably roughened. This process include the removal of unsound concrete and bond-inhibiting films, strength verification, and opening of the pore structure.

Some of the most common surface preparation methods are: brushing, grinding, scarifying, bush-hammering, steel shotblasting, sandblasting, each with advantages and disadvantages associated to several factors as the desired roughness profile of the prepared surface, cost, and processing time. As a confirmation, Chajes et al. (1996) showed that the interfacial bond strength increased when the surface is prepared using mechanical abrading. Similarly, Yao et al. (2005) presented an experimental study on the bond shear strength between FRP and concrete using a near end supported (NES) single-shear pull test; the corresponding test results emphasized the role of a careful specimen preparation that significantly affected the bond capacity, together with the amount of the removed substrate at failure. Delaney and Karbhari (2007) reported that the surface preparation influences not only the instantaneous behaviour but also the durability of the system. Unfortunately, extensive experimental data concerning the FRP to concrete bond quality are available mainly for sandblasting while for others treatments few data can be found (Mazzotti et al. 2007).

At a design level, most FRP design and construction Guidelines recommend surface preparation methods for effective applications. The International Concrete Repair Institute (ICRI) produced a guideline (Savoia et al. 2009) for concrete surface preparation for polymer overlays that carefully describes several concrete surface preparation methods stating the advantages and the disadvantages of each one. The ACI Committee 440 (2002) suggests abrasive or water blasting techniques for surface preparation to a minimum concrete surface profile CSP 3, as defined by ICRI. Nevertheless, even with roughness level not strictly complying with ICRI/ACI (1999) prescriptions, an adequate level of adhesion can be obtained (Shen et al. 2002). The Italian technical guideline DT 200 (CNR 2004) suggests that once the quality control of the substrate has been performed, the deteriorated concrete has been removed, the concrete cross-section restored, and the existing steel reinforcement has been properly treated, then sandblasting of the concrete surface to be strengthened should be performed.

Ueda and Dai (2005) observed that a large amount of the scattering of experimental results concerning bond strength is due to concrete surface conditions and preparation, even when different operators or laboratories follow the same surface treatment procedure. More recently Serbescu et al. (2013) collected a large amount of data from which they calibrated some empirical laws also concerning the effect of surface preparation.

In Iovinella et al. (2013) an extensive experimental campaign is described, where surface roughness was measured on concrete specimens treated with different surface preparation, prior to FRP application, by means of a laser profilometer and its various aspects were condensed in a simple roughness coefficient (Fig. 3.7a). Two types of tests were carried out on the strengthened specimens: conventional pull-off tests (suitable for on-site application—Fig. 3.7b) and pull-out bond tests. Results show a clear correlation between the type of surface preparation and the bond strength obtained by pull-out tests (Fig. 3.8a). The interface law (by means of the fracture energy—Fig. 3.8b), which will be introduced in the next section, is also affected by the type and effectiveness of the considered treatment. The quantitative approach allowed for proposing a simple design formula able to take into account the specific roughness considered when predicting the bond strength.

In this framework, by following the CNR (2004) approach the fracture energy can be defined as:

$$\Gamma_f = k_G \cdot k_b \cdot k_r \sqrt{f_{cm} \cdot f_{ctm}} \tag{3.7}$$

where $k_r = 0.766 + 0.08 \cdot I_R$, while all the other coefficients have the conventional meaning.

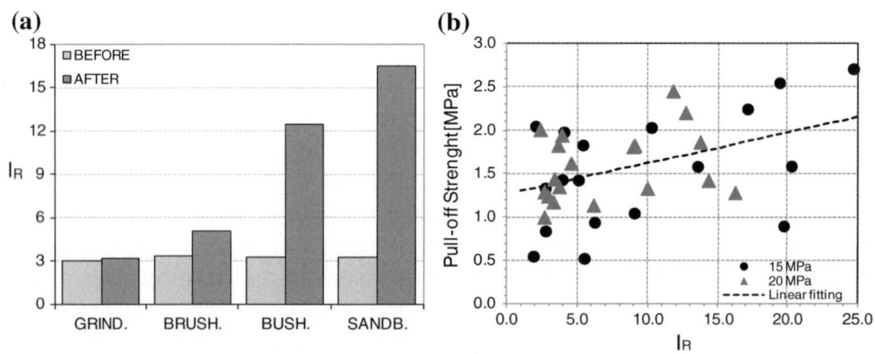

Fig. 3.7 **a** Roughness index measured from different surface preparation and **b** corresponding pull-off strength (Iovinella et al. 2013)

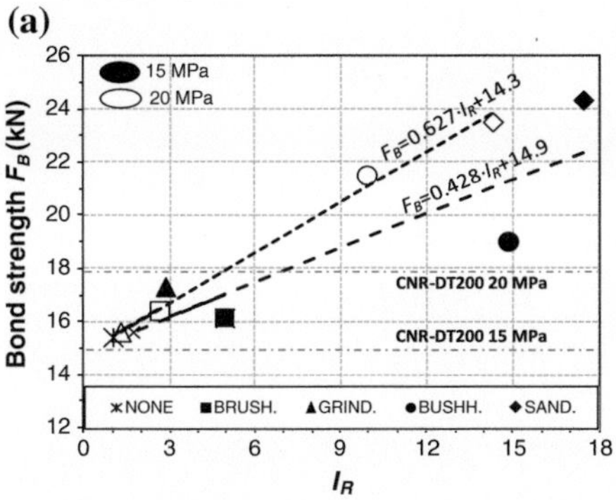

Fig. 3.8 **a** Bond strength and **b** fracture energy coefficient k_G as a function of roughness index (Iovinella et al. 2013)

Bond as a Fracture Mechanics Process

Introduction

Fracture propagation during intermediate crack-induced debonding (ICD) occurs in concrete, therefore it is reasonable to assume that the *quasi-brittle behaviour* of concrete governs the debonding process. It seems appropriate to model the debonding mechanism within the framework of fracture mechanics applied to quasi-brittle materials (Bazant and Planas 1997). In linear elastic fracture mechanics (LEFM) it is well known that a singularity in the stress and displacement fields occurs at the crack tip (Anderson 2004; Bazant and Planas 1997). In LEFM the nonlinearity of the material is neglected although in applications only a limit value of the stress can be reached and in a zone near the crack tip a stress re-distribution occurs, which is related to the nonlinearity of the material. The nonlinearity near the crack tip can be due to hardening or softening of the material. In order to overcome the inapplicability of LEFM for quasi-brittle materials such as concrete, the concept of *fracture process zone* (FPZ) is introduced. The FPZ represents a bridging zone between the cracked and uncracked regions, where progressive softening occurs. It is important to point out that in quasi-brittle materials the FPZ is most likely coincident with, or close in size to, the region of material nonlinearity. In other materials, such as steel, the softening part of the nonlinear zone is negligible and the nonlinear hardening zone is predominant. The characteristics and size of the FPZ depend on the material. In concrete, the FPZ is related to progressive damage that is associated with several complex phenomena (microcracking, void formation, etc.). As the fracture process progresses in concrete, coalescence of microcracks in the FPZ gives continuity to the already existing crack and consequently the crack propagates.

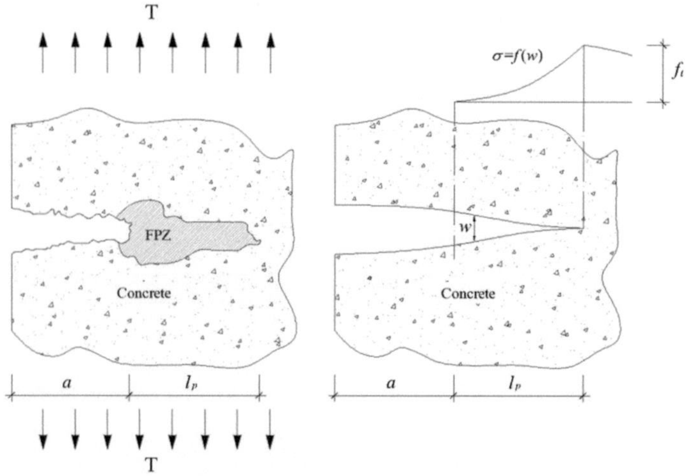

Fig. 3.9 Fracture process zone in concrete and cohesive crack model

The concept of cohesive zone or cohesive crack is associated with the work of Hillerborg et al. (1976). The cohesive crack model is a simple model for the FPZ in concrete and can be taken as a reference to compare other models (Elices et al. 2002). For a Mode-I condition, it is assumed that within the FPZ a crack tip should not be defined. Conversely, the stress σ transferred through the cohesive crack is a function of the crack opening w (Fig. 3.9). The function $\sigma = f(w)$ is characteristic of the material and is often called the *softening function*. When $w = 0$, $\sigma = f_t$, which is the tensile strength of concrete.

Several researchers attempted to study ICD phenomenon and the debonding in direct-shear tests as a Mode-II fracture problem (Anderson 2004) where the interface region is idealized to be of zero thickness with well-defined material properties (Ali-Ahmad et al. 2006; Wu and Niu 2007; Mazzotti et al. 2008). In this idealization, the interfacial crack, associated with initiation and propagation of debonding, is subject to a Mode-II loading condition. The quasi-brittle behaviour of the interfacial crack, in the spirit of the cohesive crack model, is described by introducing a cohesive material law, which relates the interface shear stress (τ) to the relative slip (s) between FRP and concrete. It is important to highlight that the cohesive material law τ-s represents the constitutive law of a fictitious material that links the FRP strip to the concrete substrate (interface). The shear stress should not be associated with the shear stress that occurs in concrete at a particular distance from the concrete surface. Although it is reasonable to assume that the stress field in concrete near the interface has an important role in the stress transfer and therefore in the debonding mechanism, the cohesive material law, herein introduced, aims to describe the interfacial debonding at the macro-scale; hence its parameters should not refer to the actual stress state in concrete at the microscopic level.

Fracture along the interface does not occur on an ideal plane parallel to the FRP strip but follows a tortuous path, which is in part controlled by the distribution of the aggregates and in part by the mixed-mode nature of the fracture process at the microscopic level. In fact, the crack continuously kinks to follow the path that requires the least amount of energy and is related to the fracture properties of the two materials (Hutchinson and Suo 1992; Gunes 2004; Gunes et al. 2009). At the macroscopic level, the microscopic mixed-mode fracture can be considered a Mode-II fracture at the FRP-concrete interface.

Direct-Shear Tests

Several analytical/numerical procedures to estimate the cohesive interfacial behaviour from the load response of direct-shear tests were developed. For example, Ali-Ahmad et al. (2006) established an experimental procedure to directly determine the Mode-II interfacial fracture law using DIC measurements. The cohesive material law for the interface, when implemented in a numerical analysis procedure, allowed for predicting the load response of concrete beams strengthened with externally-bonded FRP sheets (Wu and Yin 2003; Ali-Ahmad et al. 2007; Wu and Niu 2007). Among others, Ferracuti et al. (2006, 2007); Mazzotti et al. (2008), and Carrara et al. (2011) used a procedure similar to Ali-Ahmad et al. (2006). In those studies, the authors used strain gauge readings along the FRP surface to obtain the interfacial law. Pellegrino et al. (2008) and Pellegrino and Modena (2009) used a double-lap shear test and a small-beam test to investigate the effect of the axial stiffness of the composite on the cohesive material law and indicated the need of more research to study this aspect.

In what follows, reference will be made to the experimental work reported in Subramaniam et al. (2007). Single-lap direct-shear tests were used to evaluate the FRP-concrete debonding using the classical pull-push configuration. The tensile load was applied to the FRP sheet, while the concrete block was restrained against movement. This set-up is also referred to as the *near-end supported single-shear test* (Yao et al. 2005).

Figure 3.10 shows the specimen dimensions and the loading arrangement (Carloni and Subramaniam 2012). The Cartesian system depicted in Fig. 3.2 will be used as a reference system for the strain analysis and fracture mechanics approach that follow. Details of the test set-up and materials used can be found in Subramaniam et al. (2007, 2011). Two LVDTs were mounted on the concrete surface close to the edge of the bonded area. The LVDTs reacted off of a thin aluminium Ω-shaped plate, which was glued to the FRP surface at the beginning of the bonded area as shown in Fig. 3.10. The average of the two LVDT readings was named *global slip*. Tests were conducted in displacement control by increasing the global slip at a constant rate equal to 0.00065 mm/s, up to failure. The modality of failure of all direct-shear test specimens was associated with progressive debonding of the FRP sheet from the concrete substrate.

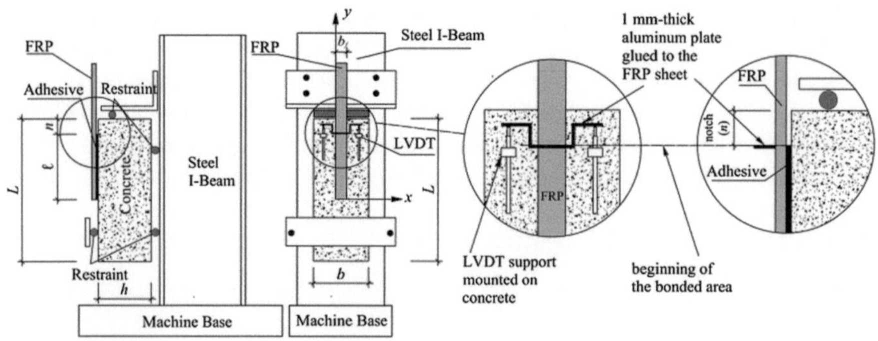

Fig. 3.10 Specimen dimensions and loading arrangement (Subramaniam et al. 2007)

The strain components on the surface of the FRP and surrounding concrete during the monotonic quasi-static tests were determined from the displacement field, which was measured using a full-field optical technique known as digital image correlation (DIC). Details about DIC can be found in Sutton et al. (1983, 2009).

The typical load-global slip response, which can be found in the aforementioned publications, showed an initial linear ascending region followed by a non-linear response. The end of the non-linear part of the response was typically marked a by a load drop that indicated that the interfacial crack has formed. The load was nominally constant after the load drop and the value of the constant load is termed *load-carrying capacity* or *bond capacity* and indicated as P_{crit}. An example of the load response, which corresponds to test W_7 in Subramaniam et al. (2007), is depicted in Fig. 3.11.

Figure 3.12 shows the variation of the axial strain ε_{yy} on the surface of the FRP along the bonded length for three points (A, B, and C) of the load response of Fig. 3.11 in the region where the load is nominally constant. The axial strain distribution along the FRP obtained from all specimens tested was nominally similar. The experimental strain values are represented by markers. The axial strain values were determined along the center line of the FRP sheet by averaging the strain across a 10 mm-wide strip for each value of y. The experimental nonlinear strain distribution along the bonded length of Fig. 3.12 was approximated using the following expression (solid line in Fig. 3.12):

$$\varepsilon_{yy} = \varepsilon_0 + \frac{\alpha}{1 + e^{-\frac{y-y_0}{\beta}}} \tag{3.8}$$

where α, β, y_0, and ε_0 were determined using nonlinear regression analysis of the strains obtained from DIC. The choice of the Eq. (3.8) is not unique (Dai et al. 2005a, b, 2006; Zhou et al. 2010; Liu and Wu 2012). The observed strain distribution along the FRP was essentially equal to zero close to the unloaded end. There was a rapid increase in strain upon approaching the loaded end. The strain leveled

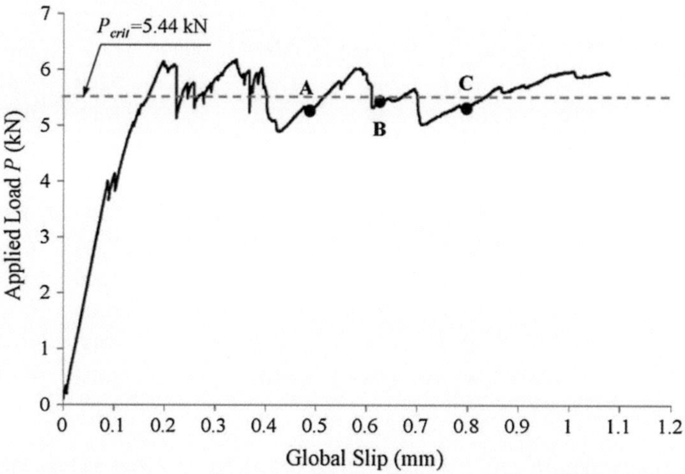

Fig. 3.11 Typical load respond (Test W_7) of a direct shear test (Subramaniam et al. 2007)

Fig. 3.12 Axial strain profiles corresponding to three points (A, B, and C) of the load response of Fig. 3.11

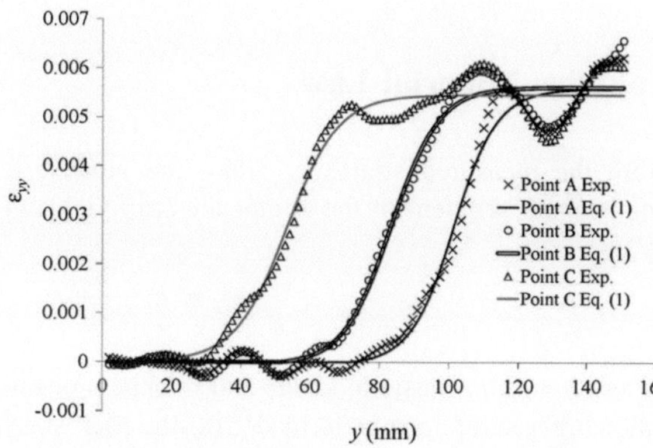

off at a value $\bar{\varepsilon}_{yy}$ which was approximately equal to 5580 με for point A. The observed strain distribution was divided into three main regions: (a) the stress-free zone (SFZ); (b) the stress-transfer zone (STZ); and (c) the fully-debonded zone (FDZ). In the FDZ, the strains were essentially constant and were found to remain unchanged with increasing the global slip, which is consistent with the observation that the load remained nominally constant (P_{crit}) after the debonding process propagated. It can be observed that a simple translation of the STZ further along the length of the FRP sheet occurred as the global slip increased while its shape remained constant. The translation of the STZ indicates self-similar crack growth. It should be noticed that the STZ does not correspond to the FPZ.

The progressive debonding of the FRP composite sheet from concrete is associated with a STZ of a fixed length L_{STZ}, which translates as the crack advances (Fig. 3.11). L_{STZ} is also termed the *effective bond length* (Chen and Teng 2001) or

Table 3.1 Fracture parameters of the tests published in Subramaniam et al. (2007)

Test	b_f (mm)	P_{crit} (kN)	L_{STZ} (mm)	ε_{yy} (με)	G_F (MPa × mm)
W_1	46	12.90	80	7200	0.874
W_2	46	12.05	76	6200	0.634
W_3	46	13.20	75	5900	0.563
W_4	38	10.09	81	6400	0.692
W_5	38	10.02	73	6200	0.652
W_6	25	5.54	80	5600	0.546
W_7	25	5.44	76	5600	0.530
W_8	25	5.36	69	6400	0.705
W_9	19	4.27	75	6400	0.686
W_10	19	4.05	78	5900	0.579

the *development length* (ACI 440.2R-08 2008). The average values of L_{STZ} are plotted in Subramaniam et al. (2007) and summarized in Table 3.1 together with the other fracture parameters, the maximum strain $\bar{\varepsilon}_{yy}$, and the load-carrying capacity.

Cohesive Material Law

From the measured strain ε_{yy} along the bonded length, the equilibrium of an infinitesimal segment of the composite strip yields (Taljsten 1996, 1997a, b):

$$\tau_{zy} = E_f t_f \frac{d\varepsilon_{yy}}{dy} \tag{3.9}$$

E_f and t_f are the elastic modulus and thickness of the composite, respectively. The following assumptions were made: (a) the FRP sheet was homogenous and linear elastic; (b) the thickness and the width of the FRP sheet were constant along the bonded length; (c) the interface was only subject to shear loading; (d) the interface between the FRP and the concrete was assumed to be of infinitesimal thickness; and (e) the concrete substrate was rigid.

The relative slip, $s(y)$, between FRP and concrete at a given location on the FRP was obtained by integrating the axial strain in the FRP up to that point.

Lu et al. (2005) commented on the possible ways to obtain the τ–s curves, observing that the violent local variations of strain measured by strain gauges entailed for substantial differences in the fracture parameters. The procedure followed by Subramaniam et al. (2007, 2011) used the strain contours obtained from DIC, which allowed to identify the fluctuations of the strain profile due to the local variations of the FRP and the substrate (Ali-Ahmad et al. 2006). The cohesive material law curves corresponding to points A, B, and C of the load response of Fig. 3.11 are shown in Fig. 3.13.

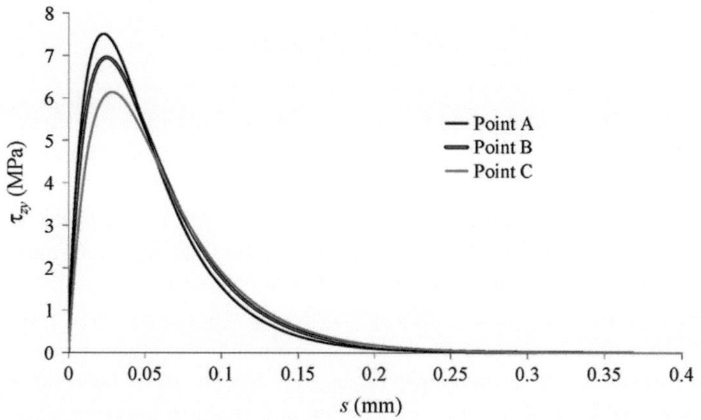

Fig. 3.13 τ_{zy} –s curves for points A, B, and C in Fig. 3.3

Several expressions of the cohesive material law are available in the literature. For example, Ferracuti et al. (2006, 2007) proposed the following relationship:

$$\tau_{zy}(s) = \bar{\tau}\frac{s}{\bar{s}}\frac{n}{(n-1)+\left(\frac{s}{\bar{s}}\right)^n} \tag{3.10}$$

where $\bar{\tau}$ the maximum is shear stress and $\bar{s}$ is the corresponding slip. Other researchers indicated these parameters as $\tau_{\max}$ and s_0, respectively. The parameter n (>2) mainly governs the softening branch of the softening curve.

Wu et al. (2012) proposed the following form of the cohesive material law (Liu and Wu 2012):

$$s'' = \frac{\alpha}{\beta}e^{-s/\alpha}\left(1 - e^{-s/\alpha}\right) \tag{3.11}$$

where s'' is the second order derivative of the slip s and is related to the shear stress if the hypotheses introduced above hold. Wu et al. (2012) used an equilibrium approach and compared the results with empirical formulas to identify the parameters α and β.

Interfacial Fracture Energy

The interfacial fracture energy G_F is the energy required to create and fully break the elementary unit area of the cohesive crack. G_F corresponds to the area under the entire τ_{zy}–s curve (Bazant and Planas 1997):

$$G_F = \int_0^{s_f} \tau_{zy}(s)ds. \quad (3.12)$$

where s_f is the slip corresponding to complete separation of the interface. The mean values of G_F for all tests can be found in the published paper (Subramaniam et al. 2007). The mean values of G_F, as well as those for the effective bond length and the maximum strain at debonding $\bar{\varepsilon}_{yy}$, were obtained from ten points of the load response within the range of global slip in which the load was nominally constant.

The relationship between the interfacial fracture energy and the fracture energy of concrete (Mode-I) is still an open discussion among researchers (Achintha and Burgoyne 2008, 2011; Carrara et al. 2011). Although the fracture process in ICD occurs in concrete, it propagates in a mortar-rich thin layer (Carloni and Subramaniam 2010) in which the mechanical and fracture properties are not easily defined. Undoubtedly, the two fracture energies are related although a convincing relationship has not been found yet. Rabinovitch (2004) successfully used the Mode-I fracture energy of concrete to study the end plate debonding (ACI 440.2R-08 2008) using the fracture mechanics concept of energy release rate. In the context of end plate debonding, the energy required to create and fully break the elementary unit area of cohesive crack should be closely related to the fracture energy of concrete as the debonding typically occurs in the concrete cover and peeling stresses are not negligible.

Taljsten (1996) obtained a relationship between the fracture energy and the load-carrying capacity in direct shear tests by considering the energy release during the advancement of the interfacial crack a by an amount da (Fig. 3.14). The *energy release rate* G per unit width b_f of the composite is obtained as:

$$G = \frac{1}{b_f}\left[\frac{\mathrm{d}}{\mathrm{d}a}(F - U_e)\right] \quad (3.13)$$

where F is the work done by the external load and is U_e the elastic energy. When debonding propagates $G = G_F$. If δ is the displacement of the point of application of the applied force P (Fig. 3.14) and C is the compliance of the system, then:

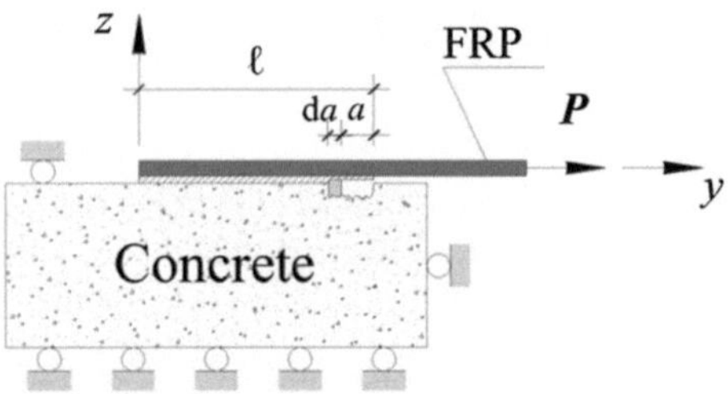

Fig. 3.14 Crack propagation in direct shear tests

$$U_e = \frac{1}{2} P^2 C \tag{3.14}$$

and when the interfacial crack propagates:

$$G_F = \frac{1}{b_f} \left[P \frac{\mathrm{d}\delta}{\mathrm{d}a} - \frac{\mathrm{d}U_e}{\mathrm{d}a} \right] = \frac{1}{b_f} \frac{P^2}{2} \frac{\partial C}{\partial a} \tag{3.15}$$

Therefore:

$$P = \sqrt{2 G_F b_f \Big/ \frac{\partial C}{\partial a}} \tag{3.16}$$

If the substrate is considered rigid and the adhesive layer is idealized as a zero-thickness layer:

$$\frac{\partial C}{\partial a} = \frac{1}{E_f b_f t_f} \tag{3.17}$$

where E_f and t_f are the elastic modulus and thickness of the composite, respectively. From Eqs. (3.16) and (3.17), the interface fracture energy G_F is related to the load-carrying capacity (Hearing 2000; Yuan et al. 2001; Liu and Wu 2012; Wu et al. 2002, 2012):

$$P_u = b_f \sqrt{2 G_F E_f t_f} \tag{3.18}$$

Equation (3.18) can be obtained through an energy balance approach (Taljsten 1996; Hearing 2000; Focacci et al. 2000; Liu and Wu 2012) and is based on the assumption that a pure Mode-II interfacial crack propagation occurs across the entire width of the composite. The theoretical load-carrying capacity under pure Mode-II was indicated as P_u in Eq. (3.18) to distinguish it from the experimental value P_{crit}.

Concluding Remarks

Contradictory resul ts (Chen and Teng 2001; Subramaniam et al. 2007; Mazzotti et al. 2008; Carloni and Subramaniam 2012) complicate the interpretation of the interfacial fracture energy G_F as a true fracture parameter. The discussion on the nature of the interfacial fracture energy as a true fracture parameter in part arises from the misuse of Eq. (3.18). From Eq. (3.18), researchers used the experimental load-carrying capacity P_{crit} to determine the fracture energy by assuming that $P_u = P_{crit}$:

$$G_F = \frac{P_{crit}^2}{2b_f^2 E_f t_f} \tag{3.19}$$

This approach led to an erroneous evaluation of G_F because the experimental value of the load-carrying capacity includes the width effect. The fracture energy cannot be directly related to the experimental load-carrying capacity of the interface. The values of G_F determined via Eq. (3.19) scale with the width of the FRP sheet. Hence, the experimental values of the fracture energy determined through Eq. (3.18) should not be considered as *true* values.

In addition to the discussion regarding the purported nature of the fracture energy, concerns remain on the applicability of the direct-shear test results to describe ICD, mainly because of the presence of a Mode-I component (peeling stresses). The Mode-I should not be confused with the one observed above at the microscopic level, but in the spirit of the macroscopic approach of the fictitious interface. The Mode-I opening is described by the relationship between the normal stress (peeling) σ_{zz} and the opening of the crack δ (Martinelli et al. 2011; Carrara and Ferretti 2013). The Mode-I interfacial fracture energy, corresponding to the area of the $\sigma_{zz} - \delta$ curve, is considerably lower than the Mode-II fracture energy (Taljsten 1996; Gunes 2004), thus even a small component of the load perpendicular to the FRP sheet could potentially reduce the load-carrying capacity of the interface. A Mode-I component is always present in the direct-shear test measurements due to the relationship between shear and moment. A limited number of experimental works reported the study of the Mode-I and mixed-mode debonding (Wan et al. 2004; Davalos et al. 2006; Alam et al. 2012). Some authors (Yao et al. 2005) recognized that the effect of a small loading angle (offset) was insignificant for relatively long bonded lengths. The presence of a Mode-I condition in beams can be explained by considering the opening of a flexural/shear crack as illustrated in Fig. 3.15 (Garden and Hollaway 1998). As the crack opens, the two faces of the crack will undergo a relative vertical displacement that will cause a mixed-mode condition for the FRP-concrete interface. Rabinovitch (2008, 2012) used a fracture mechanics approach that considered the Mode-I and Mode-II cohesive material laws and their coupling. A set of nonlinear differential equations was derived by considering a multi-layer description of the strengthened beam. A different length of the STZ for Mode-I and Mode-II can be observed in these studies. Mazzucco et al. (2012) used a similar approach to capture the coupling of the shear and pealing stresses, but introduced a contact-damage model for the adhesion between layers. Gunes et al. (2009) reported that if the strengthened beam was sufficiently strong in shear, the flexural/shear crack mouth displacement would be limited and consequently the mixed-mode nature of debonding fracture would quickly merge

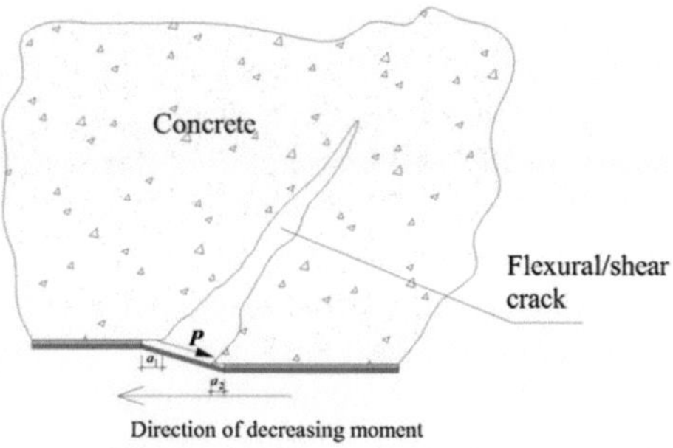

Fig. 3.15 Mixed-mode debonding propagation in beams

into a Mode-II condition. It is interesting to notice that the results published by Alam et al. (2012) showed that the effective bond length increases if the Mode-I component is significant.

Alternative approaches within the framework of fracture mechanics are available in the literature. Achintha and Burgoyne (2008, 2011), for example, studied the debonding phenomenon as a Mode-I problem by considering that the debonding often occurs in the concrete just above the interface. The Mode-I fracture energy of concrete was used in their approach. The authors observed that none of the existing studies available in the literature provided a reliable estimate of the interfacial fracture energy G_F. Gunes et al. (2009) proposed a global energy balance model to predict FRP debonding failure. The amount of energy dissipated in the system during debonding was determined by calculating the change in the potential energy of the system. The component of the energy dissipation due to the debonding process was calculated by means of the interfacial fracture energy.

Cyclic Loads

As shown in previous sections, several theoretical contributions have been proposed by researchers in recent years concerning both the behaviour of the FRP-to-concrete interface and the evaluation of the interface stresses (Ueda and Dai 2005; Dai et al. 2005a). Moreover, many experimental tests have been carried out to evaluate the bond capacity and the effective bond length (Chajes et al. 1996; Bizindavyi et al. 1999; Brosens et al. 2001; Yao et al. 2005). In particular, the influence of FRP stiffness, width, and bond length as well as concrete compressive strength and surface treatments has been investigated in depth (Brosens et al. 2001; De Lorenzis et al. 2001; Savoia et al. 2003; Guo et al. 2005; Lu et al. 2005; Faella et al. 2007a; Nigro et al. 2008). Both theoretical and experimental contributions have led to the development of design guidelines and codes (ACI 440.2R-08 2008; CNR-DT 200 2004; fib bulletin 14 2001). Such guidelines are mainly based on the results of monotonic bond tests, while many strengthened structures are subjected to fatigue loads (i.e. roads and railways bridges) or to shorter but more intense cyclic actions such as earthquakes. In particular, in these cases the FRP-concrete interface is subject to cyclic stresses which could lead to premature debonding failure.

Thus some researchers have recently begun to investigate the fatigue performance of the FRP-concrete interface (Kobayashi et al. 2003; Dai et al. 2005a, b; Bizindavyi et al. 2003; Diab et al. 2007). Nevertheless, at present, bond tests under cyclic actions performed on CFRP sheets applied on concrete blocks are not as numerous as monotonic tests, and very few contributions are available on cyclic tests performed on CFRP plates. Furthermore, few studies are available on debonding phenomena under few cycles at very high force levels, as typical occurs during earthquakes. During earthquakes, FRP instability in compression may start before debonding; however in some cases (i.e. statically determinate bridge beams and slabs), FRP laminates may be always in tension. Ko and Sato (2007) showed that the load-displacement curves recorded during cyclic tests basically traced the load-displacement curves related to monotonic counterpart tests even if plastic displacements and stiffness reduction were observed due to the partial debonding imposed by the repeated unload/reload cycles. Finally, they showed that partial debonding under cyclic loads does not affect the FRP reinforcement debonding force if adequate bond length is provided.

Compared with monotonic tests, there have been few bond tests under cyclic actions performed on CFRP sheets applied on concrete blocks. Very few contributions are available on cyclic tests performed on CFRP plates and on debonding phenomena under few cycles at high force levels, which typically occur during earthquakes. Therefore, a further series of cyclic Single Shear Tests (SSTs) bond tests under both monotonic and cyclic actions, without inversion of action sign, were performed by Nigro et al. (2011) in order to analyze both the influence of different load paths (few cycles, typical of seismic actions) and the effect of FRP bond lengths on bond behaviour between FRP reinforcement and the concrete substrate. In particular, concrete mix design was specifically designed to obtain low compressive concrete strength, to better simulate the FRP application on existing structural members that need to be strengthened (f_{cm} = 22.5 MPa), whereas sheets and plates were specifically selected to investigate the performance of reinforcements with a low or high value of axial stiffness. Moreover, to investigate the influence of the reinforcement bond length, l_b, on the interface behaviour under cyclic actions, different l_b values were assumed in the experimental program (see Fig. 3.16). Finally, different cyclic load paths were adopted to simulate a seismic event (low number of cycles) of different intensity and to evaluate the extent of the influence of cycle number on bond behaviour.

If $P_{max,M}$ is the maximum debonding load recorded during the monotonic test, the experimental outcomes of cyclic tests showed that the influence of few load-unload cycles up to 70 % of $P_{max,M}$ was negligible in terms of bond stiffness and strength for CFRP sheets both for higher and lower bond lengths than theoretical effective ones); similar results were obtained for plates, even if experimental effective bond lengths were significantly lower than theoretical ones. Moreover, a small number of load-unload cycles (i.e. a total of 40 cycles) up to 90 % of $P_{max,M}$ induced a translation of the shear stress transfer zone along the reinforcement with a

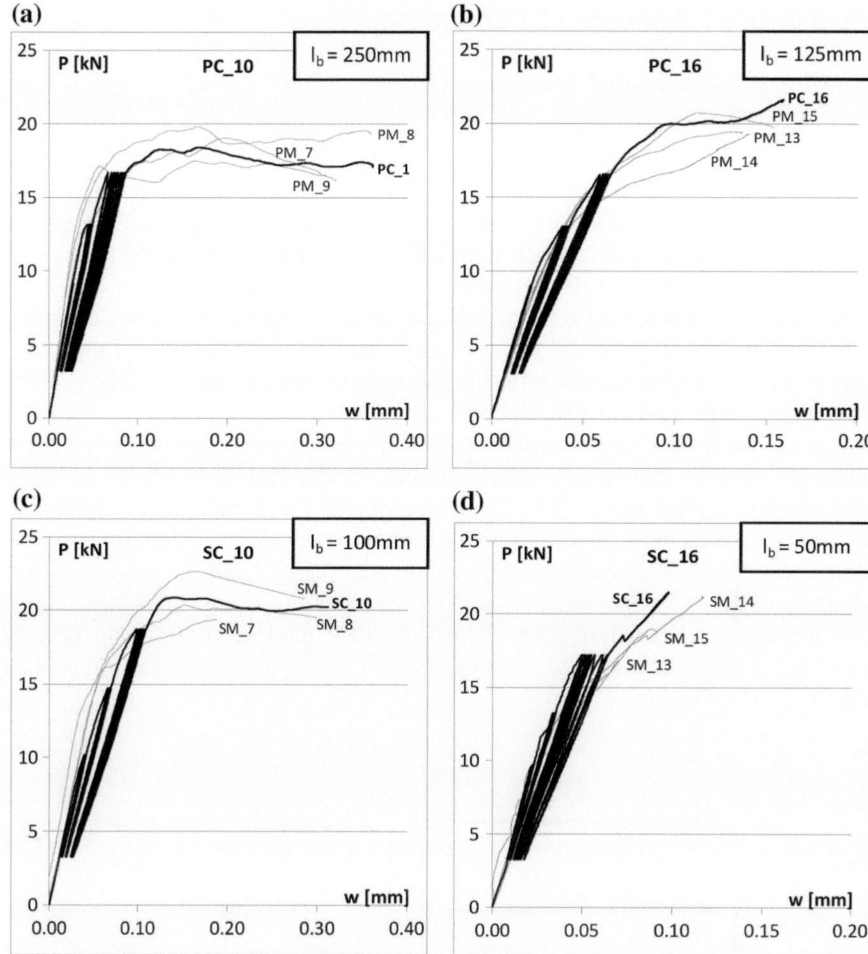

Fig. 3.16 Experimental load-displacement relationships: Plates **a** $l_b = 250$ mm, **b** $l_b = 125$ mm and Sheets, **c** $l_b = 100$ mm, **d** $l_b = 50$ mm

reduction in peak values due to interface damage; however, this phenomenon did not substantially affect debonding loads in the case of bond lengths exceeding the effective bond length. Finally, experimental tests showed that the reduction in bond length up to about 50 % of the theoretical effective bond length induced a comparable reduction in maximum debonding load on specimens subjected to monotonic or cyclic action.

Experimental Analysis of Debonding

Existing Experimental Set-Ups

Several experimental set-ups have been proposed and carried out by researchers in last years to perform bond tests, but a standard procedure has not been defined yet. The following classification of test procedure for bond test is generally assumed (Horiguchi and Saeki 1997; Chen and Teng 2003; Yao et al. 2005, Fig. 3.17): (a) double-shear pull tests; (b) double-shear push tests; (c) single-shear pull tests; (d) single-shear push tests; (e) beam (or bending) tests. These definitions are based on the loading condition in the concrete block and on the symmetry of the specimens (single or double tests refer to the presence of one or two sides of the block strengthened with the FRP reinforcement).

In the cases (a) and (c) the tensile load is applied to the external reinforcement and to the concrete element too (Fig. 3.17a, c); by contrast, in cases (b) and (d) a tensile load is applied to the FRP reinforcement and a pushing action is applied to the concrete block that, thus, is partially compressed (Fig. 3.17b, d). For both set-ups the configuration can be either symmetrical (cases a and b) or asymmetrical (cases c and d).

In the case of beam tests (Fig. 3.17e) the FRP reinforcement is not directly loaded but is however subjected to tensile stresses due to the bending action applied to the concrete element.

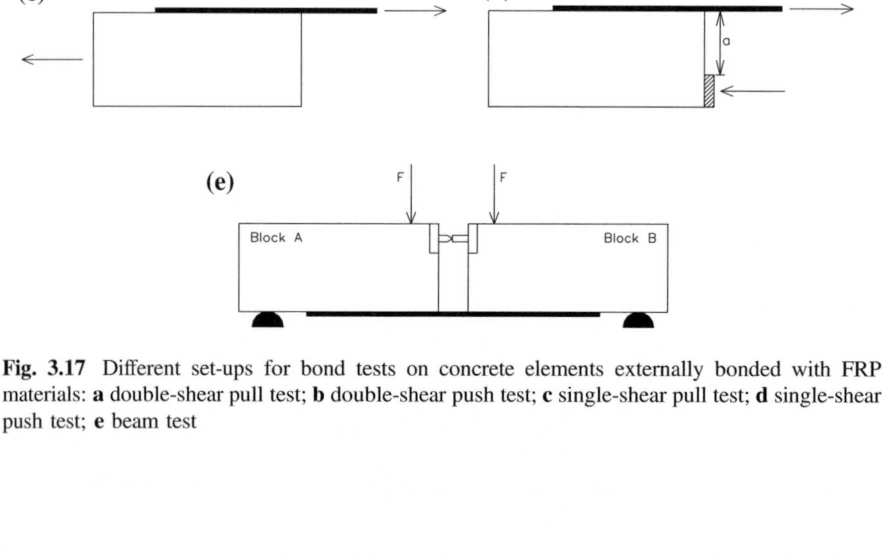

Fig. 3.17 Different set-ups for bond tests on concrete elements externally bonded with FRP materials: **a** double-shear pull test; **b** double-shear push test; **c** single-shear pull test; **d** single-shear push test; **e** beam test

The pull shear test (single or double) corresponds to a loading condition very similar to the actual one, because in existing RC elements the external FRP reinforcement is usually applied on the tension side. However, this scheme is more difficult to realize experimentally compared to the push shear test because the concrete block has to be loaded in tension. Such a loading condition is usually realized by applying tension to steel bars embedded in the block (Brosens and van Gemert 1997; Maeda et al. 1997; Ueda et al. 1999; Wu et al. 2001; Bilotta et al. 2011, Fig. 3.18a). In this scheme the set-up can be more sensitive to the geometrical inaccuracies and imperfections and thus, the repeatability or the variability of the results can be increased.

By contrast, the push shear test is more simple to realize and can give reliable predictions of bond strength, if the compressed area of concrete is not very extended (large value of the distance a in Fig. 3.17d). Indeed, suitable values of this area ensure the development of a bond failure at the concrete-FRP interface, similarly to what occurs in the pull shear test. If the loaded area of concrete is too extended the compressive stresses induced by the pushing force can limit the volume of concrete involved in the failure mechanism and, thus, lower values of debonding load can be attained due to the reduction of the fracture energy.

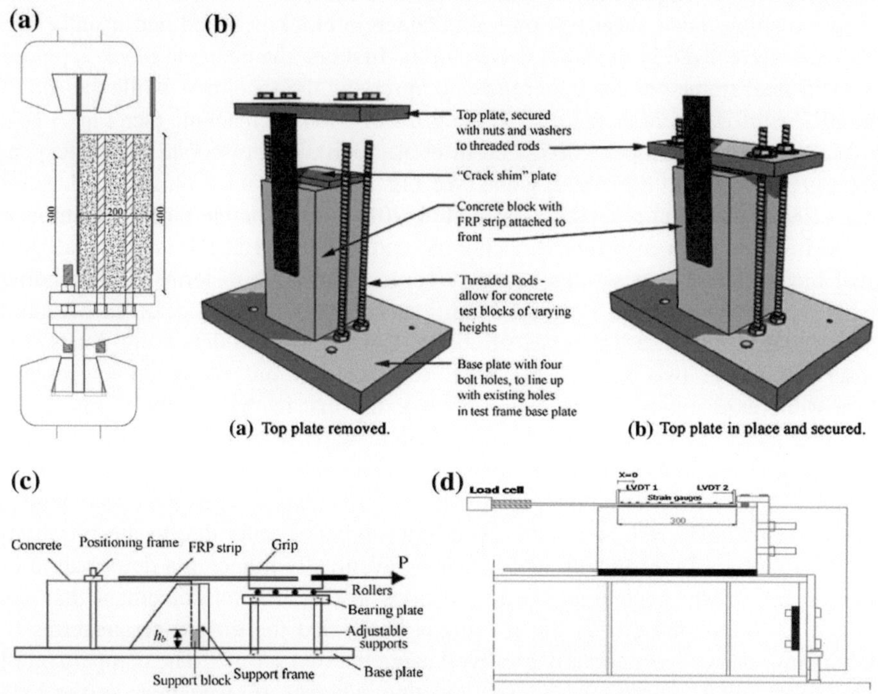

Fig. 3.18 **a** Single pull shear test by (Bilotta et al. 2011a, b); **b** Single push shear test—vertical scheme by (McSweeney and Lopez 2005); **c** Single push shear test—horizontal scheme by (Yao et al. 2005); **d** Single push shear test—horizontal scheme by (Mazzotti et al. 2005)

In general, special attention is required when a symmetrical scheme is realized and the FRP reinforcement is applied on both sides of the concrete specimen (Ceroni and Pecce 2002; Blontrock et al. 2002; Brosen and van Gemert 1997; Guadagnini et al. 2012). The most important problem to solve in this case is the alignment of the two concrete blocks for gripping the FRP reinforcement in the testing machine; indeed, some imperfections can cause additional flexural and torsional stresses in the reinforcement and, thus, can reduce the debonding load. Moreover, in the double-shear test set-ups, when one of the two bonded sides starts failing, the system loses the original symmetry and the alignment between the axis of the tensile machine and the bond surfaces; therefore, additional peeling stresses occur on one side of the specimen with the consequent reduction of the transmissible force on that side and the sudden failure of the FRP reinforcement. This phenomenon makes uncertain the definition of the effective tensile load applied to each bonded side, and very difficult the experimental monitoring of the nonlinear bond behaviour. In general, the onset of a not symmetric behaviour in a double-shear scheme may start even at the beginning of the test, due to imperfections and asymmetries in the application of the tensile loads to the FRP reinforcements, with a significant reduction of the bond capacity, especially for short bonded lengths (Yao et al. 2005). For increasing bonded lengths, the effect of misalignments or imperfections tends to be smaller.

In the single push shear test only one concrete block is tested and usually two schemes are realized in the laboratories. In the first one the concrete block is placed in a stiff steel frame with an upper plate compressing the specimen, while the end of the FRP reinforcement is clamped in the grips of a tensile machine (see Fig. 3.18a, b, McSweeney and Lopez 2005; Ceroni et al. 2008). In the second one, the specimen is placed on an horizontal plane (see Fig. 3.18c, d, Yao et al. 2005; Mazzotti et al. 2005; Matana et al. 2005; Ceroni et al. 2014), the concrete block is restrained at the free end by a mechanical anchorage and contrasted at the loaded end by a steel block. These two set-ups can give some differences in terms of debonding loads due to the different restraint conditions of the concrete blocks. In Mazzotti et al. (2013), indeed, a difference of about 10–15 % for the debonding loads was observed in the two set-ups for bond tests carried out on equal specimens. A possible explanation can be found considering that in the horizontal set-up the concrete face opposite to the FRP bonded one is prevented from transverse displacement (Fig. 3.19a); on the contrary, the same surface in the vertical set-up is completely free from restraints (Fig. 3.19b). For this reason, when the tensile load is applied to the FRP reinforcement a bending moment occurs due to the misalignment of the two forces (action and reaction), eventually inducing a deformation of the concrete prism generating peeling stresses in the FRP reinforcement that can reduce the bond strength. A 3D FE model confirmed the differences in terms of debonding loads experimentally observed using the two set-ups. The comparison of the experimental strains measured in the two schemes showed that, as the load increases, the longitudinal strains along the direction of the fibers in the vertical set-up become generally higher than strains from the horizontal one due to the additional flexural deformation.

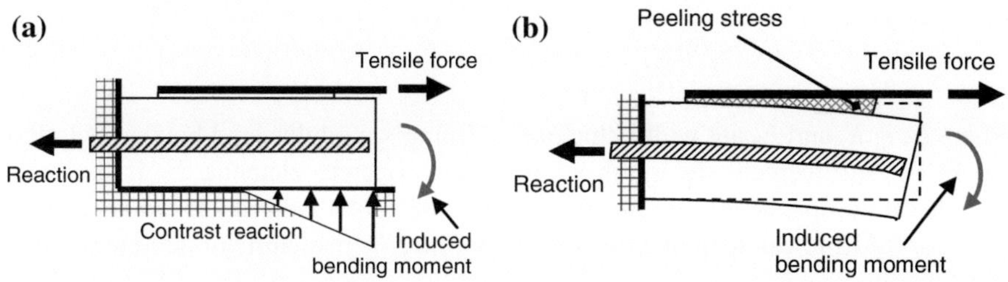

Fig. 3.19 Deformation mechanism of the concrete block in the push shear set-up: **a** horizontal scheme; **b** vertical scheme

A recent Italian round robin test has shown the reliability of the single push shear test set-up and the good repeatability of the results in terms of debonding loads (Savoia et al. 2009).

In the beam test set-up, well known for characterization of bond behaviour of steel bars in concrete elements according to the RILEM standards, the tension force is applied to the FRP reinforcement by a flexural scheme. Two blocks, placed on supports, are connected at bottom by the FRP reinforcement and above by a cylindrical hinge. Two vertical loads are generally applied (Cruz and Barros 2002; De Lorenzis et al. 2001; Ceroni et al. 2003). In some cases beam tests have been performed on a single concrete prism having a steel plate or a notch in the middlespan aimed to promote the formation of a crack in that position (Guo et al. 2005; Dai et al. 2003). However, the loading pattern of the beam test can cause a shear failure in the blocks for lower length/height ratios (Ceroni and Pecce 2006) that avoids attaining the actual bond strength.

Comparison of Experimental Results of Different Set-Ups

In order to compare experimental results coming from different set-ups, they have been compared with the theoretical expression given by Chen and Teng (2001) for the maximum debonding load:

$$N_{f,\max} = \alpha \cdot \beta_p \cdot \beta_L \cdot b_f \cdot L_e \cdot \sqrt{f_c'}; \; L_e = \sqrt{\frac{E_f \cdot t_f}{\sqrt{f_c'}}};$$
$$\beta_p = \sqrt{\frac{2 - b_f/b_c}{1 + b_f/b_c}}; \tag{3.20}$$

$$\beta_L = \sin\frac{\pi L_b}{2L_e}\ \textit{if}\ L_b \leq L_e, \qquad \beta_L = 1\ \textit{otherwise} \tag{3.21}$$

where b_f, t_f, E_f and L_b are width, thickness, Young's modulus, and bonded length of the FRP reinforcement, b_c is the width of the concrete element, f'_c is the mean cylindrical compressive strength of concrete and α is a calibration factor equal to 0.427 or 0.315 to calculate the mean value or the 5 % percentile of debonding load, respectively.

The considered experimental results have been obtained in bond tests carried out according to various set-ups:

- beam tests, named BT, (De Lorenzis et al. 2001; Cruz and Barros 2002; Guo et al. 2005; Aiello and Leone 2005; Dimande et al. 2005; Wang et al. 2005; Ceroni and Pecce 2006);
- pull shear tests, named PlST, (Brosens and van Gemert 1997; Maeda et al. 1997; Ueda et al. 1999; Wu et al. 2001; Ceroni and Pecce 2002; Aiello and Leone 2005; Boschetto et al. 2006; Yi et al. 2006);
- and push shear tests, named PsST, (Chajes et al. 1996; Takeo et al. 1997; Ueda et al. 1999; Zhao et al. 2000; Wu et al. 2001; Coronado and Lopez 2005; Lu et al. 2005; McSweeney and Lopez 2005; Pham and Al-Mahaidi 2005; Yao et al. 2005; Travassos et al. 2005; Leone et al. 2006; Savoia et al. 2009).

In the experimental database results of specimens with bonded length less than 50 mm, FRP width lower than 40 mm, and Young's modulus lower than 80,000 MPa were excluded; 448 experimental points have been collected. These limits have been fixed in order to reduce the scatter of the results, to exclude from the calibration procedure the results related to unrealistic strengthening configurations that can be strongly influenced by scale effects (too short bonded length or too low width and, thus, too low FRP-to-concrete width ratio), and to avoid the materials with elastic properties too different from those usually adopted in the experimental tests and in the practical applications. Thus, the main parameters of specimens are variable in the following ranges: concrete width, b_c = 100–500 mm, FRP width, b_f = 40–120 mm, b_f/b_c = 0.17–1, FRP thickness, t_f = 0.083-1.4 mm, number of layers, n = 1–6, Young modulus of fibers, E_f = 81380–640000 MPa, bonded length, L_b = 50–700 mm, mean compressive strength of concrete, f_{cm} = 17–62 MPa, mean tensile strength of concrete, f_{ctm} = 1.30–4.30 MPa.

In Fig. 3.20 the experimental debonding loads, distinguished according to the different set-ups (Fig. 3.20a for beam tests, Fig. 3.20b for pull tests, Fig. 3.20c for push test) are compared with the theoretical mean values given by Eq. (3.20) (i.e. the parameter α is assumed 0.427) and in Table 3.2 a summary of these comparisons is reported in terms of mean value, standard deviation, and CoV of the experimental–to-theoretical debonding load ratio, $N_{exp}/N_{f,max}$.

For the beam test (BT) the model underestimates the experimental results (the average value of $N_{exp}/N_{f,max}$ is indeed ≥1) with CoV values comparable with the push shear tests (0.21—0.27). The higher experimental loads obtained for such a

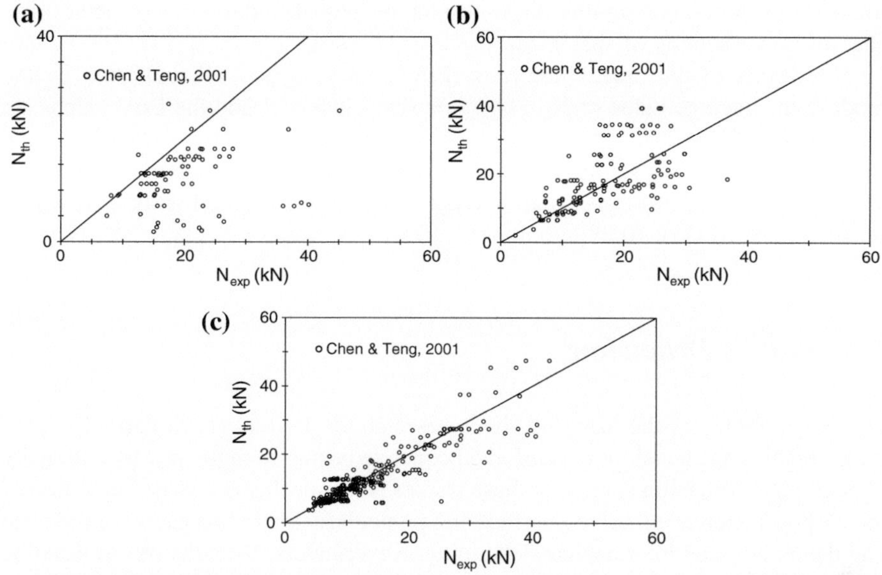

Fig. 3.20 Experimental results versus theoretical debonding load by Eq. (3.1): **a** beam test (BT); **b** pull shear test (PlST); **c** push shear test (PsST) single or double

Table 3.2 Statistical summary of experimental to theoretical debonding load ratio

	$N_{exp}/N_{f,max}$	
Beam test (61 data)	Mean	1.30
	Stand. dev	0.27
	CoV	0.21
Pull shear test (115 data)	Mean	0.98
	Stand. dev	0.36
	CoV	0.36
Push shear test (272 data)	Mean	1.01
	Stand. dev	0.23
	CoV	0.22
All results (448 data)	Mean	1.04
	Stand. dev	0.29
	CoV	0.28

set-up could be related to the length-to-height ratio of the concrete block that could not lead to a real 'debonding failure'. Indeed, the dimension of the block could activate an 'arc' resistant mechanism that allows transferring a lower tensile force to the external FRP plate compared to the 'beam' mechanism and higher stresses in the concrete strut that results in a shear failure.

For the pull shear tests (PlST) the theoretical formula overestimates the experimental results (the average value of $N_{exp}/N_{f,max}$ is indeed ≤1) and the CoV is higher

(0.36–0.42) than other set-ups; these results are probably due to the crucial influence of imperfections of specimens.

The results of the push shear tests (PsST) seem to be well replicated by the models: the average value of $N_{exp}/N_{f,max}$ is very close to 1 and the CoV values are the lowest (0.22–0.25).

If the results of all set-ups are considered, the synthesis reported in Table 3.1 shows that Eq. (3.19) furnishes an average value of $N_{exp}/N_{f,max}$ lightly ≥ 1, with a reliable value of CoV (0.28).

Intermediate Debonding

As shown above, in the last years huge research efforts have been carried out for understanding the behavior of reinforced concrete beams strengthened by externally bonded FRP. The main subject of these studies is the mechanical characterization of the FRP-to-concrete adhesive interface and particularly the investigation of the plate end debonding and intermediate debonding. Nevertheless, formulations to describe the FRP-to-concrete bond behavior are still under discussion. Various proposals have been derived from simplified mechanical models and calibrated making use of the experimental results available in the scientific literature (Teng et al. 2002). Alternatively numerical simulations allow to simulate and, thus, further investigate the debonding phenomenon. Roberts (1989) provided a simplified model for evaluating interface stresses in FRP (or even steel) strengthened beams; simplified equations for evaluating shear and normal stresses throughout the FRP-to-concrete interface have been provided by assuming linear elastic behavior of the adhesive interface. Similar relationships, even obtained under simplified hypotheses for the interface behavior, have been provided in Malek et al. (1998). The authors showed, through experimental and numerical comparisons, that such simplified formulae usually result in a close approximation of the complex stress patterns which develop throughout the FRP-to-concrete interface.

The above mentioned research papers mainly deal with interface stress distribution in the elastic range, which is an aspect of concern for serviceability conditions. Premature loss of bonding between FRP and concrete needs to be studied by considering a suitable non-linear relationship between interface stresses and strains. Holzenkaempfer (1994) proposed a bi-linear relationship between shear stresses and interface slips; based on such model Taljsten (1997a, b) determined the expressions of the ultimate bearing capacity of FRP-to-concrete joints.

In Faella et al. (2006a, b) a numerical model is implemented and validated: the study is mainly focused on debonding failure which can occur at the FRP-cut-off section (plate end debonding) or throughout the FRP-to-concrete adhesive interface (intermediate debonding). Interface slips between reinforced concrete beam and FRP laminates are considered and, consequently, a well-established non-linear shear stress-slip law is introduced. Moreover, non-linear stress-strain relationships are utilized for modeling the other structural materials and a completely non-linear

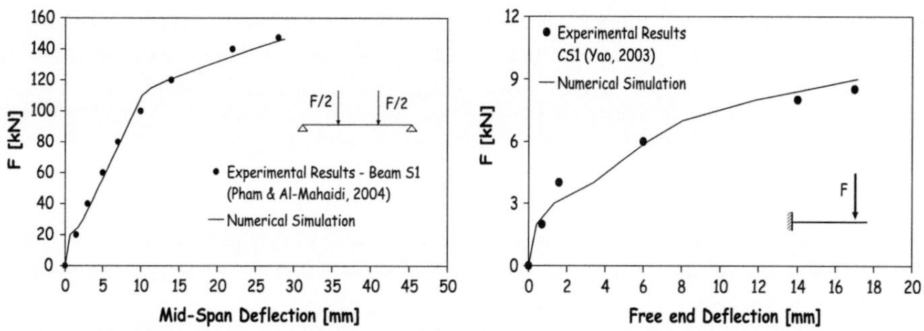

Fig. 3.21 Experimental and numerical Load-Deflection curves. (Faella et al. 2008a)

analysis procedure is obtained by means of a secant approach; such non-linear procedure allows for reproducing the whole structural behaviour up to failure which can be due to FRP tearing, concrete crushing or interface debonding. Figure 3.21 shows how the complete evolution of the displacement-vs-force curve recorded for a simply supported beams (Pham and Al Mahaidi 2004) and a cantilever beam (Yao et al. 2005) externally strengthened by FRP can be followed by the numerical procedure which provides also a good estimation of the ultimate load and displacement.

A high level of uncertainty still overshadows the mechanical understanding of intermediate debonding due to the complex interactions between several phenomena, such as cracking in concrete, steel yielding in longitudinal rebars, interface adhesion properties, the amount of reinforcement, the load condition and so on. As a result of this incomplete understanding of the mechanical reasons leading to intermediate debonding failure of FRP-strengthened RC beams, several analytical approaches have been proposed within the scientific literature and adopted by the most common design codes. Since those procedures work in rather different ways involving various parameters and adopting different relationships for defining interface properties, they generally lead to rather different predictions of the ultimate load resulting in intermediate debonding. Moreover, such procedures actually neglect or disregard the role played by several mechanical parameters in controlling the structural response of FRP-strengthened RC beams, adopting simplified expressions for deriving formulae usually calibrated on the available experimental observations.

Two different methodological paths can be followed for defining reasonably simplified design formulae based on experimental results:

- direct calibration of empirical expressions against experimental results by means of well-established mathematical procedures like least-square minimization of the overall difference between the experimental observations and the corresponding analytical values;

– validation of refined numerical models (i.e. based on finite element discretization) by means of a limited number of experimental results and extrapolation of those results by means of the above mentioned numerical procedures.

Despite of several proposals, numerical models are not yet suitable to take into account all phenomena affecting debonding. Therefore, a simplified design formula (see Chap. 2—Application 3) based on a statistically consistent procedure for determining the safety levels required for defining the so-called “characteristic” and “design” values of the maximum axial strain developed in FRP at intermediate debonding is probably more useful for design purposes.

Figure 3.22 shows a comparison between design curves obtained in Chap. 2—Application 3 and other curves according to some of the models outlined in Sect. 3. Both the model by Said and Wu (2008) and the formula adopted by ACI440-08 lead to predictions in terms of maximum axial strain in FRP at debonding which are not conservative enough to be used for design purposes. Moreover, the predictions based on the model by Teng et al. (2004) are rather close to the values obtained by the current CNR-DT 200/2004 provisions. In the case of low concrete strength (namely, for f_c < 40 MPa), both formulations look not conservative enough for design purposes.

On the other hand, the predictions obtained by applying the formulation in Chap. 2 by considering the $k_{IC,5\,\%}$ coefficient, a further safety factor $\gamma_{f,d}$ = 1.2 addressing the quality of the application, a confidence factor FC = 1 as “full knowledge” is achieved about the mechanical properties of structural materials demonstrate the higher level of conservativeness achieved by this proposed formula. Finally, although the curves representing the results of the model by Teng et al. (2003) are generally even more conservative than those obtained by the model proposed in Chap. 2, it could result in too strict provisions for a cost-effective

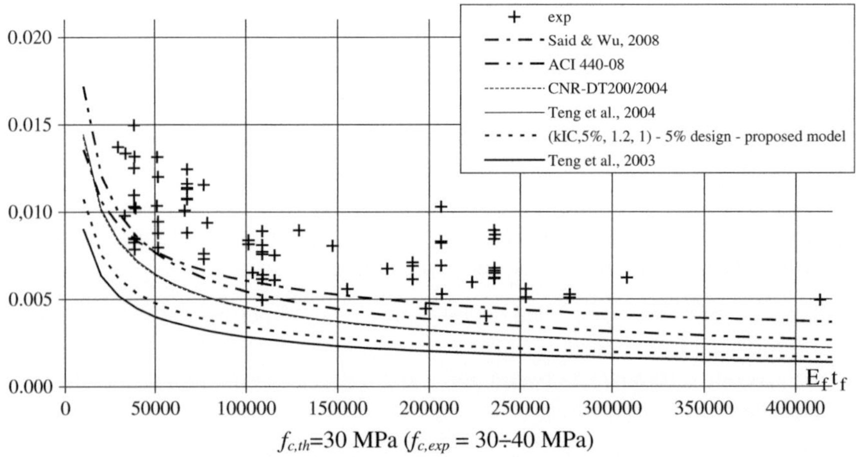

Fig. 3.22 Experimental results and code provisions

application of FRP strengthening. The two curves representing the design formula and the model by Teng et al. (2003) are rather close but the former can move upward if a unit value is also considered for $\gamma_{f,d}$, as a result of a certified application procedure allowing for higher values of the relevant mechanical properties of the adhesive-to-concrete interface.

Bond Law Identification Method

The FRP-to-concrete interface behaviour is often described by its fracture energy G_F, which is directly related to the ultimate load F_{max} observed in pull-out tests. Nevertheless, assessing the G_F value is not sufficient for reproducing the overall behaviour of the FRP-to-concrete interface for modeling problems such as, for instance, intermediate debonding in RC beams externally strengthened by FRP reinforcement (Faella et al. 2008a). Thus, an accurate local bond–slip model is of fundamental importance in modeling FRP-strengthened RC elements.

As showed in the previous sections, the pull test delivers the ultimate load of the FRP-to-concrete interface, but can also provide useful information on the local bond–slip behaviour of the interface if axial strains of the FRP reinforcement are measured with closely spaced strain gauges. Indeed, the shear-stress-relative-slip relationship, describing the FRP-to-concrete interface law, can be identified starting from the values of the strains recorded during the tests at different load levels.

Commonly, the shear stress of a particular location along the FRP-to-concrete interface can be found using a difference formula, whereas the corresponding slip can be found by a numerical integration of the measured axial strains of the FRP.

In particular the interface shear stresses $\tau_i(z)$ can be obtained by the variation of axial stresses, and thus strains, throughout the FRP by the following relationship between two strain gauges at distance Δz_i:

$$\tau_i = \frac{\varepsilon_{i+1} - \varepsilon_i}{\Delta z_i} \cdot E_f \cdot t_f \tag{3.22}$$

where E_f and t_f are FRP Young's modulus and thickness, respectively. Typical shear stress profiles assessed for sheets and plates, respectively, are reported in Fig. 3.21a, b. Note that, at the loaded end of the reinforcement, shear stresses assessed for loads close to the debonding of the reinforcement are lower than those assessed for lower loads. This indicates that in this zone of the reinforcement the shear stress-slip law is in the softening stage typical of a post-elastic behaviour.

On the other hand, assuming for the sake of simplicity that concrete strain is negligible with respect to FRP counterpart, the slip values corresponding to the shear stress values obtained by Eq. (3.21) can be calculated by integrating the axial strains measured during the test by the following relationship:

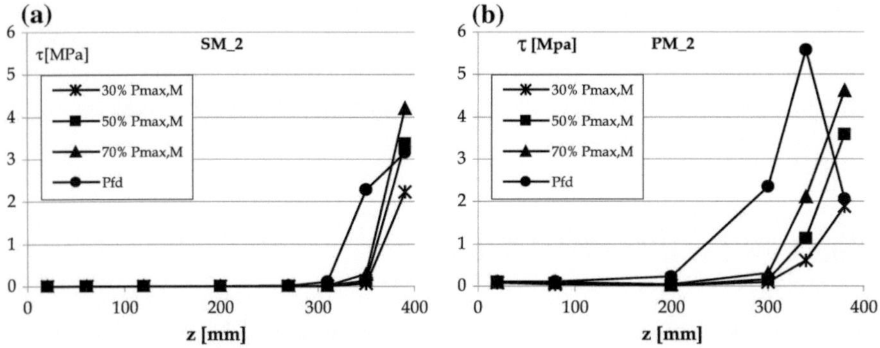

Fig. 3.23 Shear stresses assessed on sheet **a** and plate **b** reinforced specimens

$$s_{i,i+1} = \sum_{k=0}^{i} \frac{(\varepsilon_{k+1} + \varepsilon_k)}{2} \cdot (x_{k+1} - x_k) \tag{3.23}$$

Therefore, the bond law at the FRP-to-concrete interface can be obtained by calculating the shear stresses using Eq. (3.22) (considering the strains recorded by the first two gauges—e.g. at 400 and 380 mm in Fig. 3.23) and the corresponding slips using Eq. (3.23) (considering all the strain gauges applied on the FRP reinforcement). In this way the experimental interface law is obtained directly with respect to values of shear stresses and relative slips based on experimental strains (see Fig. 3.24).

Moreover, the couples of values $(\overline{s_j}, \overline{\tau_j})$ can be "directly" used to calibrate the τ-s relationship through a numerical regression, such as the least square method. This method (called DirIM in Faella et al. 2009) is very simple, but it does not often produce accurate local bond–slip curves. In particular the shear stress deduced from

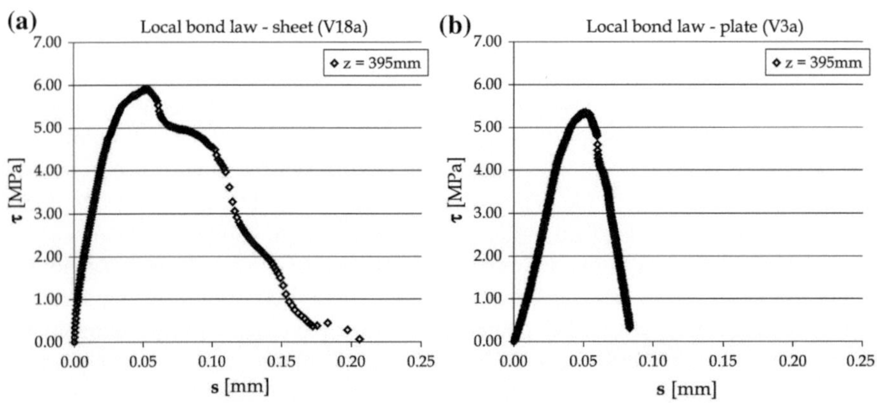

Fig. 3.24 Experimental bond law at the FRP-to-concrete interface: sheet **a** and plate **b**

axial strains can be not reliable due to sensitivity to the distance between strain gauges in the averaging procedure needed for estimating shear stresses. Consequently the method can noticeably underestimate the values of fracture energy and provide bond–slip curves, attained from different tests, substantially different.

In this regard, Ferracuti et al. (2007) presented a procedure to calibrate non-linear FRP-to-concrete interface laws from experimental results of bond tests: strains along the composite are used to obtain shear stress–slip data, whereas the maximum transmissible force is used to prescribe the value of fracture energy, G_F, of interface law. Hence, the non-linear interface law is obtained by a DirIM, taking into account a restraint on G_F in the calibrating procedure. The interface law is then used to simulate the tests and a good agreement between numerical and experimental results are showed. Nevertheless, the distribution of shear stresses cannot be directly compared with data provided by the pull out tests, because both interface shear stresses and local displacements cannot be directly measured during the usual pull-out tests.

In Lu et al. (2005), some existing bond–slip models was presented and assessed using the results of some pull tests on simple FRP-to-concrete bonded joints, leading to the conclusion that a more accurate model, unaffected by such uncertainties, is required. For this reasons three new bond–slip models of different levels of sophistication were proposed, highlighting a novel aspect in calibrating the models on the predictions of a meso-scale finite element model. Through comparisons with the test database, all three bond–slip models are shown to provide accurate predictions of both ultimate load and strain distribution in the FRP reinfocement. In particular it was showed that, while a more precise bond-slip model should consist of a curved ascending branch and a curved descending branch (see also Savoia et al. 2003 and Ferracuti et al. 2007), also other shapes such as a bilinear model can be used as a good approximation.

Among the three models proposed in Lu et al. (2005), the last is just represented by a bilinear law, identified by the following relationships for determining the three parameters τ_{max}, s_e and s_u:

$$\tau_{\max} = 1.5\beta_w f_t, \tag{3.24}$$

$$s_e = 0.015\beta_w f_t, \tag{3.25}$$

$$s_u = \frac{2.G_F}{\tau_{\max}} = \frac{2 \cdot 0.308 \cdot \beta_w^2 \cdot \sqrt{f_t}}{1.5 \cdot \beta_w \cdot f_t} = 0.41 \cdot \beta_w \cdot \sqrt{f_t} \tag{3.26}$$

where f_t is the tensile strength of the concrete and β_w is a well-known geometrical factor (as function of the ratio between the width of the FRP and the concrete member which the FRP is applied on). Slightly dissimilar expressions were suggested in literature for this factor (Lu et al. 2005; Teng et al. 2002; CNR-DT 200 2004; fib Bulletin 14 2001), but the difference among the expressions is however very small and all the mentioned equations are suitable for practical applications. Note that, for simplicity and uniformity, some relationships were formally rewritten.

It is worth observing that the relationships (23–25) were assessed by using the experimental results of bond tests performed on sheets (i.e. wet-lay-up system characterized by thickness ranging between 0.133 and 0.5 mm for one or three layers). Very few experimental data was related to tests performed on plates (i.e. preformed system characterized by thickness ranging between 1 and 2 mm) whereas the reinforcement thickness particularly affects debonding behaviour: indeed, the greater the thickness, the higher the increase in the normal and shear stresses at FRP to concrete interface and consequently the probability of premature debonding occurrence (Oehlers and Moran 1990; Tounsi et al. 2009). Even if the numerical analysis may take into account this parameter, the comparison performed with the experimental data appears clearly lacking from this point of view.

By contrast, recently (Bilotta et al. 2011b), in order to assess a design formulation to predict the plate end debonding load in RC elements strengthened with Externally Bonded Reinforcement (EBR) made of FRP materials by means of a statistical analysis, the experimental debonding loads of several bond tests available in the literature have been collected. Cured in situ (sheets) and preformed (plates) FRP systems have been distinguished to better exploit the performance of the former ones, as concerns the plate end debonding failure. Lu et al. (2005) suggested an expression for assessing the interface relationship and consequently the fracture energy value that takes into account the strength of the concrete, but completely neglects the influence of the reinforcement properties, in particular the FRP thickness. This assumption, regardless of the value of β_w, lead to same values of maximum shear stress for plates and sheets applied on the same concrete.

Thus, based on such results and the outcomes of Yao et al. (2005); Ferracuti et al. (2007); Faella et al. (2009) and Bilotta et al. (2011b), the IndIM procedure and the bilinear shape for the bond law were extensively used in Bilotta et al. (2012) with the final aim of calibrating bilinear interfaces laws, as stable as possible, and assessing their reliability for sheets and plates, separately.

Since relevant measures of shear stresses and corresponding relative strains cannot be directly obtained by pull-out tests, an alternative procedure using the experimental measures in terms of axial strain values $\varepsilon_{f,j,i}$ recorded at distance z_j under the force F_i was adopted for "indirectly" calibrating the τ-s interface relationship. The simple but effective bilinear model is taken because closed-form solutions are available for such shape of the interface law, as extensively shown in Faella et al. (2003). This model is defined by well-known relationships in which three parameters identify the bond law: the maximum shear stress, τ_{max}, the corresponding slip, s_e, and the ultimate slip, s_u, beyond which the interface shear stress can be considered null.

Therefore, for each set, q, of parameters τ_{max}, s_e and s_u a given interface law is defined, and the corresponding theoretical value $\varepsilon_{fth,j,i}$ of the axial strain developed in the FRP plate at a distance z_j under the force F_i can be evaluated. Even if numerical procedures, such as finite differences, can generally be utilized, the choice of a bilinear bond law allowed for using the closed-form solutions (Faella et al. 2003) taking also into account, if necessary, the influence of the parameter bond length, L, in the solution of the problem.

The identifying procedure was applied on a wide collection of experimental results attained by pull tests during which not only the load but also the corresponding axial strains of the FRP reinforcement were measured. Both the consistency of IndIM method and the robustness of the assumption on the bond law shape was showed by a comparison, in terms of axial strains throughout the bonded length, between theoretical predictions and the corresponding measured values. Even if the uncertainness in accurately identifying the parameter s_e indicated that the bond behaviour in the elastic stage was not perfectly approximated by a linear branch, the result obtained by assuming a bilinear law were satisfying.

Finally, several bond law relationships, identified by three parameter (i.e. the maximum shear stress, τ_{max}, the corresponding elastic slip, s_e, and the ultimate slip, s_u), have been compared. The elastic and ultimate slips, s_e and s_u respectively, are on average the same for sheets and plates, although the dispersions of the values obtained by the identifying method are somewhat high. Conversely the values of maximum shear stress, τ_{max}, obtained for sheets bond laws are always higher than those obtained for plates interface relationships, of about 30 % in average. Clearly the same differences are attained in terms of fracture energy. Such results are in agreement with the theoretical strength models available in the literature for predicting debonding of sheets and plates separately, confirming the advisability of assessing a bond law for the plates different from that for the sheets.

Existing Models and Code Formulations

Externally bonded FRP sheets are currently used to repair and strengthen existing reinforced concrete (RC) structures for shear and flexural applications. Proper design against various debonding failure modes is the key issue of this technique (Taljsten 1997a, 1997b; Bizindavyi and Neale 1999; Chen and Teng 2001; Nakaba et al. 2001; Lu et al. 2005; Ferracuti et al. 2007; Pellegrino et al. 2008). Typical failure modes include cover separation, plate end interfacial debonding, intermediate flexural crack-induced interfacial debonding, and critical diagonal crack-induced interfacial debonding, as described in the Italian guidelines CNR-DT 200 2004. Furthermore, some authors (e.g. Yuan et al. 2004) pointed out that, although there exist many experimental setups to evaluate the FRP-concrete bond strength, a standard test procedure does not exist yet. The most diffused experimental setups are represented by the so-called direct shear tests, single and double, and by the bending/beam test. In bending tests the FRP composite is bonded to the bottom of a beam subjected to flexure. Bending tests are sometimes carried out on small scale specimens where a notch or a hinge is provided in order to initiate debonding at a specific cross-section. The combination of results related to both small- and full-scale specimens is arguable due to different mechanisms and resisting contributions developing in small- and full-scale beams. For this reason, in this work the experimental tests of small-scale notched beams were discarded and only the results of full-scale strengthened RC beams subjected to bending tests were included in the database.

A wide assessment of some diffused analytical models available in literature was performed. Since it has been shown that these different kinds of approaches can lead to different results for the same amount and preparation of FRP and concrete support (Yuan et al. 2004), various test setups (single shear test, double shear test and bending test) and FRP composites preparation (pre-impregnated laminates and post-impregnated sheets) have been considered in the assessment. A comparison between experimental and analytical values of the bond strength and of the effective bond length is presented and discussed. The analytical models considered in this chapter are some of the most diffused formulations and include those adopted by the fib Bulletin 14 (2001), CNR-DT 200 (2013) and ACI 440.2R-08 (2008).

Theoretical Models

A number of analytical bond formulation have been proposed in literature by several authors, have been considered in this work and are briefly recalled for the sake of clarity.

van Gemert (1980):

$$N_f = 0.5 \cdot b_f \cdot l_b \cdot f_{ctm} \tag{3.27}$$

Tanaka (1996):

$$N_f = (6.13 - \ln l_b) \cdot b_f \cdot l_b \tag{3.28}$$

Hiroyuki and Wu 1997:

$$N_f = 5.88 \cdot l_b^{-0.669} \cdot b_f \cdot l_b \tag{3.29}$$

Maeda et al. 1997:

$$N_f = 110.2 \cdot 10^{-6} \cdot E_f \cdot t_f \cdot b_f \cdot l_e \tag{3.30}$$

Neubauer and Rostàsy 1997:

$$N_f = 0.64 \cdot k_p \cdot b_f \cdot \sqrt{f_{ctm} \cdot E_f \cdot t_f} \tag{3.31}$$

when $l_b \geq l_e$.

$$N_f = 0.64 \cdot k_p \cdot b_f \cdot \sqrt{f_{ctm} \cdot E_f \cdot t_f} \cdot \frac{l_b}{l_e} \cdot \left(2 - \frac{l_b}{l_e}\right) \tag{3.32}$$

when $l_b < l_e$.

Khalifa et al. (1998):

$$N_f = 110.2 \cdot 10^{-6} \cdot \left(\frac{f_{ck}}{42}\right)^{2/3} E_f \cdot t_f \cdot b_f \cdot l_e \tag{3.33}$$

Adhikary and Mutsuyoshi (2001):

$$N_f = b_f \cdot l_b \cdot \left(0.25 \cdot f_{ck}^{2/3}\right) \tag{3.34}$$

Chen and Teng 2001:

$$N_f = 0.315 \cdot \beta_p \cdot \beta_L \cdot \sqrt{f_{ck}} \cdot b_f \cdot l_e \tag{3.35}$$

De Lorenzis et al. 2001:

$$N_f = b_f \cdot \sqrt{2 \cdot E_f \cdot t_f \cdot G_f} \tag{3.36}$$

Yang et al. (2001) (in Lu et al. 2005):

$$N_f = \left(0.5 + 0.08 \cdot \sqrt{\frac{E_f \cdot t_f}{1000}}\right) \cdot b_f \cdot l_e \cdot 0.5 \cdot f_{ctm} \tag{3.37}$$

Dai et al. (2005a, b):

$$N_f = \left(b_f + 7.4\right) \cdot \sqrt{2 \cdot E_f \cdot t_f \cdot G_f} \tag{3.38}$$

Lu et al. (2005):

$$N_f = \beta_l \cdot b_f \cdot \sqrt{2 \cdot E_f \cdot t_f \cdot G_f} \tag{3.39}$$

Camli and Binici 2007:

$$N_f = \sqrt{\tau_f \cdot \delta_f} \cdot \sqrt{E_f \cdot t_f} \cdot b_f \cdot \tanh\left(\frac{\theta \cdot l_b}{l_e}\right) \tag{3.40}$$

Izumo (2003) (included and cited in the JCI 2003 Recommendations):

$$N_f = \left(3.8 \cdot f_{ck}^{2/3} + 15.2\right) \cdot l_b \cdot b_f \cdot E_f \cdot t_f \cdot 10^{-3} \tag{3.41}$$

Iso (2003) (included and cited in the JCI 2003 Recommendations):

$$N_f = b_f \cdot l_e \cdot 0.93 \cdot f_{ck}^{0.44} \tag{3.42}$$

Sato (2003) (included and cited in the JCI 2003 Recommendations):

$$N_f = \left(b_f + 7.4\right) \cdot l_e \cdot 2.68 \cdot f_{ck}^{0.2} \cdot E_f \cdot t_f \cdot 10^{-5} \tag{3.43}$$

The analytical models just reported were applied without considering safe and partial factors to allow the comparison with the experimental data collected within the database. The details about the notation can be found in the cited papers.

Code Formulations

fib Bulletin 14 (2001)
According to the *fib* Bulletin 14 (2001) the maximum force which can be anchored by the FRP is expressed by:

$$N_{fa,\max} = \alpha\, c_1\, k_c\, k_b\, b\, \sqrt{E_f\, t_f\, f_{ctm}} \tag{3.44}$$

where:

$$k_b = 1.06\sqrt{\frac{2 - b_f/b}{1 + b_f/400}} \geq 1 \tag{3.45}$$

$$N_{fa} = N_{fa,\max} \frac{l_b}{l_{\max}}\left(2 - \frac{l_b}{l_{\max}}\right) \quad for\ l_b < l_{b,\max} \tag{3.46}$$

The effective bond length is expressed by:

$$l_{b,\max} = \sqrt{\frac{E_f\, t_f}{c_2\, f_{ctm}}} \tag{3.47}$$

CNR-DT 200 (2004)

The Italian document CNR-DT 200 (2004) proposes a formulation similar to that of the *fib* Bulletin 14 (2001); it quantifies the maximum stress in the FRP reinforcement as a function of the fracture energy of the FRP-concrete interface:

$$f_{fdd} = \frac{k_{cr}}{\gamma_{f,d}\,\sqrt{\gamma_c}}\sqrt{\frac{2\,E_f\,\Gamma_{Fk}}{t_f}} \tag{3.48}$$

$$f_{fdd,rid} = f_{fdd}\,\frac{l_b}{l_e}\left(2 - \frac{l_b}{l_e}\right) \quad for\ l_b < l_e \tag{3.49}$$

The maximum force which can be anchored by the FRP is finally calculated multiplying the area of the composite and the stress f_{fdd}.

The factor k_{cr} distinguishes between different kinds of delamination (k_{cr} = 1 for the end delamination, k_{cr} = 3 for the intermediate delamination due to flexural cracking).

The fracture energy of the FRP-concrete interface is expressed by:

$$\Gamma_{Fk} = 0.03\,k_b\,\sqrt{f_{ck}\,f_{ctm}} \tag{3.50}$$

where:

$$k_b = 1.06\sqrt{\frac{2 - b_f/b}{1 + b_f/400}} \geq 1 \tag{3.51}$$

The effective bond length is expressed by:

$$l_e = \sqrt{\frac{E_f\,t_f}{2\,f_{ctm}}} \tag{3.52}$$

The meaning of the symbols is detailed in the *fib* Bulletin 14 (2001), ACI 440.2R-08 (2008), and CNR-DT 200 (2004).

A new version of the Italian guidelines, CNR-DT 200 R1/2013 has been recently published. It provides new equations that can improve the model accuracy. Among the others, a new equation for computing the fracture energy, which has a different values depending on the material used, the effective bond length, and the FRP-concrete strength is provided. The maximum stress f_{fdd} that can be carried by the composite preventing the end plate debonding failure is calculated as:

$$f_{fdd} = \frac{k_{cr}}{\gamma_{f,d}} \sqrt{\frac{2\, E_f\, \Gamma_{Fk}}{t_f}} \tag{3.53}$$

$$f_{fdd,rid} = f_{fdd}\, \frac{l_b}{l_e} \left(2 - \frac{l_b}{l_e}\right) \quad for\ l_b < l_e \tag{3.54}$$

The fracture energy Γ_{Fd} is computed as:

$$\Gamma_{Fd} = \frac{k_b \cdot k_G}{FC} \cdot \sqrt{f_{cm} \cdot f_{ctm}} \tag{3.55}$$

$$k_b = \sqrt{\frac{2 - b_f/b}{1 + b_f/b}} \geq 1 \tag{3.56}$$

where $k_G = 0.023$ in case of pre-impregnated laminate, and $k_G = 0.037$ in case of post-impregnated sheet. FC is an additional safety factor. In order to avoid the intermediate crack-induced debonding failure the maximum FRP stress must be less or equal to $f_{fdd,2}$:

$$f_{fdd,2} = \frac{k_q}{\gamma_{f,d}} \sqrt{\frac{E_f}{t_f} \cdot \frac{2 \cdot k_b \cdot k_{G,2}}{FC} \cdot \sqrt{f_{cm} \cdot f_{ctm}}} \tag{3.57}$$

where $k_{G,2}$ is an empirical coefficient equal to 0.10, and $k_q = 1.25$ in case of distributed load, and $k_q = 1.0$ in all other cases. The CNR-DT 200 R1/2013 computes the effective bond length, named optimum bond length, as:

$$l_e = \min\left\{\frac{1}{\gamma_{Rd} \cdot f_{bd}}\sqrt{\frac{\pi^2 \cdot E_f \cdot t_f \cdot \Gamma_{Fd}}{2}}, 200\right\} \tag{3.58}$$

$$f_{bd} = \frac{2 \cdot \Gamma_{Fd}}{s_u} \tag{3.59}$$

where $s_u = 0.25$ is the ultimate slip between the FRP and the concrete support, and $\gamma_{Rd} = 1.25$ is a modification factor.

ACI 440-2R-08 (2008)

According to ACI 440.2R-08 (2008) the maximum bond strength is calculated multiplying the maximum strain in the FRP reinforcement at the ultimate limit state by the fibre elasticity modulus, assuming perfectly elastic behaviour. The effective strain in FRP reinforcement is limited to the strain level at which debonding may occur, ε_{fd}, as defined in Eq. (3.60). The ultimate strength of the structural member is then found considering the mode of failure for an assumed neutral axis depth, as computed in Eq. (3.61).

$$\varepsilon_{fd} = 0.41\sqrt{\frac{f_c'}{nE_f t_f}} \leq 0.9\varepsilon_{fu} \tag{3.60}$$

$$\varepsilon_{fe} = \varepsilon_{cu}\left(\frac{d_f - x}{x}\right) \leq \varepsilon_{fd} \tag{3.61}$$

$$f_{fe} = E_f \varepsilon_{fe} \tag{3.62}$$

where ε_{cu} is the maximum compressive strain in the concrete, taken as 0.003; d_f and x are the depth of the FRP and the neutral axis, respectively. The maximum force which can be anchored by the FRP is finally calculated multiplying the area of the composite and the stress f_{fe}.

In case of shear or pure axial strengthening the maximum bond strength is calculated multiplying the maximum strain in the FRP reinforcement at the ultimate limit state, according to Eq. (3.63) (in case of U-Wraps or bonded face plies), by the fibre elasticity modulus, assuming perfectly elastic behaviour as in the flexural case (Eq. 3.64). k_v is an empirical coefficient limiting the ultimate strain in the reinforcement:

$$\varepsilon_{fe} = k_v\ \varepsilon_{fu} \leq 0.004 \tag{3.63}$$

$$k_v = \frac{k_1\ k_2\ l_e}{11{,}900\ \varepsilon_{fu}} \leq 0.75 \tag{3.64}$$

where k_1 and k_2 are taken equals to 1.0 (case of pure axial tension). The active bond length, i.e. the length over which the majority of the bond stress is maintained, is expressed by:

$$l_e = \frac{23{,}300}{(n_f t_f E_f)^{0.58}} \tag{3.65}$$

Assessment of Code Formulations Pellegrino

The experimental database used for the assessment of the considered FRP-to-concrete bond analytical models contains 410 specimens, 229 tested with single shear test setup (Bizindavyi and Neale 1999; Chen and Teng 2001; Yao et al. 2005; Lu et al. 2005; Toutanji et al. 2007; Ceroni and Pecce 2010), including both laminates (45 specimens) and sheets (184 specimens); 60 sheet specimens tested by double shear test setup (Chen and Teng 2001; Lu et al. 2005; Pellegrino et al. 2008); 121 specimens tested with bending test setup (Ahmed et al. 2001; Fanning and Kelly 2001; Rahimi and Hutchinson 2001; Smith and Teng 2002; Teng and Yao 2007, Pellegrino et al. 2008), including both laminates (74 specimens) and sheets (47 specimens).

In Fig. 3.25 a comparison between experimental and analytical values, obtained by fib Bulletin 14 (2001), CNR-DT 200 (2004), ACI 440.2R-08 (2008), and CNR-DT 200 R1/2013 for different experimental setups (single shear, double shear, bending test), and different composite materials (laminates and sheets) is shown.

The assessment was carried out by means of a statistical procedure. The performance of the analytical model for evaluating the maximum bond strength of the FRP-concrete interface was obtained comparing the experimental values with the corresponding analytical predictions. As usual, experimental versus theoretical bond diagrams have been built. The values above the line $P_{exp}/P_{th} = 1$ (where P is the force pulling the composite) are safe, whereas values below are unsafe. The accuracy of the various models was assessed through the use of a coefficient of variation (CoV), which measures the distance between the ratio P_{exp}/P_{th} and the optimum ideal value $P_{exp}/P_{th} = 1$.

Using the database herein described, a comparison between experimental and theoretical values for different experimental setups, namely single shear test, double shear test and bending test, was performed. A comparison between experimental and theoretical values for different composite materials, namely laminates and sheets, was made as well.

Table 3.3 summarizes the results of the statistical analysis in terms of maximum force carried by the FRP-concrete bond surface. The value of the CoV together with the indication of the corresponding mean value of the ratio between the experimental and theoretical value (Avg), are provided.

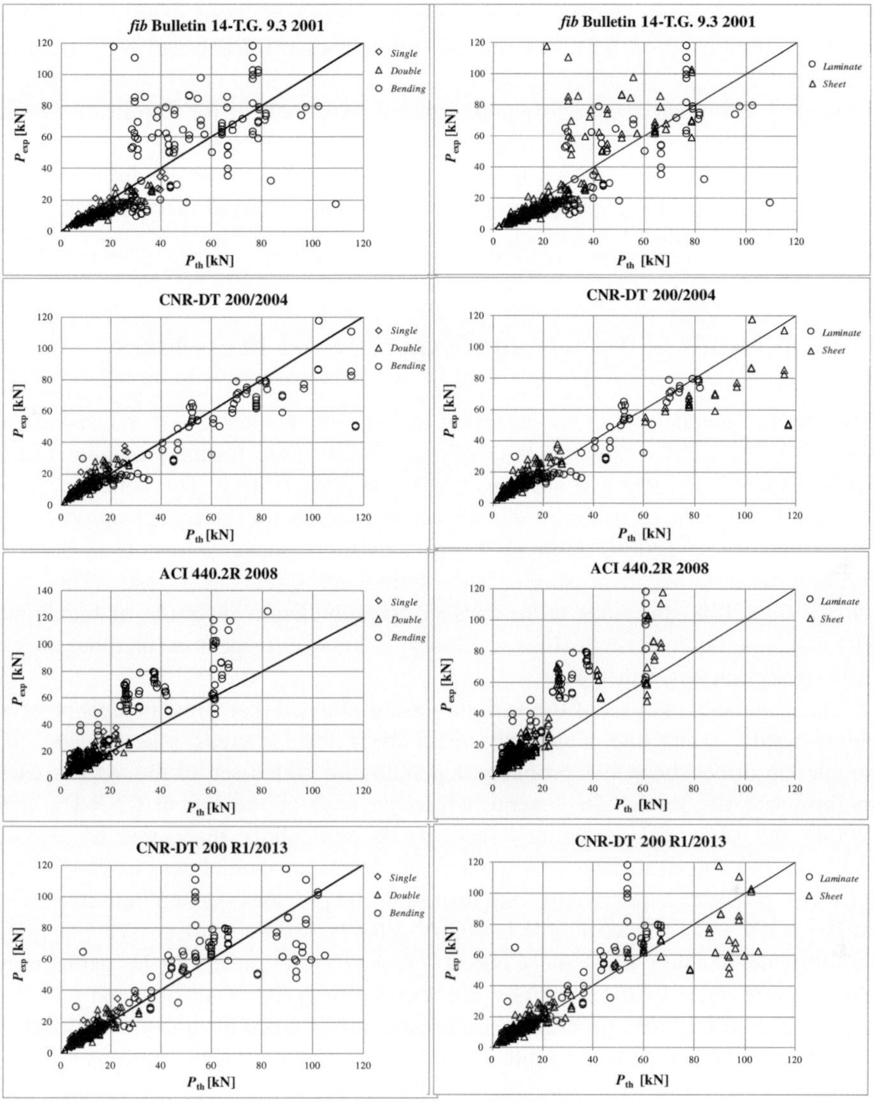

Fig. 3.25 Comparison between experimental and analytical values for different experimental setups (single shear, double shear, bending test), and different composite materials (laminates and sheets)

The analytical formulations for the effective bond length were compared to the experimental measurements for different material preparations (pre-impregnated laminates and post-impregnated sheets). Since there are few works in which the effective bond length was experimentally measured due to the practical difficulty of the procedure, the database was comprised of 48 specimens taken from

Bizindavyi and Neale (1999), Pellegrino et al. (2008), Subramaniam et al. (2007, 2011), Carloni et al. (2012), Carloni and Subramaniam (2013), and Nguyen et al. (2001).

Table 3.4 summarizes the results of the statistical analysis for the effective bond length. The value of the CoV together with the indication of the corresponding average value of the ratio between the experimental and theoretical value (Avg), the standard deviation (StD), and the percentage of the overestimated length value (OverE), are provided.

Critical Issues of Models and Experimental Procedures

Considering the analysis of the laminates and sheets together (*Sheet + Laminate* in Table 3.3), the results show that the analytical formulations for the evaluation of the FRP-concrete bond strength are sometimes non conservative. The better result in terms of coefficient of variation (CoV) was obtained with the model proposed by Camli and Binici (2006) although it provides more scattered results in case of bending tests rather than single- and double-lap direct-shear tests. The ACI 440.2R-08 (2008) provides more conservative prediction in terms of maximum FRP-concrete bond strength (Figs. 3.25 and 3.26) with respect to the other codes, thou its accuracy is rather poor.

The statistical analysis of the various set-ups showed that some models provide better results using the single-lap direct-shear test whereas others using the double-lap direct-shear test, probably depending on which set-up the authors used to formulate the model. In general, except in case of the Italian CNR-DT 200 (2004), the analytical predictions seem to be particularly inaccurate in case of full-scale bending test. The new version of the Italian guidelines, CNR-DT 200 R1/2013, provides better results with respect to the previous version both in case of single- (CoV = 0.44), and double-lap (CoV = 0.42) direct-shear test. When applied to full-scale bending tests, the CNR-DT 200 R1/2013 seems to be particularly inaccurate (CoV = 0.76), especially if compared with the result obtained with the previous version (CoV = 0.39). The same issue arises when the analytical model of the CNR-DT 200 R1/2013 is applied to the tests carried out using pre-impregnated laminate (CoV = 0.91, 2.1 times higher with respect to the previous version). This inaccuracy affects the overall results of the model (CoV = 0.55). It should be noted that 75 % of the full-scale bending test results included within the database are carried out using pre-impregnated laminate. For this reason it is not possible to evaluate whether the model proposed by the CNR-DT 200 R1/2013 is inaccurate when applied to full-scale bending tests or to pre-impregnated laminate.

Observing the effective bond length versus FRP stiffness ($E_f t_f$) diagram (Fig. 3.26h) it can be seen that the predictions are good for low values of the FRP stiffness, whereas become worse for higher values. The ACI 440.2R-08 (2008) model showed an opposite trend with respect to the other main formulations. The best result for effective bond length, in terms of CoV, was obtained by the new

Table 3.3 Main results of the statistical procedure on the maximum force carried by the FRP-concrete bond interface

Analytical model	Single		Double		Bending		Sheet		Laminate		Sheet + Laminate	
	Avg	CoV	Avg	CoV	Avg	CoV	Avg	CoV	Avg	CoV	Avg	CoV
Van Gemert (1980)	1.51	0.84	0.94	0.54	0.84	2.98	1.18	0.66	1.06	1.18	1.08	0.76
Tanaka (1996)	1.75	1.20	1.40	0.70	–	–	1.90	1.00	2.19	2.94	1.77	1.21
Hiroyuki and Wu (1997)	1.91	1.15	1.82	1.25	1.95	1.42	1.90	1.38	2.19	1.48	1.91	1.38
Maeda et al. (1997)	0.97	0.24	1.05	0.32	1.39	0.94	1.12	0.59	1.08	0.41	1.11	0.56
Neubauer and Rostásy (1997)	0.87	1.38	0.99	0.22	1.37	0.94	1.05	0.54	0.95	0.50	1.04	0.54
Khalifa et al. (1998)	1.24	0.41	1.05	0.38	1.30	0.90	1.30	0.64	1.03	0.40	1.23	0.59
fib Bulletin 14-T.G. 9.3 (2001)	0.84	0.23	0.85	0.26	1.15	0.74	0.95	0.46	0.87	0.39	0.93	0.45
Adhikary and Mutsuyoshi (2001)	0.90	0.41	0.55	0.55	0.20	0.81	0.70	0.48	0.63	0.80	0.64	0.58
Chen and Teng (2001)	1.47	1.38	1.66	0.75	2.31	1.95	1.76	1.18	1.67	1.05	1.75	1.17
De Lorenzis et al. (2001)	0.67	0.36	0.72	0.34	0.92	0.60	0.77	0.45	0.70	0.40	0.75	0.44
Izumo (2003)	0.85	0.39	0.69	0.62	0.10	0.97	0.72	0.53	0.08	1.22	0.61	0.62
Iso (2003)	1.06	0.26	0.96	0.31	1.11	0.78	1.11	0.50	0.90	0.38	1.06	0.48
Sato (2003)	0.73	0.36	0.85	0.53	0.53	0.79	0.77	0.47	0.46	0.70	0.69	0.54
Yang et al. (2003)	1.14	0.30	1.04	0.35	1.39	0.95	1.24	0.61	1.11	0.41	1.20	0.58
CNR-DT 200 (2004)	1.42	0.52	1.34	0.52	0.86	0.39	1.30	0.51	1.14	0.43	1.23	0.48
Dai et al. (2005a, b)	0.61	0.41	0.67	0.38	0.90	0.59	0.72	0.47	0.65	0.44	0.70	0.46
Lu et al. (2005)	1.00	0.18	1.17	0.32	1.62	1.18	1.22	0.68	1.14	0.57	1.21	0.67
Camli and Binici (2006)	1.01	0.31	0.84	0.37	0.62	0.55	0.93	0.35	0.80	0.51	0.87	0.40
ACI 440.2R (2008)	1.45	0.59	1.78	0.94	2.12	1.46	1.52	0.72	2.14	1.41	1.66	0.92
CNR-DT 200 R1/2013	1.30	0.44	1.22	0.42	1.14	0.76	1.19	0.37	1.41	0.91	1.24	0.55

Table 3.4 Results of the statistical analysis procedure for the FRP effective bond length carried out without distinguish between laminate and sheet composites

	OverE (%)	StD	Avg	CoV
CNR-DT 200/04	35	0.23	0.97	0.22
fib/01				
Neubauer and Rostásy (1997)				
Lu et al. (2005)	100	0.43	1.78	0.88
Chen and Teng (2001)	35	0.24	0.96	0.24
Camli and Binici (2006)				
Iso (2003)	77	0.36	1.27	0.44
Sato (2003)	0	0.20	0.63	0.41
Maeda et al. (1997)	97	4.17	5.37	5.99
Khalifa et al. (1998)				
ACI 440.2R (2008)	97	4.52	5.82	6.56

version of the Italian guidelines, CNR-DT 200 R1/2013, which improves the previous model also adopted by the fib Bulletin 14 (2001) and Neubauer and Rostasy (1997).

The results obtained show a clear influence of the test set-up on the accuracy of the analytical models. This observation is of particular importance because it highlights the needs of a standard shared test set-up. Furthermore, the poor accuracy of the analytical models when applied to full-scale RC beams show the need of further investigation on FRP strengthened RC beams.

Fib Bulletin 14 (New Formulation)

The indications provided by the most recent formulation of the fib Bulletin 14, which are still under discussion and, thus, were not considered for the previous assessment, are shown below for the sake of completeness. In the new draft of fib Bulletin 14, three types of deboding failure modes, needing specific verifications, are considered:

1. Debonding at the end anchorage zone;
2. Debonding at the end anchorage zone for Concrete Cover Delamination and Critical Diagonal Crack debonding;
3. Debonding at intermediate cracks due to shear stresses (IC Debonding);

Debonding due to unevenness of the concrete surface is also cited, but clearly no calculation is furnished, but only attention in the surface treatment. About the "*Debonding at the end anchorage zone*", under the hypothesis that in EBR strengthened concrete elements the bond behaviour at the end of the reinforcement can be assimilated to what happens in bond tests, thus assuming the interfacial shear

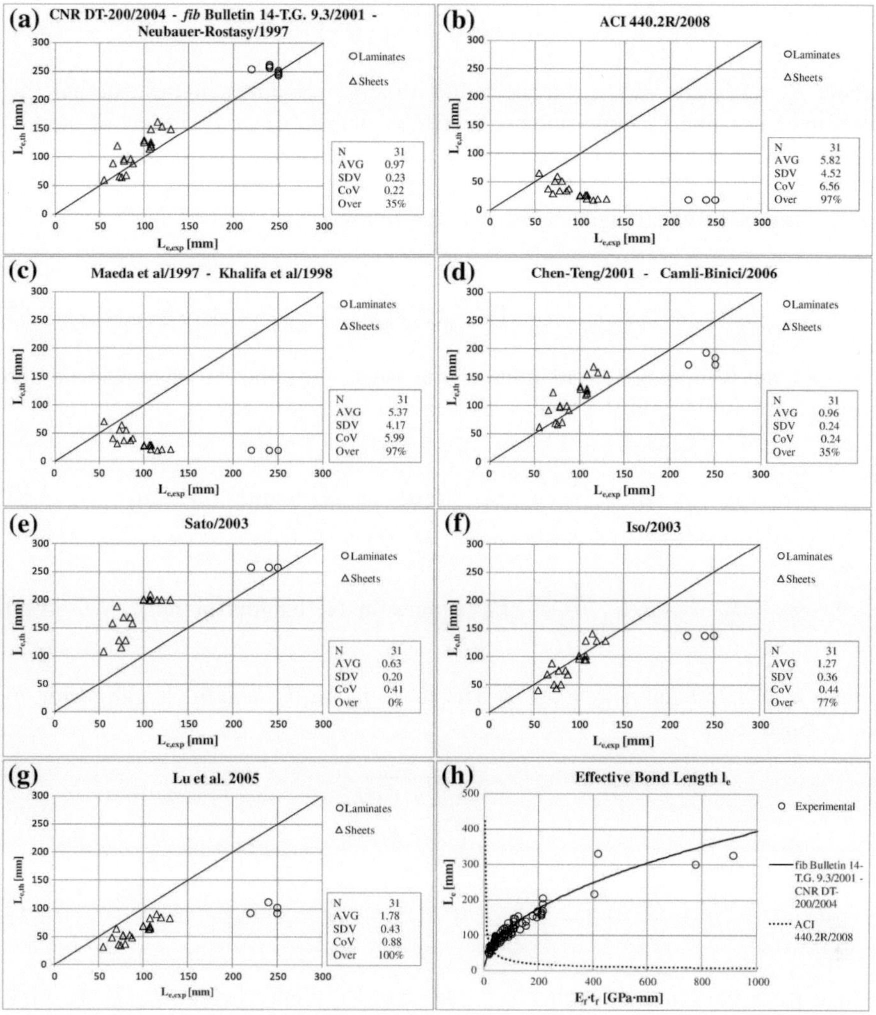

Fig. 3.26 Comparison between experimental and analytical effective bond length results **a–g**, and between experimental and analytical effective bond length as the FRP stiffness increases for the considered codes\recommendations **h**

stresses predominant and the normal ones negligible, the following well-known formulation for the maximum tensile load in the FRP reinforcement (Holzenkampfer 1994; Brosens and van Gemert 1999; Chen and Teng 2001; Teng and Yao 2007; CNR-DT 200 2004; Seracino et al. 2004a; Yao et al. 2005; Toutanji et al. 2007) is assumed:

$$F_{\max} = b_f \sqrt{2 \cdot E_f \cdot t_f \cdot \Gamma_f} \quad (3.66)$$

where b_f, t_f, E_f are width, thickness and Young's modulus of the external FRP reinforcement, and Γ_f is the fracture energy associated to the bond law of the FRP reinforcement-concrete interface. Such a general expression has been calibrated on results of pull-push bond tests and the fracture energy has been assumed depending on the compressive strength of concrete, as follows:

$$\Gamma_f = k_b^2 \cdot f_{cm}^{2/3} \quad (3.67)$$

Thus, the following formulations of the mean, f_{fbm}, and the characteristic debonding strength, f_{fbd}, in the FRP reinforcement are proposed:

$$f_{fbm} = k_m \cdot k_b \cdot \beta_L \cdot \sqrt{\frac{2 \cdot E_f}{t_f} \cdot f_{cm}^{2/3}} \text{ [forces in N, lengths in mm]} \quad (3.68)$$

$$f_{fbk} = k_k \cdot k_b \cdot \beta_L \cdot \sqrt{\frac{2 \cdot E_f}{t_f} \cdot f_{cm}^{2/3}} \text{ [forces in N, lengths in mm]} \quad (3.69)$$

where $k_m = 0.250$ and $k_{0.05} = 0.170$. The shape factor, k_b, and the length factor, β_L, are defined as:

$$k_b = \sqrt{\frac{2 - b_f/b_c}{1 + b_f/b_c}} \quad (3.70)$$

$$\beta_L = \frac{L_b}{L_e} \cdot \left(2 - \frac{L_b}{L_e}\right) < 1 \quad \text{if } L_b < L_e \text{ and } \beta_L = 1 \text{ if } L_b > L_e \quad (3.71)$$

where L_b is the bonded length, b_c is the concrete section width, f_{cm} is the mean cylinder compressive strength of concrete.

The coefficient k_m has been assessed using the experimental results collected in more than 280 bond tests (Bilotta et al. 2011) on concrete elements externally bonded with FRP plates and sheets with compressive strength f_{cm} variable in the range 15–62 MPa, Young's modulus of FRP reinforcement 82÷400 GPa, thickness of FRP reinforcement 0.083÷1.6 mm, layers of strengthening 1÷3, reinforcement—to concrete width ratio 0.15÷1. In particular, k_m has been assessed by means of a statistical procedure as the least-square coefficient minimizing the difference between theoretical values and experimental results according to the design by testing philosophy to determine capacity models suggested in (see Chap. 2, CEN, 1993, Eurocode 0). The value $k_{0.05}$ has been then calculated under the hypothesis that the debonding load has a normal distribution.

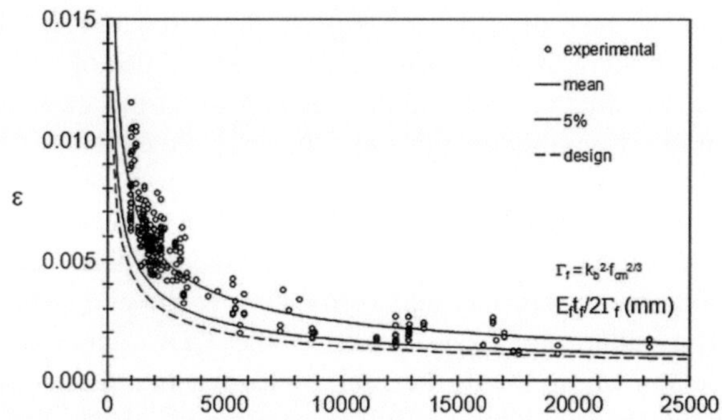

Fig. 3.27 Comparison of experimental and theoretical values of debonding strain according to Eqs. (3.3) and (3.4)

In Fig. 3.27 the distribution of experimental values of maximum strain at debonding collected by (Bilotta et al. 2011) is plotted in comparison with the mean and characteristic provisions given by Eqs. (3.68) and (3.69).

Associated to the bond model given by Eqs. (3.3) and (3.4), the effective bond length L_e may be estimated by the following expressions:

$$L_e = \frac{\pi}{k_b} \cdot \sqrt{\frac{E_f \cdot t_f}{8 \cdot f_{cm}^{2/3}}} \quad \text{for mean provision} \tag{3.72a}$$

$$L_e = 1.5 \cdot \frac{\pi}{k_b} \cdot \sqrt{\frac{E_f \cdot t_f}{8 \cdot f_{cm}^{2/3}}} \quad \text{for design provision} \tag{3.72b}$$

The failure mode indicated as "*Debonding at the end anchorage zone for Concrete Cover Delamination and Critical Diagonal Crack debonding*", is usually determined by a diagonal shear crack formed near the support end of the FRP reinforcement in the unstrengthened portion. When the distance between the FRP reinforcement end and the adjacent beam support is very small, the failure can be induced by the formation of a major shear crack intersecting the plate with the detachment of a thin layer of concrete or associated to the complete concrete cover delamination. In this case a real CDC debonding occurred due to the formation of the critical diagonal crack in the strengthened portion. If the plate end distance is increased, the CDC may fall outside the plated region, and only concrete cover separation is observed. In presence of internal steel stirrups, the shear crack can propagates below the steel stirrups and lead to the superficial debonding or to concrete cover delamination. On the contrary, the lack of internal steel shear reinforcement could induce the propagation of the diagonal crack along the height of the beam. This has been often observed in experimental bending tests where concentrated loading points are applied. Since the lacking of internal steel stirrups is one of the cause of such failure, the behaviour can be significantly enhanced by shear strengthening measures, such as the use of externally bonded FRP U-shaped jackets, that can restrain the opening-up of the shear crack as well as the separation of the strengthening plate and the concrete cover. A variety of strength models have been proposed for calculating the strength for such a failure (Janzse 1997; Oehlers and Moran 1990; Smith and Teng 2002a; Oehlers et al. 2004; Gao et al. 2005; Teng and Yao 2007; Toutanji et al. 2006); most of them are based on shear verification or combined shear-moment verification at the end section of the FRP reinforcement. The failure load associated to the CDC debonding or to the concrete cover delamination should be range between the shear strength of the concrete in the unplated beam and the full shear strength of the plated beam.

For the "*Debonding at intermediate cracks due to shear stresses (IC Debonding)*", three possible approaches are proposed: a simplified approach, an approach based on the envelope line of tensile stress and an approach based on the force transfer between the concrete and external FRP reinforcement.

In the simplified approach, similarly to what reported in (CNR-DT 200 2004), at the ultimate limit state the maximum bending moment can be calculated assuming that in the FRP reinforcement the tensile stress does not exceed the bond strength given by the following relationship:

$$f_{fbm,IC} = k_{cr,m} \cdot f_{fbm} \tag{3.73a}$$

$$f_{fbm,IC} = k_{cr,m} \cdot f_{fbm} \tag{3.73b}$$

where the values of the bond strength at the end, f_{fbm} and f_{fbm}, are defined by Eqs. 3.3 and 3.4 and the coefficients $k_{cr,m}$ and $k_{cr,d}$ can be taken equal to 2.0 and 1.5, respectively, if specific data are not available.

The approach based on the envelope line of tensile stress can be complex for design purposes, since in most cases it requires the evaluation of the tensile stress in the FRP reinforcement in adjacent cracks and the definition of the crack spacing. According to this approach, indeed, the stress variation, $\Delta\sigma_f$, in the FRP strengthening between two subsequent cracks should not exceed a suitable limit value $\Delta\sigma_R$, which corresponds to the maximum increase in tensile stress that can be transferred by means of bond stresses along the cracks spacing. The value $\Delta\sigma_R$ depends, in general, on the bond constitutive law, the distance between cracks, s_r, and the stress level, σ_f, in the FRP reinforcement under the ultimate load condition (fib bulletin 14 2001; Teng et al. 2003; Wu and Niu 2007; Oller et al. 2009; DAfStb 2012).

To avoid the calculation of the tensile stress variation in adjacent cracks, $\Delta\sigma_f$, an alternative approach can be used (Oller et al. 2009). This approach consists of a bending-shear interaction diagram related to IC debonding, obtained through the limit stress variation $\Delta\sigma_R$. This multi-linear diagram is obtained through the shear and bending moment values associated to the maximum transferred force along the crack spacing for some key points related to steel yielding in the critical cracks involved. This method directly compares the design shear force and bending moment values to the IC debonding interaction diagram, but the disadvantage of this method remains the evaluation of the crack spacing.

The approach based on the force transfer between concrete and FRP is simpler than the previous one and does not depend on the crack spacing. This approach implies the limitation of the interfacial shear stress resulting from the change of tensile force along the FRP reinforcement to a certain design shear strength.

References

Achintha, P. M. M., & Burgoyne, C. J. (2008). Fracture mechanics of plate debonding. *Journal of Composites for Construction, 12*(4), 396–404.

Achintha, P. M. M., & Burgoyne, C. J. (2011). Fracture mechanics of plate debonding: validation against experiment. *Construction and Building Materials, 25*(6), 2961–2971.

ACI (2002). *Design and construction of externally bonded FRP systems for strengthening concrete structures.* ACI 440.2R-02, American Concrete Institute, Farmington Hills, Mich.

ACI 440.2R-08 (2008). *Guide for the Design and Construction of Externally Bonded FRP Systems for Strengthening Concrete Structures.* ACI440.2R-08 American Concrete Institute, Farmington Hills, MI, 2008, 76 pp.

ACI Committee 440F (2002). *Guide for design and construction of externally bonded FRP systems for strengthening concrete structures.*

Adhikary, B. B., & Mutsuyoshi, H. (2001). Study on the bond between concrete and externally bonded CFRP sheet. In *Proceedings of the 6th International Symposium on Fiber Reinforced Polymer Reinforcement for Concrete Structures (FRPRCS-5)* (vol. 1, pp. 371–378).

Ahmed, O., van Gemert, D., & Vanderwalle, L. (2001). Improved model for plate end shear of CFRP strengthened RC beams. *Cement & Concrete Composites, 23*(1), 3–19.

Aiello, M.A., & Leone, M. (2005). Experimental bond analysis of concrete—FRP (fiber reinforced polymer) reinforced. In *Proceedings of fib Symposium "Keep Concrete Attractive*". Budapest, Hungary, May.

Alam, M. S., Kanakubo, T., & Yasojima, A. (2012). Shear-peeling bond strength between continuous fiber sheet and concrete. *ACI Structural Journal, 109*(1), 75–82.

Ali-Ahmad, M., Subramaniam, K. V., & Ghosn, M. (2006). Experimental investigation and fracture analysis of bebonding between concrete and FRP. *Journal of Engineering Mechanics, 132*(9), 914–923.

Ali-Ahmad, M., Subramaniam, K. V., & Ghosn, M. (2007). Analysis of scaling and instability in FRP-concrete shear debonding for beam-strengthening applications. *Journal of Engineering Mechanics, 133*(1), 58–67.

Anderson, T. L. (2004). *Fracture mechanics: fundamentals and applications.* Boca Raton: Florida, CRC Press.

Bazant, Z. P., & Planas, J. (1997). *Fracture and size effect in concrete and other quasibrittle materials.* Boca Raton: Florida, CRC Press.

Bilotta, A. (2010). *Behaviour of FRP-to-concrete interface: theoretical models and experimental results, Doctoral thesis, Naples.* Italy: University of Naples "Federico II".

Bilotta, A., Ceroni, F., Nigro, E., & Pecce M. (2011b). Design by testing of debonding load in RC element strengthened with EBR FRP materials. *10th International Symposium on Fiber Reinforced Polymer Reinforcement for Reinforced Concrete Structures,Tampa, Florida, USA.* April 2-4, 2011.

Bilotta, A., Ceroni, F., Nigro, E., Di Ludovico, M., Pecce, M., & Manfredi, G. (2011a). Bond efficiency of EBR and NSM FRP systems for strengthening of concrete members. Journal of

Composites for Construction, ASCE. 15(5), 757–772. October 1, 2011, ISSN 1090-0268/2011/5.

Bilotta, A., Faella, C., Martinelli, E., & Nigro, E. (2011c). Indirect identification method for bilinear bond-law relationship, J. of Composites for Construction, ASCE, doi:10.1061/(ASCE) CC.1943-5614.0000253).

Bilotta, A., Faella, C., Martinelli, E., & Nigro, E. (2012). Indirect identification method of bilinear interface laws for FRP bonded on a concrete substrate. *Journal of Composites for Construction*, 16, 171–184, ISSN: 1090-0268, doi:10.1061/(ASCE)CC.1943-5614.0000253.

Bizindavyi, L., & Neale, K. W. (1999). Transfer lengths and bond strengths for composites bonded to concrete Journal of Composites for Construction ASCE n. 153–160.

Bizindavyi, L., Neale, K. W., & Erki, M. A. (2003). Experimental Investigation of Bonded Fiber Reinforced Polymer-Concrete Joints under Cyclic Loading. *Journal of Composites for Construction ASCE, 7*(2), 127–134.

Blontrock, H., Taerwe, L., & Vanwalleghem, H. (2002). Bond testing of externally glued FRP laminates at elevated temperature. In *Proceedings of the International Symposium Bond in Concrete: from research to standards, Budapest* (pp. 648–654). ISBN 963-420-714-6.

Boschetto, G., Pellegrino, C., Tinazzi, D., & Modena, C. (2006). Bond behaviour between FRP sheets and concrete: an experimental study. In *Proceedings of the 2nd International Fib Congress. June, Naples, Italy*, CDROM.

Brosens, K. (2001). *Anchorage of externally bonded steel plates and CFRP laminates for the strengthening of concrete elements*. PhD thesis, K.U. Leuven, 2001, 225 pp.

Brosens, K., & van Gemert, D. (1997). Anchoring stresses between concrete and carbon fibre reinforced laminates. In *Proceedings. of the 3rd International Symposium on Non-Metallic (FRP) Reinforcement for Concrete Structures*, Japan Concrete Institute, Sapporo, 1, 271–278.

Buyukozturk, O., Gunes, O., & Karaca, E. (2004). Progress on under-standing debonding problems in reinforced concrete and steel members strengthened using FRP composites—Short survey. *Con-struction and Building Materials, 18*(1), 9–19.

Camli, U. S., & Binici, B. (2007). Strength of carbon fiber reinforTced polymers bonded to concrete and masonry. *Construction and Building Materials, 21*, 1431–1446.

Carloni, C., & Subramaniam, K. V. (2010). Direct determination of cohesive stress transfer during debonding of FRP from concrete. *Composite Structures, 93*(1), 184–192.

Carloni, C., & Subramaniam, K. V. (2012). Application of fracture mechanics to debonding of FRP from RC members. *ACI SP 286-10.*

Carloni, C., & Subramaniam, K. (2013). Investigation of sub-critical fatigue crack growth in FRP/concrete cohesive interface using digital image analysis. *Compos—Part B: Eng, 51*, 35–43.

Carloni, C., Subramaniam, K. V., Savoia, M., & Mazzotti, C. (2012). Experimental determination of FRP-concrete cohesive interface properties under fatigue loading. *Composite Structures, 94*, 1288–1296.

Carrara, P., Ferreti, D., Freddi, F., & Rosati, G. (2011). Shear tests of carbon fiber plates bonded to concrete with control of snap-back. *Engineering Fracture Mechanics, 79*, 2663–2678.

Carrara, P., & Ferretti, D. (2013). A finite-difference model with mixed interface laws for shear tests of FRP plates bonded to concrete. *Composites: Part B, 54*, 329–342.

Ceroni, F., Garofano, A., & Pecce, M. (2014). Modelling of bond behavior in masonry elements externally bonded with FRP materials, in press on Composite part B, Elsevier.

Ceroni, F., & Pecce, M. (2002). Bond behaviour of R.C. elements externally reinforced with FRP laminates. In *Proceedings of the International Symposium Bond in Concrete—from research to standards. Budapest* (pp. 622-629) ISBN 963-420-714-6.

Ceroni, F., & Pecce, M. (2005). Strength and ductility of RC beams strengthened with FRP sheets under monotonic and cyclic loads. *Proceedings. of fib Symposium "Keep concrete Attractive", Budapest, Hungary* (418–423).

Ceroni, F., & Pecce, M. (2006). Bond tests on concrete and masonry blocks externally bonded with CFRP. In *Proceedings of Third International Conference on FRP Composites in Civil Engineering, CICE 2006*, Miami, Florida, USA, pp. 17–20.

Ceroni, F., & Pecce, M. (2007). Bond performance in concrete elements strengthened with CFRP sheets. In *Proceedings of FRP RCS8*, July, Patrasso, Greece, CD ROM.

Ceroni, F., & Pecce, M. (2010). Evaluation of bond Strength in concrete element externally reinforced with CFRP sheets and anchoring devices. *Journal of Composites for Construction, ASCE, 14*(5), 521–530.

Ceroni, F., Pecce, M., Matthys, S., & Taerwe, L. (2008). "Bond tests on concrete elements with CFRP and anchorage systems", *Composites: Part B. Elsevier, 39*, 429–441.

Chajes, M. J., Finch, W. W., Januszka, T. F., & Thomson, T. A. (1996). Bond and force transfer of composite material plates bonded to concrete. *ACI Structural Journal*, 93(2), 208–217.

Chen, J. F., & Teng, J. G. (2001). Anchorage strength models for FRP and Steel Plates bonded to concrete. *ASCE J. of Structural Engineering, 127*(7), 784–791.

Chen, J. F., & Teng, J. G. (2003) Shear capacity of FRP-strengthened RC beams: FRP debonding. *Construction and Building Materials*, *17*, 27–41.

Chen, J. F., Yang, Z. J., & Holt, G. D. (2001). FRP or steel plate-to-concrete bonded joints: effect of test methods on experimental bond strength. *Steel Compos Structures, 1*(2), 231–244.

CNR-DT 200. (2004). Instructions for design, execution and control of strengthening interventions through fiber-reinforced composites, Council of National Research, Rome.

CNR. (2013). Guide for the design and construction of externally bonded FRP systems for strengthening existing structures, CNR-DT 200 R1/2013, National Research Council.

Coronado, C. A., & Lopez, M. M. (2005). Modelling of FRP-concrete bond using nonlinear damage mechanics. In C. K. Shield & J. P. Busel(Eds.), *Proceedings of 7th International Symposium on FRP Reinforcement for Concrete Structures*, Kansas City, Missouri.

Cruz, J. M. S., & Barros, J. A. O. (2002). Bond behaviour of carbon laminate strips into concrete by pullout bending test. In *Proceedings of the International Symposium "Bond in Concrete—from research to standards*", *Budapest* (pp. 614–621). ISBN 963-420-714-6.

DAfStb. (2012). On the strengthening of concrete parts with adhesively bonded reinforcement, German Committee for Reinforced Concrete.

Dai, J. G., Sato, Y., Ueda, T., & Sato, Y. (2005b). Static and Fatigue Bond Characteristics of Interfaces between CFRP Sheets and Frost Damage Experienced Concrete. In *Proceedings of FRPRCS-7*, (pp. 1515–1530) ACI-SP-230-86.

Dai, J., Ueda, T., Hiroki, O., & Sato, Y. (2003). Experimental study on the mix-mode fracture of FRP sheet-concrete interface. *JCI International Symposium on Latest Achievement in Technology and Research on Retrofitting Concrete Structures, Interface Mechanics and Structural Performance*, *Kyoto, Japan* (pp. 121–128).

Dai, J., Ueda, T., & Sato, Y. (2005a). Development of the nonlinear bond stress–slip model of fiber reinforced plastics sheet-concrete interfaces with a simple method. *Journal of Composites for Construction, ASCE* 9(1), 52–62.

Dai, J. G., Ueda, T., & Sato, Y. (2006). Unified analytical approaches for determining shear bond characteristics of FRP-concrete interfaces through pullout tests. *Journal of Advanced Concrete Technology, 4*, 133–145.

Davalos, J. F., Kodkani, S. S., & Ray, I. (2006). Fracture mechanics method for Mode-I interface evaluation of FRP bonded to concrete substrates. *Journal of Materials in Civil Engineering, 18* (5), 732–742.

De Lorenzis, L., Miller, B., & Nanni, A. (2001). Bond of Fiber-Reinforced Polymer Laminates to Concrete. ACI Materials Journal. 98-M29, 256–264.

Delaney, J., & Karbhari, V. (2007). Defect criticality in FRP strengthening. In *Proceedings of the 8th International Symposium in Fiber-Reinforced (FRP) Polymer Reinforcement for Concrete Structures (FRPRCS-8) CD-ROM*, *University of Patras, Greece* (pp. 3–20).

Diab, H., Wu, Z., & Iwashita, K. (2007). Experimental and numerical investigation of fatigue behavior of frp-concrete interface. In *Proceeding of FRPRCS-8*, *Patras, Greece* (pp. 16–18).

Dimande, A. O., Juvenades, L. F. P., & Figueiras, J. A. (2005). Bond characterization between concrete and fiber-reinforced polymer. In *Proceedings of* 3rd *International Conference on Composites in Construction, CCC2005*, *Lyon, France*.

Elices, M., Guinea, G. V., Gómez, J., & Planas, J. (2002). The cohesive zone model: advantages, limitations and challenges. *Engineering Fracture Mechanics, 64*, 137–163.
Faella, C., Martinelli, E., & Nigro, E. (2002). Aderenza tra calcestruzzo e fogli di FRP utilizzati come placcaggio di elementi inflessi. Parte II: modelli teorici ed elaborazioni numeriche. Atti del XIV Congresso C.T.E., Bologna.
Faella, C., Martinelli, E., & Nigro, E. (2003). Interface behaviour in FRP plates bonded to concrete: experimental tests and theoretical analyses. In *Proceedings of the International Conference on Advanced Materials for Construction of Bridges, Buildings and other Structures—III, Davos (Svizzera)*, 7–12 September 2003.
Faella, C., Martinelli, E., & Nigro, E (2006a). Formulation and Validation of a Theoretical Model for Intermediate Debonding in FRP Strengthened RC Beams. In *Proceedings of the 2nd International fib Congress, Naples, Italy*, 5–8 June 2006, Paper 0735.
Faella, C., Martinelli, E., & Nigro, E (2006b). Intermediate Debonding in FRP Strengthened RC Beams: A Parametric Analysis, In *Proceedings of the 2nd International fib Congress, Naples, Italy*, 5–8 June 2006, Paper 0993.
Faella, C., Martinelli, E., & Nigro, E., (2007a). Direct versus Indirect identification of FRP-to-concrete interface relationships, Asia-Pacific Conference on FRP in Structures. Hong Kong (China), 12–14 December 2007.
Faella, C., Martinelli, E., & Nigro, E. (2008a). Formulation and Validation of a Theoretical Model for Intermediate Debonding in FRP Strengthened RC Beams. *Composites Part B*, 39(4), 645–655. ISSN 1359-8368.
Faella, C., Martinelli, E., & Nigro, E. (2009). Direct versus Indirect Method for Identifying FRP-to-Concrete Interface Relationships. *ASCE Journal for Composites for Construction*, 13 (3), 226–233. ISSN 1090-0268.
Fanning, P. J., & Kelly, O. (2001). Ultimate response of RC beams strengthened with CFRP plates. *Journal of Composites for Construction, ASCE, 5*(2), 122–127.
Ferracuti, B., Savoia, M., & Mazzotti, C. (2006). A numerical model for FRP-concrete delamination. *Composites: Part B, 37*, 356–364.
Ferracuti, B., Savoia, M., & Mazzotti, C. (2007). Interface law for FRP-concrete delamination. *Composite Structures, 80*(4), 523–531.
fib. (2001). Externally Bonded FRP Reinforcement for RC Structures. fib Bulletin 14, Technical Report, Task Group 9.3—FRP Reinforcement for Concrete Structures, International Federation for Structural Concrete, Lausanne (CH). ISBN 978-2-88394-054-3.
Focacci, F., Nanni, A., & Bakis, C. E. (2000). Local bond-slip relationship for FRP reinforcement in concrete. *Journal of Composites for Construction, 4*(1), 24–31.
Garden, H. N., & Hollaway, L. C. (1998). An experimental study of the influence of plate end anchorage of carbon fibre composite plates used to strengthen reinforced concrete beams. *Composite Structures, 42*, 175–188.
Guadagnini, M., Serbescu, A., Palmieri, A., Matthys, S, Bilotta, A, Nigro, E, Ceroni, F, Czaderski, C, Olia, S, Szambo, Z, Balazs, G, & Mazzotti, C. (2012). Round robin test on the bond behaviour of externally bonded frp systems to concrete. In *Proceedings of CICE 2012, 6th International Conference on FRP Composites in Civil Engineering*, *Rome, Italy, CD ROM* (pp. 13–15).
Gunes, O. (2004), *A Fracture based approach to understanding debonding in FRP bonded structural members*, PhD Thesis, Massachusetts Institute of Technology, Cambridge, MA.
Gunes, O., Buyukozturk, O., & Karaca, E. (2009). A Fracture-based model for FRP debonding in strengthened beams. *Engineering Fracture Mechanics, 76*, 1897–1909.
Guo, Z. G., Cao, S. Y., Sun, W. M., & Lin X. Y. (2005). Experimental study on bond stresses-slip behaviour between FRP sheets and concrete. In Chen & Teng (Eds.), *Proceedings of the International Symposium on Bond Behaviour of FRP in Structures, BBFS 2005* (pp. 77–83).
Hearing, B. F. (2000), *Delamination of reinforced concrete retrofitted with fiber reinforced plastics*, PhD Thesis, Massachusetts Institute of Technology, Cambridge, MA.

Hillerborg, A., Modéer, M., & Petersson, P. E. (1976). Analysis of crack formation and crack growth in concrete by means of fracture mechanics and finite elements. *Cement and Concrete Research, 6*, 773–782.

Hiroyuki, Y., & Wu, Z. (1997). Analysis of debonding fracture properties of CFS strengthened member subject to tension. In *Proceedings of 3rd international symposium on non-metallic (FRP) reinforcement for concrete structures* (vol. 1, pp. 284–94).

Holzenkaempfer. (1994). Ingenieurmodelle des verbundes geklebter bewehrung fur betonbauteile, Dissertation, TU Braunschweig (in German).

Horiguchi, T., & Saeki, N. (1997). Effect of test methods and quality of concrete on bond strength of CFRP sheet. In *Proceedings of International Symposium on Non-metallic (FRP) reinforcement for concrete structures*, Sapporo, Japan, Japan Concrete Institute, Vol.1, pp. 265–270.

Hutchinson, J. W., & Suo, Z. (1992). Mixed-mode cracking in layered materials. *Advances in Applied Mechanics, 29*, 63–191.

International Concrete Repair Institute. (1997). Selecting and Specifying Concrete Surface Preparation for Sealers, Coatings and Polymer Overlays. Technical Guideline 1997 No. 03732.

Iovinella, I., Prota, A., & Mazzotti, C. (2013). Influence of surface roughness on the bond of FRP laminates to concrete, *Construction and Building Materials, 40*, 533–542.

Japan Concrete Institute (JCI). (2003). Technical report of technical committee on retrofit technology. In *Proceedings International Symposium on Latest Achievement of Technology and Research on Retrofitting Concrete Structures*.

Khalifa, A., Gold, W. J., Nanni, A., & Aziz, A. (1998). Contribution of externally bonded FRP to shear capacity of RC flexural members. *Journal of Composites for Construction, ASCE, 2*(4), 195–203.

Ko, H., & Sato, Y. (2007). Bond stress-slip relationship between FRP sheet and concrete under cyclic load. *Journal of Composites for Construction ASCE, 11*(4), 419–426.

Kobayashi, A., Matsui, S., & Kishimoto, M. (2003). Fatigue Bond of Carbon Fiber Sheets and Concrete in RC Slabs Strengthened by CFRP. In K.H. Tan (Ed.), *Proceedings of FRPRCS-6* (vol. 2, pp. 865–874).

Leone, M., Aiello, M. A., & Matthys, S. (2006). The influence of service temperature on bond between FRP reinforcement and concrete. In *Proceedings of the 2nd International Fib Congress, Naples, Italy*.

Liu, K., & Wu, Y. F. (2012). Analytical identification of bond-slip relationship of EB-FRP joints. *Composite; Part B, 43*, 1955–1963.

Lu, X. Z., Teng, J. G., Ye, L. P., & Jiang, J. J. (2005). Bond–slip models for FRP sheets/plates bonded to concrete. *Engineering Strucures, 27*, 920–937.

Maeda, T., Asano, Y., Sato, Y., Ueda, T., & Kakuto, Y. (1997). A study on bond mechanism of carbon fiber sheet. In *Proceedings of the 3rd International Symposium on Non-Metallic (FRP) Reinforcement of Concrete Structures*, Vol. 1, October.

Malek, A. M., Saadatmanesh, H., & Ehsani, M. R. (1998). Prediction of failure load of R/C beams strengthened with FRP plate due to stress concentration at the plate end. *ACI Structural Journal* 95(2), 142–152. ISSN 1090-0268.

Martinelli, E., Czaderski, C., & Motavalli, M. (2011). Modeling in-plane and-of-plane displacement fields in pull-off tests FRP strips. *Engineering Structures, 33*, 3715–3725.

Matana, M., Nanni, A., Dharani, L., Silva, P., & Tunis, G. (2005). Bond performance of steel reinforced polymer and steel reinforced grout. In *Proceedings of the International Symposium on Bond Behaviour of FRP in structures, Honk Kong* (pp. 125–132).

Mazzotti, C., Ceroni, F., & Pecce, M., (2013). Effect of test set-up of bond strength in concrete elements externally bonded with CFRP plates. In J. Barros & J. Sena-Cruz (Eds.), *Proceedings of FRPRCS11, UM, Guimarães*.

Mazzotti, C., Savoia, M., & Ferracuti, B. (2005). A New Set-Up for FRP-Concrete Stable Delamination Test. In C. K. Shield & J. P. Busel (Ed.), *Proceedings of 7th International Symposium FRP Reinforcement for Concrete Structures, Kansas City, Missouri* (pp. 165–180).

Mazzotti, C., Savoia, M., & Ferracuti, B. (2007). Mode II fracture energy and interface law for FRP—concrete bonding with different concrete surface preparations. In *Proceedings of FRAMCOS 6. FRAMCOS 6—Fracture Mechanics of Concrete and Concrete Structures. Catania, Italy*. (vol. 2, pp. 1249–1257).

Mazzotti, C., Savoia, M., & Ferracuti, B. (2008). An experimental study on delamination of FRP plates bonded to concrete. *Construction and Building Materials, 22*, 1409–1421.

Mazzucco, G., Salomoni, V. A., & Majorana, C. E. (2012). Three-dimensional contact-damage coupled modeling of FRP reinforcements—simulation of the delamination and long term process. *Computer and Structures, 110–111*, 15–31.

McSweeney, B. M., & Lopez, M. M. (2005). FRP-Concrete Bond Behavior: A Parametric Study Through Pull-Off Testing. In C. K. Shield & J. P. Busel (Eds.), *Proceedings of the 7th International Symposium FRP Reinforcement for Concrete Structures, Kansas City, Missouri* (pp. 441–460).

Miller, B., & Nanni, A. (1999). Bond Between CFRP Sheets and Concrete Congress. In *Proceedings ASCE 5th Materials Cincinnati, Ohio* (pp. 240–247).

Nakaba, K., Kanakubo, T., Furuta, T., & Yoshizawa, H. (2001). Bond behaviour between fiber-reinforced polymer laminates and concrete. *ACI Structural Journal, 98*(3), 359–367.

Neubauer, U., & Rostásy, F. S. (1997). Design aspects of concrete structures strengthened with externally bonded CFRP-plates. In *Proceedings of 7th International Conference on Structural Faults and Repair Concrete + Composites* (vol. 2, pp. 109–118).

Nigro, E., Di Ludovico, M., & Bilotta, A. (2008). Concrete interface relationships under monotonic and cyclic actions. In *Fourth International Conference on FRP Composites in Civil Engineering (CICE2008), Zurich, Switzerland* (pp. 22–24).

Nigro, E., Di Ludovico, M., & Bilotta, A. (2011). Experimental Investigation of FRP-Concrete Debonding under Cyclic Actions. *Journal Of Materials In Civil Engineering*, 23, 360–371. ISSN: 0899-1561, doi:10.1061/(ASCE)MT.1943-5533.0000173.

Nguyen, D. M., Chan, T. K., & Cheong, H. K. (2001). Brittle failure and bond development length of CFRP–concrete beams. *Journal of Composites for Construction, ASCE, 5*(1), 12–17.

Oehlers, D. J., & Moran, J. P. (1990). Premature failure of externally plated reinforced-concrete beams. *Journal of Structural Engineering—ASCE* 116(4), 978–995. ISSN 0733-9445.

Oller, E., Cobo Del Arco, D., & Marì Bernat, A. R. (2009). Design proposal to avoid peeling failure in FRP-strengthened reinforced concrete beams. *Journal of Composites for Construction, 13*(5), 384–393.

Pellegrino, C., & Modena, C. (2009). Influence of axial rigidity on FRP-concrete bond behavior: an analytical study. *Advances in Structural Engineering, 12*(5), 639–649.

Pellegrino, C., Tinazzi, D., & Modena, C. (2008). Experimental study on bond behavior between concrete and FRP reinforcement. *Journal of Composites for Construction, ASCE* 12(2), 180–189.

Pham, H., & Al-Mahaidi, R. (2004). Prediction models for debonding failure loads of CFRP retrofitted RC beams. In *Proceedings of the 2nd International Conference on FRP Composites in Civil Engineering, CICE 2004, Adelaide (Australia)* (pp. 8–10), December 2004.

Pham, H. B., & Al-Mahaidi, R. (2005). Modelling of CFRP—concrete shear-lap tests. In *Proceedings of 3rd International Conference on Composites in Construction, CCC2005, Lyon, France*.

Rabinovitch, O. (2004). Fracture mechanics failure criteria for RC beams strengthened with FRP strips—a simplified approach. *Composite Structures, 64*, 479–492.

Rabinovitch, O. (2008). Debonding analysis of fiber-reinforced-polymer strengthened beams: cohesive zone modeling versus a linear elastic fracture mechanics approach. *Engineering Fracture Mechanics, 75*, 2842–2859.

Rabinovitch, O. (2012). Dynamic debonding on concrete beams strengthened with composite materials. *International Journal of Solids and Structures, 49*, 3641–3658.

Rahimi, H., & Hutchinson, A. (2001). Concrete beams strengthened with externally bonded FRP plates. *Journal of Composites for Construction, ASCE, 5*(1), 44–56.

Roberts, T. M. (1989). Approximate analysis of shear and normal stress concentrations in the adhesive layer of plated RC beams, The Structural Engineer, 67(12), 1989. *ISSN, 229–233*, 0039–2553.

Said, & Wu. (2008). Evaluating and proposing models of predicting IC Debonding Failure. *ASCE Journal of Composites for Construction* 12(3), 284—299. June 1, 2008.

Savoia, M., Bilotta, A., Ceroni, F., Di Ludovico, M., Fava, G., Ferracuti, B., Mazzotti, C., Nigro, E., Olivito, R., Pecce, M., & Poggi, C. (2009). Experimental round robin test on FRP concrete bonding. In *Proceedings of FRP RCS9, Sydney, Australia* (pp. 13–15).

Savoia, M., Ferracuti, B., & Mazzotti, C. (2003). Non linear bond-slip law for FRP-concrete interface. In *Proceedings of the conference FRPRCS-6, Singapore*.

Serbescu, A., Guadagnini, M., & Pilakoutas, K. (2013). Standardised double-shear test for determining bond of FRP to concrete and corresponding model development. *Composites Part B Engineering, 55*, 277–297.

Shen, X., Myers, J. J., Maerz, N., & Galecki, G. (2002). Effect of surface roughness on the bond performance between FRP laminates and concrete. In B. Benmokrane, & E. El-Salakawy (Eds.), *Proceedings of the 2nd International Conference on Durability of Fiber Reinforced Polymer (FRP) Composites for Construction, University of Sherbrooke, Canada* (pp. 607–616).

Smith, S. T., & Teng, J. G. (2002). FRP-strengthened RC beams-II: assessment of debonding strength models. *Engineering Structures, 24*(4), 397–417.

Subramaniam, K. V., Carloni, C., & Nobile, L. (2007). Width effect in the interface fracture during shear debonding of FRP sheets from concrete. *Engineering Fracture Mechanics, 74*(4), 578–594.

Subramaniam, K. V., Carloni, C., & Nobile, L. (2011). An understanding of the width effect in FRP–concrete debonding. *Strain, 47*, 127–137.

Sutton, M. A., Orteu, J. J., & Shreier, H. W. (2009). *Image correlation for shape, motion and deformation measurements*. New York: Springer.

Sutton, M. A., Wolters, W. J., Peters, W. H., Ranson, W. F., & McNeill, S. R. (1983). Determination of displacements using an improved digital correlation method. *Image and Vision Computing, 1*(3), 133–139.

Takeo, K., Matsushita, H., Makizumi, T., & Nagashima, G. (1997). Bond characteristics of CFRP sheets in the CFRP bonding technique. In *Proceedings of Japan Concrete Institute* (Vol. 19, No. 2, pp. 1599–1604).

Taljsten, B. (1996). Strengthening of concrete prisms using the plate-bonding technique. *International Journal of Fracture, 82*, 253–266.

Taljsten, B. (1997a). Strengthening of beams by plate bonding. *ASCE Journal of Materials in Civil Engineering* 9(4), 206–211. ISSN 1943-5533.

Taljsten, B. (1997b) Defining anchor lengths of steel and CFRP plates bonded to concrete. *International Journal of Adhesion and Adhesives* 7(4), 319–327.

Teng, J. G., et al. (2001). FRP composites in civil engineering. Hong Kong: Elsevier.

Teng, J. G., Chen, J. F., Smith, S. T., & Lam, L. (2002). FRP Strengthened RC Structures, John Wiley & Sons Ltd., Chichester (UK), 245 pp. ISBN 0-471-48706-6.

Teng, J. G., Lu, X. Z., Ye, L. P., & Jiang, J. J. (2004). Recent Research on Intermediate Crack Induced Debonding in FRP Strengthened Beams. In *Proceedings of the 4th International Conference on Advanced Composite Materials for Bridges and Structures, Calgary, AB, Canada*.

Teng, J. G., Smith, S. T., Yao, J., & Chen, J. F. (2003). Intermediate crack-induced debonding in RC beams and slabs. *Construction and Building Materials* 17(6–7), 447–462. ISSN 0950-0618.

Teng, J. G., & Yao, J. (2007). Plate end debonding in FRP-plated RC beams-II: Strength model. *Engineering Structures, 29*, 2472–2486.

Tounsi, A., Hassaine, Daouadji T., Benyoucef, S., & Addabedia, E. A. (2009). Interfacial stresses in FRP-plated RC beams: Effect of adherend shear deformations, International Journal of Adhesion & Adhesives—Elsevier, 29(4), 2009. *ISSN, 343–351*, 0143–7496.

Toutanji, H., Saxena, P., Zhao, L., & Ooi, T. (2007). Prediction of interfacial bond failure of FRP—Concrete surface. *Journal of Composites for Construction, ASCE, 11*(4), 427–436.

Travassos, N., Ripper, T., & Appleton, J. (2005). Bond stresses characterization on CFRP-RC interfaces. In *Proceedings of 3rd International Conference Composites in Construction*, CCC2005, Lyon, France, July.

Ueda, T., & Dai, J. (2005). Interface Bond between FRP Sheets and Concrete Substrates: properties, numerical modeling and roles in member behavior. *Progress in Structural Engineering and Materials, John Wiley & Sons Ltd, 7*(1), 27–43.

Ueda, T., Sato, Y., & Asano, Y. (1999). Experimental study on bond strength of continuous carbon fiber sheet. In *Proceedings of 4th International Symposium on Fiber Reinforced Polymer reinforcement for Reinforced Concrete structure* (pp. 407–16).

Van Gemert, D. (1980). Force transfer in epoxy-bonded steel–concrete joints. *International Journal of Adhesion and Adhesives, 1*, 67–72.

Wan, B., Sutton, M. A., Petrou, M. F., Harries, K. A., & Li, N. (2004). Investigation of bond between fiber reinforced polymer and concrete undergoing global mixed mode I/II loading. *Journal of Engineering Mechanics, 130*(12), 1467–1475.

Wang, R. G., Liu, W. B., Dai, C. Q., Zhang, C. H., & Zhu, X. (2005). Study on adhesion properties of adhesive materials used for carbon fiber reinforced concrete structure. In *Proceedings of 3rd International Conference on Composites in Construction, CCC2005*, Lyon, France.

Wu, Z. S., & Niu, H. (2007). Prediction of crack-induced debonding failure in R/C structures flexurally strengthened with externally bonded FRP composites. *Doboku Gakkai Ronbunshuu E, 63*(4), 620–639.

Wu, Y. F., Xu, X. S., Sun, J. B., & Jiang, C. (2012). Analytical solution for the bond strength of externally bonded reinforcement. *Composite Structures, 94*, 3232–3939.

Wu, Z., & Yin, J. (2003). Fracturing behaviors of FRP-strengthened concrete structures. *Engineering Fracture Mechanics, 70*(10), 1339–1355.

Wu, Z. S., Yuan, H., & Niu, H. (2002). Stress transfer and fracture propagation in different kinds of adhesive joints. *Journal of Engineering Mechanics, 128*(5), 562–573.

Wu, Z. S., Yuan, H., Yoshizawa, H., & Kanakubo, T. (2001).Experimental/analytical study on interfacial fracture energy and fracture propagation along FRP-concrete interface. *ACI International SP-201-8,* 133–52.

Yao, J., Teng, J. G., & Chen, J. F. (2005). Experimental study on FRP-to-concrete bonded joints. *Composites Part B Engineering, 36*, 99–113.

Yi, W. H., Kang, D. E., Woo, H. S., Choi, K. S., Yoo, Y. C., & Keung-Hwan, K. (2006). A study on bond mechanism of fiber reinforced polymer bonded to concrete. *Proceeding of the 2nd International fib Congress, June, Naples, Italy, CDROM.*

Yuan, H., Teng, J. G., Seracino, R., Wu, Z. S., & Yao, J. (2004). Full-range behavior of FRP-to-concrete bonded joints. *Engineering Structures, 26*(5), 553–564.

Yuan, H., Wu, Z., & Yoshizawa, H. (2001). Theoretical solutions on interfacial stress transfer of externally bonded stee/composite laminates. *Structural Engineering/ Earthquake Engineering, 18*(1), 27–39.

Zhao, H. D., Zhang, Y., & Zhao, M. (2000). Research on the bond performance between CFRP plate and concrete. In *Proceedings of 1st Conference on FRP concrete structures of China* (pp. 247–53).

Zhou, Y. W., Wu, Y. F., & Yun, Y. (2010). Analytical modeling of the bond-slip relationship at FRP-concrete interfaces for adhesively-bonded joints. *Composites: Part B, 41*, 423–433.

Chapter 4
Shear Strengthening of RC Elements by Means of EBR FRP Systems

Giorgio Monti, Tommaso D'Antino, Gian Piero Lignola, Carlo Pellegrino and Floriana Petrone

Abstract In this chapter, a thorough review of some typical models adopted to represent the effect of EBR FRP systems on the shear capacity of RC elements is carried out, by comparing their advantages, their accuracy and also by highlighting their possible weaknesses. Then, more light is shed upon the so-called "variable angle strut model", commonly adopted in RC structures, but whose application in the FRP field raises some concerns. Along the same line, the formulations provided by current Codes for the evaluation of the shear capacity of FRP-RC elements are widely discussed and proved to overestimate the actual shear capacity. Having assessed some of the shortcomings of such models, an alternative approach is proposed. The final outcome is a revised model for EBR FRP shear-strengthening that accounts for the actual contribution of steel stirrups and FRP strips/sheets to the shear capacity, while at the same time checking the stress in concrete.

Keywords FRP · Strengthening · Shear · Reinforced concrete

G. Monti (✉) · F. Petrone
Sapienza University of Rome, Rome, Italy
e-mail: giorgio.monti@uniroma1.it

T. D'Antino · C. Pellegrino
University of Padua, Padua, Italy

G.P. Lignola
University of Naples Federico II, Naples, Italy

C. Pellegrino and J. Sena-Cruz (eds.), *Design Procedures for the Use of Composites in Strengthening of Reinforced Concrete Structures*, RILEM State-of-the-Art Reports 19, DOI 10.1007/978-94-017-7336-2_4

Introduction

Strengthening a reinforced concrete beam in shear is a task full of difficulty and should be treated with extreme care by any designer. Due to the brittle nature of shear, the capacity of the resisting mechanism should be correctly evaluated, both before and after the strengthening intervention.

In the assessment phase, one should be aware of the amount of transverse steel internally placed in the beam, but even more attention should be paid to any possible degradation and corrosion in steel, which may jeopardize the efficiency of the stirrups. As for the details, such as the presence of hooks to correctly anchor the stirrups, it goes without saying that these are seldom found in old structures. Old stirrups are almost always found "open" and their effective functioning mostly relies on bond to concrete. This implies that the current rules for evaluating the contribution of transverse steel are, as a matter of fact, overestimating it. However, this matter is not treated by most codes and also in this chapter we will assume that the steel stirrups are completely effective and can actually attain yielding. After all, since the actual contribution of concrete in existing structures is generally underestimated, one may accept to overestimate the contribution of steel, because the two resisting mechanisms are working in parallel and can actually supplement each other's weaknesses. This should elicit the designer to a deeper attention to the actual state of degradation of the materials before uncritically using the capacity equations.

In the strengthening phase, if one decides to apply FRP composites to provide the needed additional capacity, the question raises immediately: by applying FRP, how are the existing mechanisms affected? As a matter of fact, the new resisting mechanism is composed of three sub-mechanisms in parallel and its resulting capacity must be studied more carefully. For example, yielding of the transverse reinforcement cannot be given for granted anymore, since debonding of FRP can occur when the stirrups are still elastic. Also, the concrete strut angle that has been found in the assessment phase will certainly change and its new value will determine the strengthened shear capacity and also the failure mode. While it may seem inappropriate to mention the "failure mode", since shear is undoubtedly a brittle mechanism, it is worth however to consider that a failure where stirrups are already yielded is certainly more desirable that a failure where the stirrups are not. It can be useful to distinguish the two ensuing failures as "pseudo-ductile" and "brittle". Thus, the aim of the FRP-strengthening design is that of achieving a pseudo-ductile type of failure, where steel has yielded before FRP has debonded.

Having briefly commented on the procedure for assessment and strengthening, this chapter starts with a review of the current formulations and models in the most diffused codes, where the conceptual weaknesses outlined above are confirmed. Then, their assessment is carried out with respect to a large experimental database, where, also from the numerical standpoint, some other not fully satisfying aspects are highlighted. Finally, the chapter is concluded with the formulation of a new model for FRP shear strengthening that attempts to overcome the lacunae previously identified and some indications are given for future research work.

Current Formulations and Models on Shear Strengthening of RC Elements by Means of EBR FRP Systems

Different design approaches for strengthening the shear capacity of beams by means of FRP are currently available in codes or guidelines. Most of these approaches rely on models that quantify the nominal shear strength by simply adding up the contributions of concrete, steel, and externally bonded FRP, thus implicitly assuming that the FRP contribution is not influenced by the others. Typically, the following equation is adopted:

$$V_n = V_c + V_s + V_f \tag{4.1}$$

where V_n, V_c V_s, and V_f are the global (nominal) shear capacity of the element considered, the shear capacity contribution provided by concrete, steel and FRP, respectively. Several authors have focused their efforts in proposing equations for the contribution of FRP (V_f), recommending to refer to current guidelines for unstrengthened structures for evaluating the contributions of concrete, V_c, and steel V_s. Thus, shear in FRP-strengthened beams is not treated rigorously, since the various contributions are considered as separate and not interacting, as actually is the case. Current codes for non-strengthened structures, such as Eurocode 2 (CEN 2004), ACI 318 (ACI 2005) and *fib* Model Code (*fib* 2010) are mainly based on the well-established truss model, but present rather different approaches for shear design procedures as far as contributions of steel and concrete are concerned, and assume different shear crack angles, i.e., different angles of the compressed strut in the truss model.

The parameters having significant influence on the shear behaviour of RC members strengthened with externally bonded FRP and the role of these parameters in current design codes were analysed by Modifi and Chaallal (2011). One of these parameters is the cracking angle, for which they concluded that it should be implemented in the calculation of (V_f). Lima and Barros (2011) performed a reliability analysis by means of some collected experimental data and concluded that the experimental critical shear crack angle, θ_{cr}, depends on the existing conventional shear reinforcement in the strengthened beam and may be quite different from the values recommended by the design codes.

The assessment procedure presented in the next sections is based on the comparison between experimental and analytical results, considering the following quantities:

$$V_{R,\exp}^{ref} = V_c + V_s \tag{4.2}$$

$$V_{R,\exp}^{str} = V_c + V_s + V_{f,\exp} \tag{4.3}$$

$$V_{f,\exp} = V_{R,\exp}^{str} - V_{R,\exp}^{ref} \tag{4.4}$$

where $V_{R,\exp}^{ref}$ is the experimental shear strength of the unstrengthened reference control beam, $V_{R,\exp}^{str}$ is the experimental shear strength of the strengthened beam, whereas, V_c, V_s and $V_{f,\exp}$ are concrete, steel and FRP contributions to the global shear strength, respectively.

A brief review of some analytical models for the shear capacity of FRP strengthened elements is presented thereafter. The following models included in guidelines/recommendations are considered: the model proposed by *fib* bulletin 14 (2001), that of the Italian National Research Council (CNR DT-200 2004) and its revision of 2012, based on the work by Monti and Liotta (2007), and that of the American Concrete Institute ACI 440 (2008). Other analytical models proposed by various authors are also considered (Chen and Teng 2003a, b; Carolin and Täljsten 2005; Pellegrino and Modena 2008; Bukhari et al. 2010; Modifi and Chaallal 2011). All equations are presented using the same notation as in the original formulation. A notation list is provided after each equation or a direct reference to the relevant code is give, when necessary.

fib TG 9.3 (2001)

The analytical formulations proposed in *fib* TG 9.3 (2001) on shear strengthening of RC beams are based on regression of experimental results (Triantafillou and Antonopoulos 2000). The shear capacity of the strengthened element is calculated as follows:

$$V_{Rd} = \min(V_{cd} + V_{wd} + V_{fd}, V_{Rd,2}) \tag{4.5}$$

$$V_{fd} = 0.9 \cdot \varepsilon_{fd,e} \cdot E_{fu} \cdot \rho_f \cdot b_w \cdot d \cdot (\cot\theta + \cot\alpha) \cdot \sin\alpha \tag{4.6}$$

FRP fully wrapped configuration:

$$\varepsilon_{fd,e} = 0.17 \cdot \left(\frac{f_{cm}^{2/3}}{E_{fu} \cdot \rho_f}\right)^{0.30} \cdot \varepsilon_{fu} \tag{4.7}$$

Side or U-shaped FRP jackets:

$$\varepsilon_{fd,e} = \min\left[0.65 \cdot \left(\frac{f_{cm}^{2/3}}{E_{fu} \cdot \rho_f}\right)^{0.56} \cdot 10^{-3}; 0.17 \cdot \left(\frac{f_{cm}^{2/3}}{E_{fu} \cdot \rho_f}\right)^{0.30} \cdot \varepsilon_{fu}\right] \tag{4.8}$$

For continuously bonded shear reinforcement of thickness t_f:

$$\rho_f = 2 \cdot t_f \cdot \sin\alpha / b_w \tag{4.9}$$

For FRP reinforcement in the form of strips or sheets of width b_w at spacing s_f:

$$\rho_f = (2 \cdot t_f / b_w) \cdot (b_f / s_f) \tag{4.10}$$

where V_{cd} and V_{wd} are the design values of the shear capacity of concrete and transversal steel, respectively; V_{fd} is the design value of FRP contribution, $\varepsilon_{fd,e}$ is the design value of the effective FRP strain, ε_{fu} is the ultimate strain of the FRP, E_{fu} is the elastic modulus of the FRP in the principal fibre orientation (in GPa); f_{cm} is the cylinder compressive strength of concrete (in MPa); ρ_f is the FRP strengthening ratio, d is the effective depth of the cross section; α is the angle between principal fibre orientation and longitudinal axis of member; θ is the angle of the diagonal crack with respect to the member axis.

CNR DT-200 (2004) and CNR DT-200R1 (2012)

The Italian CNR DT-200 (2004) and CNR DT-200R1 (2012) guidelines contain provisions on FRP shear strengthening that are based on the studies of Monti and Liotta (2007) who developed a mechanics-based model, as opposed to the regression-based models at that time available. The shear capacity of the strengthened element is calculated as follows:

$$V_{Rd} = \min\left\{V_{Rd,cd} + V_{Rd,s} + V_{Rd,f}; V_{Rd,\max}\right\} \tag{4.11}$$

FRP side-bonding configuration

$$V_{Rd,f} = \frac{1}{\gamma_{Rd}} \cdot \min\{0.9 \cdot d; h_w\} \cdot f_{fed} \cdot 2 \cdot t_f \cdot \frac{\sin\beta}{\sin\theta} \cdot \frac{w_f}{p_f} \tag{4.12}$$

$$f_{fed} = f_{fdd} \cdot \frac{z_{rid,eq}}{\min\{0.9 \cdot d; h_w\}} \left(1 - 0.6 \cdot \sqrt{\frac{l_{eq}}{z_{rid,eq}}}\right)^2 \tag{4.13}$$

$$z_{rid,eq} = z_{rid} + l_{eq} \tag{4.14}$$

$$z_{rid} = \min\{0.9 \cdot d; h_w\} - l_{eq} \cdot \sin\beta \tag{4.15}$$

$$l_{eq} = \frac{s_f}{f_{fdd}/E_f} \cdot \sin\ \beta \tag{4.16}$$

FRP U-wrapped or fully-wrapped configuration

$$V_{Rd,f} = \frac{1}{\gamma_{Rd}} \cdot 0.9 \cdot d \cdot f_{fed} \cdot 2 \cdot t_f \cdot (\cot\theta + \cot\beta) \cdot \frac{w_f}{p_f} \tag{4.17}$$

U-wrapped configurations

$$f_{fed} = f_{fdd} \cdot \left[1 - \frac{1}{3} \cdot \frac{l_e \cdot \sin\beta}{\min\{0.9 \cdot d; h_w\}}\right] \tag{4.18}$$

Fully-wrapped configurations

$$f_{fed} = f_{fdd} \times \left[1 - \frac{1}{6} \times \frac{l_e \times \sin\ \beta}{\min\{0.9 \times d; h_w\}}\right] + \frac{1}{2} \times (\varphi_R \times f_{fd} - f_{fdd}) \times \left[1 - \frac{l_e \times in\beta}{\min\{0.9 \times d; h_w\}}\right] \tag{4.19}$$

$$\varphi_R = 0.2 + 1.6 \cdot \frac{r_c}{b_w} \tag{4.20}$$

$$0 \le \frac{r_c}{b_w} \le 0.5 \tag{4.21}$$

$$l_e = \sqrt{\frac{E_f \cdot t_f}{2 \cdot f_{ctm}}} \tag{4.22}$$

$$\Gamma_{Fk} = 0.03 \cdot k_b \cdot \sqrt{f_{ck} \cdot f_{ctm}} \tag{4.23}$$

$$k_b = \sqrt{\frac{2 - \frac{b_f}{b}}{1 + \frac{b_f}{400}}} \tag{4.24}$$

$$f_{fdd} = \frac{1}{\gamma_{f,d} \cdot \sqrt{\gamma_c}} \sqrt{\frac{2 \cdot E_f \cdot \Gamma_{Fk}}{t_f}} \tag{4.25}$$

where $V_{Rd,ct}$ and $V_{Rd,s}$ are the concrete and steel contributions to shear capacity according to the current building code, and $V_{Rd,f}$ is the FRP contribution to the shear capacity, h_w is the cross-section depth, b_w is the width of the member, f_{fed} is the effective FRP design strength, β is the fibre angle with respect to the member longitudinal axis, θ represents the angle of shear cracks (to be assumed equal to 45° unless a more detailed calculation is made), and w_f and p_f are FRP width and spacing, respectively, measured orthogonally to the fibre direction (for FRP strips installed one next to each other, the ratio w_f/p_f shall be set equal to 1.0), r_c is the corner radius of the section to be wrapped, l_e is the optimal bonded length, E_f and t_f are the Young's modulus and the thickness of FRP, respectively, and f_{ctm} is the mean tensile strength of concrete. Γ_{Fk} is the specific fracture energy of the FRP—concrete interface, f_{ck} is the characteristic compressive strength of concrete, k_b is a geometric coefficient depending on both width of the strengthened beam b and width of the FRP system b_f; $b_f/b \geq 0.33$ (if $b_f/b < 0.33$, the value for k_b corresponding to $b_f/b < 0.33$ is adopted). For laminate/sheet end debonding it is assumed that the provided bond length is equal to or larger than the optimal bonded length, f_{fdd} is the ultimate design strength. The partial factors are $\gamma_{f,d} = 1.20$ (for FRP debonding failure mode and strengthening system with certification of each component as well as the final product to be applied to a given support) and $\gamma_c = 1.5$ (concrete partial safety factor).

Main Differences Between DT-200R1 and the Previous DT-200

The revision R1 of CNR DT-200 published in 2012 aims, first of all, at incorporating the provisions of the Italian National Building Code update of 2008, which was issued after the first publication of DT-200 in 2004. The revised document upgrades the theory on FRP debonding, which shear strengthening refers to.

FRP Layout

Only two layouts are considered in the revised document, in particular, side-bonding application is not allowed, due to potential debonding problems occurring under seismic reverse actions. Therefore, only two possible layouts are considered: U-wrapping and full wrapping.

Design Shear Capacity

The main differences between the two versions of DT-200 are found in the evaluation of the shear capacity: the 1996 and the 2008 Italian National Building Code are based on completely different approaches for computing the design shear capacity of unstrengthened RC members. The theoretical background moved from superposition theory of different shear contributions, namely, concrete capacity, transverse reinforcement capacity, of 1996 Code, to the truss analogy (either with fixed or variable strut angle) of 2008 Code. The following equations refer to CNR DT200/2004 (upper) and Revision 1/2012 (lower):

$$\begin{aligned} V_{Rd} &= \min\left\{V_{Rd,ct} + V_{Rd,s} + V_{rd,f};\, V_{Rd,\max}\right\} \\ V_{Rd} &= \min\left\{V_{Rd,s} + V_{Rd,f};\, V_{Rd,c}\right\} \end{aligned} \tag{4.26}$$

where it can be seen that the concrete contribution $V_{Rd,ct}$ has been removed from the overlapping resisting mechanism, thus leaving only the steel transverse reinforcement and the FRP. Crushing of the concrete strut is checked in both versions.

Optimal Anchoring Length

Slight differences are found on the optimal (or effective) anchoring length. The following equations refer to CNR DT-200/2004 (left) and Revision 1/2012 (right):

$$l_{ed} = \sqrt{\frac{E_f t_f}{2 f_{ctm}}} \qquad l_{ed} = \min\left\{\frac{1}{\gamma_{Rd} f_{bd}} \sqrt{\frac{\pi^2 E_f t_f \Gamma_{Fd}}{2}};\, 200\ \text{mm}\right\} \tag{4.27}$$

where E_f is the FRP elastic modulus, and t_f is the FRP thickness; in the Revision, clear reference is made to both the specific fracture energy Γ_{Fd} and the concrete bond strength ($f_{bd} = 8\Gamma_{Fd}$) (both depending on the mean tensile strength of concrete f_{ctm}, which was explicitly appearing in the previous version); also, a partial safety factor γ_{Rd} has been introduced.

Specific Fracture Energy

Formal differences are found on the specific fracture energy. The following equations refer to CNR DT-200/2004 (left) and Revision 1/2012 (right):

$$\Gamma_{Fk} = 0.03k_b\sqrt{f_{ck}f_{ctm}} \qquad \Gamma_{Fd} = \frac{k_b k_G}{CF}\sqrt{f_{cm}f_{ctm}} \tag{4.28}$$

The first novel aspect is the introduction of the experimentally-calibrated coefficient k_G equal to 0.023 mm for preformed composites and to 0.037 mm for wet layup, while in the previous version it was directly assumed as 0.03 mm. Furthermore, the dependency on the compressive concrete strength is made to refer to the mean value f_{cm} rather than the characteristic value f_{ck} (5 % percentile). Additionally, according to the 2008 Code for existing structures, a confidence factor CF depending on the knowledge level achieved through on-site testing of the structure is added. The geometric coefficient k_b also has a different expression, as seen in the next section.

Geometric Coefficient K_b

This geometric term k_b accounts for the coverage of FRP (having width b_f) over the RC member surface (having width b). The following equations refer to CNR DT-200/2004 (left) and Revision R1/2012 (right):

$$k_b = \sqrt{\frac{2 - \frac{b_f}{b}}{1 + \frac{b_f}{400}}} \qquad k_b = \sqrt{\frac{2 - \frac{b_f}{b}}{1 + \frac{b_f}{b}}} \tag{4.29}$$

The main difference is at the denominator where $b_f/400$ is replaced by b_f/b. Nonetheless, a major difference is in the limit range of b_f/b: while in the previous version it was $b_f/b \geq 0.33$, in the revision it is $b_f/b \geq 0.25$.

Design Value of FRP Debonding Strength

This is the fundamental design parameter in FRP strengthening. The following equations refer to CNR DT-200/2004 (left) and Revision R1/2012 (right):

$$f_{fdd} = \frac{1}{\gamma_{f,d}\sqrt{\gamma_c}}\sqrt{\frac{2E_f\Gamma_{Fk}}{t_f}} \qquad f_{fdd} = \frac{1}{\gamma_{f,d}}\sqrt{\frac{2E_f\Gamma_{Fd}}{t_f}} \tag{4.30}$$

The main difference is in the probabilistic value assigned to the parameters in the equation: the specific fracture energy is defined through its design value Γ_{Fd}, rather than its characteristic value Γ_{Fk} (5 % percentile). This results in the elimination of the partial safety factor for concrete γ_c that is absorbed by that relevant to the specific fracture energy.

ACI 440 (2008)

The American Concrete Institute (ACI 440 2008) guidelines are based on the study of Khalifa et al. (1998). The total shear strength is calculated as:

$$V_n = V_c + V_s + \psi_f V_f \tag{4.31}$$

$$V_f = \frac{A_{fv} \cdot f_{fe} \cdot (\sin\alpha + \cos\alpha) \cdot d_f}{s_f} \tag{4.32}$$

$$A_{fv} = 2 \cdot n \cdot t_f \cdot w_f \tag{4.33}$$

$$f_{fe} = \varepsilon_{fe} \cdot E_f \tag{4.34}$$

Completely wrapped members

$$\varepsilon_{fe} = 0.004 \leq 0.75 \cdot \varepsilon_{frp} \tag{4.35}$$

$$k_2 = \frac{d_{fv} - L_e}{d_{fv}} \tag{4.36}$$

Two- and three-side wrapped

$$\varepsilon_{fe} = k_v \cdot \varepsilon_{frp} \leq 0.004 \tag{4.37}$$

$$k_v = \frac{k_1 \cdot k_2 \cdot L_e}{11{,}900 \cdot \varepsilon_{frp}} \leq 0.75 \tag{4.38}$$

$$L_e = \frac{23{,}300}{(n \cdot t_f \cdot E_f)^{0.58}} \tag{4.39}$$

$$k_1 = \left(\frac{f_c'}{27}\right)^{2/3} \tag{4.40}$$

$$k_2 = \frac{d_{fv} - 2L_e}{d_{fv}} \tag{4.41}$$

where ψ_f is a reduction factor, having a value of 0.95 for completely wrapped configuration or 0.85 for two- and three-side schemes; A_{fv} is the area of FRP shear reinforcement with spacing s_f; f_{fe} is effective stress in the FRP; α is the angle between principal fibre orientation and longitudinal axis of member; d_f is the effective depth of FRP reinforcement; E_f is the tensile modulus of FRP; ε_{fe} is the effective strain level in FRP reinforcement attained at failure and is the bond-reduction coefficient. ACI adopts a 45°-truss-angle analogy without variation of the shear crack angle.

Eurocode 8

Eurocode 8 is devoted to the design of structures for earthquake resistance, and in particular part 3 is devoted to the assessment and retrofitting of buildings.

Shear capacity of brittle RC members is enhanced through the application of FRP strips or sheets applied either by fully wrapping or by bonding the strengthening to the sides and the soffit of the beam (U-shaped strip); it is noted that in this code also side bonding is allowed, however, here it is not shown.

The total shear capacity, limited either by internal stirrups or externally bonded FRP, is evaluated as the sum of the shear capacity of the existing member (according to Eurocode 8 or 2) plus the contribution of FRP, $V_{Rd,f}$, determined according to the following equation:

$$V_{Rd,f} = 0.9d\, f_{fdd,e} 2t_f \left(\frac{w_f}{s_f}\right)^2 (\cot\theta + \cot\beta)\sin\beta \tag{4.42}$$

where the geometry of the FRP strips is described by their thickness t_f, their width w_f, their spacing s_f, and by their angle β, while θ gives the crack angle.

A crucial role is played by the design value of FRP effective debonding strength $f_{fdd,e}$, which is evaluated differently depending on the FRP layout. It is made to depend on the effective anchorage length L_e, on the internal lever arm z, on the FRP design debonding strength f_{fdd}, and on the FRP design tensile strength $f_{fu,w}$, which is interestingly made to depend on the radius R of the section corners rounding.

In the case of fully wrapped members it is:

$$f_{fdd,e(w)} = f_{fdd}\left[1 - 0.36\frac{L_e \sin\beta}{2z}\right] + 0.5(f_{fu,w}(R) - f_{fdd})\left[1 - \frac{L_e \sin\beta}{z}\right] \tag{4.43}$$

In the case of U-wrapped members it is simply:

$$f_{fdd,e(U)} = f_{fdd}\left[1 - 0.36\frac{L_e \sin\beta}{z}\right] \tag{4.44}$$

The FRP debonding strength, given in Eq. (A.25) of EN1998-3, has approximately the same expression as in the CNR seen above.

In any case, the total shear capacity of strengthened members should not be larger than the maximum shear strength controlled by diagonal compression failure in the web, $V_{R,\max}$ (evaluated in accordance with Eurocode 2).

Chen and Teng (2003a, b)

Chen and Teng (2003a, b) models are widely used for FRP shear-strengthened elements. They evaluate the shear capacity according to two failure modes: FRP rupture (2003b) and FRP debonding (2003a). In case of side-bonded configurations, debonding failure mode occurs, in case of wrapped configurations, rupture failure mode should be considered, whereas, for U-jacketing, the shear capacity should be evaluated according to both modes and the smallest value has to be used.

The FRP contribution to shear strength is expressed as:

$$V_f = 2f_{frp,e}t_{frp}w_{frp}\frac{h_{frp,e}(\cot\theta + \cot\beta)\sin\beta}{s_{frp}} \tag{4.45}$$

$$f_{frp,e} = D_{frp}\sigma_{frp.\max} \tag{4.46}$$

FRP debonding failure mode:

$$\sigma_{frp.\max} = \min\left\{\begin{array}{c} f_{frp} \\ 0.427 \times \beta_w \times \beta_L \times \sqrt{\frac{E_{frp}\times\sqrt{f'_c}}{t_{frp}}} \end{array}\right\} \tag{4.47}$$

$$L_{\max} = \min\left\{\begin{array}{ll} \frac{h_{frp,e}}{\sin\beta} & \text{for U-jackets} \\ \frac{h_{frp,e}}{2\sin\beta} & \text{for side plates} \end{array}\right\} \tag{4.48}$$

$$h_{frp,e} = z_b - z_t \tag{4.49}$$

$$z_t = d_{frp,t} \tag{4.50}$$

$$z_b = 0.9\cdot d - (h - d_{frp}) \tag{4.51}$$

$$\beta_L = \left\{\begin{array}{ll} 1 & if\ \lambda \geq 1 \\ \sin\frac{\pi\lambda}{2} & if\ \lambda < 1 \end{array}\right\} \tag{4.52}$$

$$\lambda = \frac{L_{\max}}{L_e} \tag{4.53}$$

$$L_e = \sqrt{\frac{E_{frp}\cdot t_{frp}}{\sqrt{f'_c}}} \tag{4.54}$$

$$\beta_w = \sqrt{\frac{2 - w_f/(s_f \cdot \sin\beta)}{1 + w_f/(s_f \cdot \sin\beta)}} \tag{4.55}$$

FRP rupture failure mode:

$$\sigma_{frp.\max} = \min\left\{\begin{array}{ll} 0.8 f_{frp} & if\ \frac{f_{frp}}{E_{frp}} \leq \varepsilon_{\max} \\ 0.8\varepsilon_{\max} E_{frp} & if\ \frac{f_{frp}}{E_{frp}} > \varepsilon_{\max} \end{array}\right\} \tag{4.56}$$

where $f_{frp,e}$ is the effective stress in the FRP intersected by the critical shear crack at the ultimate limit state, f_{frp} is FRP tensile strength, β is the angle between the principal fibre orientation and the member longitudinal axis, θ is the angle of diagonal crack with respect to the member longitudinal axis, $\sigma_{frp,\max}$ is the maximum stress in the FRP and D_{frp} the stress distribution factor. $h_{frp,e}$ is the effective height of the FRP, z_t and z_b are the coordinates of the top and bottom ends of the effective FRP, $d_{frp,t}$ is the distance from the compression face to the top edge of the FRP, h is the height of the beam, and d_{frp} is the distance from the compression face to the lower edge of the FRP (thus, $d_{frp} = h$ for U jackets). It is recommended to use $\varepsilon_{\max} = 1.5\,\%$ if other specific recommendations are not available. Further details can be found in Chen and Teng (2003a, b).

Carolin and Taljsten (2005)

The design model of Carolin and Taljsten (2005) is based on the superposition principle of the shear contributions and the strut and tie model. The FRP contribution to shear strength V_f can be expressed as:

$$V_f = \eta \varepsilon_f E_f t_f z \left(\frac{\cos\theta}{\sin\alpha}\right) \sin\beta \tag{4.57}$$

where η is a reduction factor ($\eta = 0.6$), ε_f is the critical strain in the fibre, E_f is the modulus of elasticity of the fibre, t_f is the thickness of fibre, and z is the length of the vertical tie in the truss, for steel stirrups usually expressed by the internal lever arm or $0.9d$. The variables α, β and θ are angles considering crack inclination, fibre direction and the difference between them, respectively. Further details can be found in Carolin and Taljsten (2005).

Pellegrino and Modena (2008)

This model is based on the observed experimental shear failure due to peeling of a triangular portion of concrete cover in U-wrapped and side-bonded FRP strengthened beams and takes into account the interaction between external FRP and internal transverse steel reinforcement. FRP strains are assumed equal to those of internal stirrups. The FRP shear contribution V_f is obtained from the rotational equilibrium of the forces F_f and F_c operating in the FRP and concrete surface, respectively, at failure (Figs. 4.1 and 4.2):

$$V_f = \frac{2n_f t_f L_f w_f \varepsilon_{fe} E_f h_f}{s_f} \tag{4.58}$$

where n_f is the number of layers, t_f is the thickness of FRP (one layer), w_f is the width of FRP, E_f is the elastic modulus of FRP in the principal fibre orientation, h_f is the vertical distance from the top edge of the FRP shear reinforcement to the bottom of concrete cross-section, is the spacing of FRP strips and s_f is the effective strain:

$$\varepsilon_{fe} = \frac{2f_{ct}A_c \cos^2 \varphi b_{c,v}}{n_f t_f L_f E_f \frac{h_f - l_e}{h_f} b_f} \tag{4.59}$$

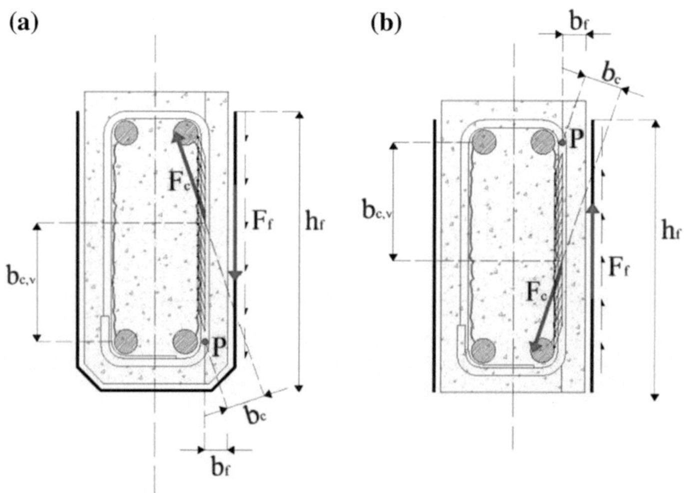

Fig. 4.1 Forces acting in the cross section of "U-jacketed" (**a**) and side-bonded beams (**b**) by Pellegrino and Modena (2008)

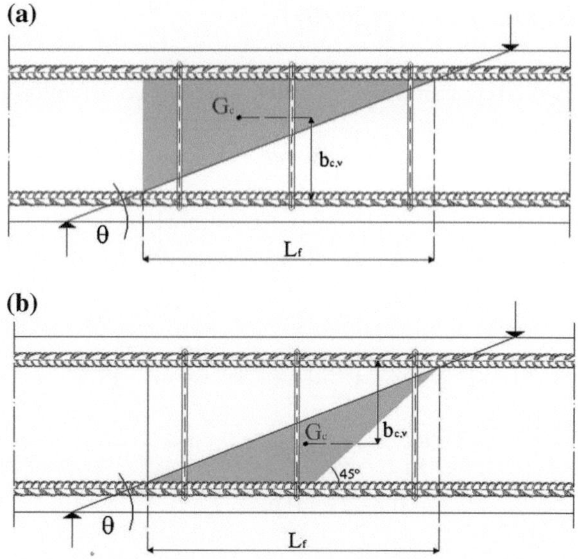

Fig. 4.2 Shape of the fracture surface of "U-jacketed" (**a**) and side-bonded beams (**b**) by Pellegrino and Modena (2008)

where f_{ct} is the tensile strength of concrete, A_c is the area of beam cross-section, L_f is the length of failure surface, l_e is the effective bond length, while the meaning of symbols $b_{c,v}$ and b_f is clear from Figs. 4.1 and 4.2; and ϕ is the angle characterizing the conventional roughness of the interface, which is assumed equal to 79°, according to the experimental results obtained by the authors.

An improved formulation for the contribution of the transversal steel is also proposed for FRP shear-strengthened beams. Variable amplitude for the diagonal crack is assumed. Furthermore, when external FRP reinforcement is applied, the maximum stress in the internal transverse steel is assumed equal to its yielding value only if the effective FRP strain is higher than the steel yield stress. This can allow, with a simplified approach, to take into account the interaction between steel and FRP. Therefore, the formulation for the steel contribution in the presence of an external FRP reinforcement is proposed as:

$$V_s = \left[\alpha \rho_v \left(1 - \frac{c}{d}\right) \cot\theta \cdot \min(\varepsilon_{fe} E_s; f_y)\right] b_w d \tag{4.60}$$

where α is taken as 0,75 to take into account that the stress level in the transverse steel varies with the diagonal crack amplitude; ρ_v is the transverse steel ratio, f_y is the yield stress of the transverse steel, c is the depth of the neutral axis, and b_w is the width of the cross section, d is the effective depth. Further details can be found in Pellegrino and Modena (2008).

Bukhari et al. (2010)

In the regression-based model the FRP contribution to shear strength is given as:

$$V_{Rd,FRP} = Cz_f b_w E_f(\cot\theta + \cot\beta)\sin\beta \tag{4.61}$$

$$C = \min\left[\rho_f \varepsilon_{fe,1}; \rho_f^* \varepsilon_{fe,2}\right] \tag{4.62}$$

For side wrap:

$$\varepsilon_{fe,1} = 0.7\left\{40.25(\rho_f E_f/f_c^{2/3})^{-0.7}\right\} \cdot 10^{-3} \leq 0.4\varepsilon_{fu} \leq 0.004 \tag{4.63}$$

For U wrap:

$$\varepsilon_{fe,1} = 0.8\left\{29.14(\rho_f E_f/f_c^{2/3})^{-0.48}\right\} \cdot 10^{-3} \leq 0.4\varepsilon_{fu} \leq 0.004 \tag{4.64}$$

$$\varepsilon_{fe,2} = \min\left\{\frac{\varepsilon_{fu}}{2}; 0.64\sqrt{\frac{f_{ct}}{E_{fd} \cdot t_f}}; 0.004\right\} \tag{4.65}$$

$$\rho_f^* = \rho_f(d_f - nl_{t\max}/3)/z_f \tag{4.66}$$

$$l_{t\max} = 0.7\sqrt{\frac{E_f t_f}{f_{ctm}}} \tag{4.67}$$

where ρ_f is FRP shear reinforcement ratio, $n = 0$ for fully wrapped sections, 1 for U-wrap and 2 for side wrap, $z_f = 0.9d$ and $l_{t\max}$ is the anchorage length required to develop full anchorage capacity. Further details can be found in Bukhari et al. (2010).

Modifi and Chaallal (2011)

The FRP contribution to shear strength is assumed as:

$$V_f = \rho_v E_f \varepsilon_{fe} b d_f(\cot\theta + \cot\alpha)\sin\alpha \tag{4.68}$$

$$\varepsilon_{fe} = 0.31\beta_c\beta_L\beta_w\sqrt{\frac{\sqrt{f_c'}}{t_f E_f}} \leq \varepsilon_{uf} \tag{4.69}$$

For U-jackets:

$$\beta_c = \frac{0.6}{\sqrt{\rho_f E_f + \rho_s E_s}} \tag{4.70}$$

For side-bonded FRP:

$$\beta_c = \frac{0.43}{\sqrt{\rho_f E_f + \rho_s E_s}} \tag{4.71}$$

and:

$$\lambda = \frac{L_{\max}}{L_e} \tag{4.72}$$

$$L_e = \sqrt{\frac{E_f t_f}{2 f_{ct}}} \tag{4.73}$$

$$\beta_w = \sqrt{\frac{2 - w_f/s_f}{1 + w_f/s_f}} \tag{4.74}$$

where ρ_v is FRP shear reinforcement ratio, E_f is the elastic modulus of FRP in the principal fibre orientation, d_f is FRP effective depth α is the angle between principal fibre orientation and longitudinal axis of member and θ is the angle of diagonal crack with respect to the member axis. It should be noted that in the case of a continuous FRP sheet, the FRP width (w_f) and the spacing (s_f) can be assumed equal to 1. ε_{fe} is the effective strain, f_c' is compressive strength of concrete, β_c is the cracking modification factor, β_L is a decreasing coefficient (FRP effective anchorage length ratio) which represents the effect of FRP sheets having an anchorage length shorter than the effective length L_e. It is assumed $\beta_L = 1$ for $\lambda \geq 1$ or $\beta_L = (2 - \lambda)\lambda$ for $\lambda < 1$. $L_{\max}$ is the maximum available bond length, calculated as $d_f / \sin\alpha$ for U-jackets or $2 \cdot d_f / \sin\alpha$ for side plates. L_e is the effective bond length in mm. Further details can be found in Modifi and Chaallal (2011).

Assessment of Current Formulations and Models on Shear Strengthening of RC Elements by Means of EBR FRP Systems

Experimental Validation

A wide experimental database, mainly derived from Pellegrino and Modena (2008) and Sas et al. (2009) was collected to assess the above mentioned analytical models for shear capacity of FRP strengthened elements.

The experimental database contains values from experimental investigations on 29 beams with T-cross-section and 186 beams with rectangular cross section, in total 215 experimental results. Material properties and geometrical parameters obtained from experimental tests and reported in the original papers have been considered for calculating the predictions of each model. Partial safety factors have not been included in the calculations of the predictions of the models. The assessment of the accuracy of codes' predictions has been done by distinguishing the strengthening schemes. In detail, the database contains:

65 control unstrengthened beams;
36 U-jacketed beams with stirrups;
27 U-jacketed beams without stirrups;
16 side-bonded beams with stirrups;
49 side-bonded beams without stirrups;
22 fully wrapped beams.

The overall shear strength of the RC beams has been considered for the comparison, to take into account the performance of the models for estimating FRP contribution when combined with the basic models on RC elements and obtain information about interaction between concrete, steel and FRP contributions to the overall shear capacity. It should be noted that the accuracy of the overall shear strength is influenced by the accuracy of both the FRP and RC design codes. The experimental values of the overall shear strength have been directly obtained from the tests included in the database, whereas predictions of overall shear strength have been obtained by combining basic model codes and models for FRP strengthened structures. Estimation of contribution of FRP (V_f) is made according to the above mentioned models for FRP strengthening, whereas contributions of concrete, steel and compressive strength of concrete strut in the truss model are estimated according to basic model codes for unstrengthened RC structures (Eurocode 2, ACI 318 code and *fib* Model Code 2010).

The angle of the shear crack θ may have significant influence on prediction of models, hence various values of this angle have been considered in this study. Since ACI 318 code recommends to use the sum of concrete and steel contributions with $\theta = 45°$, Eurocode 2 recommends to use only the steel contribution with an angle variable between $21.8° \leq \theta \leq 45°$ (for this assessment the angle that maximized the overall shear strength, i.e. when the strength of the tie in the truss model is equal to that of the concrete strut, was found), and the *fib* Model Code recommends also to use the sum of concrete and steel contributions with $\theta = 36°$, these three cases have been considered for the assessment. As recommended by the codes, the compressive strength of the inclined concrete strut in the truss model is always considered as the upper limit for the overall shear strength.

The assessment of the models considered was evaluated using the collected database, comparing the theoretical values $V_{n,theo}$ (the design value of the overall shear strength provided by the design model) with the experimental shear strength $V_{n,\mathrm{exp}}$. The main statistical measures adopted in this study are the percentage of conservative predictions, the average ratio between theoretical and experimental values (AVG), the standard deviation (STD) and the coefficient of variation (CoV). The coefficient of variation (CoV) is the ratio between the standard deviation and the average value and it aims to describe the dispersion of the variable without depending on the variable's measurement unit. When the CoV increases, the dispersion of the ratio between theoretical and experimental values increases, and the accuracy of the model decreases.

The database was divided in five groups (U-jacketed beams with stirrups; U-jacketed beams without stirrups; side-bonded beams with stirrups; side-bonded beams without stirrups and fully wrapped beams) to study the accuracy of the models for FRP shear strengthened beams combined with the basic models for RC beams for different strengthening schemes. The lack of data regarding completely wrapped beams, probably due to its rare use in practical applications, suggested to consider them as a single group regardless of the internal transverse steel.

In Table 4.1 values for the coefficient of variation (CoV) and the corresponding values of average ratio between theoretical and experimental values (AVG) obtained are summarized for various combinations of basic codes and models for strengthened structures and angles of shear crack θ.

In general, both the Italian code CNR-DT 200 (2004), based on the model by Monti and Liotta (2007), and the model of Pellegrino and Modena (2008) give good results for most cases. In particular, the model of Pellegrino and Modena (2008) gives good predictions in terms of CoV and AVG, since it takes into account the interaction between steel and FRP contributions not only proposing a formulation for the FRP contribution to shear strength but also for the steel contribution. It shows good results if combined with both Eurocode 2 (2004) and *fib* Model Code (2010). Starting from the model of Pellegrino and Modena (2008), a modification of the angle characterizing the conventional roughness of the interface from its original value $\phi = 79°$ (Pellegrino and Modena 2008) to $\phi = 75°$, is included to further improve the performance of this model in terms of CoV and AVG.

In general, combining the model of Pellegrino and Modena (both original and modified version) with Eurocode 2 for non-strengthened structures, the best predictions are obtained when the shear crack angle is assumed as $\theta = var$ i.e. for the common shear calculation procedure in Europe. Improvement in behaviour of this model with modification of angle ϕ from 79° to 75° can be also observed.

Table 4.1 Coefficient of variation (CoV) and average ratio between theoretical and experimental values (AVG) for various combinations of basic codes and models for strengthened structures and angles of shear crack θ

Model code	EC2 (2004)						fib MC10 (2010)				ACI318	
Statistical measure	θ = 45°		θ = 36°		θ = var (45 − 21.8)		θ = 45°		θ = 36°		θ = 45°	
	CoV	AVG	CoV	AVG	CoV	AVG	CoV	AVG	CoV	AVG	CoV	AVG
U-jacketing without transversal reinforcement												
fib TG 9.3 (2001)	0.23	1.14	0.20	0.96	0.56	0.66	0.61	2.32	0.54	1.99	–	–
CNR-DT 200 (2004)	0.29	1.28	0.22	1.10	0.37	0.78	0.45	1.65	0.35	1.36	–	–
ACI 440 (2008)	–	–	–	–	–	–	–	–	–	–	0.48	1.67
Chen and Teng (2003a, b)	0.61	2.35	0.61	2.35	0.61	2.35	0.79	2.97	0.79	3.97	–	–
Carolin and Täljsten (2005)	0.21	1.01	0.30	0.83	0.75	0.58	0.32	1.25	0.26	1.00	–	–
Pellegrino and Modena (2008)	0.37	1.39	0.37	1.39	0.37	1.39	0.57	1.92	0.57	1.92	–	–
Pellegrino and Modena φ = 75°	0.34	1.08	0.34	1.08	0.34	1.09	0.51	1.40	0.51	1.40	–	–
Bukhari et al. (2010)	0.47	1.76	0.44	1.63	0.37	1.34	0.67	2.67	0.65	2.39	–	–
Modifi and Chaallal (2011)	0.49	1.82	0.45	1.69	0.35	1.39	0.68	2.79	0.65	2.50	–	–
U-jacketing with transversal reinforcement												
fib TG 9.3 (2001)	0.59	1.78	0.45	1.30	0.39	0.83	0.47	1.64	0.35	1.29	–	–
CNR-DT 200 (2004)	0.62	1.82	0.49	1.33	0.38	0.84	0.37	1.35	0.28	1.05	–	–
ACI 440 (2008)	–	–	–	–	–	–	–	–	–	–	0.37	1.28
Chen and Teng (2003a, b)	0.78	3.31	0.68	2.40	0.41	1.35	0.59	2.07	0.49	1.67	–	–
Carolin and Täljsten (2005)	0.46	1.48	0.32	1.09	0.41	0.75	0.33	1.21	0.30	0.93	–	–
Pellegrino and Modena (2008)	0.45	1.54	0.41	1.40	0.31	1.13	0.43	1.50	0.36	1.26	–	–
Pellegrino and Modena φ = 75°	0.33	1.18	0.29	1.08	0.26	0.95	0.35	1.26	0.31	1.09	–	–
Bukhari et al. (2010)	0.76	2.99	0.65	2.17	0.37	1.23	0.55	1.93	0.44	1.54	–	–
Modifi and Chaallal (2011)	0.74	2.64	0.63	1.92	0.37	1.09	0.52	1.77	0.41	1.41	–	–
Side bonding without transversal reinforcement												
fib TG 9.3 (2001)	0.27	0.94	0.37	0.81	0.76	0.59	0.60	1.93	0.58	1.70	–	–
CNR-DT 200 (2004)	0.39	1.44	0.36	1.35	0.29	1.12	0.54	1.93	0.51	1.78	–	–

(continued)

Table 4.1 (continued)

Model code	EC2 (2004)						fib MC10 (2010)				ACI318	
Statistical measure	θ = 45°		θ = 36°		θ = var (45 − 21.8)		θ = 45°		θ = 36°		θ = 45°	
	CoV	AVG	CoV	AVG	CoV	AVG	CoV	AVG	CoV	AVG	CoV	AVG
ACI 440 (2008)	–	–	–	–	–	–	–	–	–	–	0.46	1.42
Chen and Teng (2003a)	0.46	1.04	0.50	0.94	0.61	0.77	0.60	1.36	0.60	1.21	–	–
Carolin and Täljsten (2005)	0.47	1.00	0.53	0.89	0.70	0.71	0.55	1.28	0.58	1.10	–	–
Pellegrino and Modena (2008)	0.26	1.16	0.26	1.16	0.26	1.16	0.45	1.53	0.45	1.53	–	–
Pellegrino and Modena φ = 75°	0.33	0.87	0.33	0.87	0.33	0.87	0.40	1.09	0.40	1.09	–	–
Bukhari et al. (2010)	0.50	1.62	0.48	1.51	0.46	1.30	0.69	2.42	0.68	2.22	–	–
Modifi and Chaallal (2011)	0.44	1.49	0.41	1.37	0.39	1.13	0.61	2.12	0.58	1.90	–	–
Side bonding with transversal reinforcement												
fib TG 9.3 (2001)	0.38	1.26	0.30	0.95	0.49	0.72	0.44	1.41	0.34	1.11	–	–
CNR-DT 200 (2004)	0.62	2.14	0.50	1.63	0.26	1.02	0.45	1.47	0.35	1.20	–	–
ACI 440 (2008)	–	–	–	–	–	–	–	–	–	–	0.23	1.07
Chen and Teng (2003a)	0.55	1.68	0.41	1.24	0.33	0.86	0.30	1.24	0.24	0.97	–	–
Carolin and Täljsten (2005)	0.74	1.23	0.67	0.94	0.48	0.84	0.49	1.07	0.57	0.81	–	–
Pellegrino and Modena (2008)	0.41	1.47	0.37	1.38	0.29	1.18	0.32	1.32	0.24	1.13	–	–
Pellegrino and Modena φ = 75°	0.22	1.05	0.22	0.98	0.30	1.10	0.23	1.09	0.20	0.94	–	–
Bukhari et al. (2010)	0.66	2.02	0.52	1.48	0.24	0.99	0.37	1.39	0.28	1.09	–	–
Modifi and Chaallal (2011)	0.68	1.98	0.55	1.45	0.36	0.95	0.44	1.40	0.36	1.12	–	–
Complete wrapping												
fib TG 9.3 (2001)	0.92	2.54	0.83	1.90	0.44	1.09	0.61	1.79	0.54	1.43	–	–
CNR-DT 200 (2004)	0.84	2.57	0.74	1.90	0.48	1.18	0.54	1.05	0.52	0.99	–	–
ACI 440 (2008)	–	–	–	–	–	–	–	–	–	–	0.51	1.05
Chen and Teng (2003b)	1.05	2.48	0.95	1.83	0.65	1.18	0.70	1.70	0.66	1.36	–	–

The simple addition of shear components considering them as independent is one of the assumptions that have been questioned by a number of authors. Many authors observed that that the maximum shear contributions of steel stirrups and FRP may not be reached simultaneously, so that their combined contribution may be less than the sum of the respective peak values of V_f and Vs (Chen et al. 2010, 2013; Pellegrino and Modena 2002, 2006; D'Antino et al. 2012). For accurate evaluation of the shear resistance, the determination of the maximum value of the combined contribution of steel stirrups and FRP strips is to be recommended, since the simultaneous use of these maximum values seems to be an unconservative approach in the evaluation of shear strength (Chen et al. 2010, 2013). Bousselham and Chaallal (2006, 2008), Pellegrino and Modena (2002, 2006, 2008) and Pellegrino and Vasic (2013) have found that the amount of steel shear reinforcement has a significant effect on the effectiveness of shear strengthening using FRP strengthening. In particular, the efficiency of the FRP strengthening technique seems to decrease with an increase in the axial rigidity ratio between the internal steel shear reinforcement and the external FRP shear reinforcement. This is probably due to the fact that the internal stirrups, in some circumstances related to premature debonding of the external FRP strengthening, cannot reach yielding as assumed by the main codes for RC elements without strengthening.

Design Comparison

A reinforced concrete member with rectangular cross section is herein considered to evaluate numerically the potential differences in the evaluation of shear capacity of rectangular RC members strengthened by means of externally bonded FRP systems, given by different formulations and models.

The models considered are:

CNR DT 200 (2004);

CNR DT 200R1 (2012), assuming for the concrete strut either a variable inclination angle or fixed and equal to 45°;

"Eurocode 8: Design of structures for earthquake resistance—Part 3: Assessment and retrofitting of buildings", assuming for the concrete strut either a variable inclination angle or fixed and equal to 45°;

ACI 440.2R-08 "Guide for the Design and Construction of Externally Bonded FRP System for Strengthening Concrete Structures".

The cross-section of the RC member has side dimensions b = 300 mm and h = 500 mm (see Fig. 4.3), it is made of C20/25 concrete and yielding stress of stirrups is 450 MPa. Stirrups D8 are considered at different spacing; a low value of 50 mm for spacing is assumed to simulate high transverse reinforcement ratio while a high value of 300 mm is assumed to simulate reduced transverse reinforcement ratio. Different U-wrap and fully wrapped FRP layouts are also considered, calling for GFRP strips having width equal to spacing (100 mm), $E_f = 80.7$ GPa, $\varepsilon_f = 3$ % and $t_f = 0.48$ mm. In the case of U-wrap, the thickness of the slab (200 mm) is

Fig. 4.3 Cross section of RC beam

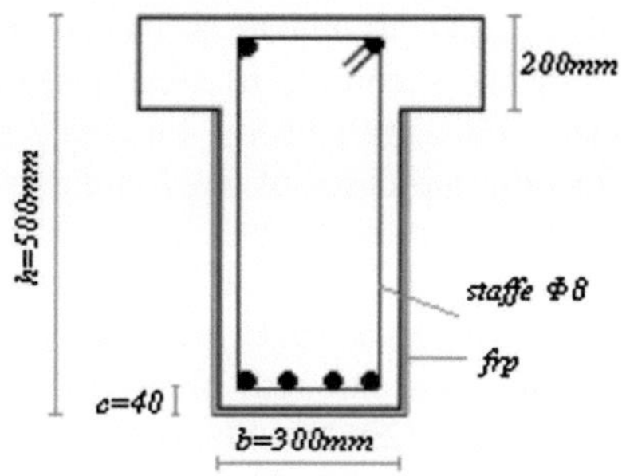

excluded from the FRP application, yielding to reduced effectiveness of the FRP on the sides of the T beam.

The differences in terms of unstrengthened shear capacity of the RC member (second bar), shear capacity after FRP strengthening (first bar), and the percentage increase are reported for each formulation and model in the following figures. A different figure is proposed for each FRP layout (U-wrapping and full wrapping, because only these two are considered in all the formulations).

Figure 4.4 shows that if the transverse reinforcement presents a low amount (i.e. stirrup spacing equal to 300 mm), according to CNR DT-200R1 (2012), the strength increment due to U-Wrapped FRP is about 147 % if a fixed strut inclination is assumed (equal to 45°), and it reduces to about 133 % in the case of variable inclination (and in particular θ = 23°); however, the absolute increment is

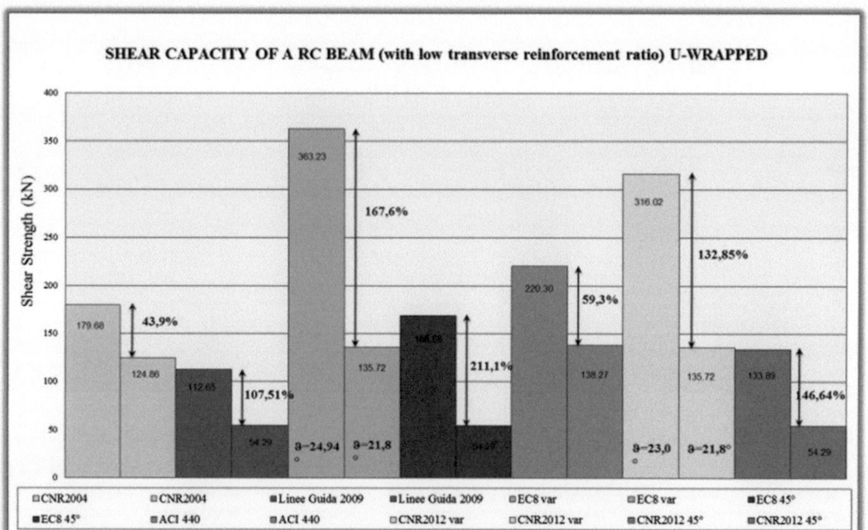

Fig. 4.4 Shear capacity of a reinforced concrete beam with large spacing of stirrups, U-wrapped with FRP

higher, like as higher is the shear capacity of the unstrengthened RC member evaluated according to a variable inclination truss model. In this sense, there is a clear similarity with the Eurocode 8 model; in fact they have similar assumptions for both the cases of fixed angle and of variable angle for the compressed struts. However, the highest shear capacity is predicted by the Eurocode 8 model. It should be noted that in Figs. 4.4, 4.5, 4.6 and 4.7 the results provided by the Italian Linee Guida (2009) are also included.

The shear strength of FRP U-wrapped member computed according to CNR DT-200R1 (2012) with fixed angle, is one of the lowest values predicted by the models, being actually lower than the shear strength of the unstrengthened member according to ACI 440 and to CNR DT 200R1 (2012) and Eurocode 8 with variable angle.

Similarly, one of the lowest percentage increments is given by ACI 440 and CNR DT-200R1 (2012) with variable inclination angle. The lowest increment is given by the original formulation of CNR DT-200 (2004), having fixed strut inclination, being almost equal to one fourth of the increment predicted by Eurocode 8 with the same fixed angle assumption.

Figure 4.5 shows the strength increment due to fully-wrapped FRP. According to CNR DT-200R1 (2012), the strength increment is about 215 % if a fixed strut angle is assumed, and shear capacity is one of the lowest when FRP strengthened member is considered; however, the percentage increment is significant. In this case, the shear strength of the FRP-strengthened members is always higher than every single prediction of unstrengthened shear capacity. The highest percentage increment is obtained with Eurocode 8 with fixed angle and it is close to 300 %.

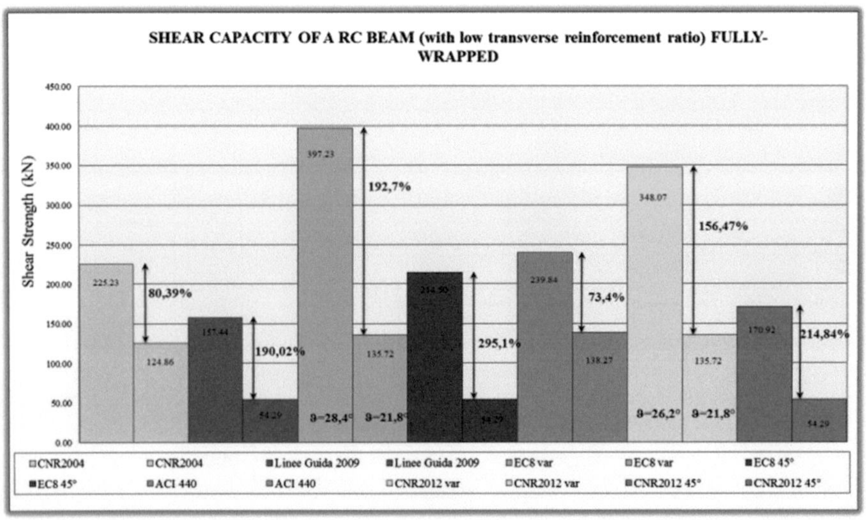

Fig. 4.5 Shear capacity of a reinforced concrete beam with large spacing of stirrups, fully-wrapped with FRP

In the case of fully-wrapped RC members, the lowest percentage increase is provided by ACI 440.

In the case of high internal transverse reinforcement ratio, the shear capacity increment obtained from FRP is expected to be reduced, mainly because the shear capacity tends to be limited by the concrete contribution being the impact of increasing the tensile side capacity, only marginal. However, not all the models perform in the same way. In this case, the differences in terms of unstrengthened shear capacity are less evident, while the shear capacity increases due to FRP strengthening are very scattered as clearly shown for each formulation and model in the following figures.

Both U-wrapped and fully-wrapped predictions (Figs. 4.6 and 4.7), especially according to CNR DT-200R1 (2012) and Eurocode 8 with variable angle, present a reduced shear percentage increase. This is totally in agreement with the previous comments on the impact of improving the tensile side of a mechanism, potentially already effective. This is confirmed by the angle θ reaching the threshold value of 45° in the FRP fully-wrapped configuration. In this case, in fact, the CNR DT-200R1 provides the same shear capacity for both variable and fixed angle, when FRP fully wrapped is considered; this implies that failure is related also to concrete strut crushing. In this case of reduced internal stirrup spacing, ACI 440 provides one of the highest shear capacities for the strengthened members. The highest capacity is provided by the original formulation of CNR DT-200 (2004), out-of-date nowadays.

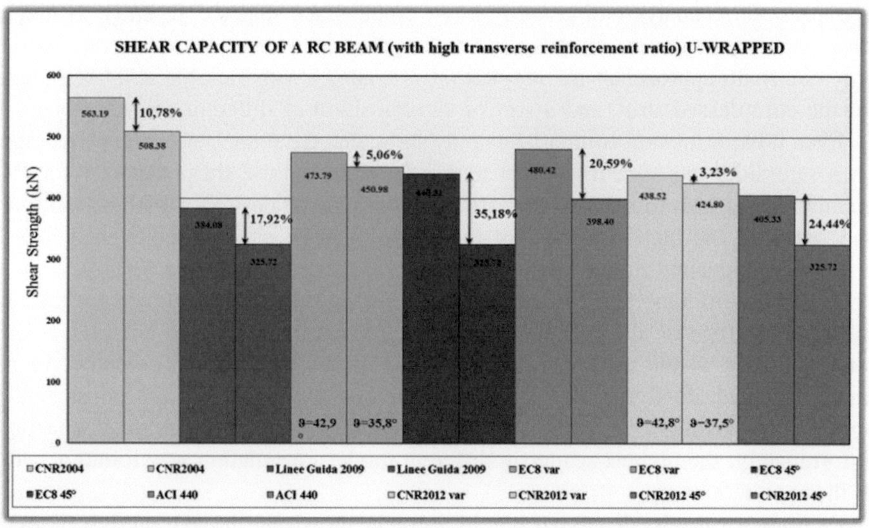

Fig. 4.6 Shear capacity of a reinforced concrete beam with small spacing for stirrups, U-wrapped with FRP

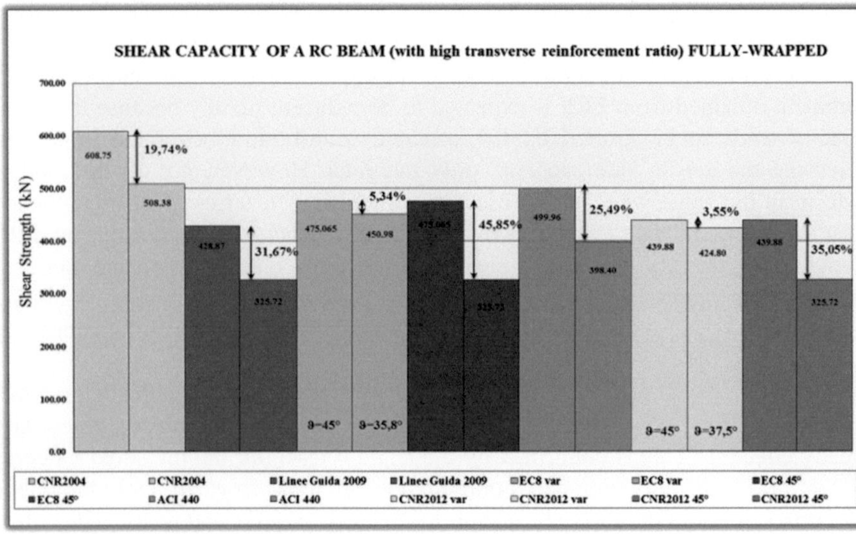

Fig. 4.7 Shear capacity of a reinforced concrete beam with small spacing for stirrups, fully-wrapped with FRP

Lacunae and Proposals of Improvement for Current Formulations and Models on Shear Strengthening of RC Elements by Means of EBR FRP Systems

The parametric analysis discussed above emphasized that the recently available shear capacity predictive models provide largely scattered predictions, even though only two main approaches are adopted: truss analogy with variable and fixed angle for the compressed strut, and a sort of superposition of different contributions.

Even if experimental comparisons may show that the models are, on the average, on a safe side, too high a level of underestimation of the shear capacity for RC members may lead to over-conservative and expensive design of FRP retrofitting, and potential too high detriment of the shear capacity predictions, even if it is a brittle and undesired failure mode.

The main differences in the predictions of the considered models are not limited to the percentage or absolute increments in shear capacity due to EB-FRP wrapping, but also to the prediction of the shear capacity of the RC member in its unstrengthened configuration. FRP-strengthening interventions are most likely applied in existing old buildings, which usually show large stirrup spacing: unfortunately, the highest scatter in different model formulation was found exactly in those situations.

Preliminary results discussed herein seem to show that the assumption of variable angle of compressed struts catches with higher accuracy the different behavior of RC members in shear, accounting also for the internal, existing, transverse reinforcement. The selection of the angle has a strong impact on the ultimate capacity of the RC member in shear, thus, in any assessment of unstrengthened and strengthened capacity, the angle variability range should be the same for both assessments. Pragmatically, some codes differentiate the assumption on the compressed strut angle to take into account the seismic behavior of the structure; in a sense, limiting the inclination angle to 45° reduces the demand on the compressed concrete strut to account implicitly for the potential reduction of concrete capacity due to flexural ductility demand. The validation of this assumption should be further investigated because it is not feasible to validate a cyclic shear model by means of monotonic static tests, as the most common experimental tests available in the scientific literature are.

Revised Model for EBR FRP Shear-Strengthening

The literature review and the assessment of design formulations discussed in the previous sections prove that structural models currently available cannot reliably predict the shear strength of EBR FRP-strengthened RC elements. In fact, most of the researchers' effort in the last two decades has been devoted to develop models for the accurate evaluation of FRP strips/sheets contribution to the shear capacity of RC elements, mainly depending on the geometry of FRP and the mechanical characteristics of both FRP and concrete. As a result, many analytical and mechanical models are available for modeling in detail the behavior of any FRP strengthening system and predicting its capacity, but no comprehensive model, capable of accounting for the interaction of all the resisting mechanisms, has been developed so far.

Trying to fill this lacuna, in this section a recently developed mechanics-based model (Petrone and Monti 2014) will be briefly illustrated: it allows evaluating the actual contribution of all resisting mechanisms to the shear capacity of FRP-strengthened RC beams/columns and predicting the failure sequence between yielding of existing transverse steel ties and debonding of FRP sheets/strips, while monitoring the corresponding compressive stress in concrete.

The mechanical model, see Fig. 4.8, derives from the general configuration of a RC cracked beam strengthened in shear with FRP, see Fig. 4.9: in its general form, the model aims at representing the mechanical behavior of EBR FRP-strengthened RC structures, for any arrangement of FRP strips/sheets and shear reinforcement.

When dealing with existing cracked structures, geometry and materials are all known quantities. Therefore, the exact solution of the mechanical model in Fig. 4.8 can be conveniently carried out by imposing a strong compatibility condition along the crack between FRP strips/sheets and steel stirrups and then by setting the equilibrium condition between compressive and tensile resultant forces.

The compatibility condition allows evaluating the actual strain of each fiber of FRP and each stirrup crossing the crack: once the strain is known, stress and corresponding tensile forces in FRP strips/sheets and steel stirrups can be easily

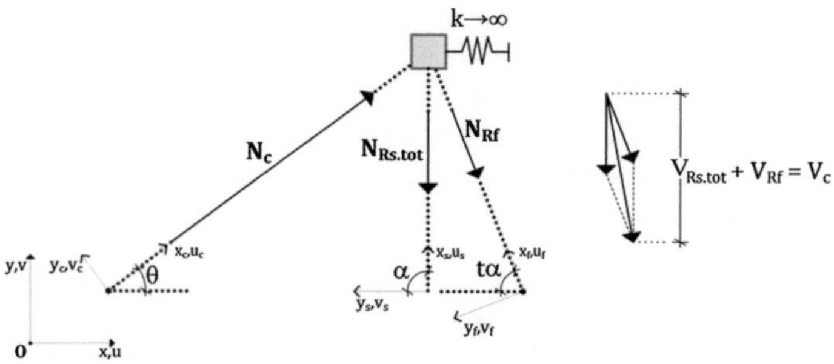

Fig. 4.8 Simplified mechanical model. From Petrone and Monti (2014)

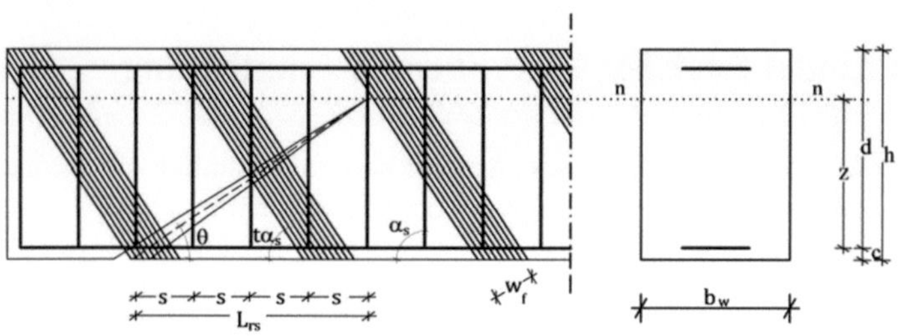

Fig. 4.9 RC beam strengthened in shear with FRP strips: general configuration. From Petrone and Monti (2014)

calculated. Then, the equilibrium condition gives the corresponding compressive force in the concrete strut.

With reference to Fig. 4.10, the compatibility condition ensures that at each location along the crack, corresponding to the x axis in the local reference system, the strain in FRP strips and stirrups varies proportionally to the crack width w_{cr}, given by:

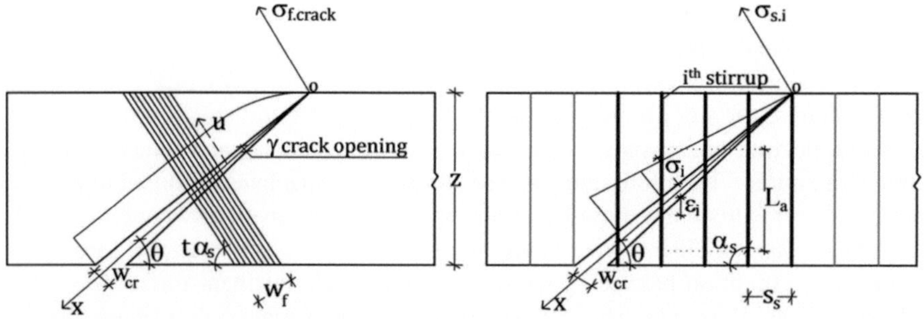

Fig. 4.10 Compatibility along the crack: FRP trip and stirrups deformation. From Petrone and Monti (2014)

$$w_{\rm cr} = 2 \cdot x \tan\frac{\gamma}{2} \tag{4.75}$$

Since w_{cr} varies with the crack opening angle γ and linearly with x, the actual contribution of FRP and steel is univocally defined by the crack opening at each location along the crack.

The evaluation of FRP strips/sheets and steel stirrups contribution to the shear capacity, expressed as function of x and γ, is achieved by making use of the formulations explained thereafter.

As for the FRP strips/sheets, the following relation holds:

$$V_{\rm Rf}(\gamma, x) = 0.9d \cdot \sigma_{\rm f,eff}(\gamma, x) \cdot 2t_{\rm f}(\cot\theta + \cot t\,\alpha_{\rm s})\frac{w_{\rm f}}{p_{\rm f}} \tag{4.76}$$

where $d = h - c$ is the effective depth of the cross-section, $\sigma_{\rm f,eff}(\gamma, x)$ is the effective stress in the FRP, calculated according to the model proposed by Monti and Liotta (2007), θ and $t\alpha_s$, see Fig. 4.8, are the slopes of the crack and the FRP strips/sheets with respect to the beam longitudinal axis, respectively, the latter expressed as function of α_s, which is the slope of the stirrups, through the factor t. Also, t_f and p_f are the thickness and the width of the single FRP sheet/strip, respectively. Being w_f the overall width of the FRP crossing the crack, the ratio w_f/p_f represents the number of FRP strips/sheets crossing the crack.

As for the steel stirrups, the following relation holds:

$$V_{\rm Rs.tot}(\gamma, x) = 00.9 \cdot z\frac{A_{s_1}}{s_s}(\cot\alpha_s + \cot\theta)\sin\alpha_s \sum_{i=1}^{n} \sigma_{s_i}(\gamma, x) \tag{4.77}$$

where $A_{s_1} = n_b\pi\phi_s^2/4$ is the cross-sectional area of one stirrup, n_b is the number of legs of the single stirrup and ϕ_s is the bar diameter; s_s is the stirrups spacing and $\sigma_{s_i}(\gamma, x) = \min\left[E_s\varepsilon_{s_i}(\gamma, x), f_{ym}\right]$ is the tensile stress in each stirrup, expressed as a function of the strain $\varepsilon_{s_i}(\gamma, x)$, given as:

$$\varepsilon_{s_i}(\gamma, x) = \frac{2 \cdot x \tan(\gamma/2)}{\sin(\alpha_s + \theta) \cdot L_a} \tag{4.78}$$

where L_a represents the “bonding length” of the stirrups. Therefore each stirrups contributes to the overall shear capacity proportionally to its actual strain, which depend on the crack opening and on its location along the crack.

Figure 4.11a, b shows how the strain and the stress of stirrups change at varying of the crack opening,γ: it is worth noticing that the strain, which is function of the crack width, see Eqs. (4.79) and (4.80), varies linearly along the x axis with w_{cr}. For

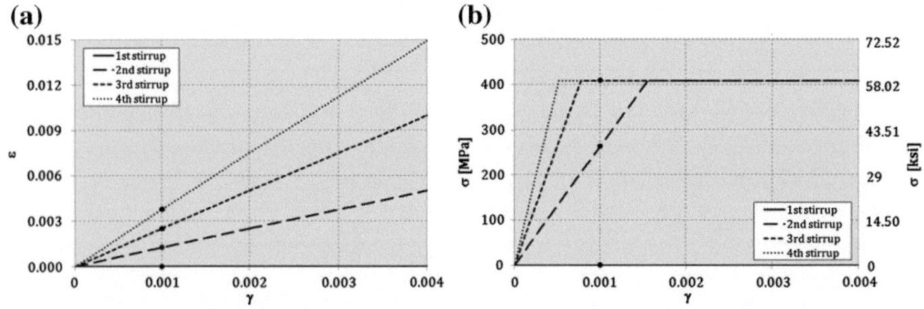

Fig. 4.11 Strain (**a**) and stress (**b**) in each stirrup as function of the crack opening, γ. Petrone e Monti (2014)

the set of parameters chosen for this analysis, that is $H = 400$ mm; $\theta = 30°$; $s_s = 100$ mm; $f_{yk} = 355$ MPa $E_s = 210$ GPa and when $\gamma = 1 \cdot 10^{-3}$, four stirrups cross the crack, including the first one, see Fig. 4.11a, b, located at the crack's tip that does not give any contribution. The second stirrup, which is the closest one to the crack's tip, is still in the linear branch and contributes to the shear capacity proportionally to its actual strain, the third and the fourth are both yielded and, even exhibiting different deformations, give the same maximum contribution to the shear capacity.

Once the stress in each stirrup is known, the contribution of all of them to the shear capacity can be easily calculated by making use of Eq. (4.77).

Figure 4.12a, showing how $V_{\mathrm{Rs.tot}}$ changes at varying of the crack opening, proves that for $\gamma = 1 \times 10^{-3}$ not all the stirrups contribute to the shear capacity with their maximum strength and demonstrates that the hypothesis of assuming all the stirrups as yielded can lead to unsafe overestimations of the shear capacity. Then Fig. 4.12b clearly shows how much steel stirrups and the FRP sheet contribute to the shear strength for an assigned configuration of the beam and a certain crack opening, see point 1 and 2 in Fig. 4.12b.

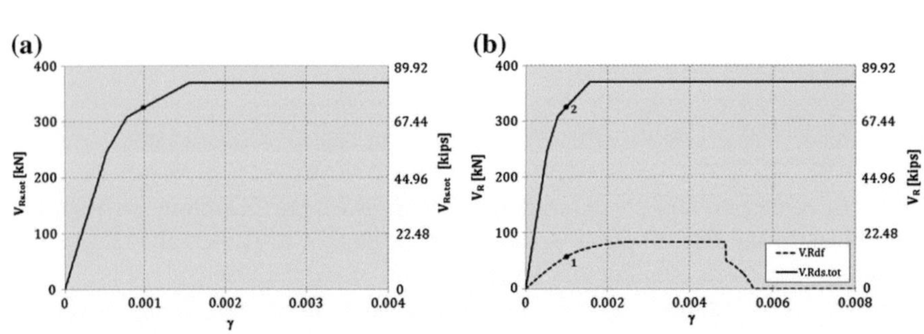

Fig. 4.12 Actual contribution of all the stirrups (**a**) and the FRP sheet (**b**) to the shear capacity at varying of γ. Petrone e Monti (2014)

The equilibrium condition of the resultant forces, see Fig. 4.8, allows evaluating the magnitude of the force in the concrete strut, that is:

$$V_{\mathrm{Rf}}(\gamma) + V_{\mathrm{Rs.tot}}(\gamma, x) = V_c(\gamma, x) \tag{4.79}$$

where $V_{\mathrm{Rf}}(\gamma)$ and $V_{\mathrm{Rs.tot}}(\gamma, x)$ have the meaning expressed above and $V_c(\gamma, x)$ is the vertical projection of axial compressive force acting on the concrete strut. The equilibrium condition expressed in Eq. (4.79) can be ensured, if the following condition is met:

$$V_c(\gamma, x) \leq V_{c,\max} \tag{4.80}$$

that can be explicitly expressed as:

$$\sigma_c(\gamma, x) \cdot b_w L_{rs} \sin^2\theta \leq f_c' \cdot b_w L_{rs} \sin^2\theta \tag{4.81}$$

where θ, as already mentioned, is the slope of the compressive strut with respect to the beam longitudinal axis, b_w is the width of the cross section, $L_{rs} = z(\cot\theta + \cot\alpha_s)$ is the length of the representative span of the shear resisting model, see Fig. 4.9, and $f_c' = 0.6 f_c(1 - f_c/250)$ is the reduced compressive strength of concrete, calculated according to Regan (1993). Equation (4.81) can be then simplified into:

$$\sigma_c(\gamma, x) \leq f_c' \tag{4.82}$$

which is the verification of a concrete section subjected to the axial force $F = V_c(\gamma, x)/\sin\theta$.

Needs for Future Work

From the content of this chapter, two main fields of research appear in need of further studies for clarifying some still unclear aspects in the assessment of the shear capacity obtained through FRP-strengthening: one is relevant to the experimental studies and the other to the development of models that are reliable and at the same time usable for design purposes.

As far as the experimental studies are concerned, it is noted that nowadays there is a wealth of information and data available and easily accessible to researchers to validate their models. The next step would be to assess the effect of the axial load on the shear capacity and, which is even more important, the effect of biaxial loading conditions. Here, we are mainly referring to columns under seismic conditions, since we consider that for beams the existing database is larger than that for columns. Many other experimental studies can be envisaged for assessing and improving the effectiveness of construction details, such as anchorages or bonding agents, but this attains to the sphere of industrial optimization of applications, certainly not to the development of fundamental understanding.

The latter has had a tremendous acceleration since the first studies dating back to the last decade, where the common approach to the description of the effects of FRP strengthening on the shear capacity was merely based on the observation of the experimental results and on their description through regression-based equations. In the following years, much has been done by many authors, whose works are reported in this chapter, to understand in depth the underlying mechanisms and thus obtaining a better control over the variables determining the efficacy of a strengthening design. Here, research has initially mainly focused on the contribution of FRP to shear capacity, but it has soon moved towards interpreting the interaction with the other co-existing resisting mechanisms, of steel and concrete.

Nowadays, it is proved that the common accepted rule of assuming the shear capacity of RC members strengthened in shear with FRP as the sum the maximum contribution of both FRP and stirrups can lead to an unsafe overestimation of the shear capacity. This issue has been pointed out by some authors, when comparing experimental shear capacity values with the theoretical ones, but without giving a convincing explanation of that. In this chapter, a model is proposed to better understand the mechanical behavior of FRP-strengthened RC beams in shear and calculate their shear capacity in a more appropriate manner. In fact, it has allowed demonstrating that there are (not peculiar) situations where the steel stirrups are still elastic when FRP debonds: in those conditions, the actual shear capacity is lower than that computed through the purely additive formulas. This has urged a coordinated research effort, currently under way, which aims at revisiting all the available experimental data and at comparing them with the predictions of this new formulation, with the purpose of adjusting the current design formulas.

References

Assemblea Generale Consiglio Superiore LL PP. (2009). Linee Guida per la Progettazione, l'Eesecuzione ed il Collaudodi interventi di rinforzo di strutture di c.a., c.a.p. e murarie mediante FRP, 24/07/2009, Rome, Italy.

American Concrete Institute. (2005). Building code requirements for structural concrete (ACI 318-05) and commentary (ACI 318R-05), Farmington Hills, Mich., USA.

American Concrete Institute. (2008). Guide for the design and construction of externally bonded FRP systems for strengthening of concrete structure (ACI 440.2R-08), Farmington Hill, Mich., USA.

Bousselham, A., & Chaallal, O. (2006). Behaviour of reinforced concrete T-beams strengthened in shear with carbon fibre-reinforced polymer—An experimental study. *ACI Structural Journal, 103*(3), 339–347.

Bousselham, A., & Chaallal, O. (2008). Mechanisms of shear resistance of concrete beams strengthened in shear with externally bonded FRP. *Journal of Composites for Construction, 12*(5), 302–314.

Bukhari, I. A., Vollum, R. L., Ahmad, S., & Sagaseta, J. (2010). Shear strengthening of reinforced concrete. *Magazine of Concrete Research, 62*(1), 65–77.

Carolin, A., & Taljsten, B. (2005). Theoretical study on strengthening for increased shear bearing capacity. *Journal of Composites for Construction, 9*(6), 497–506.

CEN. (2004). Eurocode 2: Design of Concrete Structures—Part 1-1: General Rules and Rules for Buildings, EN 1992-1-1 2004, Comité Européen de Normalisation, Brussels, Belgium.

Chen, G. M., Teng, J. G., & Chen, J. F. (2013). Shear strength for FRP-strengthened RC beams with adverse FRP-steel interaction. *Journal of Composite for Construction, 17*(1), 50–66.

Chen, G. M., Teng, J. G., Chen, J. F., & Rosenboom, O. A. (2010). Interaction between steel stirrups and shear-strengthening FRP strips in RC beams. *Journal of Composites for Construction, 14*(5), 498–509.

Chen, J. F., & Teng, J. G. (2003a). Shear capacity of FRP-strengthened RC beams: FRP debonding. *Construction and Building Materials, 17*, 27–41.

Chen, J. F., & Teng, J. G. (2003b). Shear capacity of FRP-strengthened RC beams: FRP rupture. *Journal of Structural Engineering, 129*(5), 615–625.

CNR—Italian Research Council, Advisory Committee on Technical Recommendations for Construction. (2004). Guide for the Design and Construction of Externally Bonded FRP Systems for Strengthening Existing Structures. Materials, RC and PC Structures, Masonry Structures (CNR-DT 200/2004), Rome, Italy.

CNR—Italian Research Council, Advisory Committee on Technical Recommendations for Construction. (2012). Guide for the Design and Construction of Externally Bonded FRP Systems for Strengthening Existing Structures. Materials, RC and PC Structures, Masonry Structures (CNR-DT 200R1/2012), Rome, Italy.

D'Antino, T., Pellegrino, C., Salomoni, V., & Mazzucco, G. (2012). Shear behavior of RC structural members strengthened with FRP materials: A three dimensional numerical approach. ACI SP286-05, 69–84.

Fédération Internationale du Béton. (2001). *Externally bonded FRP reinforcement for RC structures*. Task Group 9.3, Bullettin No. 14, Lausanne, Switzerland.

Fédération Internationale du Béton. (2010). *Model code for concrete structures*. Switzerland: Lausanne.

Khalifa, A., Gold, W. J., Nanni, A., & Abdel Aziz, M. I. (1998). Contribution of externally bonded FRP to shear capacity of RC flexural members. *Journal of Composites for Construction, 2*(4), 195–202.

Lima, J., & Barros, J. (2011). Reliability analysis of shear strengthening externally bonded FRP models. *Proceedings of the Institution of Civil Engineers (ICE)—Structures and Buildings, 164*(1), 43–56.

Modifi, A., & Chaallal, O. (2011). Shear strengthening of RC beams with EB FRP: Influencing factors and conceptual debonding model. *Journal of Composites for Constructions, 15*(5), 62–74.

Monti, G., & Liotta, M. A. (2007). Tests and design equations for FRP-strengthening in shear. *Construction and Building Materials, 21*, 799–809.

Pellegrino, C., & Modena C. (2002). FRP shear strengthening of RC beams with transverse steel reinforcement. *Journal of Composites for Construction, 6*(2), 104–111.

Pellegrino, C., & Modena, C. (2006). FRP shear strengthening of RC beams: experimental study and analytical modelling. *ACI Structural Journal, 103*(5), 720–728.

Pellegrino, C., & Modena, C. (2008). An experimentally based analytical model for shear capacity of FRP strengthened reinforced concrete beams. *Mechanics of Composite Materials, 44*(3), 231–244.

Pellegrino, C., & Vasic, M. (2013). Assessment of design procedures for the use of externally bonded FRP composites in shear strengthening of reinforced concrete beams. *Composites Part B Engineering, 45*(1), 727–741.

Petrone, F., & Monti, G. (2014). FRP-RC beam in shear: Mechanical model and assessment procedure for pseudo-ductile behavior. *Polymers, 2014*(6), 2051–2064.

Regan, P. E. (1993). Research on shear: A benefit to humanity or a waste of time? *Structural Engineering, 71*, 337–347.

Sas, G., Täljsten, B., Barros, J., Lima, J., & Carolin, A. (2009). Are available models reliable for predicting the FRP contribution to the shear resistance of RC beams? *Journal of Composites for Construction, 13*(6), 514–534.

Triantafillou, T. C., & Antonopoulos, C. P. (2000). Design of concrete flexural members strengthened in shear with FRP. *Journal of Composites for Construction, 4*(4), 198–205.

Chapter 5
Confinement of RC Elements by Means of EBR FRP Systems

Stavroula Pantazopoulou, Ioannis Balafas, Dionysios Bournas, Maurizio Guadagnini, Tommaso D'Antino, Gian Piero Lignola, Annalisa Napoli, Carlo Pellegrino, Andrea Prota, Roberto Realfonzo and Souzana Tastani

Abstract This chapter reviews the issues of confinement in plain and reinforced concrete under concentric compression and summarizes the state of the art regarding the available confinement models for strength and stress-strain behaviour of encased confined concrete, and the corresponding magnitude of dependable strain capacity. The mechanisms of confinement failure, the evolution of Poissons' effects under low and high confinement, and the ensuing material compaction at high confining pressures (plastification) are discussed. The effects of stress concentrations near corners, the effectiveness of layers and influence of adhesive, other scale effects and the influence of specimen morphology on mechanical behaviour are also outlined. Next, the chapter concentrates on the effects of embedded reinforcement both longitudinal and transverse. Confinement effectiveness in the

---

Electronic supplementary material The online version of this chapter (doi:10.1007/978-94-017-7336-2_5) contains supplementary material, which is available to authorized users.

S. Pantazopoulou (✉) · I. Balafas
University of Cyprus, Nicosia, Cyprus
e-mail: pantazopoulou.stavroula@ucy.ac.cy

D. Bournas
The University of Nottingham, Nottingham, UK

M. Guadagnini
University of Sheffield, Sheffield, UK

T. D'Antino · C. Pellegrino
University of Padua, Padua, Italy

G.P. Lignola · A. Prota
University of Naples Federico II, Naples, Italy

A. Napoli · R. Realfonzo
University of Salerno, Salerno, Italy

S. Tastani
Democritus University of Thrace, Xanthi, Greece

C. Pellegrino and J. Sena-Cruz (eds.), *Design Procedures for the Use of Composites in Strengthening of Reinforced Concrete Structures*, RILEM State-of-the-Art Reports 19, DOI 10.1007/978-94-017-7336-2_5

presence of combined flexure and shear (in plastic hinge regions), local effects due to rotation capacity increase, and effects of FRP confinement on overall member behaviour are discussed. Shape effects that occur in hollow or oblong sections are also considered. Furthermore, the chapter gives an outline regarding the characteristics of the international database of tests for confinement, and its calibration with the database of the available confinement models including those included in the design standards (ACI, CNR and EC8-III).

General Aspects

The effectiveness of FRPs as a passive confining system for plain concrete has been documented through extensive experimental research which began in the mid-1980s (Fardis and Khalilli 1981). The success of jacketing by wrapping with FRP layers on the lateral surface of concrete elements builds on the natural tendency of concrete to expand as damage accumulates due to internal cracking. Actually, the effective Poisson's ratio of concrete is known to increase gradually from the reference "elasticity" value which is near 0.25, to the "incompressibility" limit of $v \approx 0.5$ at peak stress, and even well beyond that limit once the material enters the state of failure past the peak point. It is now well known that concrete in compression cannot fail (i.e., v will not exceed the continuum mechanics incompressibility limit of 0.5) unless the material volume is free to expand laterally—thus, any form of lateral restraint to expansion, provided by any type of jacketing scheme with a reasonable stiffness in the hoop direction will resist expansion by mobilizing passively generated lateral confining pressures.

Initially, unreinforced cylindrical or prismatic specimens either wrapped with FRP jackets, or encased in FRP-tubes were studied in compression. In the early studies, variables of the investigation were the type of composite, the orientation of fibres, the number of layers, the arrangement of wraps in strips, the cross sectional geometry of the specimen, the chamfering radius at the corners of specimens with rectangular cross-sections, the concrete quality, and the mode of loading (monotonic or cyclic). Later, these studies were extended to include the confinement effectiveness when FRP wraps are used in jacketing of reinforced concrete members. Motivation was to explore the prospects of FRP-jacketing as a measure for seismic retrofit, which became at the end of the 1990s in response to the pressing need for fast rehabilitation of existing brittle construction after the occurrence of significant earthquakes. For this reason, most of the experimental studies in this area focused into lightly reinforced columns with widely spaced stirrups.

Although a variety of tests have been conducted where FRP jackets function as confining devices, yet most of the confinement models have been derived and calibrated with data obtained from tests of cylinders and prisms under uniaxial

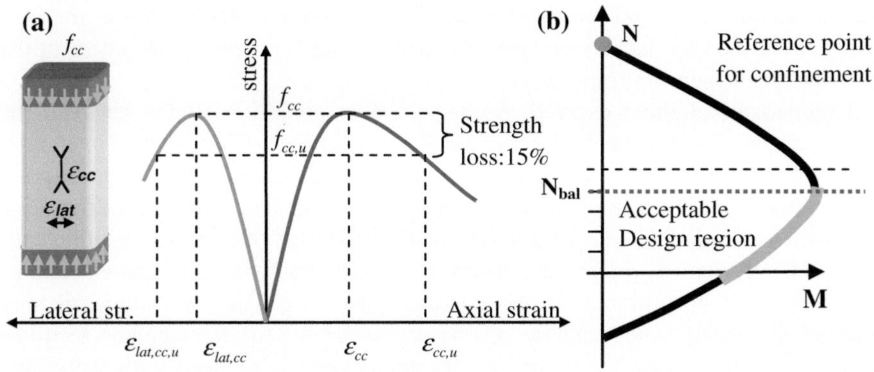

Fig. 5.1 **a** Typical axial stress versus axial and lateral strain diagram. **b** Axial load—flexural moment interaction diagram of typical column

compression. Published tests have relatively small aspect ratios (height to cross-section size around 2), dictated as a rule by the test space of compression frames. However, even if the aspect ratios of the specimens could resemble those of actual columns, it is relevant to note that, with regard to the typical axial load/flexural moment interaction diagram that typifies column behaviours (Fig. 5.1), all published tests of this type correspond to a single point, i.e. concentric compression, marked by the red circle in Fig. 5.1b, whereas the region of design interest is marked by the green line in the figure. This is an important disclaimer to be recalled when confinement models are used in order to analyse usual columns, where the axial load/flexural moment combinations lie below the point of balanced failure (a value of axial load ratio $N/(A_g f_{co}) < 0.4$).

The benefits imparted by FRP confinement in actual reinforced concrete columns concern the following aspects:

(a) Confined concrete strength, which exhibits a relatively mild increase over that of unconfined concrete owing to the passive nature of the confining action;
(b) Impressive increase of axial strain capacity (i.e. the range of deformation that may be developed without significant loss of strength). By convention, which prevailed over time, the limit of deformation capacity is usually taken as the axial strain at post-peak strength equal to 85 % of peak, and is denoted in the literature either by ε_{85} or $\varepsilon_{cc,u}$;
(c) Significant shear strength increase and clamping action that effectively confines jacketed lap splices of longitudinal reinforcement in tension, as well as lateral restraint against buckling of embedded longitudinal reinforcement in compression, enhanced bond in confined anchorages, significant enhancement of displacement and rotational ductility, and postponement of undesirable brittle modes of failure.

The main scope of this chapter focuses only on aspects (a) and (b) above, whereas (c) refers to the seismic retrofit of lightly reinforced columns. Thus (c) will be mentioned briefly and only with reference to the role of the confining pressure imparted by the FRP jackets on those mechanisms that support development of the associated full member response.

Comparison of the measured average axial compression strains between the various types of specimens tested illustrates that jackets are *most effective* in circular plain concrete specimens—here, the only occasion of jacket failure before rupture due to attainment of the nominal strain capacity of the material occurs by debonding over the overlap length of a wrap layer (if the anchorage of a layer is insufficient). Jacket effectiveness is progressively lower in specimens with: (i) Square sections, where strain concentrations at the corners induce local jacket rupture thereby limiting the axial strain capacity of encased concrete. (ii) Rectangular sections, because in these cases, the confinement effectiveness is reduced with increasing aspect ratio of the cross-section, with a negligible contribution in terms of enhancement of concrete compressive strength in wall-like cross-sections. (iii) Hollow sections; in these sections, the interior opening enables some degree of radial dilation thereby partly relieving the restraint to volumetric expansion that gives rise to passive lateral confining stress.

FRP confinement effectiveness is compromised from the ideal values reported in unreinforced jacketed column tests if the specimens are already reinforced with longitudinal and transverse steel reinforcement, particularly when transverse stirrup details are non-conforming with modern standards. Examples of this situation are lightly reinforced columns with sparse spacing and inadequate anchorage of embedded stirrups, and columns with corroded reinforcement. Jacket failure by rupture in compressed wrapped reinforced concrete members is often attributed to longitudinal bar buckling that occurs over the unsupported length between one or more stirrup spacings. Following this event, failure is characterized by disintegration of the encased concrete column, with abrupt loss of load carrying capacity. However, several counter-examples in the experimental literature, where jacketed reinforced concrete columns have reportedly attained similar levels of strain capacity as identical unreinforced jacketed specimens, underscore the fact that the occurrence of bar buckling in the presence of FRP jacketing is suppressed in some circumstances, whereas in others it prevails, thereby threatening the safety of the rehabilitation scheme.

It is therefore essential, within the framework of performance-based strengthening, that FRP jacketing design details (i.e. number of layers, arrangement, and material employed) be linked explicitly to the targeted deformation capacity of the strengthened member, by consistently sorting through all the possible modes of failure of the application. The scenarios above represent the characteristic response of FRP encased columns at different ranges of values of the determining design parameters. Note that the passive confining pressure imparted by FRP jackets is a continuous phenomenon, linearly increasing with the axial stiffness of the jacket (quantified as the product of the jacket thickness and its modulus of elasticity) and inversely proportional to the dimensions of the confined member's cross-section. Thus, a feasible design objective would be to define a critical value for the FRP confining pressure such that the retrofit scheme may fundamentally alter the structural response of lightly reinforced concrete members from brittle to ductile by suppressing or postponing undesirable modes of failure.

A valuable test-bed for evaluation of any proposal or hypothesis in the direction of actually controlling, by design, the performance of a retrofit is the collection of experiments published in the literature. Several databases have been assembled before grouping specimens from several independent investigations with reference to some important qualifying trait (e.g. reinforced versus unreinforced, square versus rectangular cross-sections, brittle existing details versus modern constructions, different aspect ratios, different wrapping materials—glass, carbon, etc.). These have been used systematically to support interpretations of experimental trends or for calibration of analysis and design models for confinement. The characteristics of the assembled database and important relevant findings are summarized in the following section.

Database of Published Tests

Testing FRP-jacketed columns under concentric compression is the established method for documentation of the axial stress, axial strain, and constitutive behaviour of confined concrete. These tests are easy to conduct and data reduction may be easily carried out in the absence of a strain gradient on the cross-section.

The available experimental literature comprises a diverse collection of specimen subgroups originating from different investigations. Typically, each investigation considers a relatively small number of specimens (up to 20 on average), so parametric influences determined through these studies only account for a small range of combinations of values for the important parameters. To obtain a spherical assessment of analytical models through calibration with the test results, the boundaries posed by the individual experimental studies need to be overcome by a collective consideration of all the available data into a single database. In this report, two unified databases are assembled by merging several individual databases found in the literature (listed in the Appendix to the Chapter, Tables 5.A and 5.B (in supplementary material), along with citations to the original experimental studies considered in assembling the database). Criterion for inclusion of any given test is the requirement that lateral strains have been measured on the FRP jacket, at least up to the onset of failure. This strain measure is required so that lateral confining pressures that are responsible for the strength and deformation capacity enhancement may be estimated and considered explicitly during calibration of the analytical models.

Figure 5.2 outlines the range of parameter values included in the available database of tests. Data are classified based on the shape of the specimen cross-section and the presence or not of internal steel reinforcement. Numbers of reported samples decrease when going from plain to reinforced concrete, or from specimens with circular to specimens with rectangular cross-sections. Reinforced specimens are intended to model columns with brittle details, which are the most likely candidates for jacketing in real retrofit applications. About 2/3 of the

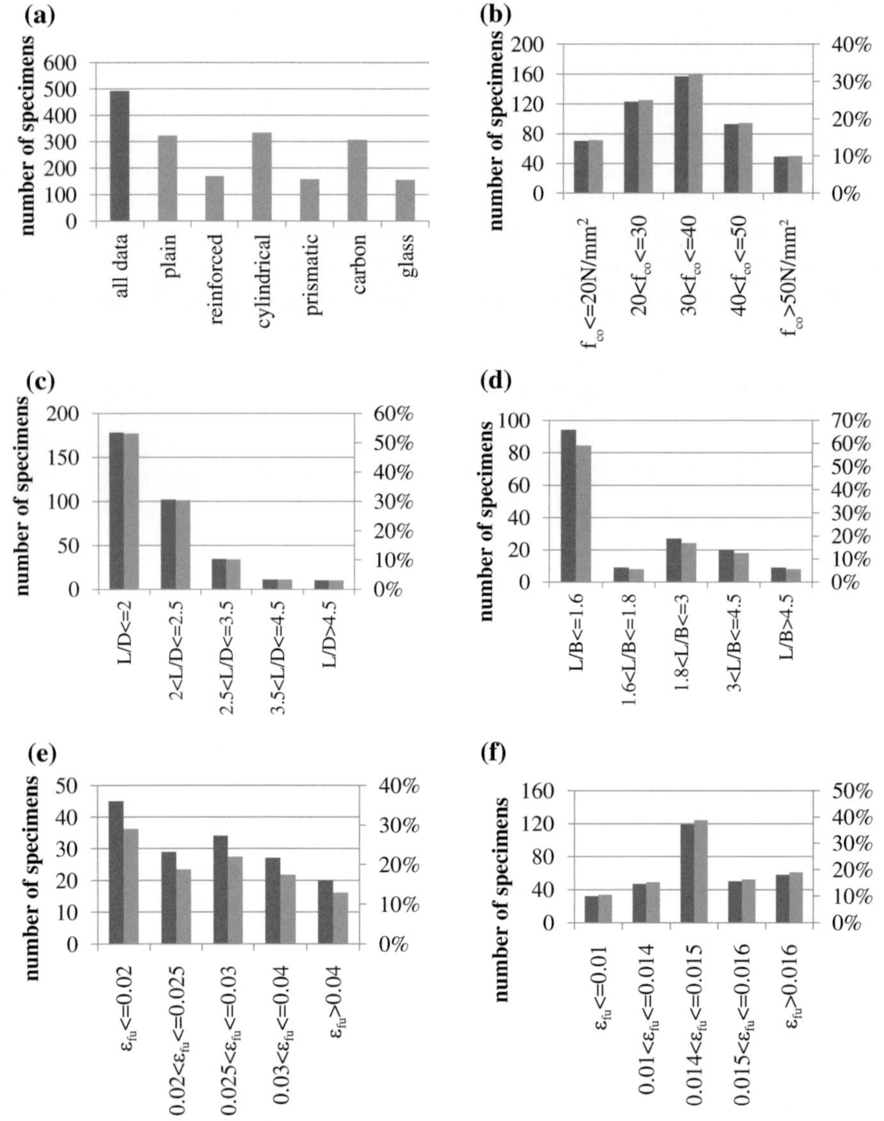

Fig. 5.2 **a** The distribution of available tests in the database. **b** Concrete strength of specimens. The axis to the right represents the percentage over the total available number, of each specimen category. **c** and **d** Aspect ratio (height to diameter/side) of available cylindrical and prismatic specimens. **e** and **f** Coupon failure strain (ε_{fu}) for GFRP and CFRP jackets. **g** and **h** GFRP and CFRP jacket thickness over the total number of tests. **i** and **j** Ratio of confined to unconfined concrete strength, normalized to lateral confining pressure at peak response, $(f_{cc}/f_{co})/\sigma_{lat}$, for GFRP and CFRP jackets respectively. **k** and **l** Ratio of confined to unconfined concrete strain at peak stress, normalized to lateral confining pressure at peak response, $(\varepsilon_{cc}/\varepsilon_{co})/\sigma_{lat}$, for GFRP and CFRP jackets respectively. **m** and **n** Confinement efficiency based on tests (k_e = ratio of FRP jacket average strain/strain capacity measured from coupon tests) for glass and carbon fiber wraps

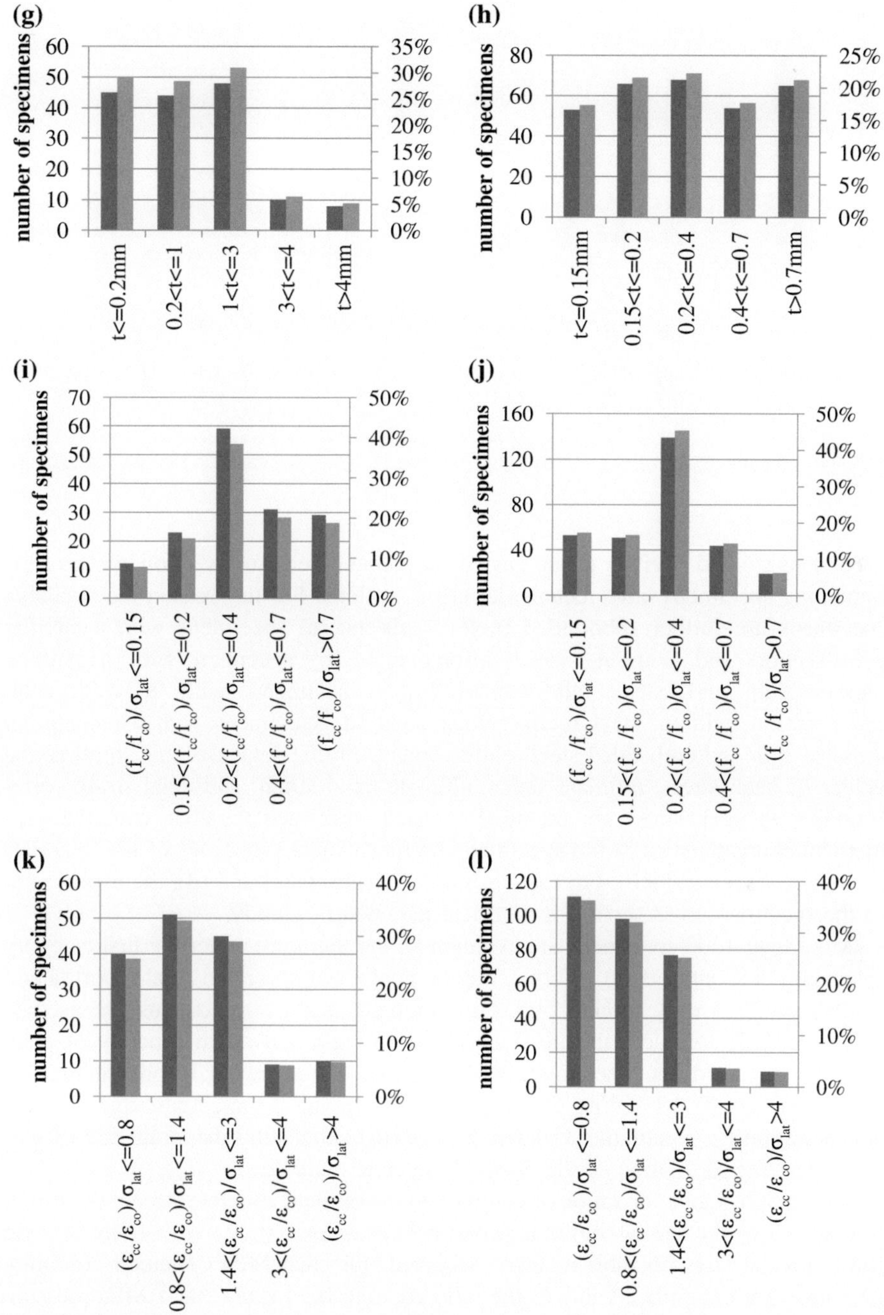

Fig. 5.2 (continued)

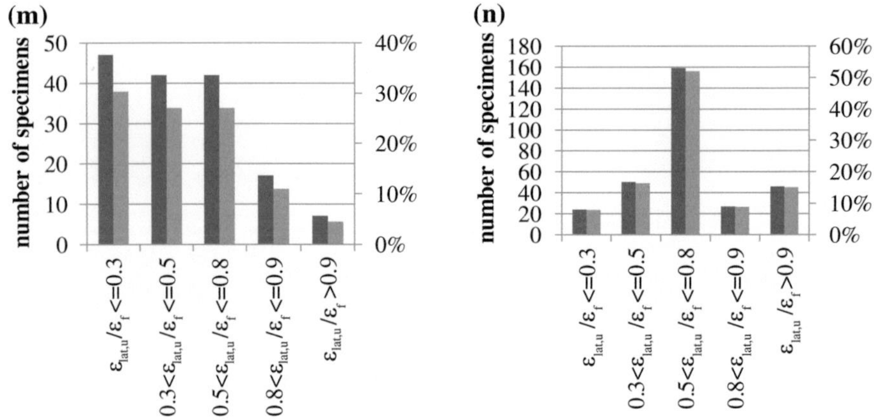

Fig. 5.2 (continued)

specimens studied in either group have a circular encased cross-section, whereas the remaining specimens have rectangular cross-sections. For this reason the available specimens are further subdivided in four subsets: (a) specimens with a circular cross-section and without steel reinforcement, (b) specimen with a circular cross-section and steel reinforcement, (c) specimens with a rectangular cross-section, without steel reinforcement, and (d) specimens with a rectangular cross-section and embedded steel reinforcement. In this database, the experimental values of peak stress, ultimate stress, ultimate axial strain, and axial strain corresponding to peak stress are included for each specimen together with the geometrical characteristics of the specimen and mechanical properties of the materials (concrete, steel and FRP). For the specimens in subsets (c) and (d), the aspect ratio of the specimen cross section is a critical parameter of study.

More than 1000 specimens are available in the literature under the first category (plain concrete, subsets (a) and (b)), but the number of specimens that report lateral strain records is only around 500. Unreinforced jacketed specimens are used to study the FRP confinement effectiveness on strength and strain capacity of plain concrete: reported specimen failures in these cases are owing to one of the following causes: (i) debonding of the jacket, (ii) rupture at the corners due to stress concentration (in rectangular cross sections), (iii) exhausting the strain capacity of the jacket due to dilation of the encased concrete material.

Figure 5.2i, j give the ratios of confined to unconfined concrete strengths, further normalized by the lateral confining pressure at peak load, $(f_{cc}/f_{co})/\sigma_{lat}$—this variable illustrates the strength ratio increase achieved, for each MPa of lateral confining pressure—for example a value of the ratio equal to 0.3 means that 1 MPa pressure will achieve 30 % confined strength increase.

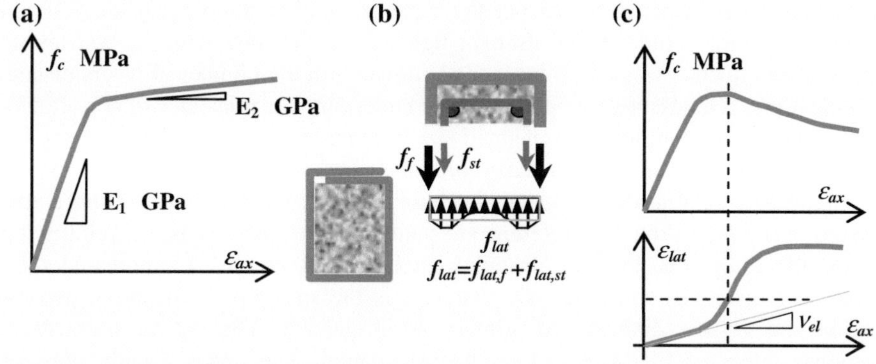

Fig. 5.3 Summary of experimental trends: **a** Behaviour of FRP-confined plain concrete. **b** Definition of effective lateral confining pressure (sum of FRP and steel contributions). **c** Evolution of average lateral strain with increasing axial strain. The ratio of $\varepsilon_{lat}/\varepsilon_{ax}$ is the apparent Poisson's ratio. The initial value is the elastic value of ν

Similarly Fig. 5.2k, l report the ratio of confined to unconfined concrete strain at peak stress, normalized by the lateral pressure, $(\varepsilon_{cc}/\varepsilon_{co})/\sigma_{lat}$. The variable illustrates the strain capacity increase for each MPa of lateral pressure.

The response of FRP-encased plain-concrete typically follows a bilinear stress—strain law, where the stiffness of the ascending branch primarily depends on the properties of concrete, whereas the slope of the hardening branch is determined by the stiffness of the jacket (Fig. 5.3a). Parameters studied in this database are the jacketing material type, the number of jacket layers, the fibre orientation in the wrap (measured by the angle of placement of the wrap weave relative to the direction of axial compression), the layout of the jacket sheets (continuous or strips), the strength of encased concrete, f_{co}, and the radius of chamfer used in the corners.

A total of 150 specimens are FRP-jacketed reinforced-concrete columns (subsets (c) and (d)). In terms of scale several specimens are simple cylinders and prisms reinforced with embedded longitudinal and transverse reinforcement, having an aspect ratio (height to cross-section height) of approximately 2 (specimen height is usually around 300–400 mm whereas a typical cross-sectional dimension is in the range of 150–200 mm; in most cases, dimensions of specimens are prescribed by limitations in the testing equipment). End effects in such short specimens prevent easy extrapolation of findings to the more general case of actual columns. But a small number of specimens are full scale columns with height to section size aspect ratio well above 2, much larger dimensions, and embedded reinforcement sizes larger than the common example. A note on matters of scale is relevant here: because the lateral confining pressure is proportional to the number of jacket layers but inversely proportional to the dimension of the column cross section, it is evident that in order to achieve the same magnitude of lateral pressure in similar specimens of different scale, different numbers of FRP layers will be needed. For example if n is the number of layers used in a 200 mm diameter column, theoretically, the number of layers required to produce the same pressure on a more realistic 400 mm diameter column would be $2n$. But, in light of the fact that with the wet lay-up process the effectiveness of the layers goes down with the number of layers placed, it is clear that scale issues are open and may undermine the relevance of constitutive laws obtained and calibrated from small size column models when these are applied to estimate the response of full scale columns.

Parameters of this experimental database include, over and above those discussed in the preceding with reference to plain concrete specimens, the reinforcing ratios, bar diameters, and the unsupported length of primary (compressed reinforcement) given by the ratio s/D_b, where s is the spacing of stirrups in the test region and D_b the diameter of primary reinforcement. The typical stress-strain response curve of FRP-jacketed reinforced concrete specimens usually presents bilinear characteristics (apparent yield and hardening plateau) reaching impressive strain magnitudes similar to those of plain specimens. A counter-example to that is the occasion of jacket failure by rupture in compressed wrapped reinforced concrete members which generally occurs at low levels of axial strain, owing to longitudinal bar buckling. This accounts for some of the scatter in the experimental values of specimens of identical dimensions, jacketed with the same material and number of layers; to organize the data, a criterion that quantifies the propensity for bar buckling owing to sparse stirrup spacing is used to limit the usable strain range of the jacket material in section "Estimation of the FRP effective ultimate strain".

Other sources of scatter in the experimental results may be owing to the following factors:

- The stiffness of the compression testing frame may affect the response of the FRP confined concrete specimen, particularly in the range of very large loads.
- The devices used for measuring both transverse and longitudinal strains, particularly regarding their ability to average the measurements over a significant gauge length (strain-gauges are particularly likely to produce non-uniform, localized strain values in this class of test specimens, especially in case of rectangular sections. Non-contact laser-scanning devices have been used recently in recording lateral strains in FRP jacketed concrete members; however, the number of available tests is still rather small).

Behaviour of FRP-Jacketed Concrete in Uniaxial Compression

The efficiency of FRP jacketing is gauged with reference to the enhancement obtained from encased concrete over that of identical unconfined concrete under compression. Observations from the database of tests support the following general trends:

(i) The quantifiable increase in compressive strength and corresponding strain capacity of confined concrete over that of unconfined concrete depends on the average lateral confining pressure exerted on encased concrete by both embedded stirrups (if any) and the outside external FRP jacket layers (Fig. 5.3b). A general expression for the average lateral confining pressure, f_{lat}, is the following:

$$f_{lat} = \frac{1}{2} \cdot \left(k_f \rho_{f,v} f_f + k_{st} \rho_{sv} f_{st}\right) = \frac{1}{2} \cdot \left(k_f \rho_{f,v} E_f \varepsilon_{lat} + k_{st} \rho_{sv} f_{st}\right) \quad (5.1)$$

where, f_f is the average axial stress in the jacket layers in directions orthogonal to axial compression, ε_{lat} is the average lateral strain developing along the perimeter of the encased column's cross-section, f_{st} the axial stress in the stirrups, ρ_{fv} and ρ_{sv} the volumetric ratios of FRP-wrap and steel stirrup reinforcement, and k_f, k_{st} the corresponding confinement effectiveness coefficients. By definition, each confinement effectiveness coefficient approximates the volume fraction of the encased core that is effectively confined by the FRP jacket and the stirrups, respectively; k_f is given in EC8-III ($k_f = a$ in Eq. (A36) of that Code). A similar expression is given for the effectiveness confinement of conventional stirrups in EC8-III ($k_{st} = a$ in Eq. A2 of the Code).

Both ε_{lat} and f_{lat} in Eq. (5.1) reflect the degree of dilation experienced by concrete normal to the direction of axial compression (Fig. 5.3b); they are related to the axial compression strain of the encased material, ε_{ax}, through an estimate of an apparent Poisson's ratio, $v(\varepsilon_{ax})$, as follows:

$$\varepsilon_{lat} = -v(\varepsilon_{ax}) \cdot \varepsilon_{ax}; \quad f_{st} = E_{st} \cdot \varepsilon_{lat} \leq f_{y,st}; \quad f_f = E_f \cdot \varepsilon_{lat} \quad (5.2)$$

Attainment of peak concrete stress coincides with attainment of an apparent Poisson's ratio in the order of 0.5–1, so, if any steel stirrups exist in the encased member, they may be assumed to have reached yielding ($f_{st} = f_{y,st}$). Based on the current literature, the enhanced confined concrete strength, f_{cc}, follows a very simple relationship with confining pressure, σ_{lat}, similar to the corresponding relationships established for conventionally confined concrete. There is a very large collection of strength models in published literature. Some of the proposed models are outlined in Table 5.1 for the sake of comparison. A far more exhaustive critical review of available models is presented by Ozbakkaloglu et al. (2013). The prevailing trend follows the mathematical form (Eq. 5.3):

$$f_{cc} = f_{co} \cdot \left[b + K_c \cdot (f_{lat}/f_{co})^a\right] \Rightarrow \bar{f}_{cc} = b + K_c \cdot \bar{f}_{lat}^{\,a} \quad (5.3)$$

where, the parameters b, K_c, and a have been calibrated against various databases, depending on the manner of calculation of the lateral confining pressure, f_{lat}; bar-quantities, $\bar{f}_{cc}$, $\bar{f}_{lat}$, in the right hand side represent the confined strength and lateral pressure values at peak response, normalized by the uniaxial compressive strength of concrete, f_{co}. Note that this general model expression for the confined concrete strength is compatible with established models for conventionally confined

Table 5.1 A sample of models for FRP confined concrete strength and strain capacity

Fardis and Kahalili (1982)	$\bar{f}_{cc} = 1 + 4.1 \cdot \bar{f}_{lat}$, $\bar{f}_{cc} = 1 + 3.7 \cdot \bar{f}_{lat}^{0.86}$ $f_{lat} = 2 f_{fu} n t_f / D$	$\varepsilon_{cc} = \varepsilon_{co} + 0.001 \frac{E_f n t_f}{f_{co} D}$
Mirmiran and Shahawy (1997)	$f_{cc} = 1 + 4.269 \cdot (f_{lat}^{0.59} / f_{co})$	N.A.
Karbhari and Gao (1997)	$\bar{f}_{cc} = 1 + 2.1 \cdot \bar{f}_{lat}^{0.87}$	$\varepsilon_{cc} = \varepsilon_{co} + 0.01 \bar{f}_{lat}$
Kono et al. (1998)	$\bar{f}_{cc} = 1 + 0.0572 \cdot f_{lat}$	$\varepsilon_{cc} = \varepsilon_{co}\left(1 + 0.14 \frac{4 t_f \cdot f_f}{D}\right)$
Spoelstra and Monti (1999)	$\bar{f}_{cc} = 0.2 + 3 \cdot \bar{f}_{lat}^{0.5}$	$\varepsilon_{cu} = 0.004 + 1.4 \frac{\rho_f f_{fu} \varepsilon_{fu}}{f_{cc}}$ $\varepsilon_{cu} = \varepsilon_{co}(2 + 1.25 \frac{E_c}{f_{cc}} \varepsilon_{fu} \bar{f}_{lat}^{0.5})$
Toutanji (1999)	$\bar{f}_{cc} = 1 + 3.5 \cdot \bar{f}_{lat}^{0.85}$ $E_1 = 10200 f_{co}^{1/3}$ $E_2 = 0.272\left(\frac{f_{co}}{\varepsilon_{co}}\right)$	$\varepsilon_{cu} = \varepsilon_{co} \cdot \left[1 + k_2 \cdot \left(\frac{f_{cc}}{f_{co}} - 1\right)\right]$ $k_2 = 310.57 \varepsilon_{fu} + 1.9$
Xiao and Wu (2000)	$\bar{f}_{cc} = 1 + 3.24 \cdot \bar{f}_{lat}^{0.8}$	$\varepsilon_{cc} = \varepsilon_{co}\left[1 + 17.4 \bar{f}_{lat}^{1.05}\right]$
Campione and Miraglia (2003)	$\bar{f}_{cc} = 1 + 2 \cdot \bar{f}_{lat}$	–
Ilki and Kumbasar (2003)	$\bar{f}_{cc} = \alpha[1 + 2.23 \cdot \bar{f}_{lat}]$ α = 8 for circular. Also, $\bar{f}_{cc} = \alpha\left[1 + 2.29 \cdot \bar{f}_{lat}^{0.87}\right]$ $f_{lat} = 0.5 k_\alpha \rho_f f_f$	$\varepsilon_{cc} = \varepsilon_{co}\left[1 + 15 \bar{f}_{lat}^{0.75}\right]$
Matthys et al. (2005)	$\bar{f}_{cc} = 1 + 2.3 \cdot \bar{f}_{lat}^{0.85}$ $\varepsilon_f^{eff} \approx 0.6 \varepsilon_{fu}$	$\varepsilon_{cu} = \varepsilon_{co} \cdot [1 + k_2 \cdot (\bar{f}_{cc} - 1)]$ $k_2 = 310.57 \varepsilon_{fu} + 1.9$
Kumutha et al. (2007)	$\bar{f}_{cc} = 1 + 0.93 \cdot \bar{f}_{lat}$	–
Tastani et al. (2006)	$\bar{f}_{cc} = 1 + 3.1 \cdot \bar{f}_{lat}$	$\varepsilon_{cc} = \varepsilon_{co} \cdot (1 + 15 \bar{f}_{lat})$ $\varepsilon_{cc,u} = \varepsilon_{cu} + 0.15 \cdot (\bar{f}_{lat} - 0.05)$
Vintzileou and Panagiotidou (2008)	$\bar{f}_{cc} = 1 + 2.8 \cdot \bar{f}_{lat}$ $\bar{f}_{lat} = 0.5 k_f \omega_w$ $\omega_w = (4 t_j / d) \cdot (f_{je} / f_{co})$ $f_{je} = f_j n^{-1/4}$ (n: number of FRP layers)	$\varepsilon_{cc,u} = \gamma_f \cdot \left(0.003 \cdot (\bar{f}_{cc})^2\right)$; $\gamma_{f-GFRP} = 1.15$ $\gamma_{f-CFRP} = 1.95$
Wu et al. (2009)	$\bar{f}_{cc} = 1 + 2.2 \cdot \bar{f}_{lat}^{0.94}$ $\bar{f}_{cc} = 1 + 3.2 \cdot \bar{f}_{lat}$ (AFRP)	N.A. $\varepsilon_{cu} = \varepsilon_{co} \cdot (1 + 9.5 \bar{f}_{lat})$ (AFRP)
Lam and Teng (2003)	$\bar{f}_{cc} = 1 + 3.3 \cdot \bar{f}_{lat}$ for $\bar{f}_{lat} \geq 0.07$ $\bar{f}_{cc} = 1$ for $\bar{f}_{lat} < 0.07$ $\varepsilon_f^{eff} = k_e \cdot \varepsilon_{fu}$ $k_{e\text{-}CFRP}$ = 0.586, $k_{e\text{-}GFRP}$ = 0.624 and $k_{e\text{-}AFRP}$ = 0.85	$\varepsilon_{cc} = \varepsilon_{co}\left[1.75 + 6.5 \cdot \left(\frac{E_f}{f_{co}/\varepsilon_{co}}\right)^{0.8} \left(\frac{\varepsilon_{f,rupt}}{\varepsilon_{co}}\right)^{1.45}\right]$
Teng et al. (2009)	$\bar{f}_{cc} = 1 + 3.5 \cdot (\rho_k - 0.01) \cdot \rho_\varepsilon \geq 1$ $\varepsilon_{je} = 0.586 \varepsilon_{fu}$ $\rho_k = 2 E_j t_j \varepsilon_{co} / (d \cdot f_{co})$ $\rho_\varepsilon = \varepsilon_{je} / \varepsilon_{co}$	$\varepsilon_{cu} = \left[0.0033 + 0.6\left(\frac{E_f}{f_{co}}\right)^{0.8} \cdot \left(\varepsilon_{f,rupt}\right)^{1.45}\right]$

(continued)

Table 5.1 (continued)

Albanesi et al. (2007)	$\bar{f}_{cc} = 1 + 3.609\bar{f}_{lat}$	$\varepsilon_{cu} = \varepsilon_{co} \cdot \left(1 + 18\bar{f}_{lat}\right)$
Jiang and Teng (2007)	$\bar{f}_{cc} = 1 + 3.5\bar{f}_{lat}$	$\varepsilon_{cc} = \varepsilon_{co} \cdot \left(1 + 17.5\bar{f}_{lat}^{1.2}\right)$
Girgin (2009)	$\bar{f}_{cc} = \bar{f}_{lat} + \sqrt{s + m\bar{f}_{lat}}$; $s = 1$ $m = \begin{cases} 2.9 \text{ for } f_{co} < 20\,\text{MPa} \\ 6.34 - 0.076 f_{co} \text{ for } 20 < f_{co} < 82\,MPa \\ 0.1 \text{ for } 82 < f_{co} < 108\,MPa \end{cases}$	N.A.
Wu and Zhou (2010)	$\bar{f}_{cc} = \bar{f}_{lat} + \sqrt{\left(\frac{16.7}{f_{co}^{0.42}} - \frac{f_{co}^{0.42}}{16.7}\right) \cdot \bar{f}_{lat} + 1}$	N.A.
Fahmy and Wu (2010)	$E_2 = m_2 \cdot \left(245.61 f_{co}^{m_1} + 0.6728 E_f\right)$ $f_{cc} = f_{co} + k_1 f_{lat}$; $k_1 = a f_{lat}^{-0.3}$ *if* $f_{co} \leq 40\,\text{MPa}$: $a = 4.5,\ m_1 = 0.5,\ m_2 = 0.83\ i$ otherwise: $a = 3.75,\ m_1 = 0.2,\ m_2 = 1.73$	$\varepsilon_{cu} = \frac{f_{cc} - f_{co}}{E_2}$

concrete, the most well-known being that of Richart et al. (1928) which used: $b = 1$, $K_c = 4.1$, and $\alpha = 1$. Alternatives to the general expression of Eq. (5.3) are models derived by adapting to the experimental trends the confinement model of Mander et al. (1988), according to which:

$$\bar{f}_{cc} = 2.25\sqrt{1 + 7.94\bar{f}_{lat}} - 2\bar{f}_{lat} - 1.254 \tag{5.4}$$

Strain capacity enhancement of confined concrete is more difficult to estimate and presents significant scatter; a number of alternative expressions have been proposed to estimate the axial strain associated with attainment of peak stress, ε_{cc}. Most of these expressions built on the original form proposed by Richart et al. (1928), which was subsequently calibrated by several other researchers and was eventually adopted in EC8-III (Eqn. A7).

$$\varepsilon_{cc} = \varepsilon_{co} \cdot \left[1 + 5 \cdot \left(\frac{f_{cc}}{f_{co}} - 1\right)\right] = \varepsilon_{co} \cdot \left[1 + 5 \cdot \left(\bar{f}_{cc} - 1\right)\right] \tag{5.5}$$

Substitution of Eqs. (5.3) into (5.5) (and taking the commonest $b = 1$) leads to:

$$\varepsilon_{cc} = \varepsilon_{co} \cdot \left[1 + 5 \cdot \left(b + K_c \cdot \bar{f}_{lat}^{a} - 1\right)\right] \approx 0.002 + 0.01 \cdot K_c \cdot \bar{f}_{lat}^{a} \tag{5.6}$$

For the post-peak "strain capacity", $\varepsilon_{cc,u}$, which is the axial strain sustained by the compressed specimen at a residual axial compressive strength equal to 85 % of peak, expressions have been mostly calibrated from tests. A lower bound expression for $\varepsilon_{cc,u}$ is similar to that derived for conventionally confined concrete, and has been obtained after calibration with the available experimental data:

$$\begin{aligned} \varepsilon_{cc,u} &= \varepsilon_{c,u} + \lambda \cdot \left(\frac{k_f \rho_{f,v} E_f \varepsilon_{lat} + k_{st} \rho_{st} f_{st}}{f_{co}} - 0.1 \right) \geq \varepsilon_{c,u} \\ \Rightarrow \quad \varepsilon_{cc,u} &= \varepsilon_{c,u} + \zeta \cdot (\bar{f}_{lat} - 0.05) \geq \varepsilon_{c,u} \end{aligned} \tag{5.7}$$

where, $\varepsilon_{c,u}$ is the strain capacity (at 85 % residual post-peak axial compressive strength) of unconfined concrete (in the range from 0.0035 to 0.004), whereas the second term in the right-hand side of Eq. (5.5) is the enhancement in strain capacity over that of unconfined concrete. Equation (5.7) recognizes that a minimum confining pressure ($\bar{f}_{lat} > 0.05$) would be necessary in order to have a noticeable effect on the strain capacity. A calibrated value for $\lambda = 0.075$ (i.e., $\zeta = 0.15$) has been proposed in the literature (Tastani and Pantazopoulou 2007; Tastani et al. 2013).

More than 100 models have been published in the literature attempting to quantify the strength and strain capacity of FRP-encased concrete. An exhaustive list of most available models have been presented by Ozbakkaloglu et al. (2013); a few samples of those expressions are listed in Table 5.1 for illustration of concepts. The majority of the models have been obtained either after calibration of the parameters of basic analytical expressions with the international test database, or they are directly a result of best fit of the tests. Usually the former are referred to as "analytical models", whereas the latter are reported as "design-oriented" approaches.

(ii) The effective strain in the confining jacket, ε_{lat} used in the estimation of the confining pressure exerted by the jacket (Eq. 5.1) at the instant of attainment of peak compressive strength of the encased concrete is not well understood. This is the reason behind the great diversity of published models proposed for estimation of a single phenomenon, namely the strength and deformation capacity enhancement of FRP-confined concrete (see for example, Table 5.1). Experimental results of axial loading tests on FRP-jacketed concrete column specimens have shown that the FRP strain level at failure is often lower than the characteristic FRP tensile strain capacity (Tastani and Pantazopoulou 2003; De Lorenzis and Tepfers 2003; Lignola et al. 2008; Realfonzo and Napoli 2011; De Luca et al. 2011). This fact may be responsible for unconservative overestimation of the member's confined strength. Indeed, the average absolute error of confinement models shows a remarkable attenuation when an effective strain value—that is lower than the nominal strain capacity of the jacket—is considered (De Lorenzis and Tepfers 2003). Many possible reasons for this phenomenon have been suggested by different authors.

To appreciate the degree of variability obtained in the data, the phenomenological Poisson's ratio (i.e. the ratio of reported average lateral strain value, ε_{lat}, to the associated axial strain value, ε_{cc}) at peak stress is plotted in Fig. 5.4. Based on

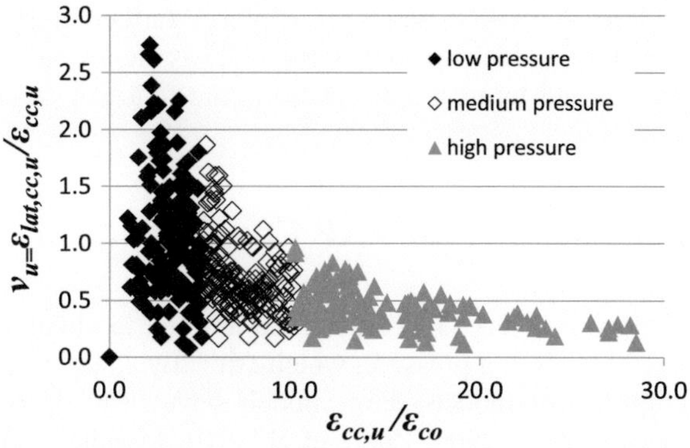

Fig. 5.4 Apparent Poisson ratio versus axial strain ductility at failure

the available test data, it appears that for low levels of confinement, evidenced by axial strain ductilities up to 5 (i.e., $\varepsilon_{cc,u}/\varepsilon_{co} < 5$) the value of v_u ranges from around 0.5 for circular, to 1 for rectangular sections. For higher confinement levels which enabled development of strain ductilities higher than 10, the values of v_u are lower, ranging from around 0.3 for circular, to 0.5 for rectangular sections.

As stated in the preceding, parameter ε_{lat} does not have a uniform value over the perimeter of the compressed member's cross-section. Peak local strain values occur at the corners where the jacket fibres change orientation. It has been shown that local deformation, $\varepsilon_{f,ch}$, which occurs at the points of chamfer, is related to local failure phenomena (rupture at the corner); the average strain value (defined as the normalized dimensional change of the cross-section), $\varepsilon_{f,ch}^{ave}$, has been linked to $\varepsilon_{f,ch}$ through simple kinematic considerations (Tastani et al. 2006).

The value of ε_{lat}, which is used in Eq. (5.1) is a mean jacket strain value that is associated with the prevailing mode of failure from among the following basic mechanisms of jacket resistance:

(a) Precipitation of debonding along the lap length of the exterior FRP jacket layer, which initiates at a strain of $\varepsilon_f{}^{deb}$;

(b) The dilation limit beyond which the nominal strain capacity of the jacket material would be overcome—this in the case of rectangular sections would occur at the corners due to strain concentration at these points, $\varepsilon_f{}^{dil}$;

(c) Cover failure and jacket rupture owing to buckling of compressed reinforcement, $\varepsilon_f{}^{buckl}$.

Thus, the effective strain ε_{lat}^{eff} is the least jacket strain value obtained from comparison of the above alternative scenarios for the FRP jacket and cannot exceed the nominal deformation capacity of the material, ε_{fu} (Tastani et al. 2006, 2013; 2007).

$$\varepsilon_f^{eff} = min\left\{ \varepsilon_f^{deb},\ \varepsilon_f^{dil},\ \varepsilon_f^{buckl} \right\} \le \varepsilon_{fu} \tag{5.8}$$

Recent work in this area explores further the issue of effective strain of the jacket, ε_f^{eff}, which appears in Eq. (5.1) as the average lateral strain, ε_{lat}. The relevant models and terms are discussed in the following section.

Estimation of the FRP Effective Ultimate Strain

The terms of Eq. (5.8) are needed in order to obtain a dependable estimation of the lateral confining pressure, which critically affects the relevance and performance of confinement models. Few models have been proposed for the individual terms (Tastani et al. 2006; Zinno et al. 2010; Lignola et al. 2012). Whereas few studies determine the prevailing value of ε_f^{eff}, through systematic consideration of the failure modes, an alternative approach focuses on determining the strain efficiency factor, k_e through collective evaluation of test databases that include specimens where lateral strains have been reported—note that the ratio of $k_e = \varepsilon_f^{eff}/\varepsilon_{fu}$ represents the usable strain range of the material that may actually be exploited in the jacket application as a fraction of its nominal strain capacity; in the following paragraphs both of these approaches will be summarized for completeness (see for example Fig. 5.2m, n).

(a) Failure strain due to FRP debonding/detachment, $\varepsilon_f^{\,deb}$

The strain at the initiation of layer detachment from its substrate as a result of bond failure ε_f^{deb}, is controlled by the bond characteristics of the substrate, that is the shear strength $f_{a,u}$ of the adhesive and the associated slip value $s_{a,u}$. For elastic response of the FRP and assuming that the elastic work for shear deformation of the adhesive (hypothesis for uniform distribution of bond stresses along the lap length of the exterior ply, L_f) is equal to the strain energy accumulated by the jacket along L_f, it follows that (Fig. 5.5):

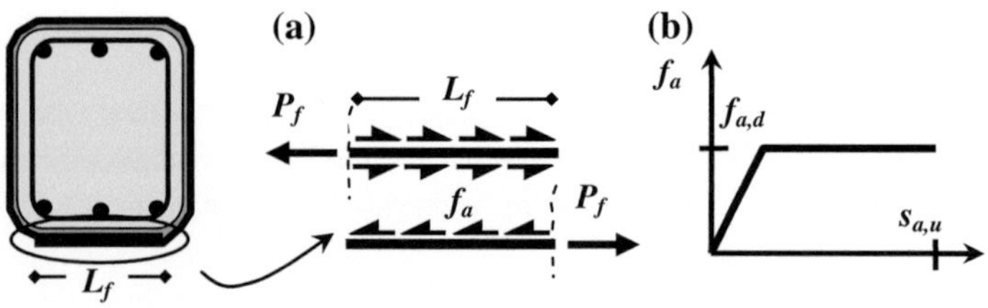

Fig. 5.5 Detailed definition of jacket debonding length. **a** Overlap of a single ply for anchorage; **b** Adhesive properties and stress transfer in the overlap length

$$t_f \cdot \int_0^{L_f} \left(\int_0^{\varepsilon_f^{deb}} f_f(\varepsilon_f) \cdot d\varepsilon_f \right) dx = \int_0^{L_f} \left(\int_0^{s_{a,u}} f_a(s_a) \cdot ds \right) dx$$
$$\Rightarrow \quad \varepsilon_f^{deb} = \sqrt{\frac{f_{a.u}}{E_f} \frac{s_{a,u}}{t_f}} \tag{5.9}$$

Hence, the ε_f^{deb} fundamentally depends on the bond of the resin between successive layers (Fig. 5.5b) as well as on the jacket stiffness, $E_f \cdot t_f$. The minimum required anchorage length $L_{f,min}$ that is necessary for the development of bond strength $f_{a,u}$, is defined from force equilibrium along the anchorage (i.e. $t_f\ E_f\ \varepsilon_f^{deb} = L_{f,min} \cdot f_{a,u}$) by using Eq. (5.10):

$$L_{f,min} = \sqrt{s_{a,u} E_f t_f / f_{a,u}} \tag{5.10}$$

In the absence of detailed information, typical values of the adhesive bond—slip law may be used for the calculation of the ε_f^{deb} (i.e., $f_{\alpha,u}$ = 5 MPa and $s_{\alpha,u}$ = 1 mm; for E_f = 170 GPa and t_f = 0.2 mm Eqs. (5.9) and (5.10) result in 85 mm of required anchorage length, and a debonding strain of 0.012). Figure 5.6 summarizes the strain values at failure of a selected group of specimens in the database. Points in Fig. 5.6a below the equal value line correspond to failures by dilation of the encased concrete core, whereas points above the equal line denote failures by jacket debonding; measured jacket strains at column failures are lower than the available strain capacity of the jacket material as evidenced in Fig. 5.6b, which plots data from plain specimens (empty squares and circles correspond to the corresponding cross sectional shapes) and the same trend is also seen in reinforced specimens (solid squares and circles correspond to the associated cross sectional shapes).

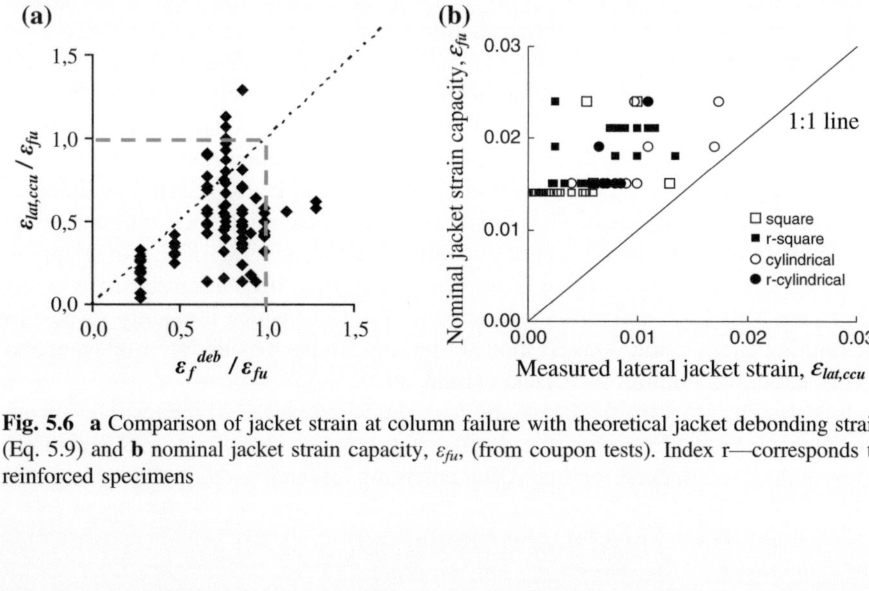

Fig. 5.6 **a** Comparison of jacket strain at column failure with theoretical jacket debonding strain (Eq. 5.9) and **b** nominal jacket strain capacity, ε_{fu}, (from coupon tests). Index r—corresponds to reinforced specimens

Thus, based on the data, FRP jacket layer detachment is a rare mode of failure in the specimens retained in the database. Regarding the minimum anchorage length, the experimental practice has shown that detachment failure can be averted if the last layer has anchorage length equal to the side dimension of the structural member. Thus, Eq. (5.10) can be used for design of FRP strengthening by introducing a safety factor $\gamma_{Lf} \geq 2$.

(b) Effective lateral strain at jacket debonding, ε_f^{eff}

Considering premature rupture of the jackets at the corners owing to stress concentrations and combined possible debonding at the free ends of the wrap layers, Zinno et al. (2010) proposed the following expression for the effective strain:

$$\varepsilon_f^{eff} = min\left\{\frac{e^{2\gamma L}-1}{1+e^{2\gamma L}}\tau_u;\ \frac{1}{t_a\gamma}\sigma_{nu}\right\}\cdot\frac{1+n}{E_f tn\gamma} \tag{5.11}$$

where τ_u and σ_{nu} are, respectively, the interfacial shear and normal stress capacity, L is the overlapping length and n is the number of FRP plies. Then,

$$\gamma = \sqrt{\frac{G_a}{t_a}\cdot\left(\frac{n+1}{E_f nt}\right)\Big/\left(1+t\frac{G_a}{t_a}\cdot\left(\frac{5+3n}{12G}\right)\right)} \tag{5.12}$$

where t and t_α are, respectively, the FRP single ply and adhesive/resin thicknesses and G and G_α are the corresponding shear moduli, respectively. An example of the provisions of the model is shown in Fig. 5.7a.

Clearly there is a theoretical background for the concept of an efficiency factor for the FRP jacket strain. However, providing a recommended value for this parameter is hampered by the limited understanding of its dependencies to many physical aspects of the problem.

According to the model proposed by Zinno et al. (2010) the effective FRP strain does not simply depend on the total thickness and elastic modulus of the jacket. Rather, the experimental evidence conclusively points to a greater dependence on the number of layers rather than the individual layer thickness. To reflect this finding the Greek code for Retrofitting (KAN.EPE 2012) adopted already from its 2008 edition, an efficiency factor $k_e = n^{-1/4}$ for $n \geq 4$, where n is the number of wrap layers according to the work of Vintzileou and Panagiotidou (2008, see model in Table 5.1). Additional findings are: (i) the dependence on elastic modulus is non-linear, (ii) the values of shear modulus could determine FRP ultimate strain variations if they are below approximately 15 GPa; and (iii) if the overlapping length exceeds a minimum value the ultimate FRP strain becomes insensitive to this parameter. This last observation concurs with Eq. (5.8) which implicitly associates the limiting strain to the weakest mode of failure; stronger modes are irrelevant and do not affect the limiting FRP jacket strain.

It is also clear that the thickness and mechanical properties of the resin layers are crucial parameters. However, the details of the resin adopted in tests are rarely given in the experimental reports. As an emerging research need, it will be essential

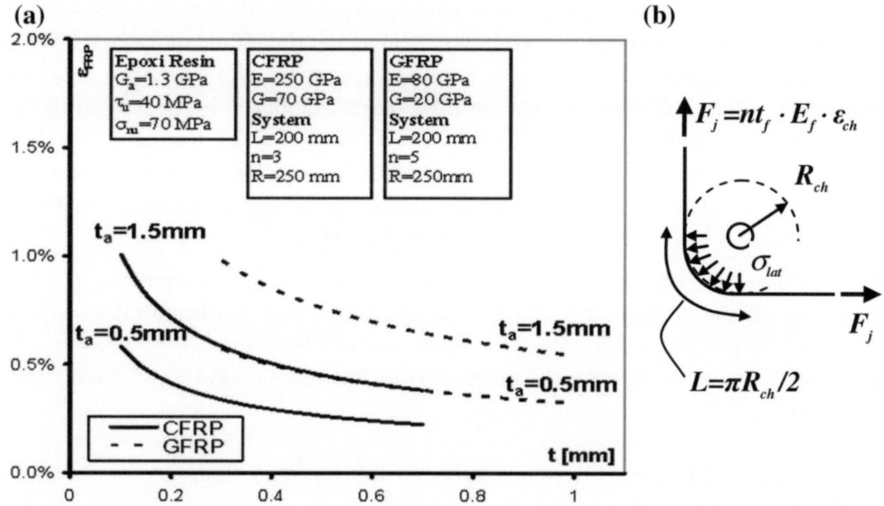

Fig. 5.7 **a** Influence of FRP thickness on FRP hoop strain. **b** Stress concentration on FRP along the chamfered corner of specimen

if this kind of data will be also collected and published in the future experimental works on FRP concrete confinement. The model described provides the least estimate for the first and second terms of Eq. (5.8); when combined with models accounting for other premature failure modes it may be used to quantify the effective FRP strain for jacketing design. It is recalled that this value is, in many cases, much lower than the value given by the flat coupon tests of the FRP material (Eq. 5.8).

(c) Effective lateral strain at jacket rupture due to concrete core dilation, ε_f^{dil}

Figure 5.4 illustrates that there is a systematic relationship between the FRP effective strain at peak confined concrete stress and concrete axial strain—referred here as the phenomenal or apparent value of the Poisson's ratio, $v = \varepsilon_f^{dil}/\varepsilon_{cc,u}$. For the estimation of a limiting effective strain associated with FRP rupture due to the dilation of the encased core, ε_f^{dil}, the other two variables (i.e., v_u and $\varepsilon_{cc,u}$) of the above ratio need to be defined. From the arrangement of the points in Fig. 5.4, it is clear that the data may be classified according with the magnitude of confinement in three discrete categories of confinement, namely lightly confined specimens ($f_{lat} < 0.2\, f_{co}$), moderately confined specimens ($0.2\, f_{co} < f_{lat} < 0.5\, f_{co}$), and highly confined specimens ($0.5\, f_{co} < f_{lat}$). Note the corresponding range in the values of developed strain ductilities: (a) Lightly confined cases developed strain ductilities $\mu_{\varepsilon cc,u} = \varepsilon_{cc,u}/\varepsilon_{co} < 5$, indicating a limited confined axial strain capacity because of the light confinement, and an associated apparent v_u value in the range from 0.5 and 1 (i.e., intensely dilative behaviour). (b) Moderately confined specimens exhibited adequate compression strain ductilities (in a range of 5–15) due to the noticeable confining pressure, with an apparent v_u value around 0.5 (i.e., controlled dilation).

(c) Highly confined specimens demonstrated large axial strain ductility (between 15 and 30) and very limited dilation, with the mean value of v_u in the range of 0.3. This response also occurred in moderately confined specimens ($f_{lat} \approx 0.4 f_{co}$). Most of specimens in the third category had an FRP jacket composed by over 4 layers.

For practical design, v_u should be taken equal to 0.5 for low and moderate confining pressures ($f_{lat} < 0.5 f_{co}$), which conservatively yields an axial compression strain ductility up to 10 (i.e. $\varepsilon_{cc,u}/\varepsilon_{co} < 10$), whereas for high pressure ($f_{lat} > 0.5 f_{co}$ and $\varepsilon_{cc,u}/\varepsilon_{co} > 10$) the recommended value is $v_u = 0.3$.

(d) Effective FRP strain at failure due to lateral stress concentration, $\varepsilon_{f,ch}$
Rounding off the corners is meant to mitigate lateral stress concentrations which occur due to the change in direction of the jacket fibers around the corners. By establishing force equilibrium the magnitude of this pressure is (Fig. 5.7b, Tastani et al. 2013):

$$\sigma_{lat} = (2\sqrt{2}/\pi) \cdot (nt_f/R_{ch}) \cdot E_f\varepsilon_{f,ch} = 0.9(nt_f/R_{ch}) \cdot E_f\varepsilon_{f,ch} \tag{5.13}$$

This pressure is proportional to the jacket stiffness (nt_fE_f), and to the lateral local FRP strain. It follows that for a given magnitude of FRP strain $\varepsilon_{f,ch} < \varepsilon_{fu}$, increasing the jacket stiffness (nt_fE_f) increases the intensity of the local stress concentration, which however is relieved by increasing parameter R_{ch}.

The local strain $\varepsilon_{f,ch}$ at specimen corners and in the middle distance between successive stirrups is defined through the magnitude of maximum buckled bar lateral deflection w (from its original straight axis) assuming polar coordinates as, $\varepsilon_{f,ch} = w/R_{ch}$. As the bar is bent outwards it imposes compressive lateral stresses on the concrete cover, which are undertaken by the FRP jacket by developing tensile forces (Fig. 5.7b). By using this definition of strain $\varepsilon_{f,ch}$ in Eq. (5.13) it follows:

$$\sigma_{lat} = 0.9 \cdot E_j \cdot (nt_f/R_{ch}^2) \cdot w \tag{5.14}$$

From Eq. (5.14) it is evident that the increase of parameter R_{ch} results in lower stress intensity values at the corners and thus it is possible for the FRP material to develop its strain capacity due to outwards bending of the compression bars without local concrete crushing of the cover due to bar bending in the lateral direction.

(e) Strain Efficiency Factor based of Experimental Evidence, k_e
Barring any undesirable failures implied by the last two terms of Eq. (5.8), namely failures by debonding and due to stress-concentrations in corners, Realfonzo et al. (2011) carried out a statistical analysis of the strain efficiency factor, k_e, in order to define the usable value of ε_f^{eff} from the nominal FRP strain reported from flat coupon tests. The statistical analysis was performed by considering a specialized database of tests published in the literature—for which the jacket ultimate strain $\varepsilon_{f,u}$ was available, i.e. 98 and 52 test specimens for CFRP and GFRP systems, respectively.

The evaluation considered the average value (k_{em}), the standard deviation (σ_{ke}), the coefficient of variation (CV), and the skewness (γ_1) and the kurtosis (γ_2); these variables were computed from the test results. Note that parameter γ_1 is a measure of the asymmetry of the k_e distribution, while γ_2 provides a measure of its "peakedness". In particular, when the k_e distribution is characterized by a tail to the right, the value of γ_1 is negative; if it has a tail to the left the skewness is positive. A positive value of γ_2, instead, implies that the distribution has an acute peak around the mean and fat tails; conversely, a negative value identifies a distribution with a wide peak and thin tails.

Table 5.2 summarizes results from the statistical analysis for k_e. Initially all data sets were considered, i.e. disregarding the dependence on the compression strength of the unconfined concrete, f_{co} (i.e. 150 test sets for CFRP and GFRP confined specimens comprising 98 specimens for CFRP confined cylinders and 52 specimens for GFRP confined specimens). The analysis was then performed by considering the dependence of k_e on f_{co} values. In this case, three different strength ranges were considered, labelled as low strength concrete (LSC), normal strength concrete (NSC) and high strength concrete (HSC). For the last two groups the analysis was conducted considering CFRP and GFRP confined specimens together (42 specimens for NSC; 16 for HSC). More emphasis was given to data in the LSC concrete range, since such types of concretes are generally found in the types of existing buildings where retrofitting by FRP confinement would be mostly needed. Within this group, the statistical study was performed considering the data belonging to only CFRP ($N_C = 62$), to only GFRP ($N_G = 29$) and to both types of FRP systems ($N_{tot} = 91$).

By focusing on the results obtained for LSC concrete it is observed that, within this range, CFRP and GFRP systems give rise to slightly different values of k_{em} (0.63 and 0.68, respectively), which are associated with a significant reduction of the data scatter especially in the case of the glass fibres; instead, disregarding the type of fibre, a mean value of the strain efficiency factor equal to 0.65 has been evaluated.

This result is illustrated in Fig. 5.8a where the probability density function of k_e—by assuming a Gaussian distribution—is plotted. The figure shows that the shape of the k_e distribution is sensitive to the type of FRP system; in particular, higher scatters are computed for GFRP systems so that a greater uncertainty in the prediction of the experimental data is associated with the use of glass fibers. Figure 5.8b shows the dependence of k_e on the FRP confining stiffness (k_{conf}). It is noted that the best fit curve is obtained by considering only the experimental specimens having $k_{conf}/f_{co} > 5$ (as recommended by Matthys et al. 2005). The best-fit curves proposed by Matthys et al. 2000, 2005 are also plotted. The mean value of k_e was equal to 0.65 when considering all the 91 specimens belonging to the LSC and MSC range (Table 5.2).

From the results it is concluded that the experimental trends are marked by significant scatter, with the strain efficiency k_e decreasing slightly with increasing confining stiffness, regardless of the concrete strength range considered.

(f) Evaluating the Strain Efficiency Factor from Constitutive Models for the Jacket

Another model for the strain efficiency factor was proposed recently by Lignola et al. (2012). Here the effects of the three-dimensional state of stress in the jacket and of the curvature at the rounded corners of rectangular cross-sections are considered.

Table 5.2 Results of statistical analysis for the definition of strain efficiency factor, k_e

	CFRP	GFRP	ALL	LSC			NSC	HSC
				($f_{co} \leq 40$ MPa)			($40 < f_{co} \leq 60$ MPa)	($f_{co} > 60$ MPa)
	$n = 98$	$n = 52$	$n = 150$	CFRP	GFRP	ALL	ALL	
				($n = 62$)	($n = 29$)	($n = 91$)	($n = 42$)	($n = 16$)
k_{em}	0.6	0.59	0.6	0.63	0.68	0.65	0.58	0.41
σ_{ke}	0.17	0.25	0.2	0.15	0.17	0.16	0.21	0.22
CV	0.28	0.42	0.33	0.23	0.25	0.24	0.37	0.54
γ_1	−0.57	−0.48	−0.59	−0.21	−2.22	−0.92	−0.3	0.11
γ_2	0.54	−1.07	−0.15	0.66	4.41	1.93	−1.04	−1.06

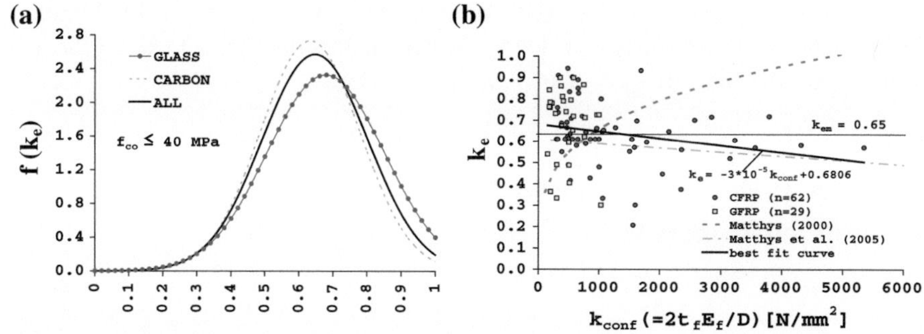

Fig. 5.8 **a** k_e Normal distribution for LSC range; **b** dependence of k_e on the FRP confining stiffness

The model is derived from first principles (accounting for equilibrium, compatibility, constitutive relationships and a 3D failure criterion (Tsai–Hill) for FRP composites). Results for ε_f^{eff} are obtained iteratively. To avoid iteration required by the original algorithm, in a more recent version it is proposed that bond between concrete and the FRP jacket be neglected, providing a direct closed form solution:

$$k_e = (=\beta) = \frac{\left[1 - \nu_{TL}\nu_{LT} + \nu_{TL}\cdot\nu_{LT}\frac{t_f}{R} + \nu_{LT}\frac{t_f}{R}\right]}{\sqrt{1+\frac{f_\theta}{f_r}\frac{t_f}{R}+\left(\frac{f_\theta}{f_r}\frac{t_f}{R}\right)^2+\nu_{TL}\left[\left(\frac{f_\theta}{f_z}\right)^2\left(1-\frac{t_f}{R}\right)^2-\left(1-\frac{t_f}{R}\right)\frac{f_\theta}{f_r}+\left(1-\frac{t_f}{R}\right)\frac{t_f f_\theta^2}{R f_r f_z}\right]}} \quad (5.15)$$

where $\nu_{TL} = \nu_{LT}\cdot E_T/E_L$ is the orthotropic Poisson's ratio and E_L and E_T are the transverse and longitudinal elastic moduli of the jacket, respectively. The longitudinal (circumferential) and transverse (vertical or radial) strengths of the jacket are f_θ and $f_r \approx f_z$, respectively. These properties have a major impact on the value of k_e, and should be recorded during the tests. In the absence of values for the orthotropic properties, average values were adopted (Lignola et al. 2012) leading to the

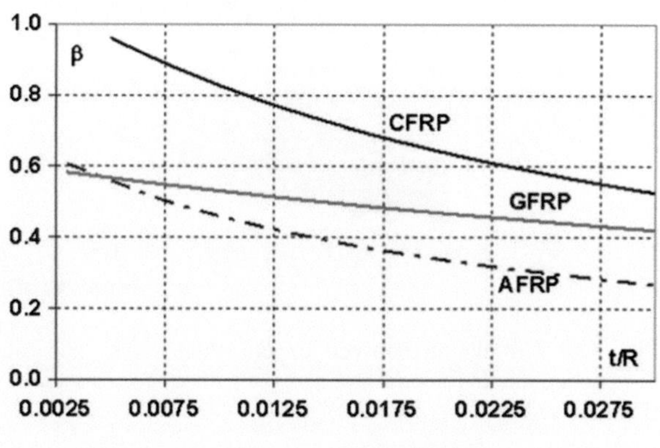

Fig. 5.9 Theoretical average estimated efficiency factors for CFRP, GFRP, and AFRP

illustrative results shown in Fig. 5.9. Although the properties of the FRP systems used in tests are rarely listed in detail, clearly such properties appear to have an influence on the performance of FRP wrapping onsite.

Reduced Jacket Effectiveness Due to Pore Compaction of Confined Concrete

Equations (5.6) and (5.7) suggest a linearly increasing relationship between confined concrete strain capacity and lateral confining pressure. This relationship performs adequately up to the range of moderate confinements with some tendency for overestimation, which deteriorates for higher confining pressures (Fig. 5.10a).

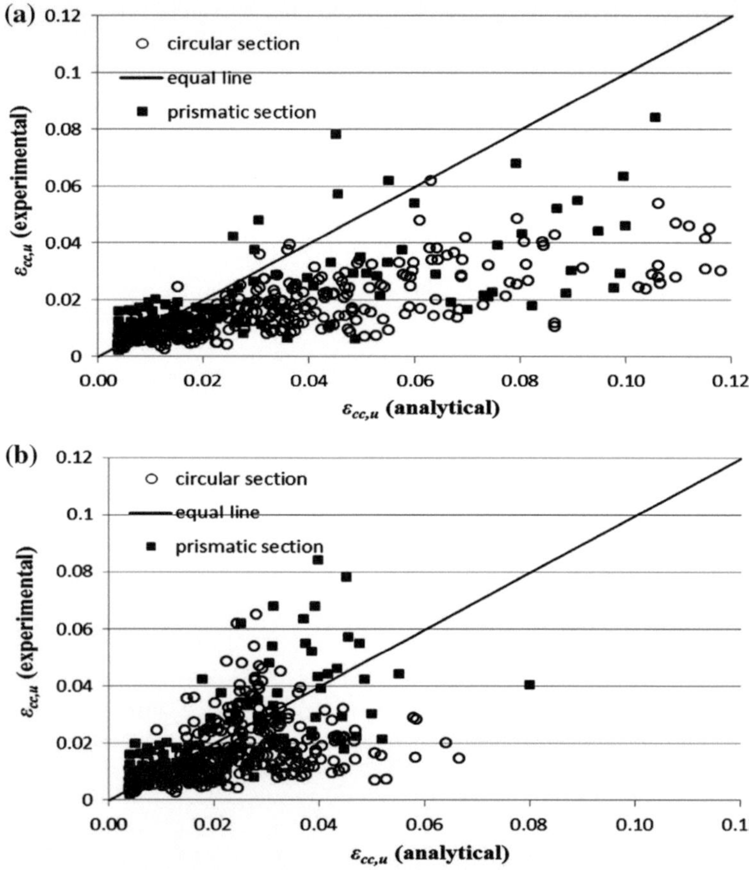

Fig. 5.10 Correlation between experimental values and analytical calculations adopting the compaction model for axial strain $\varepsilon_{cc,u}$ **a** without consideration of compaction, **b** when compaction is considered

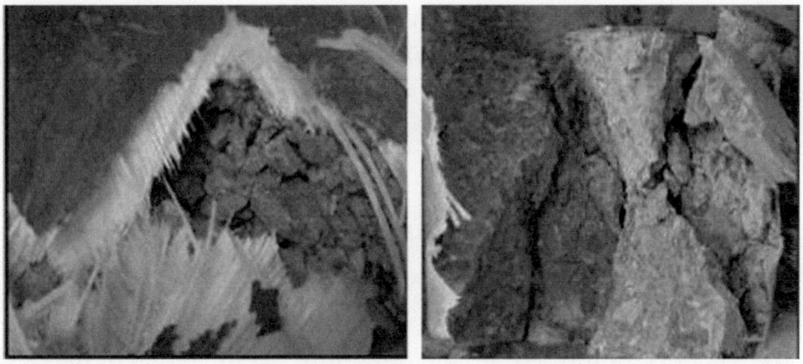

Fig. 5.11 Comparison of highly and lightly confined concrete appearance after jacket rupture

From analysis of the available strain records at failure of highly confined specimens it follows that they did not exhibit volumetric dilation but contraction (also seen at moderate pressures for lower concrete qualities).

Note that a Poisson's ratio of v_u = 0.5 or more in the lightly confined cases indicates that the lateral strain is higher than the axial strain owing to the formation of discrete cracks parallel to the compressive load; a lower value of v_u = 0.3 or less in the highly confined cases indicates that the axial compressive strain is significantly greater than the lateral expansion; thus the specimen demonstrates volumetric compaction which is accompanied with pulverization, i.e. collapse of the concrete micro-structure at advanced level of deformation. This behaviour has been already recognized as compaction of the porous concrete matrix—see the differences in the fracture patterns of highly confined and lightly confined examples in Fig. 5.11.

Tastani et al. (2013) proposed an amendment to the value of effective confining pressure, given by Eq. (5.1) to correct for the reduced passive action owing to compaction. This is done by modifying the confining effectiveness coefficients, k_f and k_{st}, to reflect the lower values of mobilized lateral pressure when compaction occurs. Therefore:

$$k_i^{conf^*} = k_i^{conf} \Big/ \left(1 + \varepsilon_A^P / \varepsilon_{co}\right), \quad i = f,\ st \tag{5.16}$$

where $\varepsilon_A{}^P = 2\varepsilon_{lat}{}^P$ is the area plastic strain that is considered as a damage index of the cross section ($\varepsilon_{lat}{}^P$ is the plastic component of the ultimate lateral strain). According to the above, Eq. (5.7) is reformulated as follows (where $\overset{*}{\varepsilon}_{lat,ccu} = \varepsilon_{lat,ccu} - \varepsilon_{lat}{}^P$):

$$\varepsilon_{cc,u} = \varepsilon_{cu} + 0.075 \cdot \left[\left(k_f^{conf*} \rho_{fv} E_f \overset{*}{\varepsilon}_{lat,ccu} + k_{st}^{conf*} \rho_{sw} f_{yst}\right) \Big/ f_c - 0.1\right] \tag{5.17}$$

Comparison between Figs. 5.10a, b, where with the introduced correction it is shown that overestimation of strain capacity is substantially reduced with the data-points clustered near the equal value line, demonstrates that compaction of the microstructure of the material at high levels of confinement ought to be considered in confinement models, particularly when estimating strength and deformation capacity of the confined member for design applications.

Confinement Models from Design Codes (CNR, ACI, EC8-III, Fib)

Some of the models described in the literature were adopted in Design Code documents such as the CNR, ACI, EC8-III and in some relevant fib Bulletins (e.g. fib Bulletin 14 2001). For the sake of completeness of presentation the corresponding design expressions are outlined briefly in the following:

(a) *fib Bulletin 14* (2001): The model is based on an iterative procedure that leads to the evaluation of the peak strain ε_{cc} and peak axial strength f_{cc}:

$$\begin{aligned} \varepsilon_{cc} &= \varepsilon_{cc,Mander}\left[\frac{E_{\sec}\left(E_{co}-E_{\sec,u}\right)}{E_{ec,u}\left(E_{co}-E_{ec}\right)}\right]^{1-\frac{E_{\sec}}{E_{co}}} \\ f_{cc} &= E_{ec,u}\varepsilon_{cc} \end{aligned} \tag{5.18}$$

where $\varepsilon_{cc,Mander}$ is defined according with Eq. (5.5) (i.e. originating from Richart et al. 1928). In the case of rectangular columns the confinement pressure is the minimum of the confinement pressures developed in the two principal directions x and y. The confinement pressure is defined from:

$$\begin{aligned} &f_{li} = \rho_{fi}k_f E_f \varepsilon_{fu}, \\ &\text{where } k_f = 1-\frac{(b-2R_{ch})^2+(h-2R_{ch})^2}{3A_g(1-\rho_l)} \end{aligned} \tag{5.19}$$

and i indicates the direction (x or y) considered. The confined concrete axial compressive strength and associated strain may also be approximated from:

$$\begin{aligned} \frac{f_{cc}}{f_{co}} &= 0.2+3\sqrt{\frac{f_{lat}}{f_{co}}} \\ \frac{\varepsilon_{cc}}{\varepsilon_{co}} &= 2+1.25\frac{E_{co}}{f_{co}}\varepsilon_{fu}\sqrt{\frac{f_{lat}}{f_{co}}} \end{aligned} \tag{5.20}$$

(b) *ACI Committee 440.2R-08* (2008): This model extends to the case of FRP confinement the model proposed by Mander et al. (1988) for the case of internal steel confinement (f_{lat} is the confining pressure). For circular cross-sections the confined strength and corresponding axial strain are given by:

$$\frac{f_{cc}}{f_{co}} = 2.25\sqrt{1 + 7.9\frac{f_{lat}}{f_{co}}} - 2\frac{f_{lat}}{f_{co}} - 1.25$$
$$\varepsilon_{cc} = \frac{1.17(5f_{cc} - 4f_{co})}{E_c} \tag{5.21}$$

(c) *CNR-DT 200/2004* (2004): Italian recommendations do not provide a closed form expression either for the confined stress-strain response or for the ultimate axial strain. The design compressive strength for confined concrete is given by:

$$\frac{f_{cc,d}}{f_{c,d}} = 1 + 2.6\left(\frac{f_{f,eff}}{f_{c,d}}\right)^{2/3} \tag{5.22}$$

where, $f_{c,d}$ is the design strength of un-confined concrete, and $f_{f,eff}$ is the effective confining pressure given by:

$$f_{f,eff} = k_{eff} f_{lat} \tag{5.23a}$$

$$f_{lat} = 0.5\rho_f E_f \varepsilon_{fd,red} \tag{5.23b}$$

Parameter $\varepsilon_{fd,red}$ is the reduced design strain of the composite, and k_{eff} is the coefficient of effectiveness computed as a product of a vertical, k_V, horizontal, k_H, and inclined, k_α, effectiveness coefficients:

$$\varepsilon_{fd,red} = min\{\eta_a \varepsilon_{fk}/\gamma_f; 0.004\} \tag{5.23c}$$

$$k_{eff} = k_H k_V k_\alpha \tag{5.23d}$$

(d) *KAN.EPE* 2012 (*Greek code for Retrofitting*): The Greek recommendations distinguish the confined concrete strength and the associated strain for circular and prismatic cross-sections through the safety factors γ_{FRP} depending on whether CFRP or GFRP jacketing material is used, respectively, as follows:

$$f_{cc,d} = f_{c,d} \cdot (1.125 + 1.25 \cdot k_f \cdot \omega_{wd})$$
$$\varepsilon_{cc} = \gamma_{FRP} \cdot 0.0035 \cdot (f_{cc,d}/f_{c,d})^2 \tag{5.24a}$$

where $f_{c,d}$ is the unconfined compressive design strength of concrete (equal to 60 % of the characteristic uniaxial concrete strength), $f_{cc,d}$ is the design confined concrete compressive strength, γ_{FRP} = 1 and 2 for CFRP and GFRP, respectively, whereas k_f is the coefficient of confinement effectiveness (see Eq. 5.19), and ω_{wd} is the mechanical ratio of confining reinforcement. In calculating its value, note that it is taken equal to the product of the volumetric ratio of the jacket (t_j is the jacket thickness and d the size of the cross section for continuous jackets), $f_{j,eff}$ is the effective jacket stress which is taken equal to the nominal strength of the jacket material, and n the number of plies used in the jacket.

$$\begin{aligned}\omega_w &= (4t_j/d)\cdot(f_{j,eff}/f_{co})\\ f_{j,eff} &= f_j n^{-1/4}\end{aligned} \tag{5.24b}$$

(e) *EC 8-III* (2005): The Eurocode 8 (Assessment & Retrofit) recommends the following Eqs. (5.25a) and (5.25b) for the definition of the confined concrete strength and the associated strain (ε_{cc}) as well as the strain capacity (ε_{cu}), where $\varepsilon_{f,eff}$ is lower than the jacket strain at rapture (ε_{fu}).

$$f_{cc} = f_{co}\left[1 + 3.7\left(\frac{k_{eff}\rho_f f_{f,eff}}{f_{co}}\right)^{0.86}\right] \tag{5.25a}$$

$$\begin{aligned}f_{f,eff} &= k_{eff}\frac{2t_f}{D}E_f\varepsilon_{f,eff}\\ \varepsilon_{cc} &= \varepsilon_{co}\cdot\left[1 + 5\cdot\left(\frac{f_{cc}}{f_{co}} - 1\right)\right]\\ \varepsilon_{cu} &= 0.004 + 0.5\frac{k_{eff}\rho_f f_{f,eff}}{f_{cc}}\end{aligned} \tag{5.25b}$$

where for circular section $k_{eff} = 1$, for rectangular $k_{eff} = \frac{2R_{ch}}{D}$ and for strips $k_{eff} = \left(1 - \frac{s_f}{2D}\right)^2$

Confined Strength Formulae for Jacketed Reinforced Concrete Columns—Considering the Interaction with Reinforcement

The interaction between internal steel reinforcement and external FRP is particularly important when considering practical retrofit applications. The first few analytical models available in the literature (e.g. fib 2001; ACI 2002; CNR 2004) were mainly calibrated through experimental studies that do not consider the influence of the existing steel reinforcement on the structural behaviour of the FRP confined element. These models correlated the increase of strength and ductility with the confinement pressure provided with the FRP strengthening only, hence neglecting the eventual contribution of the internal transverse steel reinforcement. Only some of the most recent analytical models (Kawashima et al. 2000; Li et al. 2003; Harajli et al. 2006; Tastani et al. 2006; Ilki et al. 2008; Pellegrino and Modena 2010) estimate the total confinement pressure as the sum of the confinement pressure due to the external FRP jacketing, $f_{lat,f}$, and the confinement pressure due to the internal transverse steel reinforcement (stirrups or spirals), $f_{lat,st}$, whenever such is present.

For example, the models by Megalooikonomou et al. (2011) and Pellegrino and Modena (2010) take explicitly into account the interaction between FRP and steel contemporary confinement actions and its influence on the effectiveness of the FRP confinement. This interaction can be particularly important since internal steel reinforcement configuration influences concrete cracking patterns at failure and therefore the efficiency of FRP strengthening technique in terms of ultimate capacity and ductility.

A brief review of some of the analytical models that explicitly account for embedded steel reinforcement is given in this section to provide context of the relevant issues.

(a) Li et al. (2003)

This model is only applicable to circular columns confined through CFRP with or without internal steel reinforcement. The confined concrete strength is given by:

$$f_{cc} = f_{co} + f_{lat} \cdot tg\left(45^\circ + \frac{\phi}{2}\right) \tag{5.26a}$$

where f_{lat} is the total confinement pressure calculated as the sum of the contribution of the internal steel and the external FRP. The friction angle ϕ and the peak axial strain are defined by:

$$\phi = 36^\circ + 1 \cdot \left(\frac{f_{co}}{35}\right) \leq 45^\circ \tag{5.26b}$$

$$\varepsilon_{cc} = \varepsilon_{co}\left[1 + \alpha \cdot tg^2(45^\circ + \phi/2)\frac{f_{lat}}{f_{co}}\right] \tag{5.26c}$$

where ε_{co} = 0.002 and α = 2.24. Finally, the stress-strain relation is given by:

$$f_c(\varepsilon_c) = f_{cc}\left[-\left(\frac{\varepsilon_c}{\varepsilon_{cc}}\right)^2 + 2\left(\frac{\varepsilon_c}{\varepsilon_{cc}}\right)\right] \tag{5.26d}$$

(b) Harajli et al. (2006)

This model is valid for both circular and rectangular columns. The peak compressive strength of the element confined is:

$$f_{cc} = f'_{co} + k_1 f_{lat,f} + k_1 f_{lat,st} A_{cc}/A_g \tag{5.27a}$$

$$f_{lat,f} = \left(\frac{k_{ef} k_{vf} \rho_f E_f}{2}\right)\varepsilon_{lat} \tag{5.27b}$$

$$f_{lat,st} = \left(\frac{k_{es} k_{vs} \rho_{st}}{2}\right) f_{yt} \tag{5.27c}$$

$$k_1 = 1.25\left(\frac{f_{lat,f} + f_{lat,st}A_{cc}/A_g}{f_{co}}\right)^{-0.5} \tag{5.27d}$$

The peak axial strain is computed as:

$$\varepsilon_{cc} = \varepsilon_{co}\left[1 + k_2\left(\frac{f_{cc}}{f_{co}} - 1\right)\right] \tag{5.27e}$$

$$\text{where } \varepsilon_{co} = 0.002 \text{ and } k_2 = \left(\frac{25800e^{1.17h/b}}{\left(\rho_f E_f\right)^{0.83}}\right)\varepsilon_{lat} + 2.0 \tag{5.27f}$$

(c) Tastani et al. (2006, **2008)**
This model computes the ultimate strength of an FRP confined concrete element as given by Eqs. (5.3) and (5.6) for peak strength and the associated strain at the peak point. Strain capacity is given by Eq. (5.7) after amending for the effects of compaction due to collapse of the material structure according with Eq. (5.16). The effective lateral average strain in the jacket is obtained from Eq. (5.8) considering both the risk of debonding, stress concentrations at the corners in rectangular cross-sections, and the risk of bar buckling (Fig. 5.12). Lateral confining pressure explicitly accounts for both contributions of the FRP jacket and the embedded steel reinforcement; each mechanism of confinement has an effectiveness coefficient defined according with the codes (see section "Confinement models from design codes (CNR, ACI, EC8-III, fib)" and Eq. (5.19) for k_f).

The contribution of longitudinal compression reinforcement in resisting axial load is limited by considerations about bar buckling; here the methodology involves the following steps:

Step 1. For an axially compressed, FRP-confined reinforced concrete member under axial compressive strain of ε_c, the bar axial compressive stress, f_s, and tangent

Fig. 5.12 FRP jacket rupture due to bar buckling

modulus E_h, are estimated from the bar stress strain relationship (state determination). The critical normalized unsupported length is calculated as $r = 1.5\sqrt{E_h/f_s}$ and the value is checked against the available s/D_b value (where s is the spacing of stirrups and D_b the bar diameter). When $r > s/D_b$ there is no risk for bar buckling for the given strain magnitude, whereas if $r < s/D_b$ it is assumed that the compressive reinforcement has reached a state of instability.

Step 2. If the check above confirms bar instability, then it is assumed that the apparent Poisson's ratio ν_u is equal to 0.5 for circular and 1 for square cross sections (imminent attainment of peak strength). Denote the axial strain magnitude at this point as $\varepsilon_c = \varepsilon_{s,crit}$ and the corresponding steel stress, $f_{s,crit}$ (from the stress-strain law of the steel material, at a compressive strain equal to $\varepsilon_{s,crit}$). Calculate the lateral strain in the jacket as $\varepsilon_{lat}^{buckl} = \nu_u \cdot \varepsilon_{s,crit}$.

Step 3. Calculate the ensuing redistribution of load from the longitudinal reinforcing bars to the encased concrete core. The overload during stress transfer from the unstable bars to the encased core is:

$$\Delta f_{axial} = (f_{s,crit} - f_{s,res}) \cdot \frac{A_s}{A_{gross} - A_s} = \rho_{s,gross} \cdot (f_{s,crit} - f_{s,res}) \tag{5.28a}$$

where $f_{s,crit}$ is the steel stress at the strain value associated with the onset of instability (determined above) and $f_{s,res}$ is the residual capacity of the buckled reinforcement at an axial strain value $\varepsilon > \varepsilon_{s,crit}$. (Based on work by Monti and Nuti (1992), the asymptotic post-buckling residual stress $f_{s,res}$ is $6f_y/(s/D_b)$; over a plastic hinge length of l_p the value of s used in estimating the residual post-buckling capacity of the bar is taken equal to l_p in order to acknowledge the increased unsupported length of the bar.)

Step 4. The available strength reserve of the encased concrete core is:

$$\begin{aligned} &\Delta f_{cc} = 3.1 \cdot \Delta f_{lat} \\ &\Delta f_{lat} = k_f^{conf} \rho_{fv} E_f (\varepsilon_{cc,u} - \nu_u \varepsilon_{s,crit}) \frac{R_{ch} + 0.5D_b}{b - 2R_{ch}} \geq 0 \end{aligned} \tag{5.28b}$$

where $\varepsilon_{cc,u}$ is the strain capacity of the encased concrete (Eq. 5.17), R_{ch} is the radius of chamfering at the corners, ρ_{fv} the volumetric ratio of the FRP jackets, and b the side of the cross-section.

Step 5. If the overload (Δf_{axial}) exceeds the strength reserve of the core (Δf_{cc}) failure by buckling will be instantaneous. On the other hand, if the strength reserve of the core exceeds the overload, then the risk of failure due to buckling is suppressed to occur at higher strain levels, thereby enabling the concrete core to develop its full strain capacity before failure, i.e. $\varepsilon_{cc.u}$ as specified by Eq. (5.17).

(d) Ilki et al. (2008)

The authors of the reference work proposed two separate models for the evaluation of the peak strength and strain in case of internal steel reinforcement or external CFRP confinement. The CFRP contribution is given by:

$$\left[\frac{f_{cc}}{f_{co}}\right]_{CFRP} = 1 + 2.54\frac{f_{lat}}{f_{co}}$$
$$\left[\frac{\varepsilon_{cc}}{\varepsilon_{co}}\right]_{CFRP} = 1 + 19,27\frac{b}{h}(LTF)\left(\frac{f_{lat}}{f_{co}}\right)^{0,53} \tag{5.29a}$$

where

$$f_{lat} = \frac{1}{2}k_{\alpha}\rho_f E_f \beta \varepsilon_{fu} \tag{5.29b}$$

The internal steel reinforcement contribution is given by:

$$\left[\frac{f_{cc}}{f_{co}}\right]_{ITR} = 1 + 4.54\frac{f_{lat}}{f_{co}} \quad \text{and} \quad \left[\frac{\varepsilon_{cc}}{\varepsilon_{co}}\right]_{ITR} = 1 + 5\left(\frac{f_{cc}}{f_{co}} - 1\right) \tag{5.29c}$$

Hence the peak strength and corresponding strain at peak stress of an FRP confined RC element are given by:

$$\left[\frac{f_{cc} - f_{co}}{f_{co}}\right]_{TOTAL} = \left[\frac{f_{cc}}{f_{co}} - 1\right]_{CFRP} + \left[\frac{f_{cc}}{f_{co}} - 1\right]_{ITR} \tag{5.29d}$$

$$\left[\frac{\varepsilon_{cc} - \varepsilon_{co}}{\varepsilon_{co}}\right]_{TOTAL} = \left[\frac{\varepsilon_{cc}}{\varepsilon_{co}} - 1\right]_{CFRP} + \left[\frac{\varepsilon_{cc}}{\varepsilon_{co}} - 1\right] \tag{5.29e}$$

(e) Pellegrino and Modena (2010)
This model takes into account the contribution of the internal steel and external FRP and the interaction mechanisms between steel reinforcement and external FRP strengthening both in case of circular and rectangular confined columns. The value of the peak stress and the corresponding axial strain are given by:

$$\frac{f_{cc}}{f_{co}} = 1 + k_1 \cdot \frac{P_u}{f_{co}} \tag{5.30a}$$

$$\frac{\varepsilon_{cc}}{\varepsilon_{cu}} = 0.55 + 1.5(2R_{ch}/b) \text{ for } 2R_{ch}/b < 0.3 \text{ (for } 2R_{ch}/b \geq 0.3) \tag{5.30b}$$

The ultimate stress and ultimate strain are given by:

$$\frac{f_{cu}}{f_{cc}} = 0.55 + 1.5(2R_{ch}/b) \text{ for } 2R_{ch}/b < 0.3 \text{ (for } 2R_{ch}/b \geq 0.3) \tag{5.30c}$$

$$\frac{\varepsilon_{cu}}{\varepsilon_{co}} = 2 + B\left(\frac{P_u}{f_{co}}\right) \tag{5.30d}$$

where R_{ch} is the radius of chamfering at the corners of the cross section, b is the least sectional dimension of the column and $P_u = f_{lat,f} + f_{lat,st} \cdot A_{cc}/A_g$, where $f_{lat,f} = 0.5k_f \rho_f E_f \varepsilon_f^{eff}$ and $f_{lat,st} = 0.5k_s \rho_{st} f_{y,st}$. Coefficient B assumes different values when internal steel reinforcement is either present or absent. The effective hoop FRP strain ε_f^{eff} is $\varepsilon_f^{eff} = k_e \cdot \varepsilon_{fu}$ where k_e is the efficiency coefficient assuming different expressions for plain and reinforced columns:

$$k_e = 0.25 + 0.25(2R_{ch}/b)\text{: Without steel reinforcement} \tag{5.30e}$$

$$k_e = \gamma C^{-0.7} \leq 0.8, C = E_{y,long} \cdot \rho_{y,long}/(E_f \cdot \rho_f)\text{: With steel reinf.} \tag{5.30f}$$

The parameter C takes into account the negative effects of the stress concentrations in the FRP, due to buckling of vertical bars, by means of the ratio between the rigidity of the internal longitudinal steel and that of the external FRP.

(f) Kawashima et al. (2000)
This model distinguishes between columns with circular and rectangular cross-section taking into account the effect of the internal steel reinforcement. The stress-strain relation is given by:

$$f_c = \begin{cases} E_c \varepsilon_c \left\{ 1 - \frac{1}{n} \left(\frac{\varepsilon_c}{\varepsilon_t} \right)^{n-1} \right\} & 0 \leq \varepsilon_c \leq \varepsilon_t \\ f_t + E_g(\varepsilon_c - \varepsilon_t) & \varepsilon_t \leq \varepsilon_c \leq \varepsilon_{cu} \end{cases} \quad \text{if} E_g < 0 \tag{5.31a}$$

$$f_c = E_c \varepsilon_c \left\{ 1 - \frac{1}{n} \left(1 - \frac{E_g}{E_c} \right) \left(\frac{\varepsilon_c}{\varepsilon_t} \right)^{n-1} \right\} \quad \text{if } E_g > 0 \tag{5.31b}$$

$$E_g = -0.658 \frac{f_{co}^2}{\rho_{CF} \varepsilon_{CFt} E_{CF}} + 0.078 \sqrt{\rho_{CF}} E_{CF} \tag{5.31c}$$

$$\frac{f_t}{f_{co}} = 1 + 1.93 \rho_{cf} \varepsilon_{cf} E_{cf}/f_{co} \tag{5.31d}$$

$$\varepsilon_t = 0.00343 + 0.00939 \rho_{CF} \varepsilon_{CF} E_{CF}/f_{co} \tag{5.31e}$$

The ultimate axial strain is computed as:

$$\varepsilon_{cu} = 0.00383 + 0.1014 \left(\frac{\rho_{CF} \varepsilon_{CF} E_{CF}}{f_{co}} \right)^{\frac{3}{4}} \left(\frac{f_{CF}}{E_{CF}} \right)^{\frac{1}{2}} \tag{5.31f}$$

If the strengthened element includes steel reinforcement the values of E_g, f_t and ε_t become:

$$E_g = -0.658 \frac{f_{co}^2}{\rho_{CF}\varepsilon_{CFt}E_{CF} + 0.098\rho_s f_{yh}} + 0.078\sqrt{\rho_{CF}}E_{CF} \tag{5.31g}$$

$$\frac{f_t}{f_{co}} = 1 + 1.93\frac{\rho_{cf}\varepsilon_{cf}E_{cf}}{f_{co}} + 2.2\frac{\rho_s f_{yh}}{f_{co}} \tag{5.31h}$$

$$\varepsilon_t = 0.00343 + 0.00939\frac{\rho_{CF}\varepsilon_{CF}E_{CF}}{f_{co}} + 0.0107\frac{\rho_s f_{yh}}{f_{co}} \tag{5.31i}$$

Lacunae and Proposals of Improvement for Current Formulations and Models on Confinement of RC Elements by Means of EBR FRP Systems

New Proposals for Predictive Models

In order to improve the accuracy of some current confinement's models, such as the ones considered by CNR-DT 200 (2004) and ACI 440.2R (2008), Realfonzo and Napoli (2011, 2013) developed new relationships for estimating the strength (f_{cc}) and the ultimate strain capacity ($\varepsilon_{cc,u}$) of concrete confined by FRP. The relationships were found by best-fitting experimental results collected in the database of Realfonzo and Napoli (2011). The proposed formulations are as follows:

Strength Models

For the best-fit analyses, the predictive model proposed by Wu and Wang (2009) was considered, expressed by:

$$\bar{f}_{cc} = \frac{f_{cc}}{f_{co}} = 1 + \alpha \cdot k_s \cdot \rho^{\gamma} \cdot \bar{f}_{lat}^{\beta} \tag{5.32}$$

In Eq. (5.32) α, β, and γ are three unknown parameters to be calibrated through the best-fit analysis with the purpose of minimizing the difference between the predicted and experimental strength capacity; $\bar{f}_{cc}$ is the compression strength of FRP confined concrete normalized with respect to the strength of unconfined concrete (f_{co}); k_s is the shape factor defined according to CNR-DT 200 (2004); ρ is the radius corner ratio as defined by Wu and Wang (2009); $\bar{f}_{lat}$ is the normalized confining pressure exerted by the jacket, f_{lat}/f_{co}:

$$f_{lat} = 2 \cdot t_f \cdot E_f / D \cdot (k_e \cdot \varepsilon_{fu}) = k_{conf} \cdot \varepsilon_f^{eff} \quad (5.33)$$

In Eq. (5.33), depending on the shape of the cross section, D is the cylinder diameter or the side of the square, while the product ($k_e \varepsilon_{fu}$) represents the actual ultimate hoop strain reached in the FRP jacket (ε_f^{eff}). By focusing the study on concrete cylinders, Eq. (5.32) simplifies into the following equation:

$$\bar{f}_{cc} = 1 + \alpha \cdot \bar{f}_{lat}^{\beta} \quad (5.34)$$

as, in this case, the coefficient k_s and ρ become equal to 1. Several indices on the "success of fit" can be used to find the parameters α and β that yield the lowest scatter between the experimental ($\bar{f}_{cc}^{exp}$) and theoretical ($\bar{f}_{cc}^{th}$) values of the normalized strength. For these analyses, both the mean square error (MSE) method and the mean absolute percentage error (MAPE) method were considered. According to these criteria, the following errors have to be minimized:

$$(E_{rr})_m^{MSE} = \sum_{i=1}^{n_s} E_i / n_s, \quad \text{with} \quad E_i = \left(\bar{f}_{cc,i}^{exp} - \bar{f}_{cc,i}^{th} \right)^2 \quad (5.35)$$

$$(E_{rr})_m^{MAPE} = \sum_{i=1}^{n_s} |E_i| / n_s, \quad \text{with} \quad E_i = \left(\frac{\bar{f}_{cc,i}^{exp} - \bar{f}_{cc,i}^{th}}{\bar{f}_{cc,i}^{exp}} \right) \cdot 100 \quad (5.36)$$

where n_s is the number of experimental test sets considered in the analysis; they were extracted from the general database as defined in Realfonzo and Napoli (2011).

The values of α and β introduced in Eq. (5.34) which minimize the errors $(\mathrm{Err})_m$ computed through Eqs. (5.35) and (5.36) are listed in Table 5.3. The table also reports, for each analysis, the α values obtained in case of $\beta = 1$ (i.e. using a linear law for $\bar{f}_{cc}$). Results summarized in Table 5.3 refer to best fit analyses performed by considering only test specimens belonging to the LSC group (i.e. to test specimens with $f_{co} \leq 40$ MPa), which represent the category of primary interest in the practical

Table 5.3 Values of α and β obtained through best-fit analysis (data sets with $f_{co} \leq 40$ MPa)

Type of fiber	k_{em}	MSE					MAPE				
		Linear		Non linear			Linear		Non linear		
		α	$(\mathrm{Err})_m$	α	β	$(\mathrm{Err})_m$	α	$(\mathrm{Err})_m$	α	β	$(\mathrm{Err})_m$
CFRP	0.63	3.75	0.229	3.37	0.78	0.179	3.8	13.72	3.67	0.96	13.69
$n_s = 135$	0.65	3.66	0.229	3.31	0.78	0.179	3.71	13.72	3.59	0.96	13.69
GFRP	0.68	3.3	0.257	3.05	0.78	0.209	3.08	13.29	3.00	0.96	13.23
$n_s = 53$	0.65	3.46	0.257	3.17	0.78	0.209	3.23	13.29	3.14	0.96	13.23
ALL $n_s = 188$	0.65	3.61	0.238	3.27	0.78	0.188	3.59	13.88	3.56	0.98	13.87

applications. In such analyses, the average values of the strain efficiency factor (k_{em}) adopted for the estimate of the FRP confining pressure were assumed equal to 0.65 (i.e. the average value obtained by considering all 91 specimens) or, alternatively, equal to 0.63 and 0.68 for CFRP and GFRP systems, respectively. Results of the best fit analyses performed by considering all test sets, i.e. regardless of the f_{co} value, can be found in Realfonzo and Napoli (2011).

From Table 5.3 it is observed that the values of α and β are sensitive to the type of method adopted for the best-fit analysis (MAPE or MSE); also, although the best fitting of experimental data is always given by nonlinear relationships, linear laws can be still assumed through the MAPE method, since similar errors were obtained in this case with linear and non-linear relationships.

Figure 5.13 shows the experimental ($\bar{f}_{cc} - \bar{f}_{lat}$) values and the corresponding theoretical relationships; in the figure the nonlinear and linear laws obtained from MSE and MAPE are plotted, respectively. The best-fit laws obtained from MSE and MAPE methods for the LSC range are shown in Fig. 5.14a. In the figure, the relationships adopted by CNR-DT 200 (2004), ACI 440.2R (2008), and fib Bulletin 14 (2001)—described in section "Confinement models from design codes (CNR, ACI, EC8-III, fib)"—are also plotted for comparison:

$$\bar{f}_{cc} = 1 + 2.6\bar{f}_{lat}^{2/3} \tag{5.37a}$$

$$\bar{f}_{cc} = 1 + 3.3\bar{f}_{lat} \tag{5.37b}$$

$$\bar{f}_{cc} = 0.2 + 3\bar{f}_{lat}^{1/2} \tag{5.37c}$$

The experimental $\bar{f}_{cc}$ values ($\bar{f}_{cc}^{exp}$) are compared to the theoretical ones ($\bar{f}_{cc}^{th}$) in Fig. 5.14b. The bisector corresponds to the equal value line; therefore, points falling in the lower part of the graph indicate conservative estimations, whereas points falling over the line represent unconservative results.

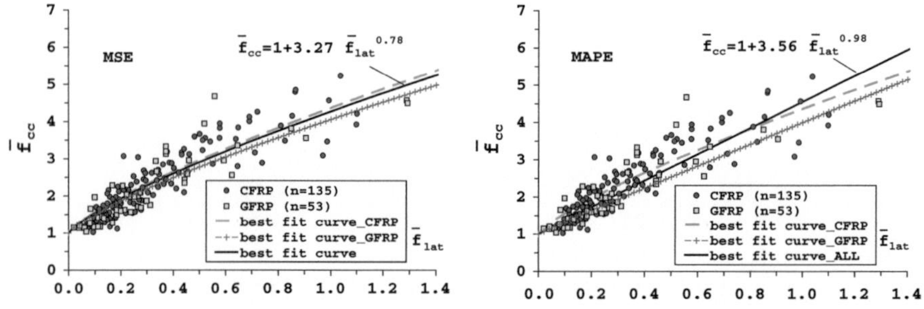

Fig. 5.13 Experimental values versus theoretical $\bar{f}_{cc} - \bar{f}_{lat}$ curves (LSC test sets only)

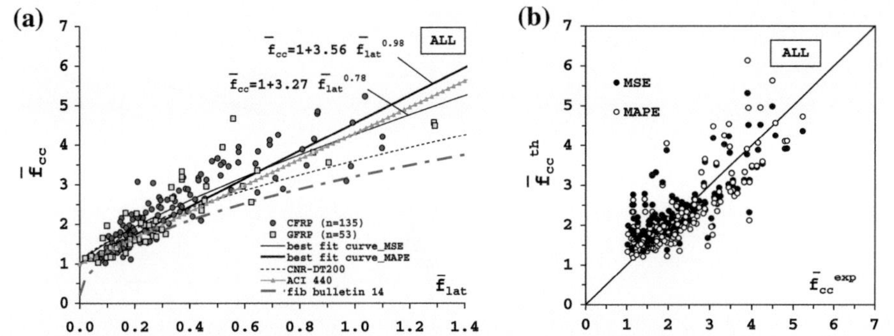

Fig. 5.14 **a** Experimental values versus theoretical $\bar{f}_{cc} - \bar{f}_{lat}$ curves, and **b** $\bar{f}_{cc}^{th}$ versus $\bar{f}_{cc}^{exp}$ comparisons (LSC test sets only)

From Fig. 5.14 it is concluded that the MAPE method leads to more conservative predictions of $\bar{f}_{cc}$ than the MSE one. Also, comparisons between the best fit curve obtained by using the MAPE method and the three code relationships confirm that:

- the CNR-DT 200 (2004) law (Eq. 5.37a) generally leads to slightly unconservative values of $\bar{f}_{cc}$ for normalized confining pressures approximately lower than 0.4, while it is more conservative than the MAPE one for higher values of $\bar{f}_{lat}$;
- the ACI formulation (Eq. 5.37b) and the MAPE law provide very similar values for confined concrete strength;
- the "practical formula" suggested by fib (Eq. 5.37c) is the most conservative relationship over the $\bar{f}_{lat}$ range.

Finally, by focusing on data sets relative to specimens with concrete strengths $f_{co} \leq 40$ MPa (category of primary interest in the practical applications) and based on the results obtained in Table 5.3, the two following general relationships can be assumed as suitable predictive strength models:

$$\bar{f}_{cc} = 1 + 3.27 \cdot \bar{f}_{lat}^{0.78} \text{ (MSE)} \tag{5.38a}$$

$$\bar{f}_{cc} = 1 + 3.59 \cdot \bar{f}_{lat} \text{ (MAPE)} \tag{5.38b}$$

The linear relationship expressed by Eq. (5.38b) can be used in place of the nonlinear best-fit one since the computed values of $(Err)_m$ are almost the same.

Strain Models

Lam and Teng (2003) consider a member sufficiently confined when the ultimate value of the "effective" normalized confining pressure $\bar{f}_{lat}$ provided by the FRP jacket exceeds the limit of 7 %:

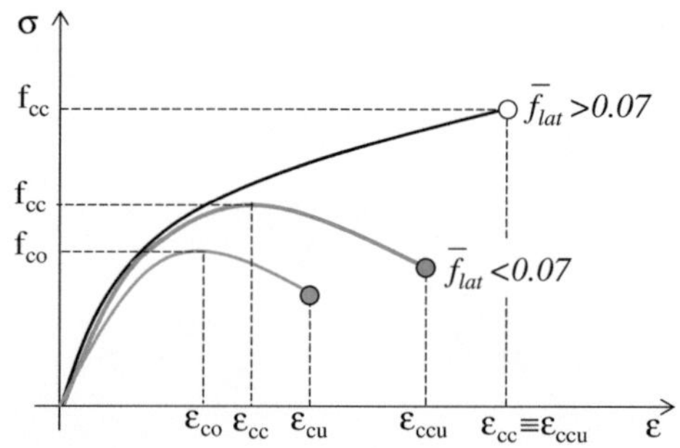

Fig. 5.15 Typical stress–strain behaviour of FRP-confined concrete

$$\bar{f}_{lat} = \frac{f_{lat}}{f_{co}} = \frac{k_{conf} \cdot (k_e \cdot \varepsilon_{fu})}{f_{co}} = \frac{k_{conf} \cdot \varepsilon_f^{eff}}{f_{co}} = 2\frac{t_f \cdot E_f}{D} \cdot \frac{\varepsilon_f^{eff}}{f_{co}} \tag{5.39}$$

Figure 5.15 depicts the qualitative trend of the stress-strain curve exhibited by an FRP confined concrete in compression. Over the 7 % threshold the stress-strain behaviour has the characteristic monotonically bilinear ascending curve and the ultimate strain $\varepsilon_{cc,u}$ is attained at the achievement of the peak compressive strength, f_{cc} (i.e. $\varepsilon_{cc,u} \equiv \varepsilon_{cc}$); instead, when the confining pressure is below such threshold the constitutive curve shows a post-peak softening branch and the maximum strength f_{cc} is reached before the FRP rupture (therefore $\varepsilon_{cc,u} > \varepsilon_{cc}$).

Realfonzo and Napoli (2013) calibrated three of the familiar strain models; these were the models of the Italian Guidelines CNR-DT 200 (2004), the model by Teng et al. (2009), and the model by De Lorenzis and Tepfers (2003):

$$\bar{\varepsilon}_{cc,u} = 1.75 + 7.5 \cdot \bar{f}_{lat}^{0.5} \tag{5.40a}$$

$$\bar{\varepsilon}_{cc,u} = 1.75 + 6.5 \cdot \bar{f}_{lat}^{0.8} \cdot \bar{\varepsilon}_f^{eff\,0.65} \tag{5.40b}$$

$$\bar{\varepsilon}_{cc,u} = 1 + 26.2 \cdot \bar{f}_{lat}^{0.8} \cdot k_{conf}^{-0.148} \tag{5.40c}$$

In Eqs. (5.40a)–(5.40c) $\bar{\varepsilon}_{cc,u}$ and $\bar{\varepsilon}_f^{eff}$ are given by:

$$\bar{\varepsilon}_{cc,u} = \frac{\varepsilon_{cc,u}}{\varepsilon_{co}} \qquad \bar{\varepsilon}_f^{eff} = \frac{\varepsilon_f^{eff}}{\varepsilon_{co}} \tag{5.40d}$$

i.e., they are obtained by normalizing the values of $\varepsilon_{cc,u}$ and ε_f^{eff} with respect to the strain ε_{co} (assumed equal to 0.2 %) corresponding to the attainment of peak stress (f_{co}) for unconfined concrete (Fig. 5.15).

Note that the constant 1.75 in Eqs. (5.40a) and (5.40b) yields $\varepsilon_{cc,u} = 0.0035$ for $\bar{f}_{lat} = 0$ (i.e. the ultimate value commonly accepted for unconfined concrete). On the other hand Eq. (5.40c) yields $\varepsilon_{cc,u} = \varepsilon_{co}$ when no FRP confinement is provided, which is the axial strain at peak stress of unconfined concrete. The framework of the three models is generalized as follows:

$$\bar{\varepsilon}_{cc,u} = 1.75 + C_1 \cdot \bar{f}_{lat}^{\;C_2} \tag{5.41a}$$

$$\bar{\varepsilon}_{cc,u} = 1.75 + C_1 \cdot \bar{f}_{lat}^{\;C_2} \cdot \bar{\varepsilon}_f^{eff\,C_3} \tag{5.41b}$$

$$\bar{\varepsilon}_{cc,u} = 1 + C_1 \cdot \bar{f}_{lat}^{\;C_2} \cdot k_{conf}^{\;C_3} \tag{5.41c}$$

the C_i coefficients (C_1, C_2, C_3) were estimated by using the mentioned database. In particular, in Realfonzo and Napoli (2013) the selected three strain models were recalibrated through best-fit analyses and then compared with the relationships proposed in literature. As done for the strength models, coefficients C_1, C_2, C_3 that yield the lowest scatter between experimental ($\bar{\varepsilon}^{exp}_{cc,u}$) and theoretical values ($\bar{\varepsilon}^{th}_{cc,u}$) of the normalized ultimate strain were calibrated by using the MSE and MAPE methods. Table 5.4 provides the outcomes of the best-fit analyses performed by restricting the range of interest to test sets with $f_{co} \leq 40$ MPa.

For the three strain models found in the literature and described by Eqs. (5.41a)–(5.41c), Table 5.4 reports the computed values of the mean error $(Err)_m$, the standard deviation of the error (σ_{Err}), and the coefficient of variation ($CV_{(Err)}$) resulting by evaluating the errors E_i according to both MAPE and MSE; for the three "predictive" models given by Eqs. (5.41a)–(c) it provides the C_i coefficients calibrated by minimizing the errors $(Err)_m$ with both MAPE and MSE methods and the resulting values of σ_{Err} and $CV_{(Err)}$, all highlighted in bold. In detail: model 1*α* refers to the relationship proposed by CNR-DT 200 (2004) and model 1*b* is the recalibrated one through the best-analysis; model 2*α* is that proposed by Teng et al. (2009), while relationship 2*b* is the corresponding recalibrated one; similarly, model 3*α* refers to the formulation proposed by De Lorenzis and Tepfers (2003) and relationship 3*b* is the best-fit one. The study was carried out more times by considering: (a) results from tests on CFRP confined specimens ($n_s = 88$); (b) results from tests on GFRP confined specimens ($n_s = 42$); (c) results from tests on specimens confined with any FRP type ($n_s = 130$). Following the recommendations by Lam and Teng (2003) the best-fit analyses were extended to only test sets for which the $\bar{f}_{lat}$ values, computed by using $k_{em} = 0.60$ in Eq. (5.33) were greater than 7 %. Conclusions drawn from Table 5.4 are as follows:

- regardless of the used strain model, the errors computed in the estimation of $\bar{\varepsilon}_{cc,u}$ are significantly higher than those obtained in the estimate of $\bar{f}_{cc}$; this is also due to the wider range of percent variation of $\bar{\varepsilon}_{cc,u}$ when compared to $\bar{f}_{cc}$;
- the errors significantly increase when test sets referring to GFRP systems are included;
- comparing each strain model (1*α*, 2*α* and 3*α*) with the re-calibrated one (1*b*, 2*b*, and 3*b*) a rather significant reduction of the mean errors is observed mainly in

Table 5.4 Results of the best-fit analyses and comparison among the strain models

Type of fiber	Model		MSE						MAPE					
			C1	C2	C3	$(Err)_m$	$\sigma_{(Err)}$	$CV_{(Err)}$	C1	C2	C3	$(Err)_m$	$\sigma_{(Err)}$	$CV_{(Err)}$
CFRP	1	α	7.50	0.50	–	15.05	25.69	1.71	7.50	0.50	–	33.68	22.26	0.66
$n_s = 88$		b	13.19	0.60	–	7.66	14.47	1.89	14.43	0.85	–	30.20	28.56	0.95
$k_{em} = 0.63$	2	α	6.50	0.80	0.65	12.82	43.59	3.40	6.50	0.80	0.65	32.02	37.04	1.16
		b	8.32	0.56	0.29	7.17	12.83	1.79	8.60	0.80	0.32	29.08	25.09	0.86
	3	α	26.20	0.80	−0.148	21.02	29.52	1.40	26.20	0.80	−0.148	40.81	16.30	0.40
		b	20.97	0.58	−0.05	7.54	14.14	1.87	39.29	0.85	−0.12	29.48	28.17	0.96
GFRP	1	α	7.50	0.50	–	48.76	80.70	1.66	7.50	0.50	–	55.63	34.74	0.62
$n_s = 42$		b	11.48	0.25	–	30.62	44.93	1.47	7.71	0.57	–	55.27	30.86	0.56
$k_{em} = 0.68$	2	α	6.50	0.80	0.65	116.81	277.51	2.38	6.50	0.80	0.65	104.56	127.88	1.22
		b	44.27	0.25	−0.73	28.37	45.99	1.62	16.15	0.61	−0.41	49.84	30.59	0.61
	3	α	26.20	0.80	−0.148	59.33	91.17	1.54	26.20	0.80	−0.148	55.92	24.41	0.44
		b	2.68	0.11	0.21	27.71	38.02	1.37	26.19	0.69	−0.14	55.64	31.87	0.57
ALL	1	α	7.50	0.50	–	25.94	53.06	2.05	7.50	0.50	–	40.70	28.40	0.70
$n_s = 130$		b	12.34	0.46	–	16.61	34.58	2.08	12.07	0.79	–	39.13	31.79	0.81
$k_{em} = 0.65$	2	α	6.50	0.80	0.65	42.93	146.56	3.41	6.50	0.80	0.65	54.10	80.16	1.48
		b	9.77	0.41	0.01	16.77	35.15	2.10	9.75	0.68	−0.01	38.45	28.11	0.73
	3	α	26.20	0.80	−0.148	33.31	60.34	1.81	26.20	0.80	−0.148	45.22	20.24	0.45
		b	10.85	0.40	0.02	16.54	34.36	2.08	8.52	0.67	0.06	38.61	32.61	0.84

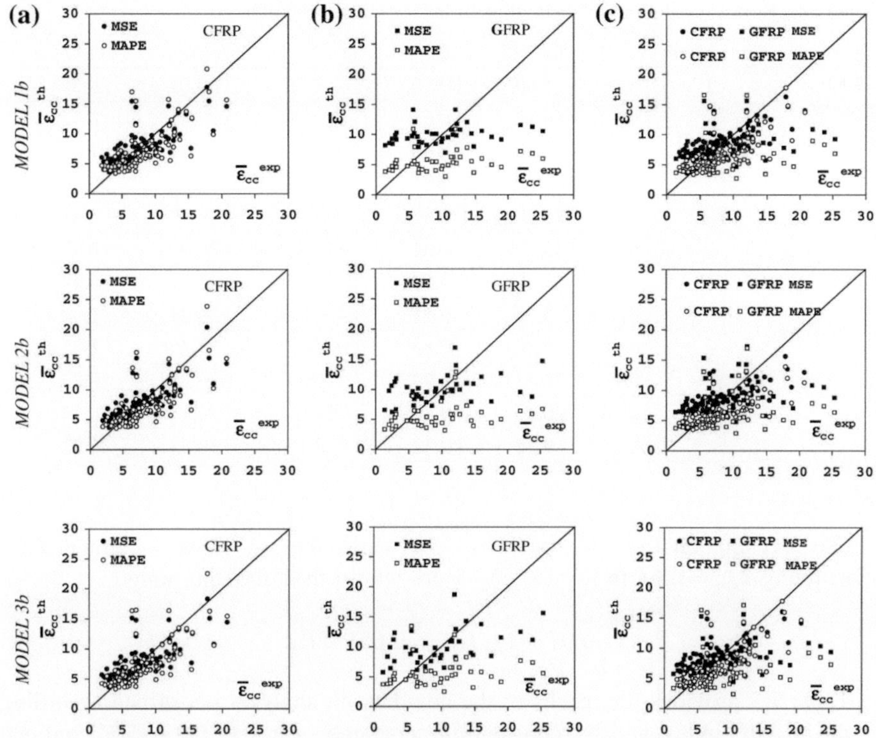

Fig. 5.16 $\bar{\varepsilon}_{cc,u}^{th}$ versus $\bar{\varepsilon}_{cc,u}^{\exp}$ comparisons for: **a** CFRP; **b** GFRP, and **c** any type of FRP systems

the case of GFRP systems, although for some models such reduction is associated to higher $CV_{(Err)}$ values.

Further comparison among the strain models best-fitted with MAPE and MSE (models 1*b*, 2*b*, and 3*b* in Fig. 5.16) mainly in case of GFRP systems, and disregarding the used predictive model, illustrates that the best-fit analyses performed with MAPE provide more conservative values of the normalized ultimate strain with respect to those obtained through MSE. Furthermore, as already evidenced by other authors (Lam and Teng 2003) a large scatter is always observed for GFRP wrapped specimens (for more detail see Realfonzo and Napoli 2013). To further improve the accuracy of the strain models given by Eqs. (5.41a)–(c)—given the lower reliability with respect to the strength models—in a second step Realfonzo and Napoli (2013) performed the best-fit analyses once again, by considering the following restrictions:

(1) Application of the MAPE index only generally yields more conservative values (Fig. 5.16).

(2) Use of test sets belonging to the LSC and MSC range only ($f_{co} \le 40$ MPa), being of more interest in practical applications;

Table 5.5 Results of the refined best-fit analyses for ($f_{co} \leq 40$ MPa)

Type	Model		C_1	C_2	C_3	$(Err)_m$	$CV_{(Err)}$
CFRP	1	α	7.50	0.50	–	30.64	0.54
$n_s = 86$		b	14.12	0.83	–	27.31	0.92
$k_{em} = 0.65$	2	α	6.50	0.80	0.65	29.64	1.22
		b	8.40	0.81	0.36	26.26	0.96
	3	α	26.20	0.80	0.148	39.87	0.41
		b	35.61	0.82	−0.11	26.61	0.97
GFRP	1	α	7.50	0.50	–	46.49	0.44
$n_s = 34$		b	14.39	0.55	–	39.24	0.90
$k_{em} = 0.65$	2	α	6.50	0.80	0.65	51.30	1.25
		b	31.42	0.50	−0.41	35.85	0.85
	3	α	26.20	0.80	−0.148	54.17	0.34
		b	9.36	0.42	0.061	38.53	0.92

(3) Focusing on the 120 specimens for which the $\bar{f}_{lat}$ values—computed by substituting $k_{em} = 0.65$ in Eq. (5.33)—were within the following range:

$$0.07 \leq \bar{f}_{lat}(k_{em} = 0.65) \leq 1.2 \tag{5.42}$$

Table 5.5 provides the results of the new best-fit analyses performed according to the MAPE index and by considering separately test sets for CFRP confined specimens ($n_s = 86$) and those for GFRP confined ones ($n_s = 34$). In particular: for the models given by Eqs. (5.40a)–(d) (models 1α, 2α, and 3α), the table reports the C_i coefficients, the values of the mean error $(Err)_m$, and the coefficient of variation $CV_{(Err)}$; for the models given by Eqs. (5.41a)–(c) (models 1b, 2b, and 3b) it highlights in bold the recalibrated values of the C_i coefficients and the resulting values of $(Err)_m$ and $CV_{(Err)}$.

By observing the results of Table 5.5 any of the three relationships may be selected as suitable strain model for CFRP/GFRP confined members. However, with the aim to preserve the structure of the relationship actually adopted by CNR-DT 200 (2004) the two following formulations were proposed by Realfonzo and Napoli (2013):

$$\bar{\varepsilon}_{cc,u} = 1.75 + 14.12 \cdot \bar{f}_{lat}^{0.83} \ (CFRP\,systems) \tag{5.43a}$$

$$\bar{\varepsilon}_{cc,u} = 1.75 + 14.39 \cdot \bar{f}_{lat}^{0.55} \ (GFRP\,systems) \tag{5.43b}$$

Confinement of Wall-like Cross-Sections

Wall-like columns (i.e. column elements with a cross sectional aspect ratio more than 3) represent a solution often adopted in reinforced concrete (RC) building structures in regions of high seismicity, due to their intrinsic ability to resist lateral load and their high stiffness in the long-side direction. In concrete columns with circular cross-section, the effectiveness of confinement is optimal, as the column cross-section is entirely and uniformly restrained by the FRP laminate. In the case of concrete columns with prismatic cross-section (that is square or rectangular), the confining pressure provided by the FRP jacket is high at the corners and low along the flat sides, and the cross-section is only partially confined (Mander et al. 1988; Tastani and Pantazopoulou 2008; De Luca et al. 2011).

Specifically, wall-like columns are of particular interest as, in this case, the effectiveness of the FRP laminate may be low. In fact, formulae provided by Codes and Guidelines to predict ultimate stress and strain of FRP confined concrete cannot be usually applied to the case of wall-like columns as they are limited to elements with cross-sections with a side-aspect-ratio not larger than 2.0.

Based on triaxial plasticity solid mechanic models (e.g. Willam and Warnke, 1975) calibrated experimentally (e.g. Elwi and Murray 1979) a simplified confinement model was provided for wall-like cross-sections (Lignola et al. 2011a, b). The confining stress field is assumed only parallel to the longer side of the cross-section, thus neglecting the confinement in the shorter direction. The average confining pressure (f_{lat}) is obtained with reference to Fig. 5.17a as:

$$f_{lat} = 2t_{FRP}E_f\varepsilon_{FRP}/b \tag{5.44a}$$

assuming cross-section height $h >$ base b. Assuming zero stress for the minimum principal stress, the confining pressure f_{lat} equal to the intermediate principal stress,

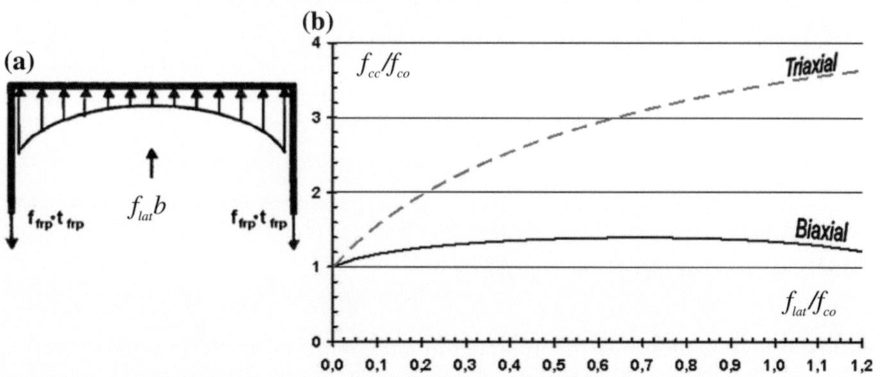

Fig. 5.17 **a** Free body diagram to evaluate the average confining pressure in concrete. **b** "Biaxial" and "Triaxial" confinement

and f_{cc} the maximum principal stress, the following approximate equation is derived for the solution describing the concrete ultimate strength surface:

$$\frac{f_{cc}}{f_{co}} = 1 + 1.42\frac{f_{lat}}{f_{co}} - 1.40\left(\frac{f_{lat}}{f_{co}}\right)^2 + 0.30\left(\frac{f_{lat}}{f_{co}}\right)^3 \tag{5.44b}$$

where $f_{lat}/f_{co} < 1.3$, otherwise it would result that $f_{lat} > f_{cc}$. Figure 5.17b compares the results from the general "triaxial" axisymmetric model (supported by confined concrete cylinder tests) with those of the "Biaxial" confinement model considered herein. Recall that Eq. (5.44b) is the closed form solution for Willam and Wranke's failure criterion under "Triaxial" axisymmetric stress (confinement). It is also noted that in all cases of "Biaxial" confinement, the increase in concrete strength, f_{cc}/f_{co}, is always less than 1.5 (see the biaxial failure envelope by Kupfer and Gerstle 1973).

Figure 5.18 presents results from the proposed model used to correlate relevant experimental results reported by Tan (2002), Maalej et al. (2003), Prota et al. (2006), and De Luca et al. (2013) both in terms of strength and lateral strain in the jacket near failure. The figure plots results obtained using for strength of the FRP material the nominal values provided by the manufacturers from flat coupon tests. Most of the experimental tests suggested that the failure of these walls and corresponding bulging of the FRP laminates occurred at fiber strains far below the ultimate values provided by the manufacturers (Prota et al. 2006). Although the proposed results fall far below the "triaxial" theoretical confinement curve (as expected) they deviate from the proposed "biaxial" theoretical confinement curve, mainly in terms of the estimated abscissa (i.e. the f_{lat}/f_{co} value), rather than in terms of the ordinate (i.e. the strength increment, f_{cc}/f_{co}, which is almost compatible and in all cases less than 1.5). In this case the average overestimation of the proposed model is about 39 % with a coefficient of variation equal to 49 %. In any case the analytical points are totally inconsistent with respect to the tests.

To explore the discrepancy in the estimated lateral pressure, the effective ultimate strain recorded during experimental tests were used in the model (see Fig. 5.18b). The average overestimation of the proposed model is about 5.7 % with a coefficient of variation equal to 10 %. The confinement model for wall-like cross-section can also be seen as a lower bound for confinement of square and rectangular cross-sections (Lignola et al. 2013a) and is used in the next subsection to assess behaviour of confined non-circular hollow cross-sections (Lignola et al. 2009).

Confinement of Hollow Sections

Hollow bridge sections are incorporated in tall columns (mainly in bridge piers) to maximize the structural efficiency of the strength-mass and stiffness-mass ratios. Hollow columns resist the high moment and shear demands through reduced

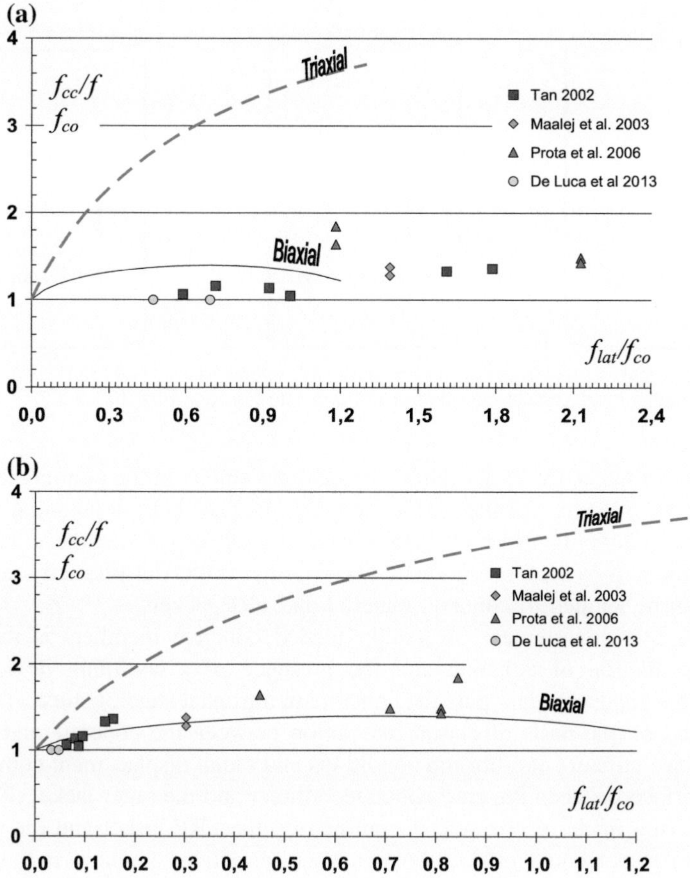

Fig. 5.18 **a** Calculated versus Experimental strength estimates (using nominal value of FRP strain capacity). **b** Results obtained using the effective FRP ultimate strain

self-weight and moderated bearing demands on foundations. The effectiveness of FRP jacketing as a confinement device in this type of element is modelled by Lignola et al. (2008) and Giamundo et al. (2014); the model can be extended to the case of solid cross-sections. An index of relative confinement stiffness is defined as $E_f t_f/(R_e - R_i)$ in order to account for the presence of the void in the cross-section (where R_e and R_i are outer and inner radii, respectively). The larger the hole, the higher is the deformability of the element thus resulting, for a similar level of dilation, in different stress paths: in the case of a solid section, the dilation of concrete is restrained by the FRP wraps and this interaction causes a strength enhancement, while in the case of thin walls, the excessive deformability allows for gaining smaller strength improvements (compared to solid sections) even though significant ductility enhancement is achieved. In this case the state of stress becomes mostly circumferential.

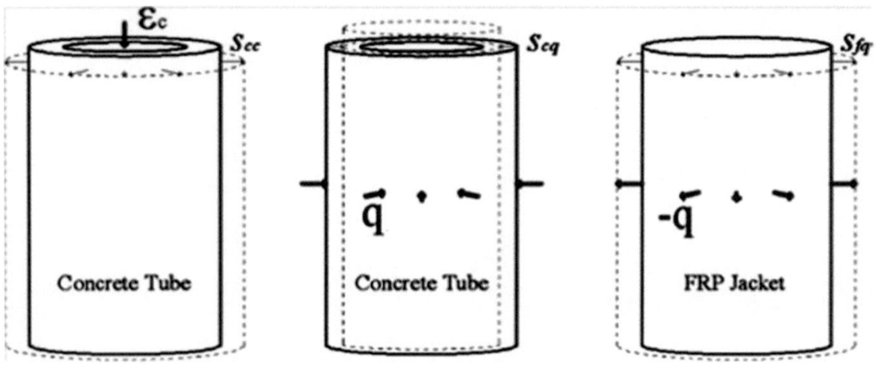

Fig. 5.19 Radial displacement contributions of concrete tube and FRP Jacket

The model traces the evolution of stresses and strains in the confinement wraps and concrete, allowing for the multiaxial state of stress and the potential failure of the external wrapping to be evaluated at each load step. Once the complete 3D stress strain step-by-step evolution is known, any multiaxial plasticity and failure criteria can be applied to concrete material and FRP jackets.

The passive confinement on axially loaded concrete members is due to the transverse dilation of concrete and the presence of a confining device which opposes this expansion and puts the concrete in a triaxial state of stress. The model is proposed on the basis of elastic interaction between the concrete and the confining device utilizing equilibrium conditions and radial displacement compatibility at the interface between the core concrete cylinder and the outer jacket (Fig. 5.19). The thick ring made of concrete is confined by the FRP jacket and the confining stress field is not equal in the two transverse directions (namely radial and circumferential). The effect of confinement is evaluated in each point of the section with the effective confining pressures which differ in the two orthogonal directions, resulting in a large computational effort. In the model proposed by Lignola et al. (2008) this aspect is solved by adopting the triaxial plasticity solid mechanic model (e.g. William and Warnke 1975) calibrated experimentally (e.g. Elwi and Murray 1979) to account for these different confining components.

The model allows for the axial and lateral stress versus strain curve to be evaluated. A solid section and hollow sections with different $\beta = R_i/R_e$ ratios, but the same relative confinement stiffness $E_f t_f/(R_e - R_i)$, can be compared.

A concrete specimen with an external radius R_e = 200 mm and unconfined concrete strength of 30 MPa is considered. The specimen was confined with a Carbon FRP wrap made of uniaxial fibers with a nominal thickness of 0.666 mm. The CFRP is applied by manual wet lay-up and the elastic modulus and ultimate uniaxial strength are 230 GPa and approximately 2000 MPa, respectively. The relative confinement stiffness (related to a solid section) is approximately 766 MPa (this means that when increasing the hole size, thickness t_f is reduced to keep the stiffness constant).

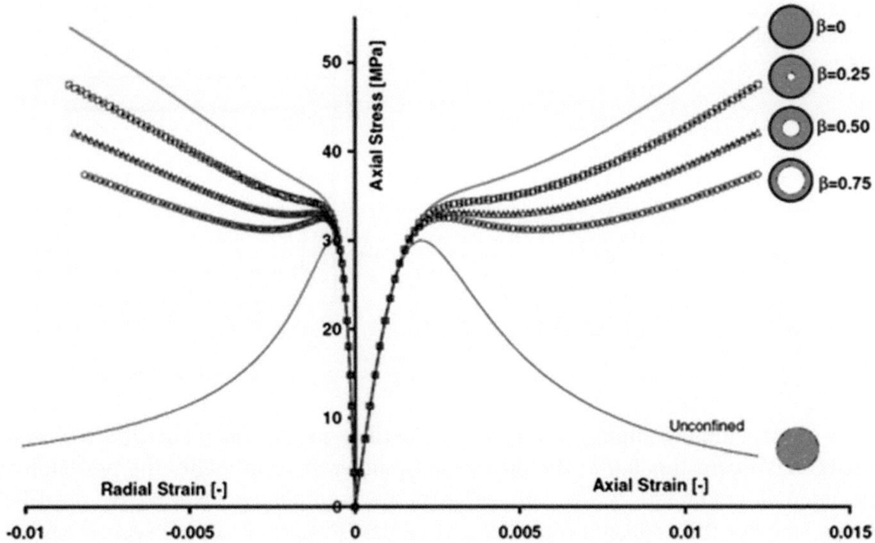

Fig. 5.20 Confined concrete: axial stress versus axial and radial strain

In Fig. 5.20 a comparison is given of the unconfined concrete stress strain curve with the confined concrete curves for different β ratios, but the same relative confinement stiffness $E_f t_f/(R_e - R_i)$. Compared to the solid section ($\beta = 0$), the effect of confinement is smaller as the hole size increases. The behaviour changes from a hardening post peak to an almost perfectly plastic behaviour in the case of very thin walls $\beta = 0.75$). Also the strength enhancement is smaller as the β ratio increases.

Generally, hollow members fail due to two premature mechanisms: compressed bars buckling and unrestrained concrete cover spalling. FRP confinement, especially in rectangular cross-sections, is not able to change the failure mode, but bar buckling is delayed and compressed concrete attains higher strains, thus resulting in higher load carrying capacity of the member.

In non-circular hollow cross-sections the effect of confinement on the walls composing the cross-section is reduced. The arch-shaped paths of the confining stresses rapidly changes in a straight distributed confinement stress field moving away from the corner, and this result is similar to what can be seen in tests of wall-like cross-sections. However, despite the similar path of confining stress directions, the effect of the confining device on a single wall-like column is higher than that exerted by adjacent walls (through the corners) and by a single wall which is part of the hollow cross-section.

Two alternatives have been proposed accounting for these two aspects (Lignola et al. 2009). In the first option each wall forming the hollow cross-section is considered separately accounting for the presence of the two adjacent walls on the opposite short sides. This very simple approach is rather good to assess strength enhancements, while it fails to predict the post-peak behaviour of the hollow

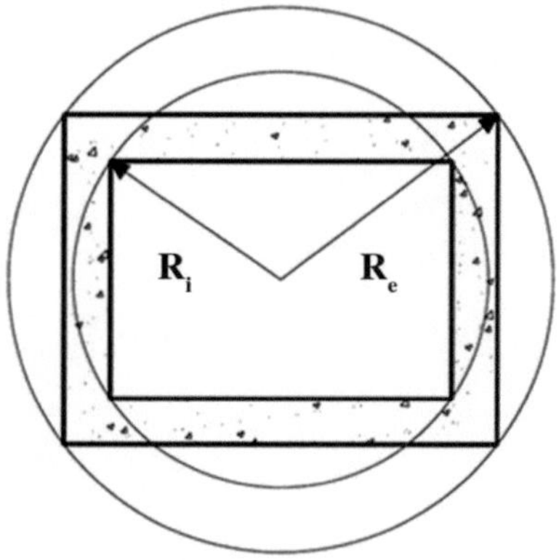

Fig. 5.21 Circular crown circumscribing the hollow rectangular cross-section

cross-section thus resulting in inaccurate ductility predictions. The effect of confinement is overestimated in the post-peak branch where in reality the presence of the internal void reduces the efficiency of the confinement exerted by the FRP wraps, and for this reason a more refined model is required. In the second option, instead, the confinement of the whole hollow cross-section forming the structural member is considered. The commonly accepted approach for dealing with confined concrete in noncircular cross-sections is to find an equivalent pressure in an equivalent circular cross-section (Teng et al. 2002). For hollow members Lignola et al. (2011b) suggested using an equivalent circular crown circumscribing the hollow section (internal R_i and external R_e radii are half diagonal inner and outer dimensions of the section, respectively), Fig. 5.21.

Bar Buckling in FRP Strengthened RC Columns

Interaction between stirrups (if available), external FRP jacketing and embedded longitudinal reinforcement has been the motivating premise in some of the models presented already in section "Confined strength formulae for jacketed reinforced concrete columns—Considering the interaction with reinforcement". Here, buckling of column longitudinal reinforcement is revisited due to its role in limiting the effectiveness of the jackets in lightly reinforced RC columns, a finding that renders this consideration a critical design issue in strengthening of RC columns through FRP jacketing. It was stated that although bar buckling cannot be suppressed altogether, yet, in the presence of FRP confinement, significant delay of the phenomenon can be achieved with a commensurate increase in the available deformation capacity of the member. Evidently, the lateral confining pressure exerted by the jackets provides additional restraining to the vertical steel rods against outwards bulging, thereby postponing buckling, especially when steel stirrups are spaced widely apart. However, if the rigidity of the external FRP jacketing is not enough to contrast buckling of vertical bars, stress concentrations in the FRP jacket at the cross-section corners can occur and cause its premature failure. In Pellegrino et al.

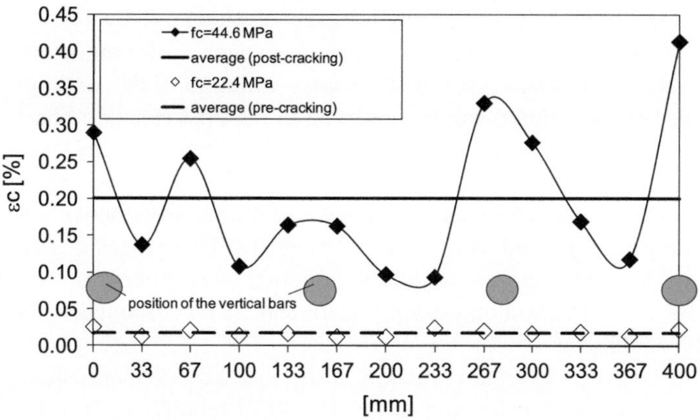

Fig. 5.22 Failure of axially loaded RC columns tested at the University of Padova

(2004) and Tinazzi et al. (2003) FRP confined circular concrete columns with internal steel reinforcement have been subjected to compressive tests. The experimental program consisted of monotonic simple compression tests on 80 × 35 cm columns with various amounts and types (Glass and Carbon) of FRP wraps. Transverse steel reinforcement consisted of 8 mm diameter stirrups, with 20 cm spacing whereas longitudinal reinforcement comprised eight 14 mm diameter bars distributed along the circumference of the column and almost equally spaced. Precise hoop strain measurements, performed by a row of strain gauges installed at the level corresponding to the middle height of the column, was arranged to deeply study the trend of hoop FRP strains in columns with steel reinforcement. Figure 5.22 shows the failure of some the axially loaded RC columns tested at the

Fig. 5.23 Strain variation along the row of circumferential strain gauges (400 mm long) in the pre-cracking (f_c = 22.4 MPa) and in the post-cracking (f_c = 44.6 MPa) phases for a CFRP confined specimen with a circular cross-section and internal steel reinforcement

University of Padova. In Fig. 5.23 strain variations along the row of circumferential strain gauges (total length of the row of strain gauges = 400 mm) for a specimen confined with CFRP wraps are shown. Intense variation of the strain profile along the perimeter of the column can be observed in the post-cracking phase with high strain concentrations mainly near the location of the vertical steel bars. The main vertical cracks approximately correspond to longitudinal bars' position due to buckling of these bars (observed after failure of all the specimens) bearing on the FRP wrapping near the middle of the column height. These measurements illustrate and confirm the prevailing notion that the internal steel influences the hoop strain profile in the FRP, the resulting crack pattern and, as a consequence, the efficiency of the FRP confinement.

It should be noted that the reduction in the FRP confinement efficiency due to bar buckling is taken into account in the analytical model proposed by Pellegrino and Modena (2010) by means of the parameter C that reduces the coefficient of efficiency, k_e, in columns with both external FRP and internal steel, when the mechanical steel ratio increases with respect to mechanical FRP ratio.

Assessment of Current Formulations and Models on Confinement of RC Elements by Means of EBR-FRP Systems

Performance and Correlation of Models

The assessment of the models described above was performed by comparing the analytical results with the experimental measurements. The average value (Avg) of the ratio between the experimental and analytical result and the coefficient of variation (CoV) were used to evaluate the accuracy of each model. Tables 5.6, 5.7, 5.8 and 6.9 show the average ratio between theoretical and experimental peak stress and ultimate strain (Avg) and the coefficient of variation (CoV) for each model considered. Circular and rectangular columns with and without steel reinforcement are distinguished.

The model of Pellegrino and Modena (2010), in which the contribution of the internal steel reinforcement is considered and the interaction between external FRP and internal steel reinforcement is taken into account, shows, for the four column typologies, the lowest values of the coefficient of variation (CoV) and average values (Avg) close to 1 and slightly conservative. The models of Harajli et al. (2006) and Ilki et al. (2008), in which the contribution of the internal steel reinforcement is considered (but the interaction between external FRP and internal steel reinforcement is not taken into account), generally show a rather good performance for the prediction of both peak stress and ultimate strain in case of both circular and rectangular cross section columns.

Table 5.6 Theoretical versus experimental statistical analysis for FRP confined circular columns without internal steel reinforcement

Models for circular columns	Columns without steel reinforcement			
	Peak stress (191 specimens)		Ultimate axial strain (175 specimens)	
	Avg	CoV	Avg	CoV
Fardis and Khalili (1981)	1.337	0.439	1.109	0.849
fib "practical formula" (2001)	1.034	0.174	2.737	2.666
ACI 440.2R-02 (2002)	0.811	0.231	0.474	0.564
Ilki and Kumbasar (2003)	0.987	0.150	1.416	0.590
Lam and Teng (2003)	0.941	0.154	0.794	0.355
CNR DT200 (2004)	0.830	0.235	–	–
Matthys et al.(2006)	1.062	0.166	1.283	0.700
Kawashima et al. (2000)	1.288	0.386	0.813	0.325
Li et al. (2003)	0.946	0.149	0.777	0.334
Harajli et al. (2006)	0.952	0.179	0.740	0.562
Tastani et al. (2006)	0.853	0.183	1.184	0.939
Ilki et al. (2008)	0.973	0.148	1.753	0.943
Pellegrino and Modena (2010)	0.991	0.142	0.983	0.323

Table 5.7 Theoretical versus experimental statistical analysis for FRP confined circular columns with internal steel reinforcement

Models for circular columns	Columns with steel reinforcement			
	Peak stress (54 specimens)		Ultimate axial strain (35 specimens)	
	Avg	CoV	Avg	CoV
Fardis and Khalili (1981)	1.251	0.370	0.970	0.701
fib "practical formula" (2001)	0.973	0.206	2.066	1.678
ACI 440.2R-02 (2002)	0.807	0.266	0.279	0.741
Ilki and Kumbasar (2003)	0.980	0.178	1.031	0.480
Lam and Teng (2003)	0.888	0.192	0.663	0.450
CNR DT200 (2004)	0.820	0.266	–	–
Matthys et al.(2006)	1.006	0.185	1.011	0.487
Kawashima et al. (2000)	1.320	0.427	0.771	0.414
Li et al. (2003)	0.916	0.181	0.590	0.498
Harajli et al. (2006)	0.925	0.221	0.708	0.622
Tastani et al. (2006)[a]	0.844 (0.756)	0.183 (0.257)	1.261 (1.071)	0.457 (0.766)
Ilki et al. (2008)	0.972	0.173	1.595	0.992
Pellegrino and Modena (2010)	0.996	0.146	0.981	0.398

Note numbers in parenthesis have been computed considering the effects of bar buckling as described in the respective model by Tastani et al. (2006) and (2008) described above

Table 5.8 Theoretical versus experimental statistical analysis for FRP confined rectangular columns without internal steel reinforcement

Models for rectangular columns	Columns without steel reinforcement			
	Peak stress (109 specimens)		Ultimate axial strain (73 specimens)	
	Avg	CoV	Avg	CoV
Fardis and Khalili (1981)	–	–	–	–
fib "practical formula" (2001)	0.688	0.381	0.782	0.447
ACI 440.2R-02 (2002)	1.058	0.199	–	–
Ilki and Kumbasar (2003)	1.013	0.311	1.367	0.800
Lam and Teng (2003)	0.772	0.281	1.201	0.905
CNR DT200 (2004)	1.068	0.188	–	–
Matthys et al.(2006)	–	–	–	–
Kawashima et al. (2000)	2.195	1.845	1.418	1.243
Li et al. (2003)	–	–	–	–
Harajli et al. (2006)	1.143	0.280	0.626	0.472
Tastani et al. (2006)	0.871	0.145	2.501	1.053
Ilki et al. (2008)	1.309	0.544	1.984	1.266
Pellegrino and Modena (2010)	0.953	0.159	0.985	0.347

Table 5.9 Theoretical versus experimental statistical analysis for FRP confined rectangular columns with internal steel reinforcement

Models for rectangular columns	Columns with steel reinforcement			
	Peak stress (121 specimens)		Ultimate axial strain (41 specimens)	
	Avg	CoV	Avg	CoV
Fardis and Khalili (1981)	–	–	–	–
fib "practical formula" (2001)	1.303	0.611	0.300	0.740
ACI 440.2R-02 (2002)	0.941	0.204	–	–
Ilki and Kumbasar (2003)	0.827	0.287	0.566	0.530
Lam and Teng (2003)	0.867	0.229	0.366	0.670
CNR DT200 (2004)	0.959	0.202	–	–
Matthys et al.(2006)	–	–	–	–
Kawashima et al. (2000)	2.288	1.625	0.591	0.526
Li et al. (2003)	–	–	–	–
Harajli et al. (2006)	1.017	0.207	0.323	0.729
Tastani et al. (2006)	0.940 (0.924)	0.075 (0.147)	0.923 (0.99)	0.696 (0.592)
Ilki et al. (2008)	1.109	0.261	0.731	0.614
Pellegrino and Modena (2010)	0.963	0.173	0.995	0.287

Effectiveness of Confinement in Members Under Moment and Axial Load Combinations

Extension of Confinement Models Calibrated to Concentric Compression Tests for the Case of Moment-Axial Load Combination

It was mentioned in the introduction that confinement models are usually calibrated with tests conducted on axially loaded members whereas in reality RC members are usually subjected to axial load coupled with bending moment, with the axial load taking on values that are relatively low compared to the axial capacity of the member (usually lower than 30 % of its crushing capacity). This means that confinement mechanisms can be strongly influenced by the strain gradient effect owing to bending moment, which reduces and limits the depth of compression zone in the cross-section, totally altering the passive confinement due to the lateral expansion of concrete. In fact, concrete in the tensile zone of cross-section may even present a reverse lateral behavior than anticipated, being in contraction instead of expansion.

In any case every reliable numerical procedure to estimate the cross-sectional behaviour under a combination of flexure and compression should include appropriate models for compressed bar buckling and concrete cover spalling (Tastani et al. 2006; Di Ludovico et al. 2010). For this reasons, a dependable stress–strain behavior for all materials composing the retrofit solution would be necessary. Even if the cross-section is not fully compressed and the benefits of confinement on concrete can be uncertain, one of the major improvements in member behaviour due to FRP wrapping is highlighted considering that, in unstrengthened columns, when steel reinforcement reaches the buckling stress in compression, as it pushes outward the surrounding concrete, the concrete cover spalls out. Confinement is able to delay bar buckling and to let the compressive concrete strains grow to higher values, thus resulting in higher load carrying capacity of the column and in significant ductility enhancement (Tastani and Pantazopoulou 2003, 2006; Pellegrino and Modena 2010; Lignola et al. 2014).

The strength increase in confined concrete due to FRP wrapping turns into a load carrying capacity increase mainly in columns loaded with small eccentricity (from Fig. 5.24a it appears that close to the state of pure bending the effect of concrete strength enhancement is not relevant because failure moves to tension side and, at lower levels of axial load, i.e. at higher eccentricities, also the influence of reinforcement buckling on the element behaviour is less significant until very high levels of rotational ductility).

Lignola et al. (2013b) performed analyses considering six values for the confinement parameter $k = f_{cc}/f_{co}$. In particular the values used were: 1 (i.e. unconfined), 1.25, 1.5, 1.75, 2, and 3. Figure 5.24 shows that the increase in material strength has almost no beneficial effect on the flexural capacity at low levels of axial load, whereas the effect becomes noticeable and more prominent at higher axial loads. At maximum axial load (concentric compression crushing) the strength

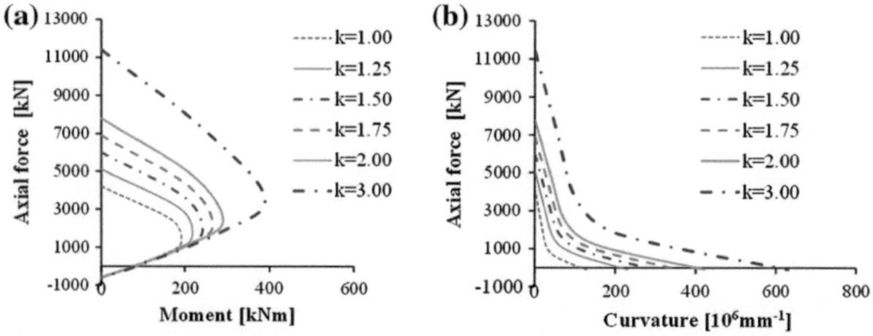

Fig. 5.24 **a** Axial load-Bending moment interaction diagram for a rectangular cross-section. **b** Axial load-ultimate curvature interaction domains for a rectangular cross-section

increase is almost proportional to the concrete strength increase (i.e. only the steel reinforcement contribution is not sensitive to confinement). The maximum bending capacity (associated with the so-called balanced failure) increases and it is achieved at increasing values of axial load, as the confinement index, *k*, increases. Figure 5.24b plots the axial load–ultimate curvature interaction diagram, illustrating the coupled effect of *k* on concrete ultimate strain and member deformation capacity. The increase of *k* impacts favourably the estimated curvature capacity at low axial load levels; this effect diminishes with increasing axial load.

Confinement of Lap Splices

After successfully retrofitting old reinforced concrete columns with FRP jacketing an anticipated result is increased deformation demand in the lap-splice regions, which, in old construction, usually occurs in the lower part of the column clear height. A design issue for the retrofit is therefore the development capacity of lap-splices confined by jacketing. FRP jackets are wrapped orthogonal to the anticipated splitting plane which is assumed to run parallel to the bars. Using a frictional model to interpret the bond action, Tastani and Pantazopoulou (2008) defined the development capacity of a confined lap-splice of length ℓ_s from $F = \mu \cdot f_{lat} \cdot p \cdot \ell_s$, where μ is the coefficient of friction at the steel-concrete interface and f_{lat} is the pressure exerted upon the lateral surface of the bar by the cover, transverse stirrups and FRP jacket. The average design bond stress f_b is given by Eq. (5.45), where N_b is the number of bars (or pairs of spliced bars) laterally restrained by the transverse pressure and p the length of the likely splitting path (p is equal to the length of the red line in Fig. 5.25a. Note that $f_{lat}{}^{f}$ and f_{lat}^{st} are obtained from Eq. (5.1) when considering the likely plane of splitting failure through the lap. Here f_{lat}^{c} represents transverse confining pressure exerted by the concrete cover.

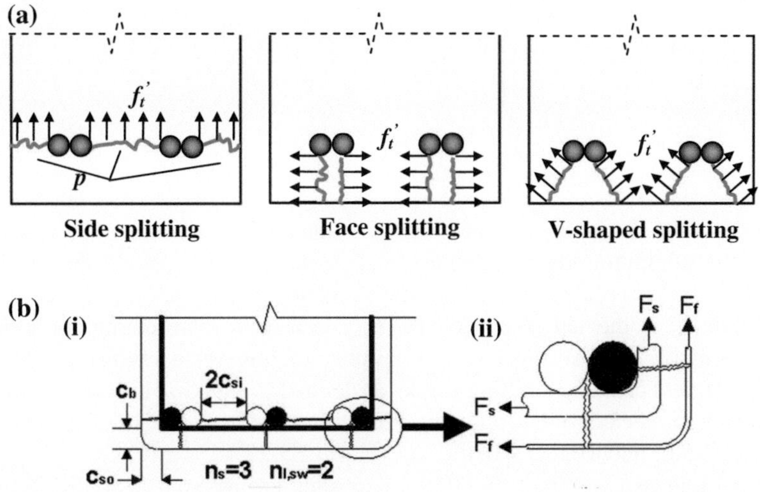

Fig. 5.25 **a** Cover crack paths in lap-spliced bars: side, face and V-shaped splitting. **b** (i) Definition of c_b, c_{si}, and c_{so}. (ii) Forces induced by stirrups and TRM or FRP jackets against side and face splitting cracks

$$f_b = \frac{2\mu}{\pi D_b}\left(f_{lat}^{c} + \frac{f_{lat}^{st} + f_{lat}^{f}}{N_b}\right) = \frac{2\mu}{\pi D_b}\left(pf_t' + \frac{k_{st}^{anch} A_{st} f_{y,st}}{N_b\, s} + \frac{2k_f^{anch} n\, t_f E_f \varepsilon_f^{eff}}{N_b}\right) \tag{5.45}$$

The $\varepsilon_f{}^{eff}$ used in Eq. (5.45) in order to calculate the lateral pressure exerted by the FRP normal to the lap is the surface strain value associated with attainment of bond strength along the bar. To obtain this, the following points are considered (Tastani and Pantazopoulou 2013): (a) Based on experimental evidences, bar slip associated with peak bond stress and bar development capacity is in the range of 0.1–0.3 mm. (b) From the results of Lura et al. (2002) it has been shown that at any point along the bar the slip is about twice the radial displacement of the internal bar boundary imposed by the displacing ribs, thus, the associated radial displacement at the bar surface is $u_{r,o}$ = 0.05–0.15 mm. (c) The corresponding hoop strain equals the $u_{r,o}$ divided by the radius of the internal boundary, $\varepsilon^{ho} = u_{r,o}/(D_b/2)$. Thus, the hoop strain at the outside boundary of the cover, where the FRP jacket is installed is: $\varepsilon_f{}^{eff} = u_{r,o}/(c + D_b/2) = 2u_{r,o} \cdot [D_b(1 + 2(c/D_b))]^{-1} = \varepsilon^{ho} \cdot (1 + 2(c/D_b))^{-1}$. For example, for $D_b = 20$ mm and $c = D_b$, ε^{ho} = 0.005–0.015 and ε_f^{eff} = 0.0017–0.005. This range of values is well below the strain capacity of the jacket itself.

The new Model Code by fib (2012) which is adopted from the recommendations drafted by TG 4.5-Bond, proposes the following semi-empirical expression (Eq. 5.46a) for the calculation of bond strength between lap-spliced bars and concrete at splitting bond failure:

$$f_b = \frac{D_b}{4l_s}\left[24.2\left(\frac{\ell_s}{D_b}\right)^{0.55}(f_{co})^{1/4}\left(\frac{c_d}{D_b}\right)^{1/3}\left(\frac{c_{\max}}{D_b}\right)^{0.1}\left(\frac{20}{D_b}\right)^{0.2}(1+K_{tr,s})\right] \quad (5.46a)$$

$$K_{tr,s} = k_s\left(\frac{A_{sw}\,n_{l,sw}}{s_h}\right) = \frac{10}{D_b n_s}\left(\frac{A_{sw}\,n_{l,sw}}{s_h}\right) \quad (5.46b)$$

where D_b is the diameter of lapped bars; f_c is the compressive strength of concrete (MPa); the minimum cover is defined as $c_d = min(c_{so}, c_b, c_{si})$ and the maximum cover $c_{max} = max(c_{so}, c_b, c_{si})$ where the variables c_{so}, c_b, and c_{si} are defined in Fig. 5.25b; ℓ_s is the lap splice length; $K_{tr,s}$ is a term representing the effect of confinement by steel stirrups; $n_{l,sw}$ = number of transverse reinforcing bar legs (stirrups or ties) crossing splitting cracks as defined in Fig. 5.25b. Use of the above equations is subjected to the following limitations:$20/D_b \leq 1.0$; $1.0 \leq c_d/D_b \leq 3.0$; $c_{max}/c_{min} \leq 5.0$, and $10K_{tr,s} \leq 0.4$.

TRM (note that in Chap. 9 TRM composites are referred to as FRCM composites) and FRP jackets normal to a lap splice of embedded reinforcement is shown through tests to provide additional resistance against splitting cracks over that supported by transverse steel reinforcement, enabling the lap splice to develop higher tensile force, as illustrated in Fig. 5.25b. To account for this contribution on the local bond strength of lap splices, a new parameter $K_{tr,f}$ is added to the transverse reinforcement parameter $K_{tr,s}$ provided by the stirrups.

The proposed modified term $K_{tr,t}$, which accounts for the total confinement applied by both the contribution of stirrups and FRP/TRM jackets, is expressed as:

$$K_{tr,t} = K_{tr,s} + K_{tr,f} = (\text{Linear Function of } k_s(A_{sw}/s_h)) + k_f 2nt_f \quad (5.47)$$

where k_s is a calibration factor for steel transverse reinforcement as defined in Eqs. (5.46a) and (5.46b); k_f is a calibration factor accounting for the effectiveness of FRP or TRM jackets; n = number of layers of fibre sheet or textile layer; and t_f is the thickness of a single fibre sheet or textile layer. The proposed factor k_f is given by Eq. (5.48):

$$k_f = k_s \frac{E_f\,\varepsilon_{f,ef}}{E_s\,\varepsilon_{sw}} \quad (5.48)$$

where E_f is the elastic modulus of the jacket in the fibre (circumferential) direction; ε_{sw} = average effective strain of the stirrups in the circumferential direction and $\varepsilon_{f,eff}$ = average effective strain of the jacket in the fibre direction. The proposed parameter $K_{tr,j}$ takes into account all the characteristics of the jacket, namely: the area of external FRP or TRM reinforcement ($2nt_f$) in the splice region; the modulus of elasticity of the jacket FRP/TRM material (E_f); and the average effective strain of the jacket in the circumferential direction ($\varepsilon_{f,eff}$). It should be also noted that application of Eq. (5.48) is based on the assumption that the jacket's height is at least equal to the lap length. The yield stress of the transverse reinforcement was removed from the initial bond model (Eqs. 5.46a and 5.46b) as it was found to have no effect on the lap splice bond strength. This effect was not measurable in most of the test results (included in the relevant databases) because splitting bond failures of the lap splice preceded yielding of stirrups. Nevertheless, the percentage of activation of the transverse steel reinforcement in the circumferential direction against splitting cracks, which is quantified by the average effective strain of stirrups ε_{sw}, was determined in Bournas and Triantafillou (2011) and was taken approximately equal to 0.135 %.

The only term in the proposed modified bond model which is still to be addressed is the average effective strain of the jacket $\varepsilon_{f,eff}$, which was found to depend on the lap length to bar diameter ratio ℓ_s/D_b. According to the experimental results presented by Bournas and Triantafillou (2011) the average bond strength along the splice length was increased by 54 and 23 % for FRP confined specimens, and by 45 and 18 % for TRM confined specimens with respect to their unconfined counterparts, for the lap lengths of twenty and forty bar diameters, respectively. Based on these enhancements of bond strength and using Eqs. (5.46a, 5.46b)–(5.48) the average strain at which the composite jackets were activated, $\varepsilon_{f,eff}$, was determined and plotted for different values of lap length to bar diameter, ranging from 15 to 45; results are shown in the diagrams of Fig. 5.26a, assuming a linear fit. Of course these diagrams should be used with care, as they are based on limited test results. More tests on FRP or TRM confined columns with different lap lengths are necessary in order to provide the best fit to the above diagrams.

Figure 5.26a illustrates a high activation strain (high $\varepsilon_{f,eff}$) of FRP or TRM jackets for short lap lengths and a lower one as the lap length increases. A key point here is that in the case of short splitting cracks (short lap splices) the activation of the jacket's part outside the lap length ℓ_s is enhanced (Fig. 5.26b), a fact quantified by the increased jacket effective strain $\varepsilon_{f,eff}$. In this way it is possible to estimate the effect of jacket confinement on RC members with lap splices as a function of lap

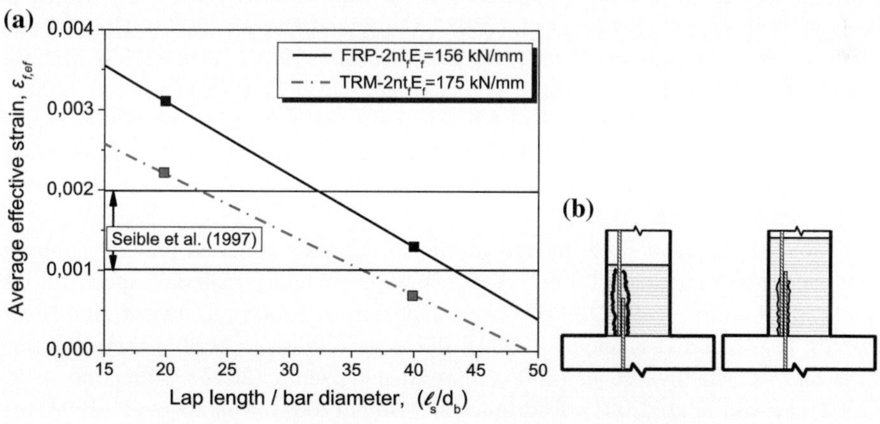

Fig. 5.26 **a** Diagrams for the evaluation of the average effective strain of FRP or TRM jackets according to the modified Model Code (2010) model. **b** Activation of FRP or TRM jacket against longitudinal splitting crack propagation in cases of short and long lap splice lengths

splice length and jacket stiffness, contrary to current practice, which is based on the work of Seible et al. (1997), who proposed a fixed value for the circumferential jacket strain corresponding to the onset of splitting in the range 0.001–0.002 (included in Fig. 5.26a). It seems that using these strains (0.001–0.002) in a jacket design may be quite conservative for low values of ℓ_s/D_b, yielding an unrealistically large number of layers. The design equations proposed in this study for the effective strain in an FRP or TRM jacket at lap splice failures are summarized below for the two modified models:

$$\varepsilon_{f,eff} = 0.0049 - 9 \times 10^{-5}(\ell_s/D_b) \text{ for FRP jackets} \tag{5.49a}$$

$$\varepsilon_{f,eff} = 0.0037 - 7.5 \times 10^{-5}(\ell_s/D_b) \text{ for TRM jackets} \tag{5.49b}$$

It is worth noting that circumferential FRP strains at the base of rectangular columns with lap splices have actually been measured experimentally by a few researchers, namely Harajli and Dagher (2008) and ElGawady et al. (2010). The former reported a value of 0.00135 for a lap splice length equal to 30 bar diameters, whereas the latter reported values in the range 0.0013–0.0023 for a lap splice length equal to 36 bar diameters. Those independently measured strains are in reasonable agreement with the modified equation, which predicts strains equal to 0.0022 and 0.0017 for the Harajli and Dagher (2008) and the ElGawady et al. (2010) test results, respectively.

Appendix: List of Sources for Database Specimens

Characteristics and experimental results for 219 FRP confined circular columns without steel reinforcement (Table 5.A-(i), in supplementary material) shown in Arduini et al. (1999), Berhet et al. (2005), Carey and Harries (2005), Harmon and Slattery (1992), Harries and Carey (2003), Kono et al. (1998), Li et al. (2003), Matthys et al. (1999), Micelli et al. (2001), Miyauchi et al. (1997), Pellegrino et al. (2004), Pessiki et al. (2001), Picher et al. (1996), Rochette and Labossiere (2000), Rousakis (2001), Shahawy et al. (2000), Silva and Rodrigues (2006), Teng and Lam (2004), Tinazzi et al. (2003), Toutanji (1999), Wang and Wu (2008), Watanabe et al. (1997) are included in the database (a).

Characteristics and experimental results for 77 FRP confined circular columns with steel reinforcement (Table 5.A-(ii), in supplementary material) shown in the works of Arduini et al. (1999), Carey and Harries (2005), Demerse and Neale (1999), Esafhani and Kianoush (2004), Ilki et al. (2008), Li et al. (2003), Lin and Liao (2004), Matthys et al. (2006), Parretti and Nanni (2002), Pellegrino et al. (2004), Pessiki et al. (2001), Rodrigues and Silva (2001), Tinazzi et al. (2003) are included in the database (b).

Characteristics and experimental results for 135 FRP confined rectangular columns without steel reinforcement (Table 5.A-(i), in supplementary material) shown in the works of Braga et al. (2004), Campione (2006), Chaallal et al. (2003), Harajli 2006, Harries and Carey (2003), Ilki and Kumbasar (2003), Mirmiran et al. (1998), Mukherjee et al. (2004), Parvin and Wang (2002), Pessiki et al. (2001), Rochette and Labossierre (2000), Rousakis et al. (2007), Wang and Wu (2008) are included in the database (c).

Characteristics and experimental results for 156 FRP confined rectangular columns with steel reinforcement (Table 5.A-(ii), in supplementary material) shown in the works of Braga et al. (2004), De Paula and Da Silva (2002), Esfahani (2004), Feng et al. (2002), Harajli et al. (2006), Hosseini and Fadaee (2004), Ilki et al. (2008), Maalej et al. (2003), Parretti and Nanni (2002), Pessiki et al. (2001), Prota et al. (2006), Tastani et al. (2006) are included in the database (d).

References

ACI Committee 440 (2002), Guide for the design and construction of externally bonded FRP systems for strengthening concrete structures, ACI 440.2R-02 (July 2002), American Concrete Institute, Farmington Hills, Michigan.

Albanesi, T., Nuti, C., & Vanzi, I. (2007). Closed form constitutive relationship for concrete filled FRP tubes under compression. *Construction and Building Materials*, *21*(2), 409–427.

American Concrete Institute (2008). Guide for the design and construction of externally bonded FRP systems for strengthening of concrete structure. ACI 440.2R-08, Farmington Hill, Mich., USA.

Arduini, M., Di Tommaso, A., Manfroni, O., Ferrari, S., & Romagnolo, M. (1999). Il confinamento passivo di elementi compressi in calcestruzzo con fogli di materiale composito. L'Industria Italiana del Cemento, 836–841 (in Italian).

Berthet, J. F., Ferrier, E., & Hamelin, P. (2005). Compressive behaviour of concrete externally confined by composite jackets. Part A: Experimental study. *Construction and Building Materials, 19*(3), 223–232.

Bournas, D. A., & Triantafillou, T. C. (2011). Bond strength of lap spliced bars in concrete confined with composite materials. *ASCE Journal of Composites for Construction, 15*(2), 156–167.

Braga, F., Laterza, M., Gigliotti, R., Dragonetti, G., & Nigro, D. (2004). Prove di compressione ciclica su pilastri in c.a. confinati con staffe e/o con tessuti in fibra di carbonio. XI National Congress "L'Ingegneria sismica in Italia" (in Italian).

Campione, G. (2006). Influence of FRP wrapping techniques on the compressive behaviour of concrete prisms. *Cement and Concrete Composites, 28*, 497–505.

Campione, G., & Miraglia, N. (2003). Strength and strain capacities of concrete compression members reinforced with FRP. *Cement and Concrete Composites*, *25*(1), 31–41.

Carey, S. A., & Harries, K. A. (2005). Axial behaviour and modeling of confined small, medium, and large-scale circular sections with carbon fiber-reinforced polymer jackets. *ACI Structural Journal, 102*(4), 596–604.

Chaallal, O., Shahawy, M., & Hassa, M. (2003). Performance of axially loaded short rectangular columns strengthened with carbon FRP wrapping. *Journal of Composites for Construction, 7* (3), 200–208.

CNR—Italian Research Council, Advisory Committee on Technical Recommendations for Construction (2004). Guide for the Design and Construction of Externally Bonded FRP Systems for Strengthening Existing Structures. Materials, RC and PC Structures, Masonry Structures (CNR-DT 200/2004), Rome, Italy.

De Lorenzis, L., & Tepfers, R. (2003). Comparative study of models on confinement of concrete cylinders with fiber reinforced polymer composites. *ASCE Journal of Composites for Construction, 7*(3), 219–237.

De Luca, A., Nardone, F., Lignola, G. P., Prota, A., & Nanni, A. (2013). Wall-like reinforced concrete columns externally confined by means of glass FRP laminates. *Advances in Structural Engineering, 16*(4), 593–603.

De Luca, A., Nardone, F., Matta, F., Nanni, A., Lignola, G. P., & Prota, A. (2011). Structural evaluation of full-scale FRP-confined reinforced concrete columns. *ASCE Journal of Composites for Construction, 15*(1), 112–123.

De Paula, R. F., Da Silva, M. G. (2002). Sharp edge effects on FRP confinement of RC square columns. *3rd International Conference on Composites in Infrastructure*.

Demers, M., & Neale, K. W. (1999). Confinement of reinforced concrete columns with fibre-reinforced composite sheets—an experimental study. *Canadian Journal of Civil Engineering, 26*, 226–241.

Di Ludovico, M., Lignola, G. P., Prota, A., & Cosenza, E. (2010). Non linear analysis of cross sections under axial load and biaxial bending. *ACI Structural Journal, 107*(4), 390–399.

EC8-III (2005). *EN 1998-3 (2005): Eurocode 8: Design of Structures for Earthquake Resistance—Part 3: Assessment and Retrofitting of Buildings*. Brussels: European Committee for Standardization-CEN.

ElGawady, M., Endeshaw, M., McLean, D., & Sack, R. (2010). Retrofitting of rectangular columns with deficient lap splices. *ASCE Journal of Composites for Construction, 14*(1), 22–35.

Elwi, A. A., & Murray, D. W. (1979). A 3-D hypoelastic concrete constitutive relationship. *ASCE Journal of Engineering Mechanics Division, 105*(4), 623–641.

Esfahani M. R., Kianoush M. R. (2004). Axial compressive strength of reinforced concrete columns wrapped with FRP. *1st Conference on Application of FRP Composites in Construction and Rehabilitation of Structures*. May 4, Tehran, Iran.

Fahmy, M. F. M., & Wu, Z. (2010). Evaluating and proposing models of circular concrete columns confined with different FRP composites. *Composites, Part B, 41*(3), 199–213.

Fardis, M. N., & Khalili, H. (1981). Concrete encased in fibreglass-reinforced-plastic. *Journal of the American Concrete Institute, 78*(6), 440–446.

Fardis, M. N., & Khalili, H. (1982). Concrete encased in fibreglass-reinforced-plastic. *ACI Structural Journal*, Title No. 78–38, Nov.–Dec., 1981, 440–446.

fédération internationale du béton (2001). Externally bonded FRP reinforcement for RC structures. Bullettin No. 14, Lausanne, Switzerland.

fédération internationale du béton (2012). *fib* Model Code 2010 Final Draft, Bulletin 65, International Federation for Structural Concrete, Lausanne, Switzerland.

Feng, P., Lu, X. Z., & Ye, L. P. (2002). Experimental research and finite element analysis of square concrete columns confined by FRP sheets under uniaxial compression. *17th Australian Conference on the Mechanics of Structures and Materials (ACMSM17)* (pp. 71–76). Gold Coast, Australia, 12–14 June.

Giamundo, V., Lignola, G., Prota, A., Manfredi, G. (2014). Analytical evaluation of FRP wrapping effectiveness in restraining reinforcement bar buckling. *ASCE Journal of Structural Engineering, 140*(7).

Girgin, Z. C. (2009). Modified failure criterion to predict ultimate strength of circular columns confined by different materials. *ACI Structural Journal, 106*(6), 800–809.

Harajli, M. H., & Dagher, F. (2008). Seismic strengthening of bond-critical regions in rectangular reinforced concrete columns using fiber-reinforced polymer wraps. *ACI Structural Journal, 105*(1), 68–77.

Harajli, M. H. (2006). Axial stress–strain relationship for FRP confined circular and rectangular concrete columns. *Cement and Concrete Composites, 28*, 938–948.

Harajli, M. H., Hantouche, E., & Soudki, K. (2006). Stress-strain model for fiber-reinforced polymer jacketed concrete columns. *ACI Structural Journal, 103*(5), 672–680.

Harmon, T. G., Slattery, K. T. (1992). Advanced composite confinement of concrete. *1st International Conference on Advanced Composite Materials in Bridges and Structures* (pp. 299–306). Sherbrooke, Québec, Canada.

Harries, K. A., & Carey, S. A. (2003). Shape and "gap" effects on the behaviour of variably confined concrete. *Cement and Concrete Research, 33*, 881–190.

Hosseini, A., & Fadaee, S. (2004). Behaviour of high strength square concrete columns strengthened with carbon fiber reinforced polymers (CFRP). *1st conference on application of FRP composites in construction and rehabilitation of structures*, Tehran, Iran.

Ilki, A., & Kumbasar, N. (2003). Compressive behaviour of carbon fibre composite jacketed concrete with circular and non circular cross-section. *Journal of Earthquake Engineering, 7*(3), 381–406.

Ilki, A., Peker, O., Karamuk, E., Demir, C., & Kumbasar, N. (2008). FRP retrofit of low and medium strength circular and rectangular reinforced concrete columns. *ASCE Journal of Materials in Civil Engineering, 20*(2), 169–188.

Jiang, T., & Teng, J.G. (2007). Analysis-oriented models for FRP-confined concrete: a comparative assessment. *Engineering Structures, 29*(11), 2968–2986.

KAN.EPE (2012). Greek Code for Retrofitting of Reinforced Concrete Structures. www.oasp.gr.

Karbhari, V. M., & Gao, Y. (1997). Composite jacketed concrete under uniaxial compression—verification of simple design equations. *Journal of Materials in Civil Engineering, 9*(4), 185–193.

Kawashima, K., Hosotani, M., Yoneda, K. (2000). Carbon fiber sheet retrofit of reinforced concrete bridge piers. *International Workshop on Annual Commemoration of Chi-Chi Earthquake, Vol. II-Technical Aspect, National Center for Research on Earthquake Engineering* (pp. 124–135), Taipei, Taiwan.

Kono, S., Inazumi, M., & Kaku, T. (1998). Evaluation of confining effects of CFRP sheets on reinforced concrete members. *2nd International Conference on Composites in Infrastructure ICCI'98* (pp. 343–355), 5–7 January, Tucson, Arizona, 343–355.

Kumutha, R., Vaidyanathan. R., & Palanichamy, M. S. (2007). Behaviour of reinforced concrete rectangular columns strengthened using GFRP. *Cement and Concrete Composites, 29*, 609–615.

Kupfer, H., & Gerstle, K. H. (1973). Behavior of concrete under biaxial stresses. *ASCE Journal of Engineering Mechanics, 99*(4), 853–866.

Lam, L., & Teng, J. G. (2003). Design-oriented stress-strain model for frp-confined concrete. *Construction and Building Materials, 17*, 471–489.

Li, Y., Fang, T., Chern, C. (2003). A constitutive model for concrete cylinder confined by steel reinforcement and carbon fibre sheet. *Pacific Conference on Earthquake Engineering.*

Lignola, G. P., Giamundo, V., Prota, A., Manfredi G. (2013b). FRP wrapping of RC members under combined axial load and bending. *11th International Symposium on Fiber Reinforced Polymer Reinforcement for Concrete Structures FRPRCS11 2013*, Guimarães, Portugal, June 26–28, 2013.

Lignola, G. P., Nardone, F., Prota, A., & Manfredi, G. (2012). Analytical model for the effective strain in FRP-wrapped circular RC columns. *ELSEVIER Composites: Part B, 43*(8), 3208–3218.

Lignola, G. P., Nardone, F., Prota, A., De Luca, A. & Nanni, A. (2011b). Analysis of RC hollow columns strengthened with GFRP. *ASCE Journal of Composites for Construction, 15*(4), 545–556

Lignola, G. P., Prota, A. & Manfredi G. (2013a). Simplified modeling of FRP wrapping of rectangular cross-sections. *2nd International Conference on Smart Monitoring, Assessment and Rehabilitation of Civil Structures SMAR 2013*. Istanbul, Turkey September 2013. Paper ID 158.

Lignola, G. P., Prota, A., Manfredi, G., & Cosenza, E. (2008). Unified theory for confinement of RC solid and hollow circular columns. *ELSEVIER Composites: Part B, 39*(7–8), 1151–1160.

Lignola, G. P., Prota, A., Manfredi, G., & Cosenza, E. (2009). Non linear modeling of RC hollow piers confined with CFRP. *ELSEVIER Composite Structures, 88*(1), 56–64.

Lignola, G. P., Prota, A., Manfredi, G. & Cosenza E. (2011a). Modeling of RC wall-like columns FRP confinement. *1st Middle East Conference on Smart Monitoring, Assessment and Rehabilitation of Civil Structures SMAR 2011*, Dubai, UAE, Paper ID 125:1–8.

Lin, J. L., & Liao, C. I. (2004). Compressive strength of reinforced concrete column confined by composite material. *Composite Structures, 65*, 239–250.

Lura, P., Plizzari, G., & Riva, P. (2002). 3D finite-element modelling of splitting crack propagation. *Magazine of Concrete Research, 54*(6), 481–493.

Maalej, M., Tanwongsval, S., & Paramasivam, P. (2003). Modelling of rectangular columns strengthened with FRP. *Cement and Concrete Composites, 25*, 263–276.

Mander, J. B., Priestley, M. J. N., & Park, R. (1988). Theoretical stress-strain model for confined concrete. *ASCE Journal of Structural Engineering, 114*(8), 1804–1826.

Matthys, S. (2000). Structural behaviour and design of concrete members strengthened with externally bonded frp reinforcement. Doctoral thesis, Ghent University, pp. 1–367.

Matthys, S., Toutanji, H., Audenaert, K., & Taerwe, L. (2005). Axial load behavior of large-scale columns confined with fiber-reinforced polymer composites. *ACI Structural Journal, 102*(2), 258–267.

Matthys, S., Toutanji, H., & Taerwe, L. (2006). Stress-strain behaviour of large-scale circular columns confined with FRP composites. *ASCE Journal of Structural Engineering, 132*(1), 123–133.

Matthys, S., Taerwe, L., & Audenaert, K. (1999). Tests on axially loaded concrete columns confined by fiber reinforced polymer sheet wrapping. *4th International Symposium on Fiber Reinforced Polymer Reinforcement for Reinforced Concrete Structures* (pp. 217–228).

Megalooikonomou G. K., Monti G., & Santini S. (2011). Constitutive model for FRP and tie—confined concrete. *3rd International Conference on Computational Methods in Structural Dynamics and Earthquake Engineering (COMPDYN 2011)*. Corfu, Greece.

Micelli, F., Myers, J. J., & Murthy, S. (2001). Effect of environmental cycles on concrete cylinders confined with FRP. *CCC2001 International Conference on Composites in Construction*. October 10–12, Porto, Portugal.

Mirmiran, A., Shahawy, M., Samaan, M., El Echary, H., Mastrapa, J. C., & Pico, O. (1998). Effect of column parameters on FRP-confined concrete. *ASCE Journal of Composites for Construction, 2*(4), 175–185.

Miyauchi, K., Nishibayashi, S., Inoue, S. (1997). Estimation of strengthening effects with carbon fiber sheet for concrete column. *3rd International Symposium (FRPRCS-3) on Non-Metallic (FRP) reinforcement for Concrete Structures* (pp. 217–224). 14–16 October, Sapporo, Japan.

Monti, G., & Nuti, C. (1992). Nonlinear cyclic behavior of reinforcing bars including buckling. *Journal of Structural Engineering*, *118*(12), 3268–3284.

Mukherjee, A., Boothby, T. E., Bakis, C. E., Joshi, M. V., & Maitra, S. R. (2004). Mechanical behaviour of fiber reinforced polymer-wrapped concrete columns-complicating effects. *ASCE Journal of Composites for Construction, 8*(2), 97–103.

Ozbakkaloglu, T., Lim, J. C., & Vincent, T. (2013). FRP-confined concrete in circular sections: Review and assessment of stress–strain models. *Elsevier Engineering Structures, 49*, 1068–1088.

Parretti, R., & Nanni, A. (2002). Axial testing of concrete columns confined with carbon FRP: effect of fiber orientation. *Proc. ICCI 2002*, Hardback, Netherlands.

Parvin, A., & Wang, W. (2002). Concrete columns confined by fiber composite wraps under combined axial and cyclic lateral loads. *Composite Structures, 58*, 539–549.

Pellegrino, C., & Modena, C. (2010). Analytical model for FRP confinement of concrete columns with and without internal steel reinforcement. *ASCE Journal of Composites of Construction, 14*(6), 693–705.

Pellegrino, C., Tinazzi, D., Modena, C. (2004). Sul confinamento di elementi in c.a. soggetti a compressione. *AICAP National Congress 2004*, 26–29 Maggio, Verona, Italy (in Italian).

Pessiki, S., Harries, K. A., Kestner, J. T., Sause, R., & Ricles, J. M. (2001). Axial behaviour of reinforced concrete columns confined with FRP jackets. *ASCE Journal of Composites of Construction, 5*(4), 237–245.

Picher, F., Rochette, P., & Laboissiére, P. (1996). Confinement of concrete cylinders with CFRP. *1st Conference on Composites in Infrastructure ICCI'96* (pp. 829–841). 15–17 January, Tucson, Arizona.

Prota, A., Manfredi, G., & Cosenza, E. (2006). Ultimate behaviour of axially loaded RC wall-like columns confined with GFRP. *Composites: Part B, 37*, 670–678.

Realfonzo, R., & Napoli, A. (2011). Concrete confined by FRP systems: confinement efficiency and design strength models. *Composites: Part B, 42*(4), 736–755.

Realfonzo, R., & Napoli, A. (2013). Confining concrete members with FRP systems: predictive vs design strain models. *Composite Structures, 104*, 303–319.

Richart, F. E., Brandtzaeg, A., & Brown, R. L. (1928). A Study of the Failure of Concrete Under Combined Compressive Stresses. University of Illinois Engineering Experimental Station, Champaign, Ill, Bulletin 185

Rochette, P., & Labossiére, P. (2000). Axial testing of rectangular column models confined with composites. *ASCE Journal of Composites for Construction, 4*(3), 129–136.

Rodrigues, C. C., Silva, M. (2001). Experimental investigation of CFRP reinforced concrete columns under uniaxial cyclic compression. *FRPRCS5 International Conference* (pp. 783–791). Cambridge.

Rousakis, T. (2001). Experimental investigation of concrete cylinders confined by carbon FRP sheets, under monotonic and cyclic axial compressive load. *Research report*. Chalmers University of Technology, Göteborg, Sweden.

Rousakis, T. C., Karabinis, A. I., & Kiousis, P. D. (2007). FRP-confined concrete members: axial compression experiments and plasticity modelling. *Engineering Structures, 29*(7), 1343–1353.

Seible, F., Priestley, M. J. N., Hegemier, G. A., & Innamorato, D. (1997). Seismic retrofit of RC columns with continuous carbon fiber jackets. *ASCE Journal of Composites for Construction, 1*(2), 52–62.

Shahawy, M., Mirmiran, A., & Beitelmann, T. (2000). Tests on modelling of carbon-wrapped concrete columns. *Composite Part B: Engineering, 31*, 471–480.

Silva, M. A. G., & Rodrigues, C. C. (2006). Size and relative stiffness effects on compressive failure of concrete columns wrapped with glass FRP. *ASCE Journal of Materials in Civil Engineering, 18*(3), 334–342.

Spoelstra, M. R., & Monti, G. (1999). FRP-confined concrete model. *ASCE Journal of Composites for Construction, 3*(3), 143–150.

Tan, K. H. (2002). Strength enhancement of rectangular RC columns using FRP. *ASCE Journal of Composites for Construction, 6*(3), 175–183.

Tastani, S., & Pantazopoulou, S. J. (2003). Strength and deformation capacity of brittle r.c. members jacketed with FRP wraps. *FIB Symposium on "Concrete Structures in Seismic Regions"*, Athens, May 6–8.

Tastani, S. P., & Pantazopoulou, S. J. (2013). Reinforcement-Concrete Bond: State Determination along the Development Length. *ASCE Journal of Structural Engineering, 139*(9), 1567–1581.

Tastani, S. P., & Pantazopoulou, S. J. (2008). Detailing procedures for seismic rehabilitation of reinforced concrete members with fiber reinforced polymers. *Elsevier Engineering Structures, 30*(2), 450–461.

Tastani, S. P., Balafas, I., Dervisis, A., & Pantazopoulou, S. J. (2013). Effect of core compaction on deformation capacity of FRP-jacketed concrete columns. *Elsevier Construction and Building Materials, 47*, 1078–1092.

Tastani, S. P., Pantazopoulou, S. J., Zdoumba, D., Plakantaras, V., & Akritidis, E. (2006). Limitations of FRP jacketing in confining old-type reinforced concrete members in axial compression. *ASCE Journal of Composites for Construction, 10*(1), 13–25 (erratum, October 2013).

Teng, J. G., Chen, J. F., Smith, S. T., & Lam, L. (2002). *FRP-strengthened RC structures*. UK: John Wiley and Sons Ltd.

Teng, J. G., Jiang, T., Lam, L., & Luo, Y. Z. (2009). Refinement of a design-oriented stress-strain model for FRP-confined concrete. *Journal of Composites for Construction, 13*(4), 269–278.

Teng, J. G., & Lam, L. (2004). Behaviour and modeling of fiber reinforced polymer-confined concrete. *ASCE Journal of Structural Engineering., 130*(11), 1713–1723.

Tinazzi, D., Pellegrino, C., Cadelli, G., Barbato, M., Modena, C., & Gottardo, R. (2003). An experimental study of RC columns confined with FRP sheets. Structural Faults and Repair. *10th International Conference*, 1–3 July, London, UK.

Toutanji, H. (1999). Stress-strain characteristics of concrete columns externally confined with advanced fiber composite sheets. *ACI Materials Journal, 96*(3), 397–404.

Vintzileou, E., & Panagiotidou, E. (2008). An empirical model for predicting the mechanical properties of FRP-confined concrete. *Elsevier Construction and Building Materials, 22*(5), 841–854.
Wang, L. M., & Wu, Y. F. (2008). Effect of corner radius on the performance of CFRP-confined square concrete columns: Test. *Engineering Structures, 30*(2), 493–505.
Watanabe, K., Nakamura, H., Honda, Y., Toyoshima, M., Iso, M., Fujimaaki, T., Kaneto, M., & Shirai, N. (1997). Confinement effect of FRP sheet on strength and ductility of concrete cylinders under uniaxial compression. *3rd International Symposium (FRPRCS-3) on Non-Metallic (FRP) Reinforcement for Concrete Structures* (pp. 233–240). 14–16 October, Sapporo, Japan.
Willam, K. J., & Warnke, E. P. (1975). Constitutive model for the triaxial behavior of concrete. *International Association for Bridge and Structural Engineering Proceedings, 19*, 1–30.
Wu, H., Wang, Y., Yu, L., & Li, X. (2009). Experimental and computational studies on highstrength concrete circular columns confined by aramid fiber-reinforced polymer sheets. *ASCE Journal of Composites for Construction, 13*(2), 125–134.
Wu, Y. F., & Wang, L. M. (2009). Unified strength model for square and circular concrete columns confined by external jacket. *ASCE Journal of Structural Engineering, 135*(3), 253–261.
Wu, Y.-F., & Zhou, Y.-W. (2010). Unified strength model based on Hoek-Brown failure criterion for circular and square concrete columns confined by FRP. *Journal of Composites for Construction, 14*(2), 175–184.
Xiao, Y., & Wu, H. (2000). Compressive behavior of concrete confined by carbon fiber composite jackets. *Journal of Materials in Civil Engineering*, 12(2), 139–146.
Zinno, A., Lignola, G. P., Prota, A., Manfredi, G., & Cosenza, E. (2010). Influence of free edge stress concentration on effectiveness of FRP confinement. *ELSEVIER Composites: Part B, 41*(7), 523–532.

Chapter 6
Special Problems

Francesca Ceroni, Marisa Pecce, Christian Carloni, Thorsten Leusmann, Harald Budelmann, Emidio Nigro, Antonio Bilotta, Joaquim Barros, Inês Costa, Gian Piero Lignola, Annalisa Napoli and Roberto Realfonzo

Abstract This chapter gives an overview on the state-of-the-art about verifications of reinforced concrete structures using Externally Bonded (EB) Fibre Reinforced Polymers (FRP) under particular loading condition. Focus is mainly put on flexural strengthening, nowadays the most common application field for composite materials in structural engineering. The items discussed in this chapter are:

- Serviceability limit states;
- Fatigue behaviour;
- Effects of fire and high temperature;
- Long term behaviour;
- Anchoring systems;
- Mechanically Fastened Systems.

F. Ceroni (✉) · M. Pecce (✉)
University of Sannio, Benevento, Italy
e-mail: ceroni@unisannio.it

M. Pecce
e-mail: pecce@unisannio.it

C. Carloni
Department of Architecture, University of Hartford, Hartford, USA

T. Leusmann · H. Budelmann
Institute for Building Materials, Concrete Structures and Fire Protection, TU Braunschweig, Germany

E. Nigro · A. Bilotta
University Federico II, Naples, Italy

J. Barros · I. Costa
University of Minho, Mino, Portugal

G.P. Lignola
University of Napoli Federico II, Naples, Italy

A. Napoli · R. Realfonzo
University of Salerno, Salerno, Italy

C. Pellegrino and J. Sena-Cruz (eds.), *Design Procedures for the Use of Composites in Strengthening of Reinforced Concrete Structures*, RILEM State-of-the-Art Reports 19, DOI 10.1007/978-94-017-7336-2_6

Keywords FRP · Serviceability · Fatigue · Anchoring systems · Durability

Serviceability Limit States

Introduction

Both excessive cracking and excessive deformations in reinforced concrete elements may lead to drawbacks in service. *Appearance, tightness and durability* are normally considered as reasons for crack control. For durability considerations, the crack width in the vicinity of reinforcement is more influential than the crack width on the surface of the element. Cracking analysis of reinforced concrete elements externally bonded with FRP can be carried out considering the same principles of reinforced concrete sections.

In particular the cracks may decrease the durability performances, functionality and appearance of the structure or may endanger the integrity of the bond interface between FRP EBR and concrete.

The limitation of tensile stresses in concrete is an adequate measure to reduce the probability of cracking in tension. The limitation of compressive stresses in concrete aims to avoid or limit excessive compression, producing irreversible strains, longitudinal cracks (parallel to the compressive strains) and nonlinear creep phenomena.

In calculating the stress conditions, account shall be taken of whether the section is expected to crack under service loads. Moreover the effects of creep, shrinkage, relaxation of pre-stressing steel and differential temperatures should be taken into account.

The limitation of tensile stresses in the steel is an indirect method to control the cracks conditions in RC elements, since it is aimed to warrant an appropriate safety margin below the yield strength and, thus, prevent uncontrolled, large, permanently open cracks due to inelastic deformations of steel bars.

Finally, the deformability limit state has to be considered, because the limitation of deflections in RC elements is generally applied since excessive deflections may restrict the normal use of the structure, induce damage to not load-bearing members or negatively influence the appearance.

For RC elements, stresses are calculated using section properties corresponding to either the un-cracked or the fully cracked condition depending on the loading conditions. Both concrete and steel are assumed to be elastic, both in tension and in compression. When an external FRP reinforcement is present, it has to be considered as linear elastic. Thus, the calculation of stresses in concrete and steel follow the same rules used for RC elements under serviceability loading conditions, with the only difference of taking into account the external reinforcement in the analysis of the un-cracked and/or the fully cracked section.

Few test results are available for verification of serviceability limit states of RC elements strengthened with FRP EBR; most of tests available in literature have been carried out with reference to ties elements (Matthys 2000; Sato et al. 2002; Ueda et al. 2002; Zhang et al. 2003; Ceroni et al. 2004) and the codes do not always include design formulas.

Moreover, an increase in the load-carrying capacity by applying FRP laminates is not accompanied by a proportional increase in system rigidity, either in global terms (with reference to the values of deflections), or in local terms (with reference to the transfer of stresses at the FRP-concrete interface). Furthermore an increase of the ultimate capacity of an RC member due to the FRP strengthening does not correspond necessarily to a proportional increase of the service load; hence verification of serviceability could be significant in element design.

Since FRP materials have high tensile strength and Young's moduli, but small cross-sectional areas, in general their effect on the limitation of deflections can be negligible, unless considering very low reinforced section.

A more significant influence on the cracking pattern due to the local bond transfer produced by the external reinforcement can be observed (Ceroni and Pecce 2004).

Clearly, the use of pre-stressed FRP systems can be more and more useful to reduce the width of existing cracks, limit the deflections and control the stresses in critical sections of structural RC members (Kim et al. 2008).

Stress Limitation

Load combinations for verifications at SLS as specified in Eurocode 2 (EN 1992-1-2 2004) should be applied. Partial safety factors for the materials, γ_M, are taken equal to 1.0, except if specified otherwise.

Under the hypothesis that the section remains plane, the strains in the steel and in the FRP reinforcement are related to the concrete strains thanks to a linear relation. Due to the assumption of linear elastic behaviour for all materials, the stresses are obtained multiplying the strains by the elastic modulus. In particular, the stress in the FRP is obtained from the following relationship:

$$\sigma_f = E_f \cdot \varepsilon_f \tag{6.1}$$

where E_f is the mean value of the modulus of elasticity of the FRP reinforcement.

Moreover, existing strain at the time of FRP installation shall be accounted for and the principle of superposition can be used.

If the maximum tensile stress in the concrete is lower than its tensile strength, the section is un-cracked and fully active; on the contrary the section should be treated as cracked.

The modular ratios $\alpha_s = \frac{E_s}{E_c}$ and $\alpha_f = \frac{E_f}{E_c}$ have to be defined to transform the actual section into a homogenized all-concrete section. These values shall be set to account for creep as well as short and long-term conditions. In particular, under permanent loading conditions, a reduced value of the elastic modulus of concrete can be used to take into account the creep effect, while under not-permanent loading conditions the ratio of the effective moduli can be used.

For calculating the strain and stress distribution along the section a linear elastic analysis both of the un-cracked (state 1) and cracked section (state 2) should be carried out.

To this aim the cracking moment, M_{cr}, for a rectangular section with base b and height h, can be evaluated as:

$$M_{cr} \approx f_{ctm} \cdot \frac{b \cdot h^2}{6} \tag{6.2}$$

where f_{ctm} is the mean tensile strength of concrete [MPa]. If M is the rate of maximum moment in the element under the service loading conditions applied after the strengthening and M_0 is the moment applied before the strengthening, the stresses induced in the materials by the overall moment $M_k = M_0 + M$ can be evaluated adding the contributes of both moments.

If M_o is higher than M_{cr}, the analysis can be developed referring to the cracked inertia (state 2) of the section for calculating both the stresses in concrete and steel reinforcement under M_o and the stresses in concrete, steel and FRP reinforcement under M; the total stress in concrete and steel is the sum of the two contributions.

On the contrary, if M_0 is lower than the cracking moment M_{cr} and the total moment M_k is greater than M_{cr}, the inertia of the un-cracked section has to be used for calculating the stresses in concrete and steel reinforcement under M_0, while the cracked inertia has to be used for calculating the stresses in concrete, steel and FRP reinforcement under M.

The case of both values M_0 and M_k lower than M_{cr} can be considered not significant for the usual applications.

When cracking has to be avoided, a limit state of decompression can be assumed for verifications and corresponds to have a zero stress at the extreme fibre of the concrete section.

The compressive stresses in the concrete should be limited to $0.6{\cdot}f_{ck}$ under the characteristic combination of loading and $0.45{\cdot}f_{ck}$ under the most unfavourable quasi-permanent load combination.

About the limitations of the tensile stress in the steel reinforcement, under the characteristic combination of actions the limit $0.8{\cdot}f_{yk}$ should not be exceeded.

The stress limit above introduce are in agreement with the indications of Eurocode 2.

The FRP stress under service load should be limited in order to avoid excessive creep or creep rupture of the FRP:

$$\sigma_f \leq \eta \cdot f_{fk} \tag{6.3}$$

for the most unfavourable quasi-permanent load combination.

The FRP stress limitation coefficient, $\eta < 1$, depends on the type of FRP and should be obtained through experiments. The Italian Guidelines (CNR DT 200/R1 2012) suggest the following ranges depending on the type of fibres: 0.75–0.95 for internal environment, 0.65–0.85 for external environment, and 0.50–0.85 for aggressive environment. The lower values are for glass fibres and the upper for carbon fibres. The same values are furnished also in ACI 440.2R-08.

Control of Cracking Phenomena

Limitation of Longitudinal Cracks

Application of the FRP reinforcement substantially changes the cracking scenario of the element, since tension stiffening phenomena develop not only at the steel-concrete interface, but also at the FRP-concrete interface. In RC elements strengthened with FRP, the crack width is thus generally smaller than for un-strengthened elements either considering the same service load or considering the same tension level in the steel, due, in both cases, to the additional tension stiffening of the external reinforcement that reduces the crack spacing. Since new cracks will appear in between existing cracks, in general, a more diffuse crack patterns, with smaller crack widths, are observed.

A real crack bridging effect due to the external FRP reinforcement has been observed in experimental tests with a further tension stiffening effect in addition to the one produced by the internal steel reinforcement (Yoshizawa and Wu 1999; Tripi et al. 2000; Matthys 2000; Ueda et al. 2002; Ceroni et al. 2004; Ceroni and Pecce 2007; Ferrier et al. 2003). Moreover as the stiffness of the FRP strengthening grows the global tension stiffening results considerably increased. In Ceroni and Pecce (2007), bending tests on beams externally bonded both with carbon and steel cords sheets were carried out and evidenced that: (1) the steel cords and carbon fibres, both impregnated with epoxy, gave very similar results when the equivalent reinforcement percentage was the same, (2) when the steel cords were bonded with cementitious grout, the tension stiffening effect was lower compared with the epoxy.

Modelling of tension stiffening in RC elements is based on many experimental tests and on a consolidated knowledge of the steel–concrete bond. On the contrary the tension stiffening effect and the cracking behaviour has not been well investigated yet for RC elements externally bonded with FRP, due also to the lack of experimental results concerning this aspect. Numerical models taking into account the tension stiffening effects of the external FRP reinforcement can be used to predict the crack width (Ceroni and Pecce 2004; Aiello and Ombres 2004; Ferretti and Savoia 2003). However for design purposes, empirical equations based on regression analysis of experimental data can be obtained to calculate directly the crack width (Tan and Saha 2008) or the crack spacing (Ceroni and Pecce 2009).

Applying the technical report (fib bulletin 14 2001) the crack width can be estimated by using the Model Code approach (Model code 90) and a specific formula is furnished for the crack spacing for taking account the external FRP reinforcement.

$$\begin{aligned} s_{rm} &= \frac{2 \cdot f_{ctm} \cdot A_{c,eff}}{\tau_{fm} \cdot u_f + \tau_{sm} \cdot u_s} \\ \tau_{s,m} &= 1.8 \cdot f_{ctm} \quad \tau_{f,m} = 1.25 \cdot f_{ctm} \end{aligned} \tag{6.4}$$

where f_{ctm} is the mean tensile strength of concrete [MPa], u_s and u_f are the perimeter of the steel bar and FRP laminates [mm] bonded to concrete, t_f the thickness of FRP [mm], $\tau_{s,m}$ and $\tau_{f,m}$ the bond stresses [MPa] along the steel-concrete and the FRP-concrete interfaces, assumed constant in s_{rm}.

According to American guidance ACI 440.2R-08, verification of cracking under service loads can be done by applying the provisions of ACI 318-11 (2013) for RC elements. The FRP external reinforcement has to be taken into account in the calculation of the inertia of the transformed section. In particular in (ACI 318-95) an empirical formula is proposed to evaluate directly the maximum crack width, w, without evaluating crack spacing:

$$w = 2 \cdot \frac{f_f}{E_f} \cdot \beta \cdot k_b \cdot \sqrt{d_c^2 + \left(\frac{s}{2}\right)^2} \tag{6.5}$$

where E_f and f_f are the modulus of Young and the tensile stress in the steel reinforcement, respectively, k_b is a bond parameter, experimentally calibrated at 1 for ribbed steel, β is the ratio of distance between neutral axis and tension face to distance between neutral axis and centroid of internal reinforcement, d_c is the concrete cover, s is the bar spacing. The presence of the external reinforcement is taken into account only in the calculation of the inertia.

Currently in the Italian Guidelines (CNR DT 200/R1 2012) no specific formulas are available for calculating crack width in RC elements externally bonded with FRP materials, but only general indications referring to well-known approach as the one proposed by Model Code or Eurocode 2.

Under a stabilized cracking condition, the indications given by Eurocode 2 (EN 1992-1-2 2004) and the new Model Code (MC 2010) for RC elements can be, indeed, extended to RC elements externally bonded with FRP materials. Verification of serviceability in the new Model Code for RC elements includes again a cover term in the crack spacing formula in order to emphasize possible deformations in the concrete cover. The cracking model kept the philosophy that the maximum crack width (called herein as design crack width) is the multiple of $2 \cdot l_{s,max}$ (slip lengths) and the average strain differences between two cracks.

According to the new Model Code, the design crack width, w_d may be calculated by:

$$w_d = 2 \cdot l_{s,\max} \cdot (\varepsilon_{sm} - \varepsilon_{cm} - \varepsilon_{cs}) \tag{6.6}$$

$$l_{s,\max} = k \cdot c + \frac{1}{4} \cdot \frac{f_{ctm}}{\tau_{bms}} \cdot \frac{\phi}{\rho_{s,ef}} \tag{6.7}$$

$$\varepsilon_{sm} - \varepsilon_{cm} = \frac{\sigma_s - \beta \cdot \sigma_{sr}}{E_s} = \frac{\sigma_s}{E_s} \cdot \left(1 - \beta \cdot \frac{\sigma_{sr}}{\sigma_s}\right) = \frac{\sigma_s}{E_s} \cdot \left(1 - \beta \cdot \frac{M_{cr}}{M_{\max}}\right) \tag{6.8}$$

where:

- σ_s is the steel stress in a crack,
- σ_{sr} is the maximum steel stress in a crack at the crack formation stage which for pure tension, defined as:

$$\sigma_{sr} = \frac{f_{ctm}}{\rho_{s,ef}} (1 + \alpha_e \rho_{s,ef}) \tag{6.9}$$

where:

$$\rho_{s,ef} = \frac{A_s}{A_{c,ef}} \qquad \alpha_e = \frac{E_s}{E_c} \tag{6.10}$$

and $A_{c,ef}$ is the effective area of concrete in tension, defined as follows:

$$A_{c,ef} = b \cdot \min\left[2.5 \cdot c, \left(\frac{d-x}{3}\right), \frac{d}{2}\right] \tag{6.11}$$

b, d and c are the width, the depth and the inferior cover of the concrete element;

- β is an empirical coefficient to assess the mean strain over $l_{s,max}$ depending on the type of loading (0.6 for short term loads, 0.4 for long term loading);
- k is an empirical parameter to take the influence of the concrete cover into consideration; according to the present knowledge, k = 1.0 can be assumed;
- η_r is a coefficient for considering the shrinkage contribution;
- ε_{sh} is the free shrinkage strain.

Note that in Eurocode 2 (EN 1992-1-2 2004), the following provision for the maximum crack spacing is provided:

$$w_k = s_{r,\max} \cdot (\varepsilon_{sm} - \varepsilon_{cm}) \tag{6.12}$$

$$s_{r,\max} = 3.4 \cdot c + 0.425 \cdot k_1 \cdot k_2 \cdot \frac{\phi}{\mu_s} \tag{6.13}$$

$$\varepsilon_{sm} - \varepsilon_{cm} = \frac{\sigma_s}{E_s} \cdot - k_t \left[\frac{f_{ctm} \cdot A_{c,eff}}{E_s \cdot A_s} + \frac{f_{ctm}}{E_{cm}} \right] \tag{6.14}$$

- k_t a factor of load duration (0.6 for short and 0.4 for long term loading);
- k_1 is a bond coefficient: 0.8 for ribbed and 1.6 for smooth steel bars;
- k_2 takes into account type loading: 0.5 for flexural and 1.0 for tensile loading.

When FRP reinforcement is present, the percentage of reinforcement normalized to the effective area of concrete in tension can be modified as follows:

$$\rho_{eff,eq} = \rho_{p,eff} + \rho_{f,eff,eq} = \frac{A_s}{A_{c,eff}} + \frac{E_f}{E_s} \cdot \frac{A_f}{A_{c,eff}} \tag{6.15}$$

being A_s, E_s and A_f, E_f the area and the Young's modulus of the internal steel and the external FRP reinforcement, respectively.

Moreover, the expression of $2l_{s,max}$ can be calculated by means of specific formulae according to the type of FRP reinforcement. In particular, Ceroni and Pecce (2009) proposed a formulation for the average and the characteristic value of the crack spacing in RC elements externally bonded with FRP sheets. The formulation is empirically based on a best-fitting procedure considering experimental results about crack spacing according to a 'design by testing' procedure suggested in Eurocode 0 (EC0, Monti et al. 2009).

The database used for assessing the crack spacing is made by experimental results of RC beams and ties externally bonded with FRP sheets carried out by the authors in different experimental programs: beams (Ceroni and Pecce 2007; Ceroni 2010) and ties (Ceroni et al. 2004). The experimental results of these tests evidenced that at the same steel stress as more cracks form and the crack spacing is reduced, crack width decreases depending on the amount of external reinforcement that influences the total load applied. Considering the full load history, the tension stiffening effect due to the fibres grows for all the types of beams after the steel yielding due to the elastic behaviour of FRP. The external reinforcement produces the highest tension stiffening effect for beams with the lowest steel reinforcement. The greater the amount of FRP is used, the greater are the effects. In the tests on RC tie-specimens carried out according to the same dimensions and set-up used by Matthys (2000), a pre-load cycle established the crack spacing of the unstrengthened element and, after application of the external FRP reinforcement, new cracks formed about in the middle. Ceroni and Pecce (2009) made several comparisons in terms of crack spacing by defining a variable δ as the ratio of the code to the experimental value. For the provision given by EC2 (EN 1992-1-2 2004), where the FRP reinforcement is taken into account by the effective reinforcement percentage, the average value of δ is 0.48 meaning that the experimental values are overestimated; this can be justified by the assumption that

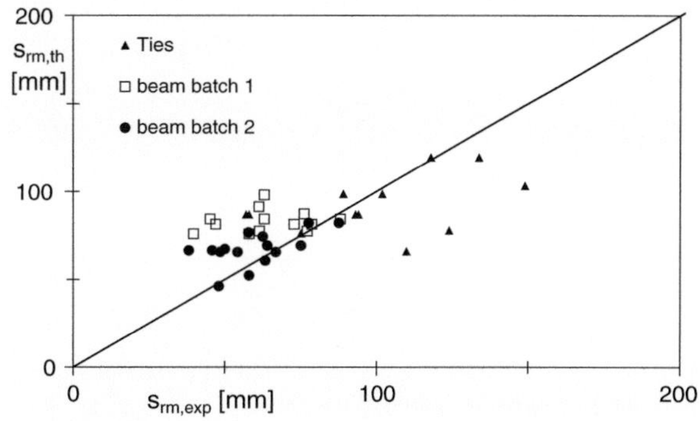

Fig. 6.1 Experimental versus theoretical average values of crack spacing given by Eq. (6.16)

such a formulation provided 'maximum' values of crack spacing. On the contrary, for the fib bulletin the mean value of variable δ is 2.14 which means a large underestimation of the experimental results.

The structure of the formulation proposed by Ceroni and Pecce (2009) for crack spacing is similar to Eq. (6.7) or (6.13) and the stiffening effect of the internal steel and the external FRP reinforcement is separately taken into account.

$$s_{rm} = s_0 + k \cdot \frac{A_{c,ef}^{\gamma} \cdot \phi^{\alpha}}{A_s^{\delta} + \left(\frac{A_f \cdot E_f}{E_s}\right)^{\beta}} \tag{6.16}$$

where s_0, k, α, β, γ and δ are parameters calibrated by the statistical procedure. In Fig. 6.1, the comparison between experimental and theoretical results given by Eq. (6.16) is reported assuming the following values of the parameters: $s_0 = 20$ mm, $k = 4$; $\alpha = 1$; $\beta = 0.75$; $\gamma = 0.5$; $\delta = 0.75$.

Equation (6.16) refers to the average value of the crack spacing, while for characteristic provision the following can be assumed:

$$s_{rm,th,k} = s_{rm,th,95\%} = 1.6 \cdot s_{rm} \tag{6.17}$$

In terms of crack width a more extended database was collected (107 data) by adding results of (Yoshizawa and Wu 1999; Matthys 2000). Equation (6.16) has been introduced in Eq. (6.12) for calculating the crack width: Fig. 6.2 shows the experimental-theoretical comparisons (mean values of δ = 0.777 and 0.715 for 64 and 107 data, respectively, CoV = 32 and 37 %).

On the contrary, the formulas of EC2 (EN 1992-1-2 2004) (Eqs. 6.12–6.14), adjusted to take into account the FRP external reinforcement, furnishes a mean value of δ = 0.67 (CoV = 50 %).

A model for considering the influence of NSM on the cracking is reported in (Zehetmaier and Zilch 2008) that based the formulation on experimental studies on NSM systems made of CFRP strips. The authors propose a modification both for

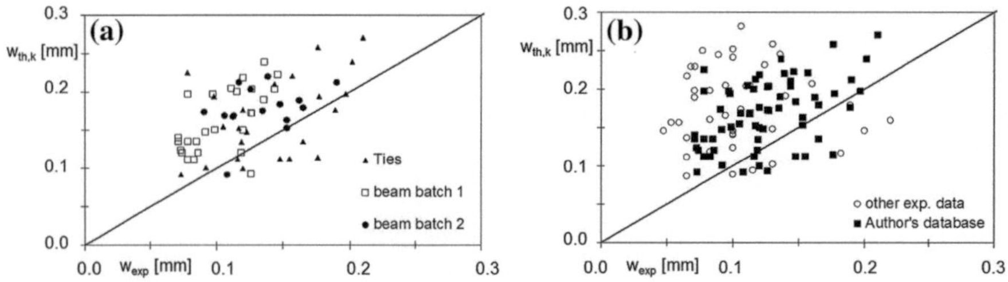

Fig. 6.2 Experimental versus characteristic theoretical values of crack width (Eqs. 6.12 and 6.16). **a** Authors' results, **b** extended database

the expression of the mean strain and the slip length in presence of NSM reinforcement.

In general, it is worth noticing that in reinforced concrete elements the spacing of internal stirrups can influence the crack spacing both with and without the FRP strengthening.

Verification of Bond Interface

Under service loading conditions, the initiation of bond interface cracks should be prevented as they may reduce the long-term integrity of the bond interface zone under e.g. cyclic loading and freeze/thaw actions. Concentrations of stresses develop especially at the end of the FRP reinforcement and the location of flexural or shear cracks.

At these locations, to avoid local debonding phenomena under the quasi-permanent load condition, the maximum principal stress, calculated based on the shear stress and the normal stress according to a linear elastic analysis (e.g. Taljsten 2004; Roberts 1989), should be smaller than the tensile strength of concrete.

According to the Italian guideline (CNR DT 200/R1 2012), the verification is developed checking that an equivalent tangential stress $\tau_{b,e}$ at the adhesive-concrete interface, under the rare or frequent load condition, is lower than the bond strength f_{bd} of the FRP strengthening–concrete interface. The equivalent tangential stress $\tau_{b,e}$ is defined through a mean tangential stress τ_m evaluated in the section at the FRP strengthening–concrete interface according to the Jourawski's formulation (Jourawski 1858) and a factor, k_{id}, that introduces by a simplified way the effect of the normal stresses.

The stresses have to be calculated under the only rate of loads applied after the adding of the FRP strengthening.

As in flexural strengthening, also in shear and torsion strengthening the externally bonded reinforcement shall not have debonding phenomena at the serviceability limit state.

Limitation of Deflections

The low axial stiffness $E_f A_f$ of FRP external reinforcement is often insufficient to reduce significantly curvatures and deflection in the strengthened beams under service load condition if the internal steel reinforcement is not yielded. On the contrary, a relevant stiffening effect is given by the external reinforcement, after the steel is yielded and is clearly proportional to the axial stiffness of the external FRP reinforcement (Ceroni et al. 2004). Also steel fabrics externally bonded to concrete beams (Balsamo et al. 2013a; Pecce et al. 2006) have evidenced an effectiveness into limiting the deflection after the yielding of internal steel reinforcement. Moreover, the bending tests reported by Balsamo et al. (2013b), evidenced that no significant difference was in using epoxy adhesive or cement-based mortar for impregnating the steel fabrics in term of deformability.

For the evaluation of deflections, the same method used for RC elements can be implemented to take into account the tension stiffening effect of both the internal and external reinforcement: (1) the most refined one is based on the double integration of the curvature, which can be determined by a cross-section analysis along the RC element; (2) a simplified calculation is based on the definition of an effective moment of inertia (Branson 1977; ACI 2005; El-Mihilmy and Tedesco 2000; Bischoff 2007) or on the calculation of a mean deflection according to the Eurocode (EN 1992-1-2 2004) and Model Code 2010 (fib bulletin 65 2012 and 66 2012) approaches.

The more refined procedure is able to take into account the tension stiffening effects in RC members due to both the internal and external strengthening based on the correspondent bond stress-slip laws (Ceroni and Pecce 2004; Aiello and Ombres 2000; Matthys 2000). Closed form equations have been also obtained assuming bi or tri-linear simplified moment–curvature responses (Razaqpur et al. 2000; Charkas et al. 2003; Rasheed et al. 2004).

The simplified formulation according to the Eurocode and Model Code approach is based on the calculation of the deflection in the FRP externally strengthened elements under the hypothesis of un-cracked and cracked section and on a 'tension stiffening coefficient' that synthesizes the tension stiffening phenomena along the element between two cracks due to both the internal and external reinforcement:

$$\alpha = \alpha_1 \cdot (1 - \zeta) + \alpha_2 \cdot \zeta \quad \zeta = 1 - \beta \cdot \left(\frac{\sigma_{sr}}{\sigma_s}\right)^2 \tag{6.18}$$

where:

- β is a coefficient taking into account the loading type (1.0 for single short-term loads and 0.5 for sustained loads or many cycles of repeated loading);
- σ_s is the tensile stress in the steel reinforcement under the service loading condition in the cracked section;

– σ_{sr} is the stress in the steel reinforcement calculated in the cracked section under the loading condition causing the first cracking (see Eq. 6.9).

The ratio σ_{sr}/σ_s may be replaced by M_{cr}/M for flexure and N_{cr}/N for pure tension, where M_{cr} is the cracking moment and N_{cr} is the cracking force and M and N are the maximum moment and normal force in the load combination considered.

The deflection at the un-cracked state, α_1, and the cracked state, α_2, can be calculated by an elastic analysis, referring to the flexural stiffness of the un-cracked section and the cracked state, respectively.

In the calculation of Eq. (6.18) the presence of the FRP external reinforcement has to be taken into account in the computation of the moment of inertia of the cracked section.

According to ACI 440.2R-08, the deflection under service loads can be calculated by applying the provisions of ACI 318-11 for RC elements. The FRP external reinforcement has to be taken into account in the calculation of the effective inertia, I_e, of the transformed section:

$$I_e = I_g \cdot \left(\frac{M_{cr}}{M_a}\right)^3 + I_{cr} \cdot \left[1 - \left(\frac{M_{cr}}{M_a}\right)^3\right] \tag{6.19}$$

where I_g and I_{cr} are the inertia of the un-cracked and cracked section, respectively, M_{cr} and M_a are the cracking and the maximum moment along the element.

In the case of NSM strengthening there are still few experimental results (Barros and Fortes 2005; Barros et al. 2007; Ceroni 2010; Balsamo et al. 2013b) to validate design formulations, but in general the same approach used for EBR systems can be assumed effective, introducing the specific modified moment of inertia. In particular, according to Ceroni (2010) and Balsamo et al. (2013b), the tension stiffening effect on the deflection provided by the NSM systems was founded to be less effective compared with the EBR technique. This was probably due to the application of the FRP reinforcement in the grooves that could lead to have higher slips along the FRP-concrete interface. In Fig. 6.3 the experimental load-deflection of the reference unstrengthened RC beams is compared with the curves of two equal beams strengthened with 2 and 3 CFRP strips, evidencing the low effect of the reinforcement on the beam stiffness.

Fatigue Behaviour

Introduction

Increasing traffic loads and aggressive environmental conditions are leading to an accelerated aging of reinforced concrete bridges and a loss in their load-carrying capacity. A well-established method to counteract effects of aging and increasing traffic loads is to strengthen bridges with fibre-reinforced polymers (FRP).

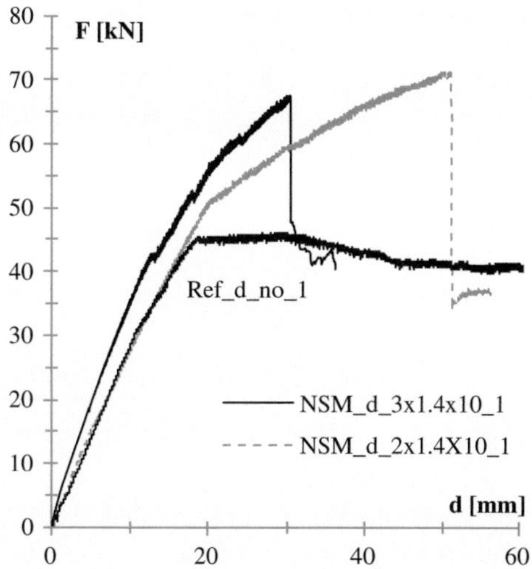

Fig. 6.3 Load-deflection curves of reference and NSM strengthened beams (Balsamo et al. 2013b)

With FRP strengthening, two objectives can be pursued: the increase of the load-carrying capacity and the extension of service life. The FRP material shows symptoms of fatigue when applied as an external reinforcement. The effect of cyclic loads on FRPs and hybrid FRPs is described in Wu (2010). Especially externally bonded FRPs (EBR) are affected by traffic induced vibrating and fluctuating loads, because cyclic loads also lead to a fatigue of concrete to FRP bond. The FRP debonding process under fatigue loading is the main objective in this chapter.

Experimental Tests

The debonding process of FRP under fatigue loading has been tested with different types of experimental tests. Commonly large scale bending tests at reinforced concrete beams strengthened with FRP under cyclic loading were carried out; see Heffernan (2004) and Kim (2008). In many cases fatigue loading leads to a failure of the interior reinforcing steel, in some cases FRP debonding could be observed followed by steel failure, see Kim (2008). If debonding is prevented, FRP strengthening leads to a higher load range or an increasing number of cycles at the strengthened RC beam, see Fig. 6.4.

According to Kim (2008), notable fatigue damage in FRP-strengthened beams is accumulated within the first cycles. The rate of damage accumulation then slows considerably until a linear damage rate is reached. Prior to fatigue failure of beams, rising damage propagation can be observed. The propagation of flexural cracks, maximum crack width, midspan deflection and steel strain show a similar trend.

These results are global measurements and do not attempt to directly evaluate the applied force in the external FRP reinforcement or the strain distribution along the

Fig. 6.4 Schematic of load range versus load cycle curves of RC beams according to Kim (2008)

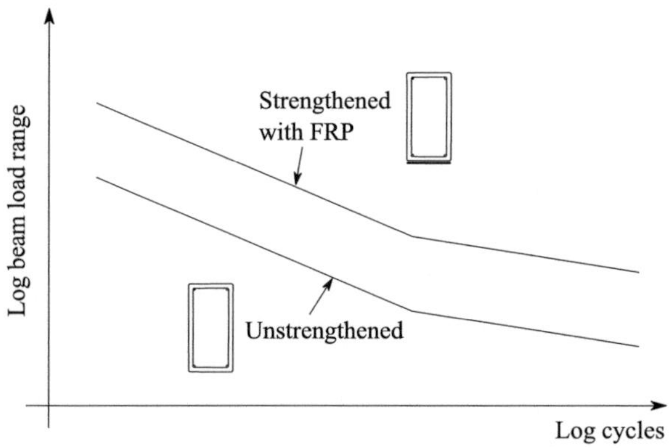

fibres. Furthermore, the applied load range is not directly related to the load-carrying capacity of the interface, but rather to the load-carrying capacity of the beam. The published experimental work on the fatigue response of strengthened beams does not provide an insight into the fatigue response of the FRP-concrete interface.

A few studies, which directly investigate the fatigue response of the FRP-concrete interface using direct-shear tests of FRP-concrete joints, have been reported in Bizindavyi et al. (2003), Budelmann (2013), Carloni (2012, 2013), Dai (2005), Diab (2009), Ferrier (2005) and Ko (2007). The maximum, F_L^O, and minimum, F_L^U, load of the fatigue cycle are defined as a percentage of the load-carrying capacity of an equivalent specimen under monotonic quasi-static conditions. The available literature suggests that the fatigue response depends on the load range, mean value, and frequency.

The most common test set-ups for direct-shear tests are described in Yao (2005). Most of the setups were also used for investigations on the fatigue response of FRP bond. Near-end supported single shear tests were carried out by Bizindavyi et al. (2003) and Carloni (2012, 2013). Far-end supported double shear tests by Ferrier (2005) and near end supported double shear tests by Budelmann (2013) and modified beam tests were carried out by Dai (2005). An extensive literature review can be found in Carloni (2013) and Kim and Heffernan (2008).

Bond Stress-Slip Behaviour Under Fatigue Loads

With an increasing number of fatigue cycles, the stiffness of the bond stress-slip response decreases (Bizindavyi et al. 2003; Carloni 2013; Ferrier 2005). This indicates that fatigue damage propagates at the interface between FRP and concrete. The translation in self-similar manner of a certain strain profile along the bond length supports this damage propagation theory such that the point, where the free strain at debonding $\bar{\varepsilon}_{yy}^{fatigue}$ is reached, propagates from the loaded end toward the

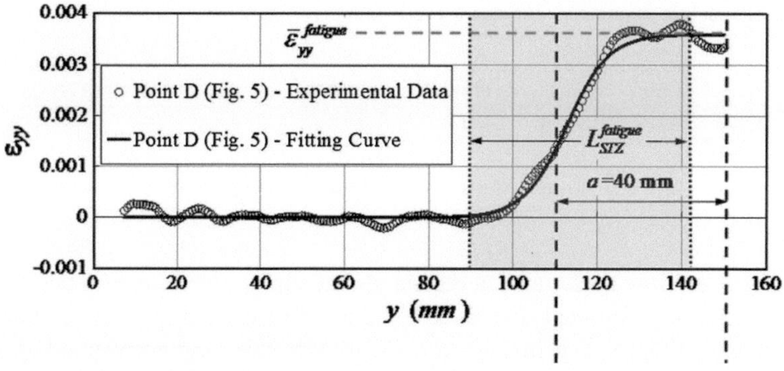

Fig. 6.5 FRP strain profile determined with DIC

other end (see Bizindavyi et al. 2003; Budelmann 2013; Carloni 2013; Ferrier 2005). A similar behaviour was observed by Hankers (1996) at externally bonded steel plates under cyclic loads.

Generally the strain profiles were measured with strain gauges. In Carloni (2013) the strain components were determined on the surface of the FRP and surrounding concrete during fatigue and monotonic tests from the displacement field, which was measured using a full-field optical technique known as digital image correlation (DIC) see (Sutton 1983, 2009).

An example of longitudinal strain distribution along the direction of the fibres determined with DIC is plotted in Fig. 6.5. Details on the strain analysis can be found in Carloni (2012, 2013) and Ali-Ahmad et al. (2006).

The nonlinear strain distribution was approximated by the following function after Ali-Ahmad et al. (2006):

$$\varepsilon_{yy} = \varepsilon_0 + \frac{\alpha}{1 + e^{\frac{y - y_0}{\beta}}} \tag{6.20}$$

where α, β, ε_0, y_0 were determined using nonlinear regression analysis of the computed strains. This approach was previously used for static tests and it can be used for fatigue as well.

The approximated strain distribution along the FRP obtained from Eq. (6.20) is also shown in Fig. 6.5. The stress transfer zone (STZ) can be identified in Fig. 6.5 as the intermediate region where the load is actually transferred from the FRP to the substrate. The length of STZ is the effective length whereas STZ itself is the fracture process zone of the interface. The length of the STZ (L_{STZ}) is termed effective bond length or anchorage length. The constant value of the strain in the fully debonded region was identified as $\bar{\varepsilon}_{yy}$.

From the best-fit strain distribution, the cohesive material law parameter $\tau_{max}^{fatigue}$ and the fracture parameter $G_F^{fatigue}$ were obtained. In Table 6.1 the average values of the fracture parameters and the value of the strain at debonding $\bar{\varepsilon}_{yy}^{fatigue}$, calculated at

Table 6.1 Fracture parameters during fatigue loading

Test #	$L_{STZ}^{fatigue}$ (mm)	$\bar{\varepsilon}_{yy}^{fatigue}$ ($\mu\varepsilon$)	$\tau_{max}^{fatigue}$ (N/mm^2)	$G_F^{fatigue}$ (N/mm)	a (mm)	P_{crit}^{Cycles} (N/mm)	g_D (mm)
DS-FT_1 [9]	50	3500	6.0	0.22	0	3.7	0.25
DS-FT_2 [9]	50	4000	6.7	0.27	68	3.6	0.30
DS-FT_3 [9]	56	4100	3.9	0.29	100	3.3	0.50

the peak of ten slow cycles after a certain global slip g_D, measured by two LVDTs, was reached, are reported for three fatigue tests carried out by Carloni and Subramaniam (2013). Ten images, processed with DIC and corresponding to the peaks of the ten slow cycles, were used to calculate the average values reported in Table 6.1. The length of the STZ during fatigue loading $L_{STZ}^{fatigue}$ appeared to be smaller than the one reported in static tests.

Carloni (2013) postulated that the sub-critical crack growth might occur in the epoxy layer rather than in a thin mortar-rich layer of concrete, as typically occurs in monotonic quasi-static tests. A similar observation was found in Carloni (2012) and in Ferrier (2005), in which the authors noticed that the fatigue performances of FRPs were greatly influenced by the physical and mechanical properties of the epoxy. This circumstance was supported by the visual analysis of the FRP sheets after failure. In fact, the debonded surface of the FRP strip was smoother in the first 40 mm close to the loaded end, which approximately corresponded to the length of the cohesive crack a during fatigue (see Carloni 2013).

Carloni (2013) used the relationship of Eq. (6.21) between the load-carrying capacity and the fracture energy of the FRP-concrete interface to prove indirectly that the load at the peak of the ten cycles P_{crit}^{Cycles}, after reaching the slip g_D, was related to the corresponding fracture energy $G_F^{fatigue}$ and, therefore, the approximation of the strain profiles with Eq. (6.20) was acceptable.

$$P_{crit}^{Cycles} = b_l \cdot \sqrt{2 \cdot G_F^{fatigue} \cdot E_f t_f} \tag{6.21}$$

P_{crit}^{Cycles} is provided in Table 6.1 for the three tests and should be compared to the applied load at the peak of the cycles P_{max}, which was equal to 4.2 kN. The calculated loads P_{crit}^{Cycles} were in good agreement with the applied load P_{max}.

Database

From the experimental test data described in Bizindavyi (2003), Budelmann (2013), Carloni (2012), Dai (2005) and Ferrier (2005) the S-N curve can be determined. For determining the number of load cycles N_{30} needed for reaching a debonded length of 30 mm, a linear increase of the debonded length is assumed. The number of load cycles needed for reaching a decoupled length of 30 mm is calculated for the tests from Bizindavyi (2003), Carloni (2012), Dai (2005) and Ferrier (2005) by linear interpolation using the bond length l_b and the number of load cycles until full decoupling. For the tests from (Budelmann 2013) the number of load cycles is taken from the strain measurements. A decoupling length of 30 mm is reached when the strain measured with the strain gauge A0 or B0 placed 30 mm from the loaded end exceeds the free strain measured with the strain gauge A1 or B1 located in the unbounded region of the CFRP-plate.

For fitting the S-N curve unified related load ranges $S_{0;i}$ at a lower load level of 0 and the corresponding number of load cycles is needed. The unified load range $S_{0;i}$ is determined in a projection analysis using the Goodman relation and Eq. (6.22).

$$S_{0,i} = \frac{\Delta F_{L,0,i}}{F_{Lb}} = \frac{\frac{F_L^O - F_L^U}{F_{Lb}}}{1 - \frac{F_L^U}{F_{Lb}}} = \frac{F_L^O - F_L^U}{F_{Lb} - F_L^U} \tag{6.22}$$

Experimental data from Bizindavyi (2003), Budelmann (2013), Carloni (2012), Dai (2005) and Ferrier (2005) with the calculated number of load cycles N_{30} and the corresponding load range S_0, as well as the width b_L, the thickness t_L, the number of layers n_L and Young's modulus E_L of the tested FRP, can be found in Tables 6.2 and 6.3. Figure 6.6 shows the S-N curve fitted to the experimental data compared with the approach given in the DAfStb-guideline (2012). The load range α is the difference between maximum and minimum load $F_L^O – F_L^U$ related to the monotonic quasi-static load-carrying capacity of the interface F_{Lb}.

A fracture mechanics based method to describe the fatigue behaviour was proposed by Diab (2009) and successively modified by Carloni and Subramaniam (2013).

In Diab (2009) the rate of debonding growth da/dN is related to the interfacial fracture energy:

$$\frac{da}{dN} = m_1 \left(\frac{G_F^{fatigue}}{G_F} \right)^{n_1} \cdot \beta \tag{6.23}$$

The coefficients m_1, n_1 and β can be determined from experimental results. In particular, β takes into account that the crack propagation rate decreases as the debonded region increases. In Eq. (6.23) an additional coefficient, related to the effect of the frequency, was considered equal to 1. $G_F^{fatigue}$ is the interfacial fracture energy during fatigue loading, see Table 6.1. Its relationship with the amplitude and mean value of the load range has not been investigated in the available literature.

Carloni and Subramaniam (2013) modified Diab's formula, as the relationship between the applied load and fracture energy during cycles was indirectly proven:

Table 6.2 Experimental data from Bizindavyi (2003)

Ref.	Test	FRP width b_L (mm)	FRP thickness t_L (mm)	FRP Young's Modulus E_L (N/mm^2)	Number of layers n_L (mm)	Bond length l_b (mm)	Maximum load F_L^O (N)	Minimum load F_L^U (N)	Bond capacity F_{Lb} (N)	Number of load cycles N_{30} (–)	Load range S_0 (–)
Biz.[a]	C1-1a	25.4	0.33	75,700	1	300	8500	0	6480	30	0.76
Biz.[a]	C1-2a	25.4	0.33	75,700	1	300	8500	0	5029	17,323	0.59
Biz.[a]	C1-2b	25.4	0.33	75,700	1	300	8500	0	5029	28,656	0.59
Biz.[a]	C1-3a	25.4	0.33	75,700	1	300	8500	0	4572	21,025	0.54
Biz.[a]	C1-3b	25.4	0.33	75,700	1	300	8500	0	4572	29,790	0.54
Biz.[a]	C1-3c	25.4	0.33	75,700	1	300	8500	0	4572	34,739	0.54
Biz.[a]	C1-3d	25.4	0.33	75,700	1	300	8500	0	4572	67,608	0.54
Biz.[a]	C1-4a	25.4	0.33	75,700	1	300	8500	0	3505	186,222	0.41
Biz.[a]	C1-4b	25.4	0.33	75,700	1	300	8500	0	3505	202,950	0.41
Biz.[a]	C2-1a	50.8	0.33	75,700	1	160	17,000	0	11,135	624	0.66
Biz.[a]	C2-1b	50.8	0.33	75,700	1	160	17,000	0	11,135	788	0.66
Biz.[a]	C2-2a	50.8	0.33	75,700	1	160	17,000	0	9185	6526	0.54
Biz.[a]	C2-2b	50.8	0.33	75,700	1	160	17,000	0	9185	26,832	0.54
Biz.[a]	C2-3a	50.8	0.33	75,700	1	160	17,000	0	7071	34,674	0.42
Biz.[a]	C2-3b	50.8	0.33	75,700	1	160	17,000	0	7071	93,133	0.42
Biz.[a]	C3-1a	50.8	0.33	75,700	1	160	17,000	0	11,460	29	0.67
Biz.[a]	C3-1b	50.8	0.33	75,700	1	160	17,000	0	11,460	60	0.67
Biz.[a]	C3-2a	50.8	0.33	75,700	1	160	17,000	0	9453	113	0.56
Biz.[a]	C3-2b	50.8	0.33	75,700	1	160	17000	0	9453	407	0.56
Biz.[a]	C3-3a	50.8	0.33	75,700	1	160	17,000	0	7453	40,022	0.44

(continued)

Table 6.2 (continued)

Ref.	Test	FRP width b_L (mm)	FRP thickness t_L (mm)	FRP Young's Modulus E_L (N/mm^2)	Number of layers n_L (mm)	Bond length l_b (mm)	Maximum load F_L^O (N)	Minimum load F_L^U (N)	Bond capacity F_{Lb} (N)	Number of load cycles N_{30} (–)	Load range S_0 (–)
Biz.[a]	C3-3b	50.8	0.33	75,700	1	160	17,000	0	7453	40,352	0.44
Biz.[a]	C4-1a	50.8	0.66	75,700	2	300	30,200	0	17496	37	0.58
Biz.[a]	C4-1b	50.8	0.66	75,700	2	300	30,200	0	17496	39	0.58
Biz.[a]	C4-2a	50.8	0.66	75,700	2	300	30,200	0	14402	190	0.48
Biz.[a]	C4-2b	50.8	0.66	75,700	2	300	30,200	0	14402	1062	0.48
Biz.[a]	C4-3a	50.8	0.66	75,700	2	300	30,200	0	11232	7650	0.37
Biz.[a]	C4-3b	50.8	0.66	75,700	2	300	30,200	0	11232	26,687	0.37
Biz.[a]	C5-1a	50.8	0.66	75,700	2	300	30,200	0	18882	38	0.63
Biz.[a]	C5-1b	50.8	0.66	75,700	2	300	30,200	0	18882	97	0.63
Biz.[a]	C4-2a	50.8	0.66	75,700	2	300	30,200	0	16825	195	0.56
Biz.[a]	C5-2b	50.8	0.66	75,700	2	300	30,200	0	16825	1430	0.56
Biz.[a]	C5-3a	50.8	0.66	75,700	2	300	30,200	0	15728	18,682	0.52
Biz.[a]	C5-3b	50.8	0.66	75,700	2	300	30,200	0	15728	31,684	0.52
Biz.[a]	G1-1a	25.4	1	29,200	1	220	11,410	0	6426	5523	0.56
Biz.[a]	G1-1b	25.4	1	29,200	1	220	11,410	0	6426	6830	0.56
Biz.[a]	G1-2a	25.4	1	29200	1	220	11,410	0	4890	8768	0.43
Biz.[a]	G1-2b	25.4	1	29,200	1	220	11,410	0	4890	9627	0.43
Biz.[a]	G1-3a	25.4	1	29,200	1	220	11,410	0	4543	112,227	0.4
Biz.[a]	G1-3b	25.4	1	29,200	1	220	11,410	0	4543	135,273	0.4
Biz.[a]	G2-1a	25.4	1	29,200	1	300	11,410	0	6477	2390	0.57

(continued)

Table 6.2 (continued)

Ref.	Test	FRP width b_L (mm)	FRP thickness t_L (mm)	FRP Young's Modulus E_L (N/mm^2)	Number of layers n_L (mm)	Bond length l_b (mm)	Maximum load F_L^O (N)	Minimum load F_L^U (N)	Bond capacity F_{Lb} (N)	Number of load cycles N_{30} (–)	Load range S_0 (–)
Biz.[a]	G2-1b	25.4	1	29,200	1	300	11,410	0	6477	7832	0.57
Biz.[a]	G2-2a	25.4	1	29,200	1	300	11,410	0	4953	21,714	0.43
Biz.[a]	G2-2b	25.4	1	29,200	1	300	11,410	0	4953	35,949	0.43
Biz.[a]	G2-2c	25.4	1	29,200	1	300	11,410	0	4953	42,308	0.43
Biz.[a]	G2-3a	25.4	1	29,200	1	300	11,410	0	4420	88,609	0.39
Biz.[a]	G2-3b	25.4	1	29,200	1	300	11,410	0	4420	245,893	0.39

[a]Bizindavyi (2003)

Table 6.3 Experimental data from Budelmann (2013), Carloni (2012), Dai (2005) and Ferrier (2005)

Ref.	Test	FRP width b_L (mm)	FRP thickness t_L (mm)	FRP Young's Modulus E_L (N/mm^2)	Number of layers n_L (mm)	Bond length l_b (mm)	Maximum load F_L^O (N)	Minimum load F_L^U (N)	Bond capacity F_{Lb} (N)	Number of load cycles N_{30} (–)	Load range S_0 (–)
Fer.[a]	A 1	16.2	1.38	67,988	1	200	12,072	0	10,284	3	0.86
Fer.[a]	A 2	16.1	1.38	67,476	1	200	11,998	0	10,220	27	0.86
Fer.[a]	A 3	16.2	1.38	67,308	1	200	12,072	0	8942	59	0.75
Fer.[a]	A 4	16.1	1.38	67,160	1	200	11,998	0	8887	118	0.75
Fer.[a]	A 5	16.2	1.38	66,770	1	200	12,072	0	8048	720	0.67
Fer.[a]	A 6	16.2	1.38	66,831	1	200	12,072	0	8048	542	0.67
Fer.[a]	A 7	16.2	1.38	66,312	1	200	12,072	0	7154	6001	0.59
Fer.[a]	A 8	16.2	1.38	66,210	1	200	12,072	0	6707	9653	0.56
Fer.[a]	A 9	16.2	1.38	65,864	1	200	12,072	0	6148	47,948	0.51
Fer.[a]	A 10	16.2	1.38	65,665	1	200	12,072	0	5813	120,251	0.48
Fer.[a]	B 1	12.8	2.57	54,380	1	200	29,113	0	17,797	2425	0.64
Fer.[a]	B 2	12.8	2.9	50,744	1	200	32,851	0	18,968	5210	0.6
Fer.[a]	B 3	12.7	2.65	51,559	1	200	29,785	0	16,222	74,385	0.56
Fer.[a]	B 4	12.7	2.61	48,905	1	200	29,335	0	14,982	207,728	0.53
Fer.[a]	B 5	12.7	2.68	–	1	200	30,122	0	13,342	573,517	0.46
Car.[b]	DS-F1	25	0.167	230,000	1	152	7590	1250	6000	255	0.75
Car.[b]	DS-F2	25	0.167	230,000	1	152	7590	1100	5100	2604	0.61
Car.[b]	DS-F3	25	0.167	230,000	1	152	7590	1100	4500	23,091	0.53
Dai[c]	S1-1	60	0.11	230	1	150	24,600	0	17,958	57	0.73
Dai[c]	S1-2	60	0.11	230	1	150	24,600	0	15,744	4615	0.64

(continued)

Table 6.3 (continued)

Ref.	Test	FRP width b_L (mm)	FRP thickness t_L (mm)	FRP Young's Modulus E_L (N/mm^2)	Number of layers n_L (mm)	Bond length l_b (mm)	Maximum load F_L^O (N)	Minimum load F_L^U (N)	Bond capacity F_{Lb} (N)	Number of load cycles N_{30} (–)	Load range S_0 (–)
Dai[c]	S1-3	60	0.11	230	1	150	24,600	0	12,546	299,162	0.51
Bud[d]	C201	50	1.4	170,000	1	1150	28,382	3820	15,958	127,862	0.49
Bud[d]	C202	50	1.4	170,000	1	1150	28,382	5117	18,005	415,550	0.55
Bud[d]	C203	50	1.4	170,000	1	1150	28,382	5819	16,267	1,634,439	0.46
Bud[d]	C204	50	1.4	170,000	1	1150	30,190	9032	16,267	156,751	0.34
Bud[d]	C205	50	1.4	170,000	1	1150	34,296	15,292	22,562	188,462	0.38
Bud[d]	C206	50	1.4	170,000	1	1150	30,190	21,729	24,942	462,744	0.38
Bud[d]	C207	50	1.4	170,000	1	1150	30,190	22,884	26,204	462,744	0.45
Bud[d]	C208	50	1.4	170,000	1	1150	30,952	5236	20,754	878	0.6
Bud[d]	C209	50	1.4	170,000	1	1150	30,952	10,436	18,374	428,569	0.39
Bud[d]	C2010	50	1.4	170,000	1	1150	30,952	10,175	18,338	428,569	0.39
Bud[d]	C401	50	1.4	170,000	1	1150	24,847	3249	13,768	200,508	0.49
Bud[d]	C402	50	1.4	170,000	1	1150	24,847	11,662	17,576	746,055	0.45
Bud[d]	C403	50	1.4	170,000	1	1150	24,847	13,626	18,862	269,850	0.47
Bud[d]	C501	50	1.4	170,000	1	1150	34,950	25,811	29,798	61,811	0.44
Bud[d]	C502	50	1.4	170,000	1	1150	34,950	25,478	29,560	61,811	0.43
Bud[d]	C503	50	1.4	170,000	1	1150	34,950	5284	16,184	746,055	0.37
Bud[d]	C504	50	1.4	170,000	1	1150	34,950	5986	17,207	269,850	0.39

[a]Ferrier (2005), [b]Carloni (2012), [c]Dai (2005), [d]Budelmann (2013)

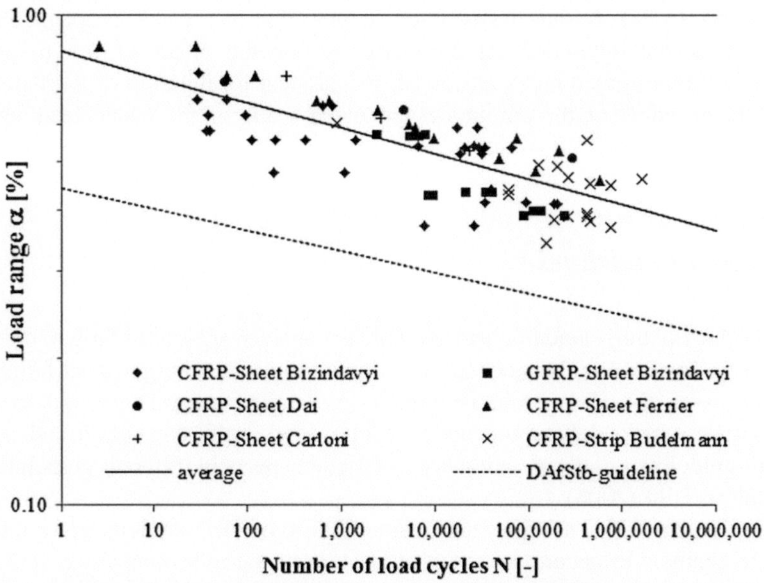

Fig. 6.6 S-N curve and experimental data from Bizindavyi (2003), Budelmann (2013), Carloni (2012), Dai (2005) and Ferrier (2005) in comparison with DAfStb-guideline (2012)

$$\frac{da}{dN} = \bar{m}_1 \left(\frac{\alpha\sqrt{\Delta P \cdot \bar{P}}}{P_{crit}} \right)^{\bar{n}_1} \bar{\beta} \tag{6.24}$$

where $\Delta P = P_{\max} - P_{\min}$ and $\bar{P} = (P_{\max} + P_{\min})/2$ are the amplitude and the mean value of the load range, respectively. P_{crit} is the monotonic quasi-static load-carrying capacity of the interface. The coefficients $\bar{m}_1$, $\bar{n}_1$, $\bar{\beta}$ can be calibrated through the experimental results. α is the frequency coefficient, which takes into account that the fracture properties during fatigue loading depend on the frequency.

With the experimental data and the projection method presented above, another formulation of Eq. (6.24) can be found:

$$\frac{da}{dN} = \frac{1}{N*} \left(\frac{(F_L^O - F_L^U)/F_{Lb}}{c \cdot (1 - F_L^U/F_{Lb})} \right)^k \tag{6.25}$$

where F_L^O and F_L^U are the maximum and the minimum values of the load range. F_{Lb} is the monotonic quasi-static load-carrying capacity of the interface. The parameters N^*, c and k can be set to 2×10^6, 0.295 and 23.1 using the S-N curve from the DAfStb-guideline (2012) for calculating the crack growth rate in mm per load cycle.

Equation (6.25) neither takes into account that the crack propagation rate decreases as the debonded region increases nor the effect of the frequency. However, it represented an extension of the classical formulation of the Paris' law in which the effect of the amplitude and mean value of the cyclic loading were explicitly considered, see Anderson (2004).

Codes and Guidelines

Most of the existing standards and guidelines consider aspects of fatigue only in a rough way. In all cases the steel stress is limited to avoid a fatigue steel failure and in some cases the FRP strain is limited. A detailed analysis method to avoid intermediate crack debonding under cyclic load only can be found in the DAfStb-guideline (2012). A summary of fatigue aspects in different guidelines can be found in Kim (2008):

ISIS Canada (2001) assumes that fatigue failure of FRP strengthened reinforced concrete beams will occur as a result of fracture of the reinforcing steel and, thus, recommends that the stress levels in the externally bonded FRP be limited such that the steel reinforcing bars do not yield.

fib bulletin 14 (2001) regards the fatigue state of strengthened reinforced concrete beams as a special design consideration. This document recommends that the steel stress range in the fatigue design of CFRP-strengthened beams be restricted to those allowed for an unstrengthened beam.

ACI Committee 440 (2008) gives a creep and fatigue limit for FRP. The magnitude of maximum applied stress including both sustained and repeated loads for CFRP should not exceed 55 % of its ultimate strength.

The Italian code for strengthening existing structures (CNR-DT 200-R1 2004) provides a conversion factor of $\eta_1 = 0.5$ for all types of FRP subjected to fatigue load to account for the potential degradation of FRP-strengthening systems.

The DAfStb-guideline (2012) considers fatigue debonding at the end anchorage of FRPs and between cracks. The steel stress range is restricted after EC2 (EN 1992-1-2 2004) as for an unstrengthened RC structure. The design concept for bond fatigue given in the DAfStb-guideline (2012) is based on the bilinear limit curve of the upper load from Fig. 6.7 and this depends on two verifications. It can be verified whether the load difference ΔF_L^O is within the elastic range $\Delta F_{L;el}$ with Eq. (6.26).

$$\Delta F_{L,el} = 0.348 \cdot f_{ct}^{1/4} \cdot F_{Lb} \geq \Delta F_L^O \tag{6.26}$$

The load difference has to be checked at the end and in the middle part of a strengthened RC structure. In the end part, the load difference ΔF_L^O has to be calculated from the plate end to the first flexure crack. In the middle part the load difference ΔF_L^O between two cracks of a theoretical crack pattern has to be

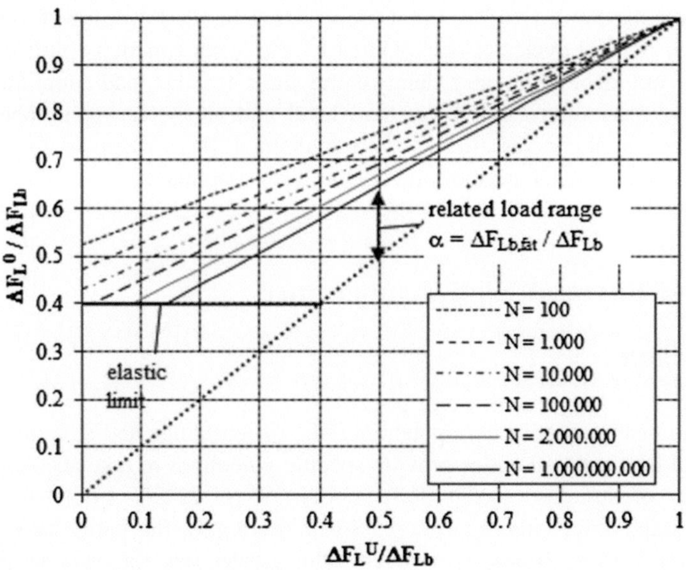

Fig. 6.7 Design concept after DAfStb-guideline (2012)

calculated. For the calculation of the crack spacing and the maximum plate force difference ΔF_{Lb} between cracks the model given in DAfStb-guideline (2012) and German Committee for Structural Concrete (2013) is used. If the condition given by the equation above is not met, the fatigue range $\Delta F_{Lb;fat}$ can be verified as follows:

$$\begin{aligned} \Delta F_{Lb,fat} &= \alpha \cdot \Delta F_{Lb} \geq \Delta F_L^O - \Delta F_L^U, \\ \alpha &= -c \cdot \frac{\Delta F_L^U}{\Delta F_{Lb}} + c, \\ c &= 0.342 \times \frac{N^{-\frac{1}{k}}}{N^*}, \\ N^* &= 2 \times 10^6, \\ k &= \begin{cases} 23.1 \; for \; N < N^* \\ 45.4 \; for \; N \geq N^* \end{cases} \end{aligned} \tag{6.27}$$

The load range between the load difference at the upper load level ΔF_L^O and the lower load level ΔF_L^U has to be smaller than the fatigue range $\Delta F_{Lb;fat}$. The fatigue range $\Delta F_{Lb,fat}$ depends on the lower load level ΔF_L^U and decreases with a rising lower load level ΔF_L^U. It also depends on the number of load cycles N.

The reduction of the fatigue range $\Delta F_{Lb,fat}$ based on an increasing number of load cycles N is described with the factor c given in Eq. (6.27). The Goodman-Smith Diagram shown in Fig. 6.7 illustrates the fatigue design concept. The horizontal line marks the elastic limit calculated with Eq. (6.26). The related load range α used in Eq. (6.27) and its dependence on the lower load level ΔF_L^U and the number of load cycles N is explained by the linear functions plotted above.

To prevent debonding and damage the limit for the load amplitude has to decrease with an increasing lower load level and an increasing number of load cycles. The design concept after DAfStb-guideline (2012) presented above shows a simple method to calculate the limit for the load amplitude.

Effects of Fire and High Temperature

Introduction

Fire is one of the most serious potential risks for buildings and structures, and for this reason international codes provide specific guidelines to take account of fire in the design of structures (ACI 318-11, EN 1991-1-2, EN 1992-1-2). In some countries (e.g. Italy, Greece, Turkey, Spain, Portugal), the earthquake, which is often follow by fire events, is an event with greater risk for physical injury and damage to objects and properties. However, also in some of these countries, national codes (NTC 2008) have recently introduced more regulation for the fire risk, by considering that human activities are in continuous development and evolution and can be more dangerous than natural events. In this chapter, the main effects of high temperatures on fibre-reinforced composite materials are summarized. In particular, some studies carried out in the last two decades on this topic are cited, mainly to show actual knowledge and future challenge about the degradation of the mechanical properties of FRP materials at high temperatures.

At first, some information about the critical temperatures for FRP are given. Then, some studies related to testing on structural members reinforced with FRP and numerical simulations of the behaviour of FRP materials at high temperatures are mentioned. Although for many applications there is a clear need for protection against high temperatures, the chapter closes with an example that shows how in some cases the protection is already self-provided by non-structural elements having other functions.

Critical Temperatures for the Mechanical Properties

As stated in previous sections, FRP are composite materials successfully applied to repair and/or strengthen RC structures. For external strengthening, the FRP plates are easily bonded on concrete using adhesive, like epoxy resins, which ensure the transfer of forces between concrete and FRP. However, degradation of mechanical properties of composites (strength, stiffness and bond) due to high temperature (Dai et al. 2013; Nigro et al. 2013), moisture absorption (Jia et al. 2005) and cycling loads (Dai et al. 2005) is a key aspect for a durable efficiency of composite materials.

Concerning high temperature, a critical condition occurs when the glass transition temperature, T_g, of the polymer matrix is achieved, due to the softening of the resin, which reduces the capacity of transfer of forces between the fibres. The precise definition of the value of T_g is still under discussion in the scientific community, because the progressive nature of the softening process makes it difficult to identify a precise temperature limit. Nevertheless, the safety check often is conservatively performed, in the temperature domain, with reference to the value of T_g properly reduced (ACI 440.2R-08 2008).

FRPs which polymerize in ordinary conditions, typical of in situ applications, are characterized by very low T_g (between 45 and 80 °C for normal and heat resistant resins, respectively). For preformed FRPs, used as internal reinforcement, is easily possible to obtain, reinforcements with T_g above 100 °C. Curing processes carried out at temperatures and pressures different from ordinary ones, allow to further increase the T_g.

Although overcoming the T_g implies a reduction in strength of the reinforcement, the drastic degradation of the resistance is reached at temperatures close to melting of the resin (temperature of crystallization, $T_c > T_g$) or even higher. The reduction of stiffness, instead, depends on the type of fibre reinforcement and it is generally negligible compared to the reduction of resistance. Therefore, the real capacity of the concrete members reinforced with FRP reinforcement, at high temperatures, can be considerably high (Nigro et al. 2011a, b, 2013).

Review of Experimental Studies

Recent experimental studies showed that the softening of the resin which begins when T_g is achieved, involves a drastic reduction of the adhesion properties (Bisby et al. 2005). Hence, the efficiency of the strengthening system for existing structures, which mainly depends on the effectiveness of the bond between FRP and concrete, is strongly affected by the temperature.

Some experimental tests (Deuring 1993) showed similar problem when conventional steel strengthening are used without mechanical anchoring. The comparison between steel and FRP strengthening systems showed that FRP, in particular sheets, without protection behave better than steel plates because of the lower heat conductivity and their smaller weight. Clearly, FRP externally strengthened RC beams or slabs need the protection with additional insulation in order to avoid the debonding between FRP sheets or laminates and concrete support. Consequently some researches were devoted to study the performances of FRP strengthened elements protected by different insulation systems in order to individuate the minimum requirements to obtain satisfactory performances in fire (Bisby et al. 2005).

No information is apparently available on the specific mechanical or bond properties of typical FRP systems used for strengthening RC structures after short-term exposure to elevated temperatures, although limited information is available on the residual mechanical properties of specific FRP materials used in the marine, aerospace, and automotive industries after exposure to high temperature (Mouritz and Mathys 1999; Mouritz and Gibson 2006; Bai et al. 2007). Basic research has been reported on the post-heating residual performance of FRP wrapped concrete cylinders (Cleary et al. 2003; Saafi and Romine 2002), although the data presented in these studies do not elucidate the specific performance of the FRP systems. Only one limited study is available on the high temperature performance of the FRP-to-concrete bond loaded in shear (Gamage et al. 2005), although residual properties are not addressed. Several large-scale standard fire tests have also been performed (Deuring 1993; Blontrock et al. 2001; Kodur et al. 2007), but the data do not address the specific performance of the FRPs either during or after high-temperature exposure.

A significant research effort over the past decade has demonstrated that appropriately designed and adequately insulated FRP strengthened RC beams, slabs, and columns, are capable of achieving adequate fire endurances (Kodur et al. 2007). However, these researches have not provided much insight into the specific performance of the FRP strengthening systems or the bond between the FRP systems and the substrate concrete, either during exposure (i.e., at high temperature) or once they have cooled to room temperature (i.e., residual performance). Information is thus required before defensible strengthening limits and allowable thermal exposures can be suggested for FRP strengthening systems, particularly in cases where the FRP system is required to be effective during or after a fire (Porter and Harries 2005). Information is also needed to develop economical fire insulation schemes for FRP strengthened members, even in cases where the FRP is not required to be structurally effective during a fire.

Numerical Modeling

Probably, in order to extend the results of experimental tests to different cases, using a numerical model could be appropriate. Indeed, the behaviour of structures exposed to fire is usually described in terms of fire resistance but, in real buildings, structural members are part of a continuous assembly, and building fires often remain localized, since the fire affected region of the structure receiving significant restraint from cooler areas surrounding it. The real behaviour of these structural elements can therefore be very different from that indicated by standard furnace tests and should be investigated, as is usual within the Fire Safety Engineering approach.

Clearly, the accuracy of the thermo-mechanical analysis is dependent on the main properties required to calculate the temperature distribution (i.e. specific heat, thermal expansion and thermal conductivity) and the constitutive laws used to define the mechanical behaviour of materials (i.e. strength, stiffness and bond properties), both at ambient and elevated temperatures. For many of these parameters suggestions in technical code are not univocal for concrete and are still lacking for FRP. Thus, more research is needed to improve the accuracy of the numerical models.

Obviously, if the FRP strengthening is not directly heated by fire or other sources of heat, the performances may be better. Hence, FRPs can be successfully used to strengthen bridges, where fire is not a primary action to be considered during design (Bisby et al. 2005). Nevertheless, it should be noted that bituminous paving casting on a bridge deck can easily lead to have high temperature (e.g. 200 °C).

On this issue Nigro et al. (2013a, b) investigated the thermo-mechanical behaviour of RC bridge decks strengthened with externally bonded FRP plates. The results are summarized in the following section.

The Effects of High Temperatures in a Practical Application

In Nigro et al. (2013a, b) two possible environmental conditions leading to thermal states different from the normal ones were studied with reference to bridge decks: (a) fire exposure over the bridge deck due to an accident involving camions; (b) bituminous paving casting on a bridge deck. Indeed, in both case the temperature at the FRP-to-concrete interface can overcome the above mentioned glass transition temperature, T_g.

The relationships suggested by Italian and American codes, to evaluate the limit strain for FRP debonding at normal temperature, were modified to take into account the effect of high temperature on the debonding of FRP. Then, thermo-mechanical analyses were performed by varying the thicknesses of the slab and the protection layer in order to assess their influence on the thermal field in the structural member. Furthermore, normal resin (NR) with $T_g = 45$ °C and heat-resistant resin (HR) with $T_g = 80$ °C were considered. The results were discussed in terms of both temperatures and safety checks carried out for both ultimate and serviceability limit states (ULS and SLS).

The ULS checks were always satisfied, mainly because the flexural capacity provided by FRP can be neglected during fire or maintenance activity. By contrast, the SLS checks performed to assess the damage levels in the FRP strengthening system during these events show that constructive details and type of resin play a key role.

In the case of fire event over the bridge deck, if the strengthening is on the bottom side, for typical design load level $\eta_{fi} = 0.7$, FRP damages (i.e. debonding) were not attained for long time of fire exposure, even if the resin achieved the glass transition temperature, T_g. If the strengthening is located on the top side of the slab, the use of heat resistant (HR) resins is suggested. If normal resin (NR) is used, a protection layer on the FRP strengthening is recommended, in order to increase the maximum time of fire exposure without FRP damage.

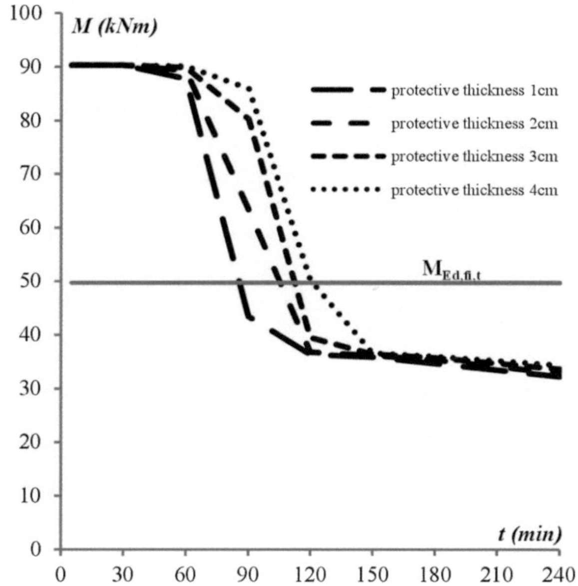

Fig. 6.8 Slab safety check in hogging moment region (fire on the bridge)—T_e = 25 °C—Slab thickness = 15 cm—(NR)

In order to show that the achievement of the T_g may be not a critical condition for the strengthening system, in Fig. 6.8 the bending moment of the FRP strengthened slab, $M_{Rd,fi,FRP}$, reduced for the effect of the fire exposure, is plotted versus the time for the 15 cm thick slab, normal resin and four thicknesses of protective layer, t_{prot}. Moreover the bending moment in fire situation $M_{Ed,fi}$ = 49.7 kNm is shown. Note that $M_{Rd,fi,FRP}$ was calculated by using the relationships suggested by CNR DT 200/2004 to evaluate the debonding strain at ambient temperature. As stated above, the relationships were modified to take into account the effect of high temperature.

Figure 6.8 shows that the time of fire exposure before the debonding of the FRP ranges between 82 and 120 min. Note that these values are higher than those related to the achievement of the T_g at the FRP-to-concrete interface (i.e. 50 and 80 min) and probably more realistic to assess the FRP damage.

In the case of bituminous paving realization, for strengthening on the bottom side, the results are quite similar to those obtained in case of fire. On the other hand, if the strengthening is located on the top side of the slab, the use of HR resins is necessary in order to avoid damages in the FRP strengthening. Otherwise thick protective layer of concrete is required also for low load levels.

The results obtained through the relationships provided by different codes (i.e. Italian and American codes) for ambient temperature and refined with the suggested model are in a good agreement (see Fig. 6.9). However, so far, more research is needed to improve the reliability of codes suggestion.

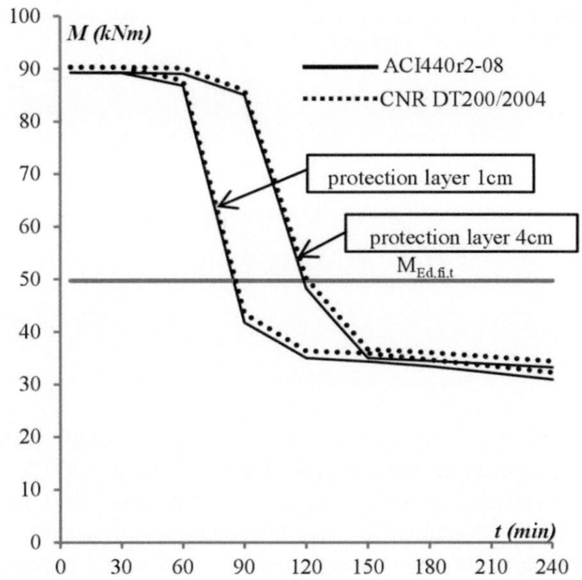

Fig. 6.9 Comparison between codes

Long Term Properties of FRP Systems

Introduction

Regarding the long term performance of a composite system, few authors have dedicated their efforts to the assessment of the performance of the intervening materials (Diab and Wu 2007; Wu and Diab 2007; Meaud et al. 2011).

Apart from changes in applied external actions, the long term performance of a CFRP-adhesive-concrete system is only expected to be affected by the creep, shrinkage or relaxation of each of the components of the system. While the creep and shrinkage of concrete have already been comprehensively studied over the years, limited information is available regarding the creep/shrinkage/relaxation of composite materials. In this chapter a resume on the collected information in this respect is provided.

FRP Relaxation

Bibliographic research has shown that FRPs are known to present low pre-stress losses, as a result of their relatively low elastic modulus (Lopez-Anido and Naike 2000), and lower stress relaxation than steel strands (Dolan et al. 2001; Sayed-Ahmed 2002). In fact, even though FRP materials are able of exhibiting an

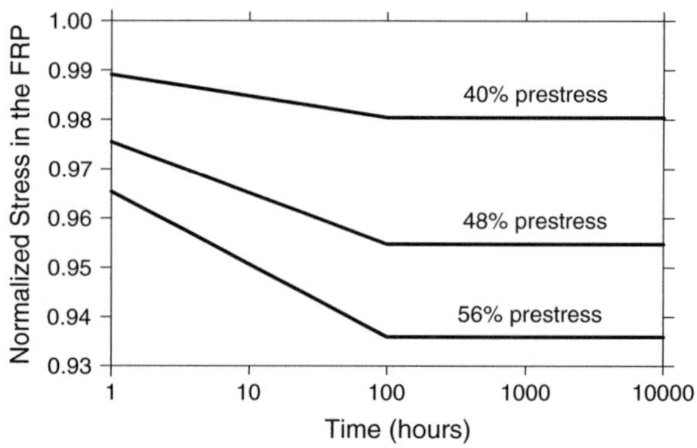

Fig. 6.10 Stress relaxation in CFRP sheets. Adapted from Wang et al. (2012)

elastic modulus, E_f, close to 200 GPa, most of the materials available reveal to have an average maximum E_f of about 160 GPa.

Wang et al. (2012) carried out relaxation tests in CFRP sheets and concluded that the relaxation loss due to sustained deformation levels ranging between 40 and 56 % of the material's tensile strength was determined to be 2.2–6.6 %, as demonstrated in Fig. 6.10. Moreover, this relaxation was mostly concentrated in the first 100 h of sustained deformation, as depicted in Fig. 6.10 and, after this period it becomes almost negligible. These authors have even suggested that the measured relaxation is primarily caused by the relaxation of the resin and straightening of fibres. The carbon fibres themselves are identified as having no relaxation whatsoever (Dolan et al. 2001).

According to Dolan et al. (2001) the relaxation losses in FRP tendons can be caused by three main sources: the relaxation of resin that bonds the fibres together, the lack of parallelism between individual fibres, and the relaxation of fibre itself. Due to these reasons, the relaxation is a characteristic attributable to the fibre type and is generally lower than 12 % over the life of the structure. In the case of CFRP tendons, relaxation losses of approximately 5 % are reported.

Based on these assumptions, it is suggested that FRP relaxation is not a relevant effect for the long-term performance of CFRP laminates. Since in FRP laminates the fibre content is particularly large when compared to FRP sheet coupons, the matrix bonding them together will take only a small portion of the applied load and, therefore, the first source of relaxation may be ignored. Regarding the alignment of the fibres, since FRP laminates are produced by machines, in opposition to FRP sheets, which are manually applied, no significant eccentricities are expected along each fibre and as a result, the second source of relaxation may also be disregarded. Finally, as carbon fibres themselves are reported to have no relaxation, the total amount of relaxation expected is even more reduced.

Adhesive Shrinkage

The volumetric shrinkage of epoxy-based adhesives is generally restricted to the shrinkage occurred during the curing process and is found to be within the range 2–7 % (Li et al. 2004; Yu et al. 2005; Khoun and Hubert 2010).

Concerning the epoxy-based adhesives used in structural applications they reveal such an insignificant shrinkage coefficient, that this parameter is often not even quantified in most materials' datasheets. However, when complete cure is achieved, the shrinkage coefficient variation becomes negligible, as suggested by (Yu et al. 2005; Khoun and Hubert 2010), and the long-term behaviour of the adhesive is no longer shrinkage dependent, but creep-dependent as it will be revealed hereafter.

Adhesive Creep

Introduction

The creep behaviour of plastics is usually fragmented in three major creep stages (ASTM 2990 2001, Majda and Skrodzewicz 2009): primary creep, secondary creep and tertiary creep. As depicted in Fig. 6.11, in the first phase the material adjusts its deformation level to the installed level of stress. This phase is followed by a stationary stage where creep gradually increases until a third phase is reached, where strain suddenly increases and fracture occurs. It is believed that this behaviour is valid under any applied stress, temperature and humidity. However, for low levels of applied stress, the time necessary to reach the tertiary creep state may be so long that it may never be achieved.

The primary function of an adhesive in structural applications is to transmit stress equally over large areas without loss of integrity (Feng et al. 2005). Structural adhesives exhibit, however, notable viscoelastic behaviour, since their deformation, ε, under a constant stress, σ, varies significantly in time. This behaviour is frequently modelled using rheological models and is usually illustrated by means of Hookean springs and Newtonian dashpots that replicate, respectively, the elastic

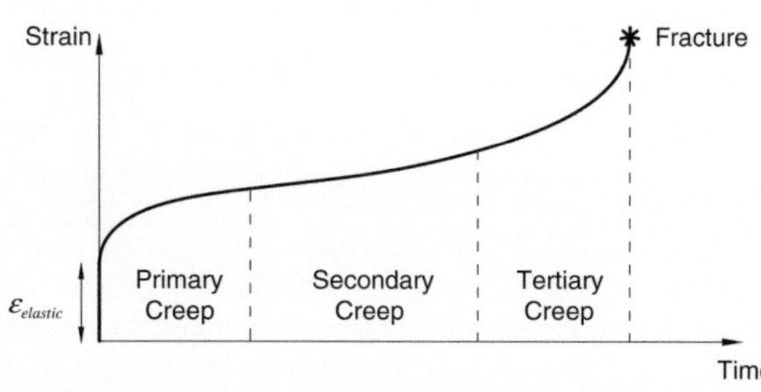

Fig. 6.11 The three stages of creep (at constant stress, temperature and humidity)

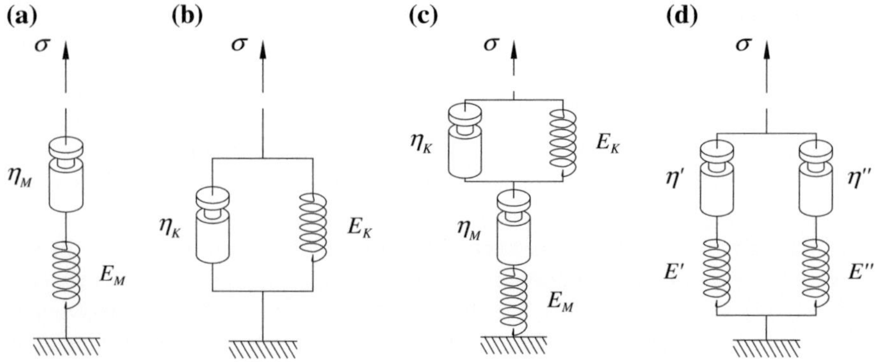

Fig. 6.12 **a** Maxwell model, **b** Kelvin model, **c** Burgers model—common configuration, **d** Burgers model—alternative configuration

and viscous components of the material's behaviour (Brinson and Brinson 2008). In Fig. 6.12, the most common rheological models are presented:

- Maxwell Model—illustrated in Fig. 6.12a, this model is a 2-parameter model that results of associating, in series, a spring with E_M elasticity and a dashpot characterized by η_M dynamic viscosity;
- Kelvin Model—depicted in Fig. 6.12b, it is also a 2-parameter model that consists in combining, in parallel, a spring of elasticity E_K and a dashpot of η_K dynamic viscosity;
- Burgers Model—schematized in Fig. 6.12a–d, this 4-parameter model can be obtained by joining in series Maxwell and Kelvin's Model (E_M, η_M, E_K and η_K in Fig. 6.12c or, in alternative, by connecting two Maxwell models in parallel (E', η', E'' and η'' in Fig. 6.12d).

It is possible to create models like the generalized Maxwell Model or generalized Kelvin Model, depicted in Figs. 6.13 and 6.14 (Brinson and Brinson 2008).

It is relatively simple to obtain the solution of each of these models and, the deduction of each equation can be found in Costa and Barros (2011).

Fig. 6.13 Generalized Maxwell fluid

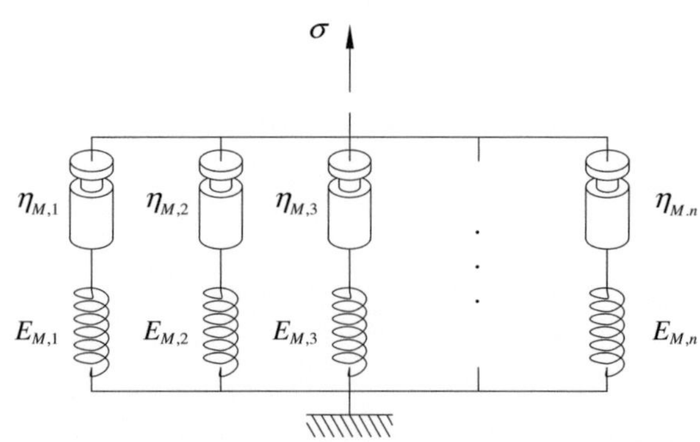

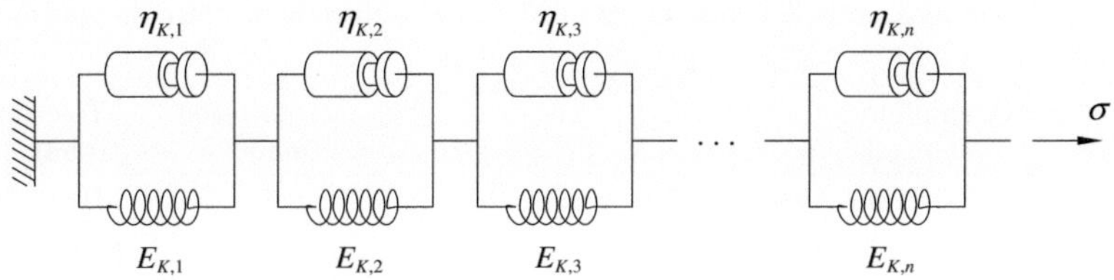

Fig. 6.14 Generalized Kelvin solid

Standards on Creep Behaviour

According to ISO 899-1 (2003) test method, load shall be applied smoothly to a standard specimen, as defined in ISO 527-2 (1993), within 1–5 s and maintained for at least 1000 h (approximately 42 days). Strains, temperature and humidity should be measured according to the schedule presented in Fig. 6.15.

ISO 527-2 (1993) recommends bone-shape specimens to be moulded using the geometry depicted in Fig. 6.16 (see also Table 6.4).

For different temperature and/or humidity level, one creep curve shall be obtained. To construct the desired creep curves, creep strains and/or creep moduli (defined in Eq. 6.28) are plotted against the logarithm of time, for every initial level of applied stress, σ, (see Fig. 6.17). Isochronous curves, which consist in Cartesian plots of stress versus strain at specific time instants, similar to those depicted on Fig. 6.17c, can also be presented.

$$E_e = 2G_r(1 + v) \tag{6.28}$$

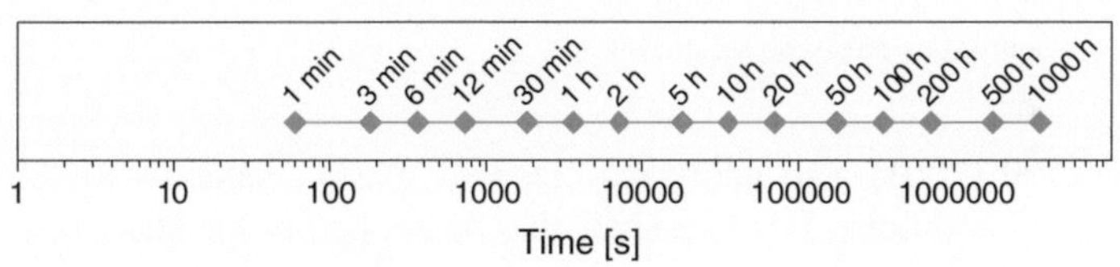

Fig. 6.15 Data acquisition schedule (ISO 899-1 2003)

Fig. 6.16 Directly-moulded specimens (Type 1A, ISO 527-2 1993)

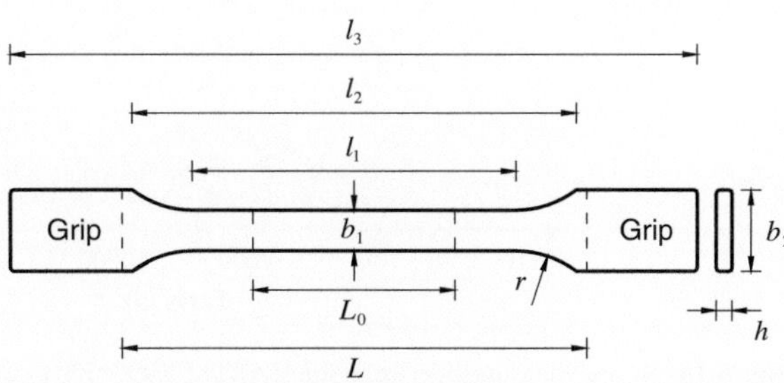

Table 6.4 Dimensions, in millimetres, of ISO 527-2 (1993) directly-moulded specimens (Type 1A)

Variable	Description	Dimension (mm)	Tolerance (mm)
b_1	Width of the narrow portion	10	±0.2
b_2	Width at the ends	20	±0.2
h	Preferred thickness	4	±0.2
L	Initial free distance between grips	115	±1
L_0	Gauge length	50	±0.5
l_1	Length of the narrow parallel-sided portion	80	±2
l_2	Distance between broad parallel-sided portions	104–113	–
l_3	Overall length	≥150	–
r	Radius	20–25	–

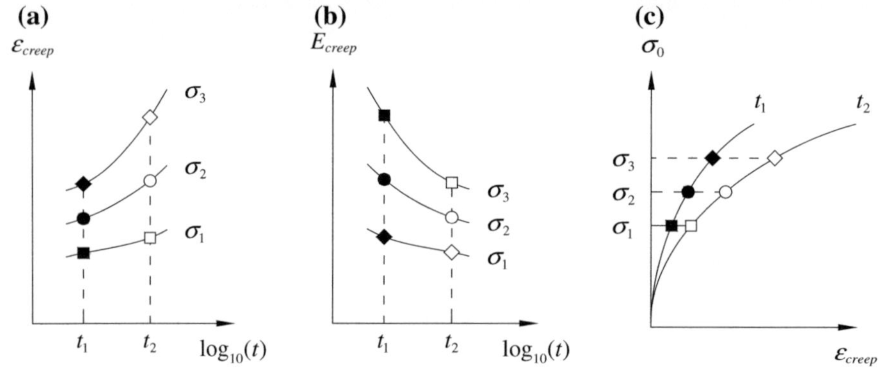

Fig. 6.17 Example of creep curves (ISO 899-1 2003): **a** creep strain curves, **b** creep modulus curves and **c** isochronous stress–strain curves

ASTM 2990 (2001) also addresses creep of plastic materials under specified environmental conditions. Loading shall also be applied in the specimen rapid and smoothly in 1–5 s, and the strains shall be recorded at the time instants shown in Fig. 6.18. To evaluate creep, the use of the normalized specimens produced according to ASTM D 638 (2003) is suggested (Fig. 6.19; Table 6.5).

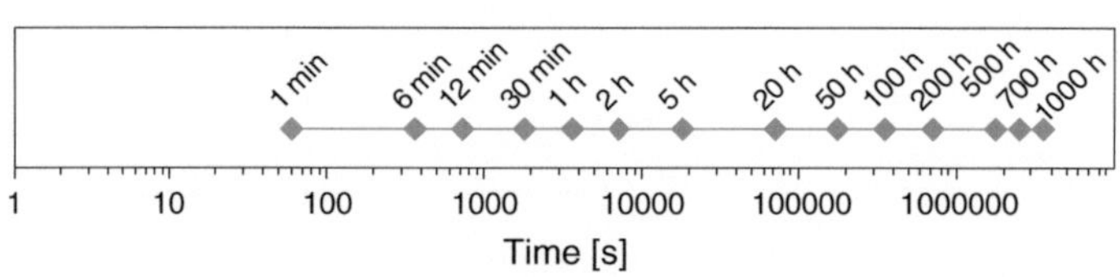

Fig. 6.18 Strain measurement schedule (ASTM 2990 2001)

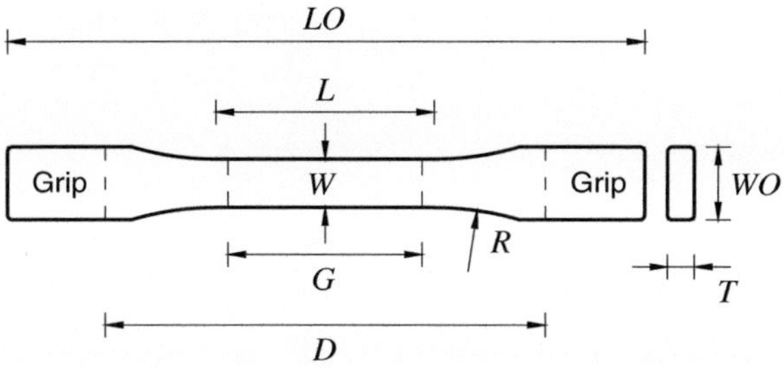

Fig. 6.19 Directly-moulded specimens (Type 1A, ISO 527-2 1993)

Table 6.5 Dimensions, in millimetres, of ASTM D 638 (2003) preferred specimens (Type I)

Variable	Description	Dimension (mm)	Tolerance (mm)
D	Distance between grips	115	±5
G	Gauge length	50	±0.25
L	Length of narrow section	50	±0.5
LO	Overall length	≥165	–
R	Radius of fillet	76	±1
T	Thickness	≤7	–
W	Width of narrow section	19	+6.4
WO	Width at the extremities	80	±2

Research on Creep Behaviour

Feng et al. (2005) suggested that it is possible to estimate the tensile creep strain, $\varepsilon_{creep}(t, T)$, by the exponential function:

$$\varepsilon(t,T) = \frac{\sigma_0}{E_0} + \sigma_0\left(\frac{1}{E_e} - \frac{1}{E_0}\right)\left(1 - e^{-(t/t^*)^{1-n}}\right) \tag{6.29}$$

where σ_0 is the applied stress level, E_0 the initial Young modulus, E_e is the equilibrium modulus given in Eq. (6.28), t^* is the relaxation time and n a coupling parameter related to moisture absorption.

$$E_e = 2G_r(1 + v) \tag{6.30}$$

where G_r is the rubbery plateau shear modulus, and v the Poisson's ratio ($v = 0.5$ since the material is in the rubbery state). Feng et al. (2005) obtained in their tests values of n ranging from 0.51 to 0.73. Majda and Skrodzewicz (2009) proposed a model purely based on Burgers Model:

$$\varepsilon(t,T)=\frac{\sigma_0}{E_0}+\frac{\sigma_0}{\eta_0}t+\frac{\sigma_0}{E_1}\left(1-e^{-t/t^*}\right) \tag{6.31}$$

where

$$t^*=\frac{\eta_1}{E_1} \tag{6.32}$$

Majda and Skrodzewicz (2009) also suggested that the coefficients of dynamic viscosity, η_0 and η_1, as well as the elastic modulus are primarily dependent on the applied stress, and:

$$\eta_0(\sigma_0)=e^{a_1-a_2\sigma_0} \tag{6.33}$$

$$\eta_1(\sigma_0)=e^{a_3-a_4\sigma_0} \tag{6.34}$$

$$E_1(\sigma_0)=a_5\sigma_0^2-a_6\sigma_0+a_7 \tag{6.35}$$

The rheological properties (E_0, E_1, η_0 and η_1) were quantified by means of nonlinear regression analysis of the experimental creep tests at four different levels of applied stress, at a constant temperature of 22 °C (Table 6.6). Later, the different a_i coefficients can be obtained applying ordinary trend lines (logarithmic/quadratic regressions) as well as by non-linear regression (Table 6.7).

Choi et al. (2007) presents the results of creep tests performed in double-shear specimens during 6 months, where creep specimens with different adhesive layer thickness and applied stress were considered. The authors suggest that the total strain due creep also follows the following exponential law:

$$\varepsilon(t,T)=\frac{\sigma_0}{E_0}+\sigma_0\frac{\phi_u(t_\infty)}{E_0}\left[1-e^{-t/t^*}\right] \tag{6.36}$$

with the values for the parameters indicated in Table 6.8.

By carrying out tensile tests with a current type of adhesive used to bond CFRP laminates to concrete in the NSM technique, Costa and Barros (2013) and Costa (2014) have concluded that after 2 days of curing, the properties of the adhesive are nearly the same as obtained for the recommended curing time (7 days).

Table 6.6 Parameters of the Burger's model determined on creep test of an epoxy adhesive at 22 °C (Majda and Skrodzewicz 2009)

σ_0 (MPa)	E_0 (GPa)	η_0 (GPa h)	E_1 (GPa)	η_1 (GPa h)	t^* (min)
15	2.232	22.4	1.173	0.48	24
20	2.232	4.12	0.788	0.23	18
25	2.232	1.49	0.896	0.07	5
30	2.232	0.36	1.380	0.06	3

Obs. The tensile strength of the adhesive is 46.6 MPa

Table 6.7 Values of a_i for simple and non-linear regressions (Majda and Skrodzewicz 2009)

Coefficient	Unit	Trend line	Nonlinear regression
a_1	ln(Pa s)	35.8	37.38
a_2	Pa-1 ln(Pa s)	27×10^{-8}	31.97×10^{-8}
a_3	ln(Pa s)	30.3	30.72
a_4	Pa^{-1} ln(Pa s)	14×10^{-8}	16.62×10^{-8}
a_5	Pa^{-1}	9×10^{-6}	12.27×10^{-6}
a_6	–	394	516.9
a_7	Pa	50.2×10^{8}	60.70×10^{8}

Table 6.8 Test results in laboratory environment −20 °C and 50 % relative humidity (Choi et al. 2007)

Specimen	Epoxy thickness (mm)	σ_0 (MPa)	Monitored FRP	$\phi_u(t_\infty)$	t^* (days)
1	0.242	0.09	Face 1	1.17	43.3
			Face 2	1.02	2.1
2	0.176	0.17	Face 1	2.89	1.1
			Face 2	2.94	1.7
3	1.500	0.17	Face 1	2.59	0.2
			Face 2	2.39	0.1

Double-shear specimen ultimate resistance: 0.56 MPa

By performing creep tensile tests with three series of specimens of this type of adhesive, these authors have verified that up to sustained stress levels of 60 % of the adhesive's tensile strength, the adhesive behaves as a classic visco-elastic material and can easily be parameterized using the modified Burgers model (Figs. 6.20, 6.21 and 6.22; Table 6.9). In the experimental tests performed, the results suggest that the properties of the adhesive tend to deteriorate with time and therefore, a special attention should be taken regarding the time between adhesive production and

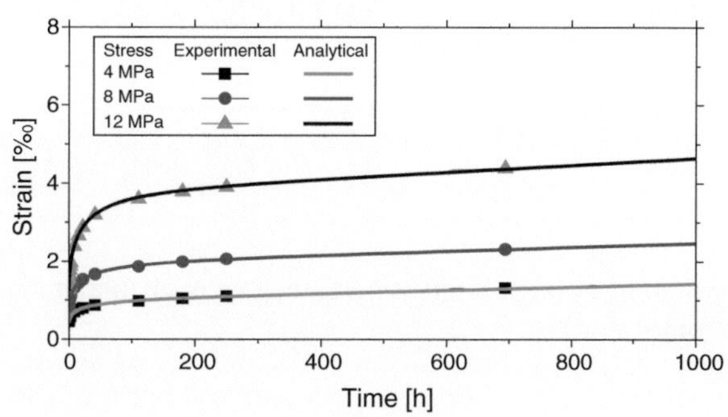

Fig. 6.20 Modified Burgers model—Series I

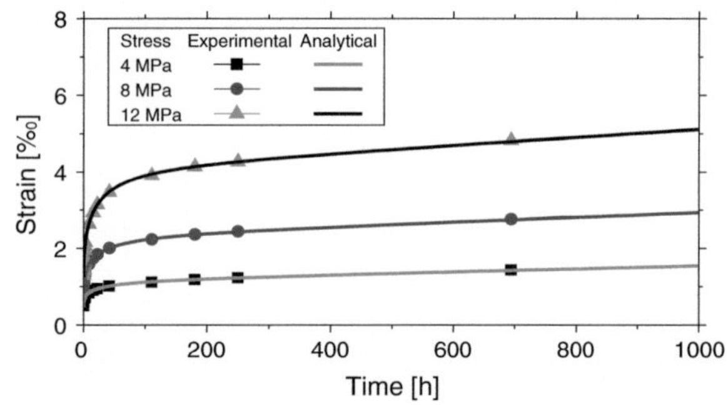

Fig. 6.21 Modified Burgers model—Series II

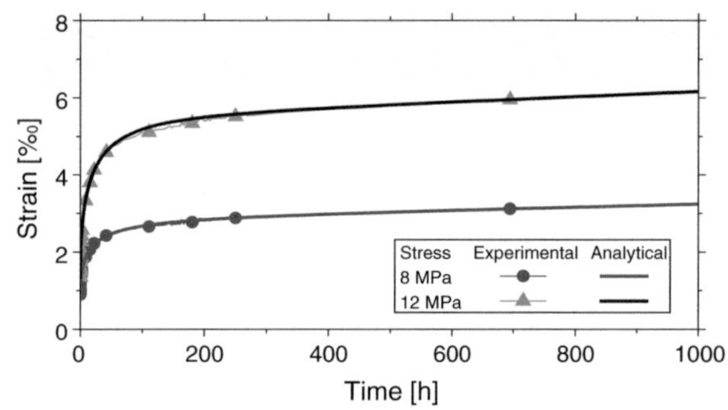

Fig. 6.22 Modified Burgers model—Series III

application. It is also noteworthy that after 1000 h of loading, the adhesive samples exhibited about 4 times the deformation at time of loading (creep modulus of roughly 25 % of the initial stiffness) without rupturing. According to the initial tensile tests performed, a maximum strain at rupture of about 3 ‰ was obtained, while during the creep tests the material was able of somehow reorganizing its internal structure to withstand almost the double of this deformation.

Based on the experimental results the following equation was proposed to estimate the tensile creep modulus (Costa and Barros 2013):

$$E_{creep}(t) = \frac{\sigma}{\varepsilon_{creep}(t)} = E_{creep} = \frac{1}{{}^{1}/_{E_M} + {}^{t}/_{\eta_M} + {}^{1}/_{E_K}\left(1 - e^{-\left(\frac{E_K}{\eta_K}t\right)^{1-n}}\right)} \tag{6.37}$$

where its good predictive performance is visible in Fig. 6.23 by considering the data from Table 6.9.

Table 6.9 Average modified Burgers equation parameters of all series tested

Parameter	E_M (GPa)	η_M (GPa h)	E_K (GPa)	t^* (h)	η_K (GPa h)	n
Series I	9.49	13,482	5.27	24.3	128	0.47
Series II	8.80	11,544	4.50	19.7	88.7	0.53
Series III	8.84	18,446	3.09	18.8	58.1	0.50
Average	9.04 (0.39) {4 %}	14,491 (3560) {25 %}	4.29 (1.10) {26 %}	20.9 (3.0) {14 %}	91.7 (35.2) {38 %}	0.50 (0.03) {6 %}

Average (Standard Deviation) {Coefficient of Variation}

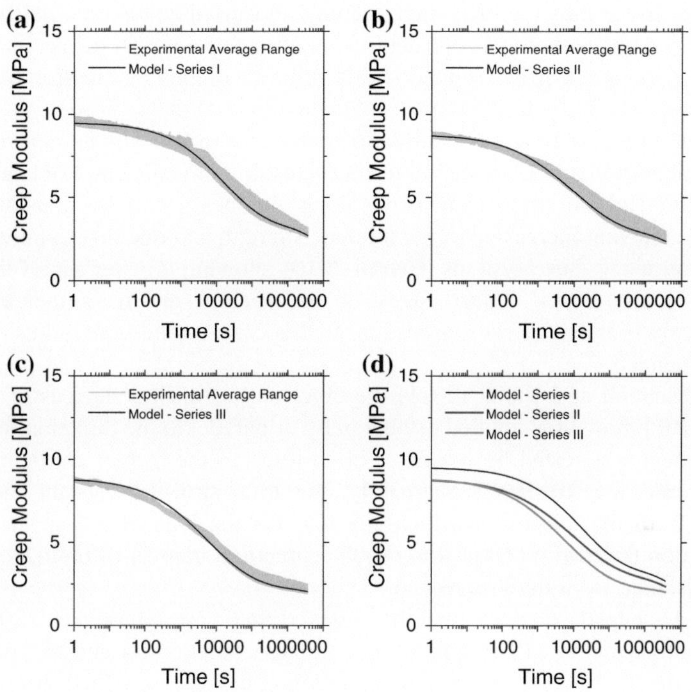

Fig. 6.23 Creep modulus curves based on the analytical results: **a** series I, **b** series II, **c** series III and **d** all analytical curves (Costa 2014)

Anchorage Systems for External Strengthening with FRP

Special Anchorage for Flexural Strengthening

For flexural strengthening the most common anchorage system is given by a FRP sheet or a FRP or steel laminate glued transversally to the strengthening direction. The good performance of such a type of anchoring was tested in Ceroni et al. (2008)

Fig. 6.24 Possible shear U-shaped anchoring devices (Ceroni 2010)

on specific bond tests. This is a reliable solution when the FRP-to-concrete width ratio is less than 1, because the transversal sheet/laminate allows extending the width of concrete covered by the FRP strengthening. In Ceroni and Pecce (2010) a simple method based on geometrical considerations is proposed in order to take into account such an enlargement of the bonded width in the theoretical expression of the debonding load.

When the FRP-to-concrete width ratio approaches to 1, a simple transversal strip becomes unhelpful and U-shaped fibre sheets result more efficient. For example one strip at the end or strips distributed along the beam can be considered (see Fig. 6.24): relevant increasing of the ultimate strength and ductility at failure can be obtained by using these systems (Ceroni 2010) allowing a better use of the tensile strength of the fibres. Sharp edges of the section are recommended to be mechanically rounded before application. In this case, a minimum radius of 30 mm is recommended.

In Al-Mahaidi and Kalfat (2011) the efficiency as anchor devices of unidirectional CFRP fabric wrap applied orthogonally or parallel to the direction of the FRP reinforcement was tested by means of bond tests. In the former configuration the load increase was of 19–28 % mainly due to a strut-tie resistant mechanism resulting from the fabric fibres inclining towards the load direction; in the latter configuration the load increase was 18–37 % due to a transfer of bond stresses to a greater distance from the loaded end.

Steel U-shaped devices as the ones tested by Blaschko (2001) and Mukhopadhyaya et al. (1998) can be an alternative solution for end anchorage. The basic scheme of tests made with special anchoring devices for CFRP strips is shown

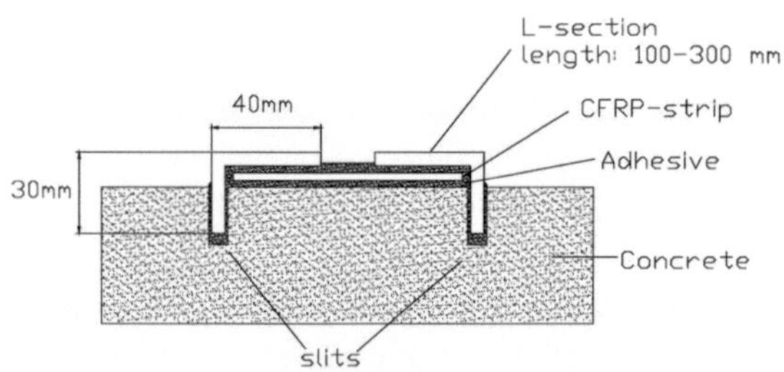

Fig. 6.25 Anchorage for CFRP strips special anchorage system (Zehetmaier 2000)

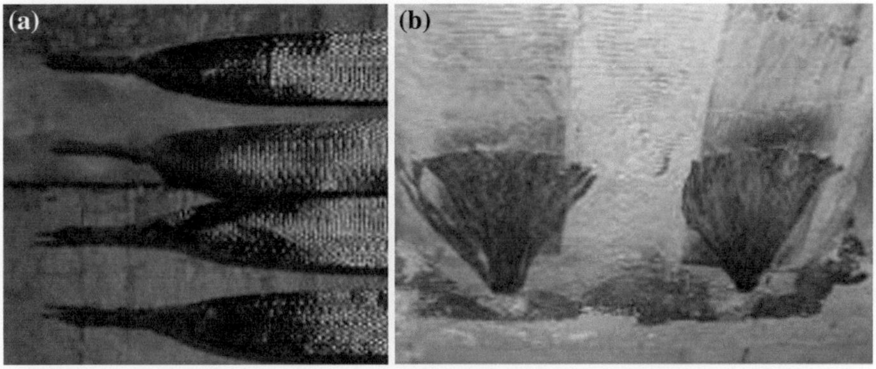

Fig. 6.26 **a** Carbon fibre fan anchor; **b** splaying of fibres outwards the hole

in Fig. 6.25 (Zehetmaier 2000). This system can be applied in case of strengthening of slabs, where no wrapping is possible, or local strengthening and as an anchorage for pre-stressed strips. A minimum concrete cover of about 20 mm is required.

In general, anchoring devices that may influence the integrity of the strengthening system should be avoided. For example, anchoring solution with bolts need of holes that could lead to interlaminar shear failure or splitting of the strip. Moreover, holes reduce the cross section of the strip. However, in case of using bolted systems, it is not adequate to drill through the strengthening strip omitting special provisions, as compressive forces can weaken the strip. Multidirectional fibres at the location of the bolts can be used in order to allow the end tabs take the full force to be anchored (Tan 2001). Bolts should be anchored in the concrete to a depth beyond the steel reinforcement.

Bolted devices could give a better performance if used to anchor prefab laminates with suitable systems provided by manufacturers.

Nail anchors made by wide ringed head nylon anchor with zinc plated hammer screw were used by (Prota et al. 2006) to mechanically anchor beams strengthened with steel tape and cementitious grout.

Fibre fan anchor systems (see Fig. 6.26) are becoming very diffuse. A hole is drilled in the concrete on the same plane of the strengthening. A glass or carbon fibres tow (see Fig. 6.26a) is forced through the impregnated fabric end into the predrilled hole and the ends are splayed outwards on the continuous sheet reinforcement with epoxy resin (see Fig. 6.26b).

Experimental studies about the effectiveness of fan shaped anchors were made by (Özdemir 2005); he founded that: (1) anchors have to be inserted at least 50 mm into the core of the concrete (depth of 130–150 mm); (2) the cross-sectional area of anchor has to be at least two times greater than the cross-sectional area of the longitudinal sheet; (3) splitting the anchor into as many smaller anchors at about a 40 mm spacing as possible.

Such type of anchoring system was tested also by (Ceroni and Pecce 2010) that achieved an increasing of ultimate strength at least of 25 % in bond tests on

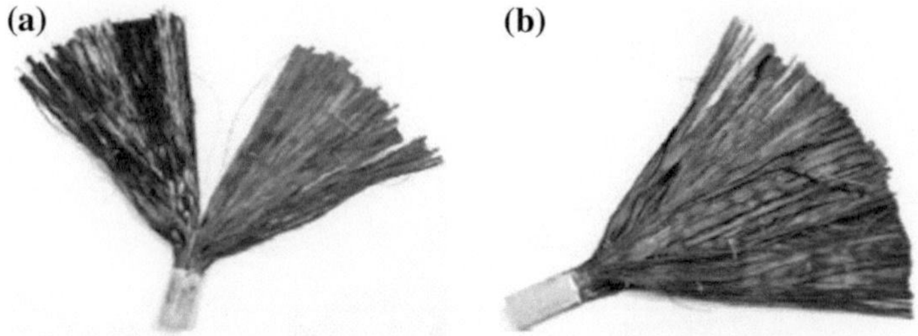

Fig. 6.27 **a** Bow-tie anchor, **b** single fan anchor (Zhang and Smith 2012a)

concrete blocks externally strengthened with carbon sheets when carbon FRP fans were used as end anchoring devices. It is worth to note that the details and accuracy of application procedures can be very influential on the strength increase.

Also (Orton et al. 2008) focused attention on the constructive detailing requested during the installation phase for warrant the efficiency of fan shaped anchor systems.

The effects of anchor splay diameter, anchor diameter and thickness of the FRP sheet have been investigated in (Niemitz et al. 2010).

In Eshwar et al. (2008) the effect of the location and the embedment of spike anchors (10 mm diameter) was experimentally investigated; the experimental results evidenced a significant improvement of strength (25–200 %). A minimum embedment depth of 50 mm in the concrete substrate is suggested, since further increase of the depth resulted ineffective. Multiple anchor spikes significantly increase the strength (+200 %).

In Zhang and Smith (2012a) the influence of FRP anchor fan configuration and dowel angle on the effectiveness of such an anchoring system for FRP reinforcement was investigated in several shear bond tests. In particular, bow-tie FRP anchors (fans oriented in opposite direction along the longitudinal FRP reinforcement) and single fan FRP anchors (fan fibres oriented only in one direction) were tested (see Fig. 6.27). Tests evidenced that single fan FRP anchors with the fan oriented in the direction of load allowed an increasing of the failure load of about 100 % compared with the not anchored specimens. The same load increase was attained using the bow-tie FRP anchors, although the fibre content was twice in this case. When the single fan anchor is positioned with the fan in the reverse direction of the applied load the efficiency into increase the failure load was lower (+60 %). Moreover, the load-slip experimental curves evidenced that the FRP fan anchors enabled a friction resistance that gave to the joint a post-peak reserve and larger ultimate slip, especially in the case of bow-tie configuration. For the single fan anchor the increasing of the angle of the anchor dowel (in the range 45°–157°, while in the basic configuration the angle is 90°) respect to the led to an increase of the failure load (25–160 %) and a decrease of the ductility of the joint due to the brittle failure of the anchor.

In Zhang and Smith (2012b) multiple single fan anchors (one, two, three) were tested in bond tests that substantially evidenced higher maximum load as the number of fan increase, even if the results were more scattered, and lower maximum slip. The concept of flexible and rigid anchors was also introduced and tested: in the flexible anchor there is no epoxy impregnating the region where the fibres of the anchor bend 90° at the junction of the fan and dowel components in order to provide larger slippage of the reinforcement. Experimental results evidenced that in the rigid anchored joints, after the plate debonding, the anchors failed immediately and no post-peak reserve of strength was observed with a consequent reduction of maximum slip that was on average only 36 % of the value attained in the flexible anchored joints. The maximum load in the rigid anchored joints was on average 32 % greater than the load in flexible ones.

The efficiency of single and bow-tie fan anchors located at the plate end or distributed along the plate into increase strength and ductility of RC slabs was proved by (Smith et al. 2013).

An alternative anchoring solution is represented by the application of a FRP bar placed transversally to the direction of the strengthening in a groove filled with resin according to the Near surface mounted bars technique (Khalifa et al. 1999, 2000). This solution can be reliable when the strengthening-to-concrete width ratio is approaching 1. Different solutions can be performed by positioning the FRP bar on a plane surface or in the corner (see Fig. 6.28). For such a technique a sufficient cover is requested for realizing the groove and installing the NSM bar inside. In (Eshwar et al. 2008) the location of the FRP bar, the grove size and the size of the anchoring bar were investigated in an experimental study. The strength was sensibly increased (also until 50 %) compared with not anchored specimens, even if the best performances were obtained when the anchor bars are located after re-entrant corner (see Fig. 6.28a) instead of before the re-entrant corner (see Fig. 6.28b), larger grooves are executed and more bars applied transversally to the reinforcement. In particular, a minimum groove size varying between 1.5 and 2.5 times the bar anchoring diameter and a minimum of 3 GFRP bars are suggested.

The use of FRP bars as anchoring system could create local concentrated stresses where fibres are turned into the groove with cutting of fibre on the corner

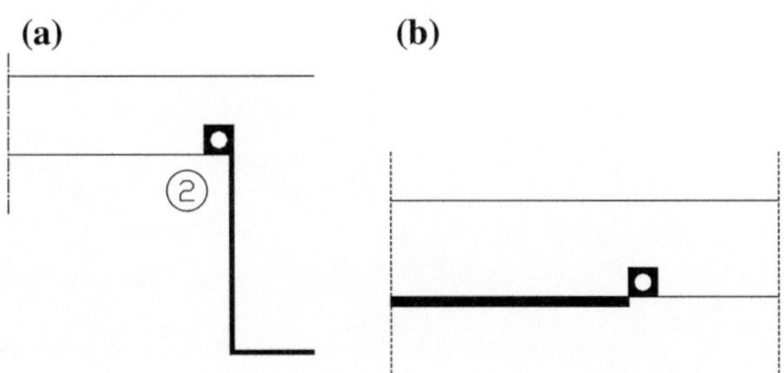

Fig. 6.28 NSM bars for plane surface anchorage for beams/slabs flexural strengthening: **a** after the re-entrant corner; **b** before the re-entrant corner

(Ceroni et al. 2008). However an increasing of ultimate strength at least of 25 % has been experimentally observed if NSM bars are used as end anchorage (Ceroni and Pecce 2010). In Eshwar et al. (2008) a minimum radius of 13 mm is suggested at the corners of the groove to limit stress concentration.

Special Anchorage for Shear Strengthening

For EBR shear strengthening anchoring systems could be particularly useful due to the reduced bond length available in order to avoid debonding. Proper anchorage can be made by systems suitably anchored in the compression zone of the strengthened section.

For shear strengthening solution similar to those proposed for flexural strengthening can be performed. Steel or FRP laminates glued or bolted transversally to the direction of the fibres can be successfully utilized when the not completely wrapped shear reinforcement configuration is adopted.

Special fan anchors can be used also in shear strengthening according to several configurations as tested by Jinno et al. (2001), Kobayshi et al. (2001) to solve the problem of passing fibres of reinforcement through the web of a 'T-section' in order to anchor fibres in the compressive zone (see Fig. 6.29).

Near surface mounted bars have been successfully applied as anchorage systems also at the end of shear strengthening (Fig. 6.30), according to the same applying procedure above described.

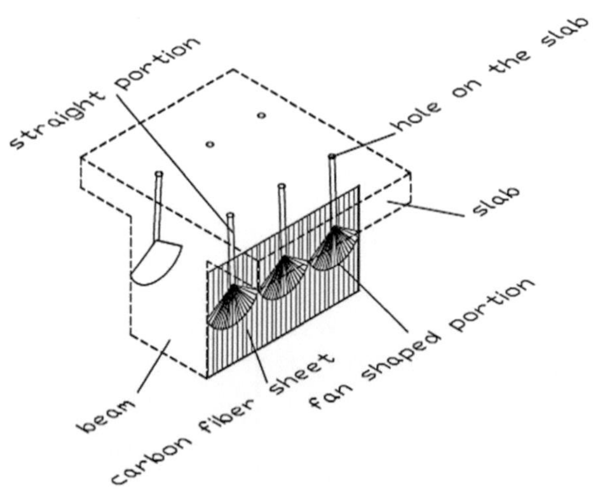

Fig. 6.29 Fan fibres anchors for shear strengthening (Jinno et al. 2001; Kobayshi et al. 2001)

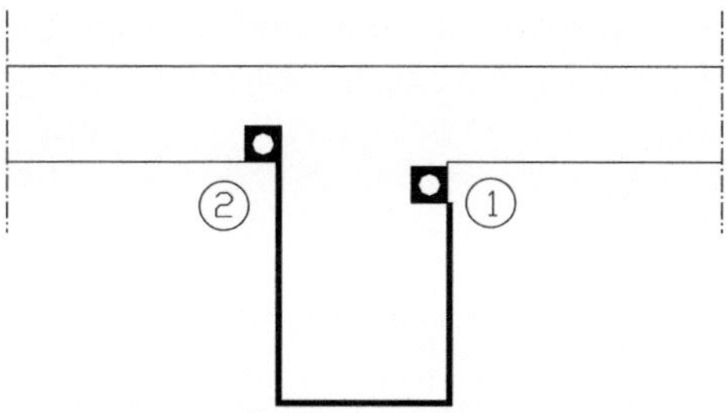

Fig. 6.30 NSM bars as anchors for U-wrap shear strengthening of an R.C. beam

Special Anchorages for Confinement

For confinement of columns where walls physically obstacle the complete wrapping of the elements, application of fan fibres (Jinno et al. 2001; Kobayshi et al. 2001) can give continuity to the reinforcement (see Fig. 6.31a).

Since the application of EBR on rectangular columns or pier walls with large aspect ratio does not actually confine the internal concrete, the EBR jacket need to be constrained on both sides along the length through the use of dowels or bolts or spike anchors. Such systems anchor the jacket to the pre-existing structure, thereby creating shorter distances which are confined between bolts (Fig. 6.31b). The positive effect of fibre anchors on the strength enhancement of wall-like columns confined with FRP was tested by (Tan 2002).

Spike anchors provide a low cost solution also to anchor the confinement jacket to the existing structure and has been tested with very good results for the attachment of FRP jackets at the reentrant corners of L-shaped cross section columns (Karantzikis et al. 2005). Experimental tests evidenced that the partial depth

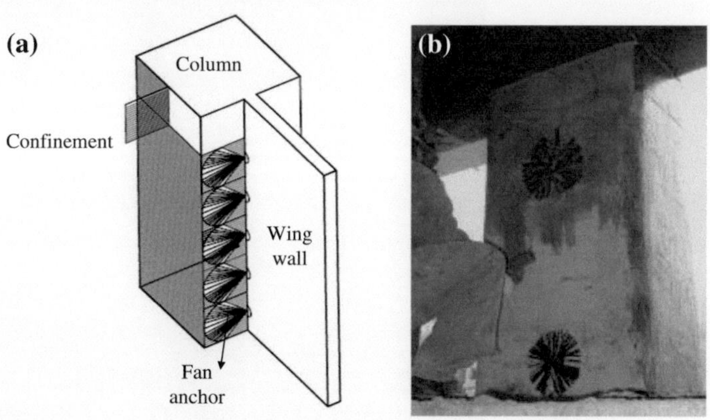

Fig. 6.31 **a** Fibre fan anchor device for confinement of columns; **b** fibre fan splayed on the confining strengthening

anchors give a good effectiveness in terms of deformability and strength increasing, while the benefit provided by full depth anchors are not justified by the high difficulty of installation.

Special Anchorages for Flexural Strengthening of Columns

The anchorage of flexural reinforcement of footing-column/wall joints should be provided. Possible solutions based on the use of steel spike anchors (Fig. 6.32) were tested by (Prota et al. 2005), where the flexural reinforcement was successively wrapped with a FRP jacket, and by (Ascione and Berardi 2011), where the FRP composite was bonded to the device steel plates bolted to each other (double lap bolted joint, see Fig. 6.33).

The flexural strengthening of columns can be realized also with NSM CFRP rods that can be anchored at the beam-column joint by wrapping the end with FRP carbon sheets (Prota et al. 2004).

In Antoniades et al. (2003, 2005) U-shaped devices made by FRP fibres or by steel or FRP laminates glued or bolted were successfully tested as anchoring

Fig. 6.32 Flexural strengthening at the footing joint of a RC column using steel spikes wrapped with a FRP jacket (Prota et al. 2005)

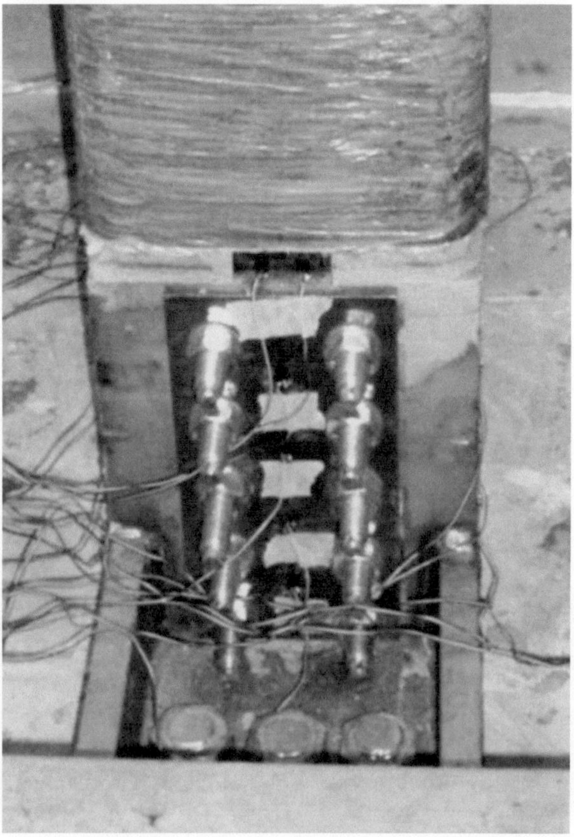

Fig. 6.33 Flexural strengthening of a RC column using anchorage device (Ascione and Berardi 2011)

systems of flexural reinforcement of column and walls at the footing zone. Effectiveness of bolted laminates into improve bond of FRP laminates in flexural strengthening of RC walls was tested by Tan et al. (2003) and Nagy-Gyorgy (2005).

Mechanically Fastened Systems

Description of the MF-FRP System

The implementation of mechanically fastened Fibre Reinforced Polymer (MF-FRP) systems for the flexural strengthening of RC members has emerged as an alternative to FRP materials adhesively bonded to the concrete substrate. The MF-FRP system consists of pre-cured FRP laminates with enhanced bearing strength that are connected to the concrete substrate by means of steel nails, anchor bolts, concrete screws, or combinations thereof.

Compared to the adhesive bonding FRP method, the benefit of MF-FRP is the speed of installation with unskilled labour, minimal or absent surface preparation under any meteorological condition and immediate use of the strengthened structures; also, the MF-FRP system is less vulnerable to fail prematurely by FRP delamination, which abruptly reduces the flexural strength gain and affects the member ductility. Some of the potential shortcomings are: possible concrete damage during anchoring and limited opportunity of installation in the presence of congested internal reinforcement in the members to be strengthened. Laboratory testing and a number of field applications have shown the effectiveness of the MF-FRP method.

FRP Strips

Unidirectional pultruded laminates currently used to be adhesively bonded onto the concrete surface are not suitable to be mechanically attached with steel anchors. These laminates are designed to have high modulus and strength in the longitudinal direction and low mechanical properties in the transverse direction of the laminate. Consequently, unless reinforcing fibres in the transverse direction of the laminate are inserted to provide adequate bearing strength, the orthotropic nature of the material causes the splitting failure of the laminate when a fastener is driven through it (Lamanna 2002). Precured laminates commercially available for strengthening with MF-FRP systems were developed in collaboration with Strongwell (USA, www.strongwell.com) and consist of a glass and carbon hybrid pultruded strip embebbed in a vinyl ester resin (see Fig. 6.34). The one shown has thickness and width of 3.2 and 101.6 mm, respectively; the 3.2 mm width is a suitable size for handling in the field. Continuous glass fibre strand mats are used to provide transverse and bearing strength, while 16–113 yield E-glass roving and 40–48 k standard modulus carbon tows are utilized to provide longitudinal strength and stiffness.

Table 6.10 summarizes the relevant mechanical properties of the FRP strips (Arora 2003). More details about the mechanical characterization of the FRP

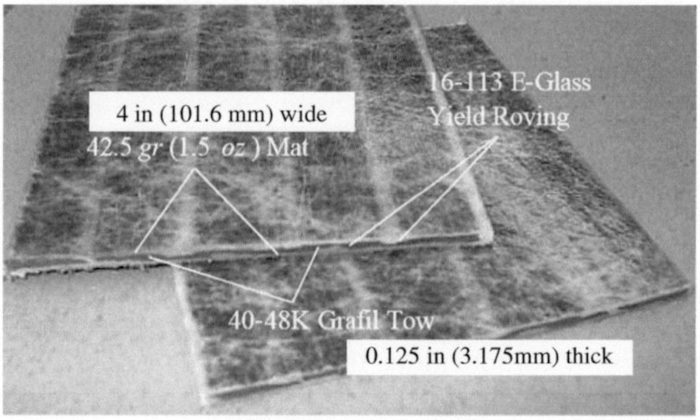

Fig. 6.34 Laminate available for the MF-FRP system

Table 6.10 Properties of FRP strips (Arora 2003)

Ultimate strength		Open-hole strength		Sustained bearing strength		Modulus of elasticity	
Mean (MPa)	COV (%)	Mean (MPa)	COV (%)	Mean (MPa)	COV (%)	Mean (GPa)	COV (%)
844	9.2	640	7.6	234	4.3	61.3	8.6

laminate in the longitudinal and transverse direction can be found in some papers (Arora 2003; Rizzo 2005; Rizzo et al. 2005a).

Fasteners

As mentioned earlier, fastener types that have been investigated for the MF FRP systems include: power-actuated fasteners (PAF), wedge bolts and wedge anchors (see Fig. 6.35).

The PAF system consists of pins embedded into base materials by means of a gunpowder charge. The effects of fastener type, washer, diameter, length, embedment depth have been investigated and discussed elsewhere (Lamanna et al. 2001; Lamanna 2002). Pre-drilling holes in the concrete is strongly recommended in order to reduce detrimental cracking phenomena. The use of the PAF system is particularly suitable when the compression strength of the concrete is less than 27 MPa. The presence of hard aggregates can prevent the fasteners to fully penetrate the concrete substrate (Bank 2004). The PAF installation requires times shorter than for wedge bolts and wedge anchors.

Wedge bolts are single-piece, heavy duty anchors that are driven into pre-drilled holes. Driving of the wedge bolt can be performed with a common rotary drill or impact wrench. As for the PAFs, the efficiency of wedge bolts is dependent on the presence of hard aggregates. Preliminary studies (Rizzo et al. 2005b) indicate that the use is not recommended for concrete with compression strength greater than 27 MPa, and with hard aggregates in the mix design.

In spite of longer installation times, wedge anchors can be used for any type of concrete; they are driven through the laminate into predrilled holes until the nut and

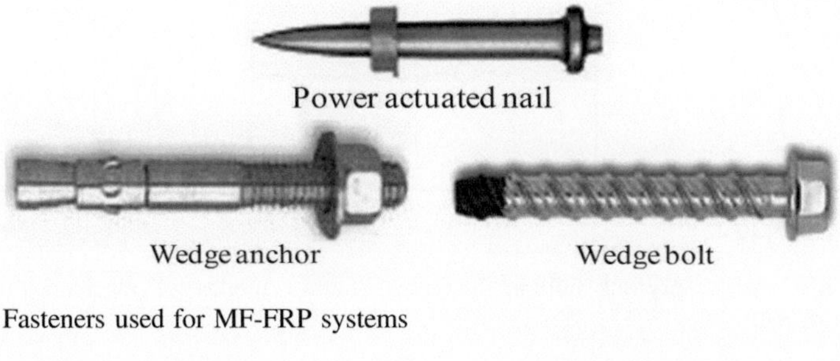

Fig. 6.35 Fasteners used for MF-FRP systems

washer are firmly secured against the laminate. The anchors are typically tightened by turning the nut with an electrical drill with torque control, according to the specifications of fastener manufactures.

Failure Modes of FRP-Fastener-Concrete Connection

The behaviour of MF-FRP connections is related to any of the components constituting the connection, which are the concrete substrate, the fastener and the FRP material. As a result, failure modes can involve the concrete, the yielding/rupture of fastener, or the FRP laminate.

The pry-out or spalling of the concrete (Fig. 6.36a) depends on the local composition of the concrete surface around the fastener. Once pry-out failure develops, the fastener rotates and the FRP laminate pulls it out of the concrete. Several factors promote the initiation of the concrete failure, such as a fastener hitting a hard aggregate during installation, low concrete strength, cracked concrete substrate conditions, short edge distance that may cause spalling, and poor fastener embedment depth (Lamanna 2002; Arora 2003).

The fastener failure usually occurs for deep embedment length, low steel strength and large edge distance (Rizzo 2005).

Four typical failure modes can involve the MF-FRP laminate (Fig. 6.36b): net-tension, cleavage-tension (or block shear), bearing and shear-out. The bearing failure is the most desirable failure mode because the connection is able to maintain its strength until significant levels of displacement (Lamanna 2002; Arora 2003). This failure is characterized by crushing on the material around the bolt-contact area followed by elongation of the hole. The other failure modes tend to develop in a more brittle way (Rosner and Rizkalla 1995). In particular, the net-tension failure is characterized by a fracture in the reduced cross section through the bolt hole, perpendicular to the direction of load. The cleavage failure consists of a crack parallel to the applied load that starts at the edge of the composite and propagates toward the bolt hole, leading to the initiation of other cracks across the net section

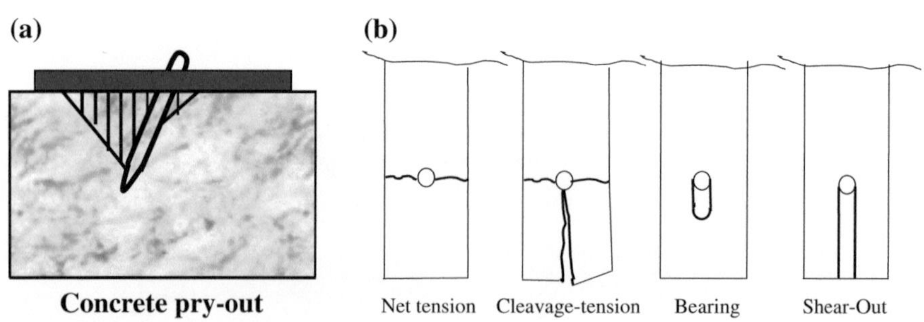

Fig. 6.36 a Concrete pry-out failure; **b** typical failure modes of mechanically fastened connections in FRP

due to in-plane stress. This failure mode is attributed to a combination of shear and tensile stress in the material. The shear-out failure, considered as a special case of bearing, is characterized by the formation of two cracks parallel to the applied load that propagate from the bolt hole toward the free edge.

State of Advancement and Current Research

The MF-FRP systems were firstly employed to strengthen RC bridges and infrastructures (Oliva et al. 2003; Rizzo 2005). After the first successful field applications, a number of experimental studies have been performed around the world with the aim of investigating the performance of RC members externally strengthened by MF-FRP laminate and quantifying the benefits obtained in terms of strength, stiffness and displacement capacity (Rizzo 2005; Lamanna et al. 2001, 2004a, b; Martin and Lamanna 2008; Lee et al. 2007).

To this aim, a recently published state-of-the-art review of the experimental research has provided compelling evidence of the effectiveness and viability of using MF-FRP laminates to rehabilitate RC beams and slabs (Brown et al. 2011). In the review, a database of collected test results was assembled to provide a valuable source of information on the performances of MF-FRP strengthened beams and one-way slabs. In a more recent paper, Martinelli et al. (2014) published an updated database of experimental results from four point-bending tests performed on MF-FRP strengthened RC members. As observed, the specimens tested so far have different sizes of the cross-section, with values of the height to-width ratios spanning from 0.5 ("slab" type) to 1.67 ("beam" type"); the clear lengths range from 1067 (small scale members) to 3505 mm (large scale members). The mechanical fastening mostly consists of shot fasteners with diameters ranging from 3.5 to 4.5 mm and lengths from 22 to 47 mm. Screw anchors were also frequently used, for which the diameters span from 4.76 to 12.7 mm and the lengths from 37 to 50.8 mm. Only in a few cases, instead, the mechanically fastening was performed by using wedge anchors (Ekenel et al. 2005; Galati et al. 2007); they were installed into the concrete by using epoxy resin as a gap filler. In these tests, the fasteners were arranged on single or multiple rows (1, 2, 4) according to aligned or staggered configurations or combinations thereof.

From the experimental investigations it has been shown that with appropriate fastener layout and FRP laminate properties, the strength increases are comparable to those of externally bonded-FRP strengthened members but with greater displacements exhibited at collapse. Also, it has been highlighted that the MF-FRP technique allows for preventing the strip delamination before concrete crushing. As an example, the experimental study performed by Napoli et al. (2010) is mentioned (see Fig. 6.37). The authors studied the effects of fastener layout and FRP strip length on flexural strength and deformability and failure mode of MF-FRP strengthened one-way slabs. Concrete screws were used to fasten the FRP strips to the concrete. At failure of specimens, the authors observed that the high

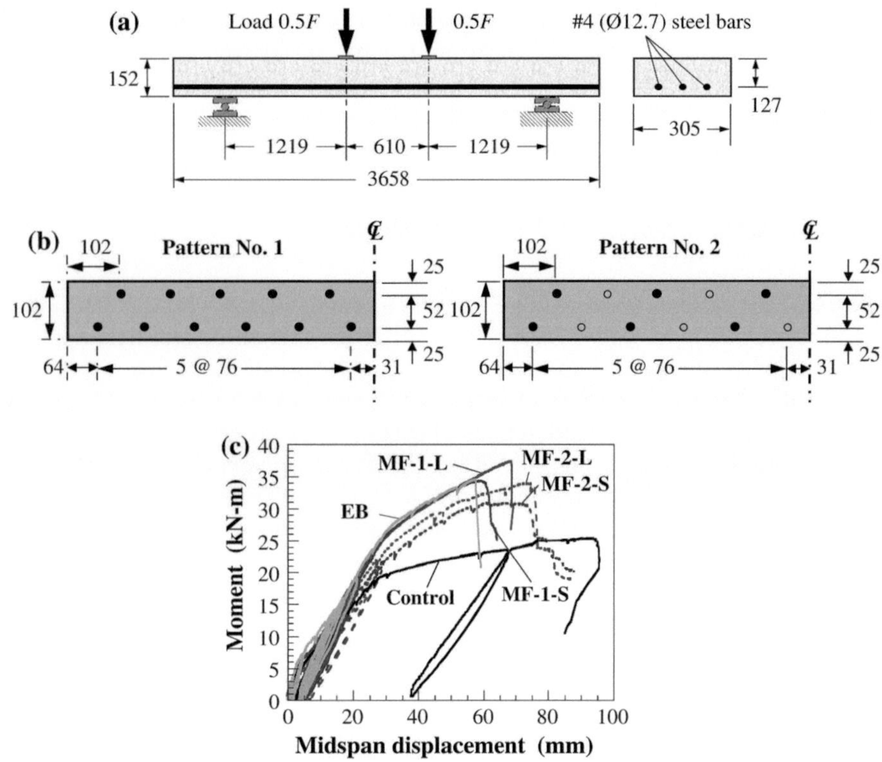

Fig. 6.37 Tests by Napoli et al. (2010): **a** specimens; **b** fastener layouts; **c** results

concentration of shear force caused spalling of the concrete cover but the FRP strip was firmly attached to the member. All MF-FRP strengthened specimens attained ultimate strength levels comparable to that of a benchmark slab strengthened with an externally bonded (EB) carbon FRP laminate, with greater deformability, as shown in Fig. 6.37c. Specimens MF-1L and MF-1S that used a larger number of anchors in the shear span (Pattern No. 1 in Fig. 6.37b), achieved marginally larger ultimate strengths but at lesser deformability than corresponding slabs MF-2L and MF-2S (using Pattern No. 2 in Fig. 6.37b).

Comparative experimental studies (Quattlebaum et al. 2005; Ekenel et al. 2006) on the static and fatigue performance of RC beams strengthened in bending by either MF- or EB-FRP laminates, showed that similar strength levels can be attained.

The feasibility of this strengthening technique was widely demonstrated experimentally in the upgrading of two-way slabs (Elsayed et al. 2009b).

Other experimental investigations examined the suitability of connecting an FRP laminate to the concrete substrate by both adhesive and mechanical anchors (EB + MF-FRP system) (Ekenel et al. 2006; Sena-Cruz et al. 2011; Ebead 2011).

One of the latest studies also showed that the MF-FRP system is viable for strengthening existing reinforced-concrete beams in shear (Johnson 2011).

Also, experimental tests have proven that the mechanically fastened composite system is an effective technique to improve the flexural capacity of corrosion damaged RC beams (El-Maaddawy 2013).

Finally, MF-FRP laminates were recently employed for enhancing the flexural capacity of timber structural members (Dempsey and Scott 2006; Schorer et al. 2008).

Design Rules

Despite the interest and promising experimental results obtained by researchers, there are no international guidelines dealing with MF-FRP till now. However, several analytical and numerical studies have been carried out through years with the aim to predict the flexural behaviour of RC members strengthened with MF-FRP systems: a state-of-the-art review has been recently published in Napoli et al. (2013). As highlighted therein, the first analytical models were based, for the sake of simplicity, on the hypothesis of "conservation of plane sections" between the concrete and the FRP, as generally accepted for EB-FRP strengthened members (Lamanna 2002; Bank et al. 2002; Bank 2004; Bank and Arora 2007; Rizzo et al. 2005b). Despite their ease of application, such models have often provided inaccurate predictions which were primarily attributed to ignoring the slip between the concrete and FRP strip. Therefore, novel proposals accounting for the concrete-FRP interfacial behaviour were recently formulated by (Lee et al. 2009; Nardone et al. 2011).

Lee et al. (2009), after observing and verifying the slip effect at the connection between FRP and concrete, introduced a FRP strain reduction factor calibrated on the basis of experimental results in order to evaluate the nominal moment capacity. However, even after this, an assumed reduction factor can lead to an incorrect evaluation of the neutral axis depth at ultimate limit state and does not provide useful information under service conditions.

Nardone et al. (2011) proposed analytical procedures for the evaluation of the flexural behaviour at both serviceability and ultimate limit states. Their analytical models account for equilibrium, compatibility and constitutive relationships of materials; in particular, they account explicitly for the slip behaviour between the concrete surface and the FRP strip due to the fasteners. The proposed models, coupled with an appropriate computation algorithm, are able to predict the fundamentals of flexural behaviour of RC members strengthened with MF-FRP strips also in terms of load versus deflection curves, failure modes, strain profiles and curvatures.

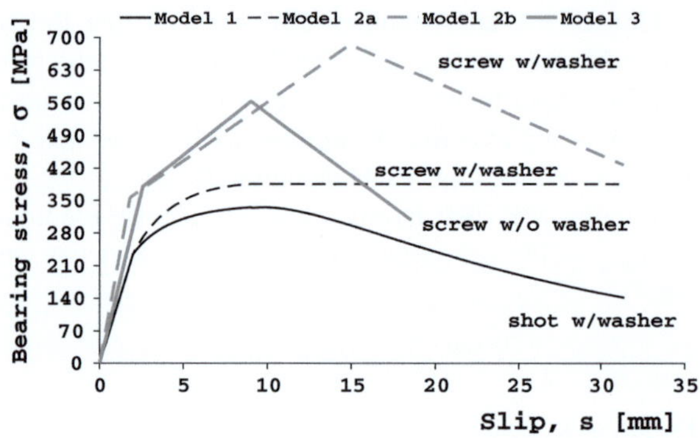

Fig. 6.38 Bearing stress-slip models

The models are capable of predicting the three failure modes experimentally found for well-designed applications, namely: bearing failure or net tension failure of the FRP laminate and concrete crushing.

The comparison between the analytical predictions and the experimental results shows a good agreement in terms of strain profiles in MF-FRP strips and moment deflection curves at serviceability limit state. Similarly, comparison of nominal flexural capacity, ultimate curvature and FRP strain at ultimate limit state show good agreement. This agreement is higher if bearing failure of many fasteners is accepted and included in the evaluations (Nardone et al. 2012). However from a design point of view, it is suggested to design for the desired sustained bearing failure of the first outermost fastener.

The knowledge of the relationship between the force acting on the fastener and the slip is fundamental in order to apply the proposed model and more research to address this fundamental parameter is needed.

To this aim, Elsayed et al. (2009a) and Realfonzo et al. (2013) recently proposed nonlinear bearing stress-slip σ-s models to describe the effect of the partial interaction between the concrete and the FRP laminate. These models, shown in Fig. 6.38, were calibrated by best-fitting experimental results of direct shear tests (DSTs) performed on MF-FRP/concrete joints with a single connector.

Two different σ-s models were found by Elsayed et al. (2009a), suitable for shot (*Model 1*) and screwed (*Model 2a*) fasteners, respectively, both with steel washers (Fig. 6.38). The first model was calibrated for MF-FRP/concrete joints having a single shot fastener with a 47 mm shank length, a 3.7 mm shank diameter and a 13 mm washer where the fastener was driven into the concrete using a powder actuated gun. The second model was for the case of a single screwed fastener with a 37 mm shank length, a 4.8 mm shank diameter and a 16 mm washer installed into the concrete using a drilling tool.

Trilinear σ-s models were proposed by (Realfonzo et al. 2013) for a single screwed fastener with (w/washer) or without (w/o washer) washer (*Models 2b* and *3* in Fig. 6.38); the screw had a 45 mm shank length, a 6 mm shank diameter, a 32 mm washer (when used) and was driven into the pre-drilled hole using a common torque wrench.

Once the first studies on the FRP-concrete interfacial behaviour were published in the literature, refined moment-deflections models were developed. Among them, the numerical model developed by Napoli et al. (2010) was based on a general algorithm formulated by implementing the differential equation of Newmark's theory for steel-concrete composite beams with linear-elastic shear connectors (Newmark et al. 1951) into an 'exact' two-node finite element (FE) developed for the analysis of partial interaction in composite beams with flexible shear connectors (Faella et al. 2008). The finite element introduced is used for nonlinear analysis through fibre discretization of the beam cross-section, and by implementing an iterative convergence procedure based on the secant value approach to account for material non-linearity, including concrete, steel and concrete-FRP interface. For the latter, the interface bearing stress-slip *Model 2a* in Fig. 6.38 and its resulting simplification in a bilinear law (Napoli et al. 2010) were successfully implemented and verified to evaluate applicability for analysis and design purposes.

A different 1-D numerical model was presented by Martinelli et al. (2014) to explicitly account for the discrete connection between FRP laminate and RC beam. Later, the model was also used by the authors to investigate the cracking process in RC members (and its implications on the structural response including the tension-stiffening effect) which is generally neglected (Martinelli et al. 2013).

Figure 6.39a depicts the considered finite element which is obtained by assembling the following three components: (a) a 1-D element that models the behaviour of an Euler-Bernoulli RC beam; (b) a rod element that simulates the mechanical behaviour of an FRP laminate; (c) two springs used as axial constraints between the FRP laminate and the RC beam. The two springs simulate the behaviour of the fasteners connecting the FRP laminate to points on the beam surface, i.e., at the points where the fasteners are actually screwed or shot. Four degrees of freedom are considered in each node to take into account the possible axial displacements of both RC beams and FRP strips, along with deflections and rotations of the former component. The stiffness matrix of the proposed FE is derived by assembling the key mechanical components of the RC beam/slab, the FRP strip and the mechanical fasteners. The nonlinear behaviour of the aforementioned components and materials is handled through well-established numerical techniques usually adopted in nonlinear FE analysis.

The procedure was validated by comparing the numerical simulations in terms of load-deflection curves with the results of some experimental tests reported in the literature. In particular, Fig. 6.39b shows a comparison for an MF-FRP strengthened slab tested by Napoli et al. (2010). In this case, the behaviour of the screws w/o washers was modelled by adopting the trilinear *Model 3* (Fig. 6.38). For comparison, the nonlinear *Model 1* for shot fasteners (Fig. 6.38) was also considered. It is shown that the numerical simulations do not significantly change with the use of the two different bearing stress-slip relationships, thus suggesting that both of them are suitable to model MF-FRP strengthened members.

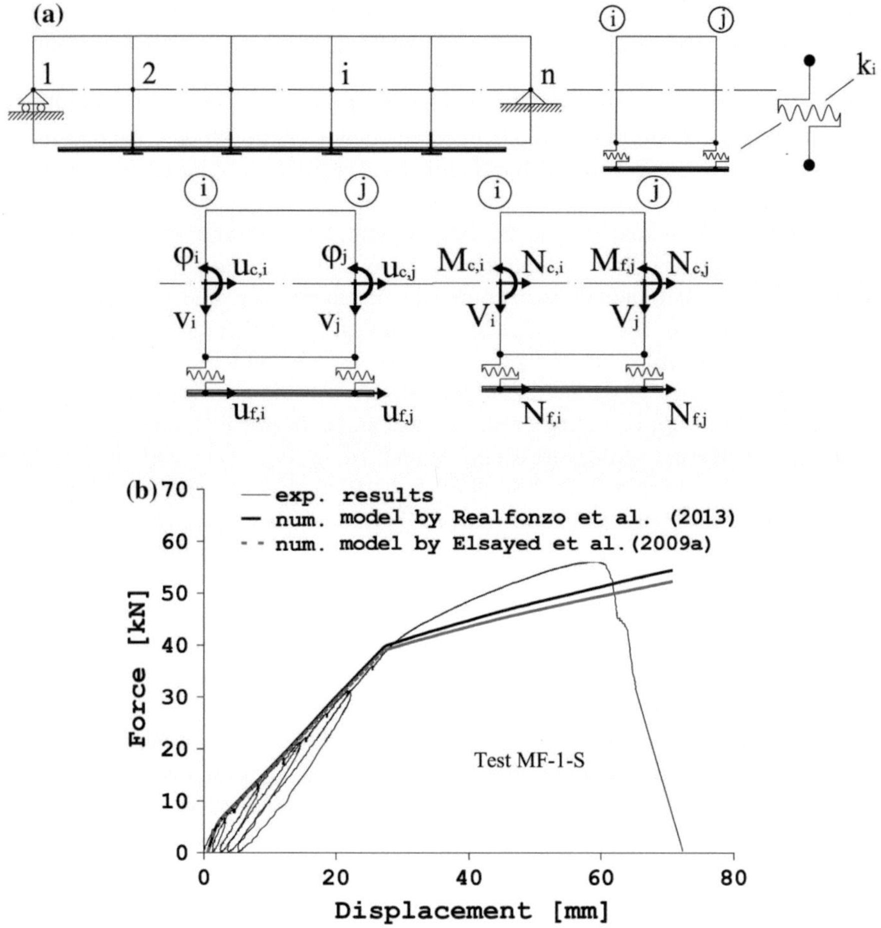

Fig. 6.39 FE model by Martinelli et al. (2014): **a** schematic and **b** validation

References

Serviceability Limit States

ACI (2005). Building code requirements for structural concrete and commentary, ACI 318R-05, ACI Committee 318, Farmington Hills, Mich.

Aiello, M. A., & Ombres, L. (2000). Load deflection analysis of FRP reinforced concrete flexural members. *ASCE Journal of Composite in Construction, 4*(4), 164–171.

Aiello, M. A., & Ombres, L. (2004). Cracking and deformability analysis of reinforced concrete beams strengthened with externally bonded carbon fiber reinforced polymer sheets. *Journal of Materials in Civil Engineering, 16*(5), 392–399.

Balsamo, A., Nardone, F., Iovinella, I., Ceroni, F., & Pecce, M. (2013a). Flexural strengthening of concrete beams with EB-FRP, SRP and SRCM. *Experimental investigation, Composites: Part B* doi:10.1016/j.compositesb.2012.10.014.

Balsamo, A., Bilotta, A., Ceroni, F., Nigro, E., & Pecce M., (2013b). Efficiency of CFRP NSM strips and EBR laminates for flexural strengthening of RC beams. In B. Joaquim & S.-C. José (Eds.), *Proceedings of FRPRCS11*, UM, Guimarães.

Barros, J. A. O., & Fortes, A. S. (2005). Flexural strengthening of concrete beams with CFRP laminates bonded into slits. *Journal Cement and Concrete Composites, 27*(4), 471–480.

Barros, J. A. O., Dias, S. J. E., & Lima, J. L. T. (2007). Efficacy of CFRP-based techniques for the flexural and shear strengthening of concrete beams. *Journal Cement and Concrete Composites, 29*(3), 203–217.

Bischoff, P. H. (2007). Deflection calculation of FRP reinforced concrete beams based on modifications to the existing Branson equation. *ASCE Journal of Composites for Constructions, 11*(1), 4–14.

Borchert, K. (2007). Bond behaviour of adhesively bonded reinforcement in service. Ph.D. Thesis, Technische Universität München (in German with English summary).

Borchert, K., & Zilch, K. (2008). Bond behaviour of NSM FRP strips in service. *Structural Concrete, 9*, 128–142.

Branson, D. E. (1977). *Deformations of concrete structures* (pp. 167–169). New York: McGraw-Hill.

Charkas, H., Rasheed, H. A., & Melhem, H. (2003). Rigorous procedure for calculating deflections of fiber-reinforced polymer-strengthened reinforced concrete beams. *ACI Structural Journal, 100*(4), 529–539.

fib bulletin No. 65. (2012). Model Code 2010-Final draft, Volume 1, International Federation for Structural Concrete, Lausanne, Switzerland. ISSN 1562-3610. ISBN:978-2-88394-105-2.

fib bulletin No. 66. (2012). Model Code 2010-Final draft, Volume 2, International Federation for Structural Concrete, Lausanne, Switzerland. ISSN 1562-3610. ISBN:978-2-88394-105-2.

Ceroni, F., Pecce, M., & Matthys, S. (2004). Tension stiffening of RC ties strengthened with externally bonded FRP sheets. *ASCE Journal of Composites for Constructions, 8*(1), 22–32.

Ceroni, F., & Pecce, M. (2004). Modelling of tension stiffening behaviour of RC ties strengthened with FRP sheets. *ASCE Journal of Composites for Constructions, 8*(6), 510–518.

Ceroni, F., & Pecce, M. (2007). Cracking behaviour of RC beams externally strengthened with emerging materials. *Construction and Building Materials, 21*(4), 736–745.

Ceroni, F., & Pecce, M. (2009). Design provisions for crack spacing and width in RC elements externally bonded with FRP. *Composites Part B, 40*(1), 17–28.

Ceroni, F. (2010). Experimental performances of RC beams strengthened with FRP materials. *Construction and Building Material, 24*, 1547–1559.

CNR DT 200/R1. (2012). Guide for the Design and Construction of Externally Bonded FRP Systems for Strengthening Existing Structures, Advisory Committee on Technical Recommendation for Construction of National Research council, Rome, Italy.

Eligehausen, R., Popov, E. P., & Bertero, V. V. (1983). Local bond stress-slip relationships of deformed bars under generalized excitations, Report no. 83/23, EERC. University of California, Berkeley.

El-Mihilmy, M. T., & Tedesco, J. W. (2000). Deflection of reinforced concrete beams strengthened with fiber-reinforced polymer (FRP) plates. *ACI Structural Journal*, *97*(5).

EN 1990. (2002). Eurocode 0: Basis of Structural Design, European Committee of Standardization.

Ferretti, D., & Savoia, M. (2003). Cracking evolution in R/C tensile members strengthened by FRP-plates. *Engineering Fracture Mechanics, 70*, 1069–1083.

Ferrier, E., Avril, S., Hamelin, P., & Vautrin, A. (2003). Mechanical behavior of rc beams reinforced by externally bonded CFRP sheets. *Materials and Structures/Matériaux and Constructions, 36*(262), 522–529.

Jourawski, D. J. (1858). Sur le résistance d'un corps prismatique et d'une Piece Composée en Bois ou on Tôle de Fer à une Force Perpendiculaire à leur Longeur, Annales des Ponts et Chausees, Memoires and Documents, 3rd Series, V. 12, Part 2, pp. 328–351

Kim, Y. J., Shi, C., & Green, M. F. (2008a). Ductility and cracking behavior of prestressed concrete beams strengthened with prestressed CFRP sheets. *ASCE Journal of Composites for Construction, 12*(3), 274–283.

Matthys, S. (2000). Structural behaviour and design of concrete members strengthened with externally bonded FRP reinforcement, Ph Doctoral thesis, Faculty of Engineering, Department of Structural Engineering, Ghent University, Ghent, Belgium.

Monti, G., Alessandri, S., & Santini, S. (2009). Design by testing: a procedure for the statistical determination of capacity models. *Construction and Building Materials, 23*(4), 1487–1494.

Pecce, M., Ceroni, F., Prota, A., & Manfredi, G. (2006). Response prediction of RC beams externally bonded with steel reinforced polymers. *ASCE Journal of Composites for Construction, 10*(3), 195–203.

Rasheed, A. H., Charkas, H., & Melhem, H. (2004). Simplified nonlinear analysis of strengthened concrete beams based on a rigorous approach. *ASCE Journal of Structural Engineering, 130* (7), 1087–1096.

Razaqpur, A. G., Švecová, D., & Cheung, M. S. (2000). Rational method for calculating deflection of fiber-reinforced polymer reinforced beams. *ACI Structural Journal, 97*(1), 175–184.

Roberts, T. M. (1989). Approximate analysis of shear and normal stress concentrations in the adhesive layer of plated RC beams. *The Structural Engineer, 67*(12), 229–233.

Sato, Y., Ueda, T., & Shoji, K. (2002). Tension stiffening effect of reinforced concrete member strengthened by carbon fiber sheet. In *Proceedings of the International Symposium "Bond in concrete", Budapest*, 20–22 November 2002, pp. 606–613.

Täljsten, B. (2004). FRP Strengthening of existing concrete structures: design guideline. Lulea University of Technology. ISSN 91-89580-03-4.

Tan, K. H., & Saha, M. K. (2008). Cracking characteristics of RC beams strengthened with FRP system. *Journal of Composites for Construction, 12*(5), 513–521.

Tripi, J. M., Bakis, C. E., Boothby, T. E., & Nanni, A. (2000). Deformation in concrete with external CFRP sheet reinforcement. *ASCE Journal of Composites for Construction, 4*(2), 85–94.

Ueda, T., Yamaguchi, R., Shoji, K., & Sato, Y. (2002). Study on behavior in tension of reinforced concrete members strengthened by carbon fiber sheet. *ASCE Journal of Composites for Construction, 6*(3), 168–174.

Yoshizawa, H., & Wu, Z. (1999). Crack behavior of plain concrete and reinforced concrete members strengthened with carbon fiber sheets. *Proceedings of Fourth International Symposium on FRP Reinforcement*, Vol. 1, pp. 767–779

Zehetmaier, G., & Zilch, K. (2008). Rissbildung und Rissbreitenbeschränkung bei Verstärkung mit CFK-Lamellen. *Bauingenieur, 83*, 19–26.

Zhang, Y., Toutanji, H., & Balagrou, P. (2003). Crack widths in R.C. beams externally bonded with CFRP sheets. *Proceedings of FRPRCS-6*, Singapore, 8–10 July 2003.

Fatigue Behaviour

ACI Committee 440. (2008). Guide for the Design and Construction of Externally Bonded FRP Systems for Strengthening Concrete Structures.

Ali-Ahmad M., et al. (2006), Experimental investigation and fracture analysis of debonding between concrete and FRP. *Journal of Engineering Mechanics ASCE, 132*(9), 914–923.

Anderson, T. L. (2004). *Fracture mechanics: fundamentals and applications*. Boca Raton: CRC Press.

Bizindavyi, L., et al. (2003), Experimental investigation of bonded fiber reinforced polymer-concrete joints under cyclic. *Journal of Composites in Construction*, pp. 127–133.

Budelmann, H., et al. (2013). *Praxisgerechte Bemessungsansätze für das wirtschaftliche Verstärken von Betonbauteilen mit geklebter Bewehrung - Verbundtragfähigkeit unter dynamischer Belastung, DAfStb Heft 593*. Berlin: Beuth Verlag.

Carloni, C., et al. (2012). Experimental determination of FRP concrete cohesive interface properties under fatigue loading. *Composite Structures, 94*(4), 1288–1296.

Carloni C., et al.. (2013). Sub-critical fatigue crack growth in FRP/concrete cohesive interface. *Composite: Part B, 51*, 35–43.

CNR-DT 200. (2004). Guide for the design and construction of externally bonded FRP systems for strengthening existing structures.

Dai, J. G., et al. (2005). Static and fatigue bond characteristics of interfaces between CFRP sheets and frost damage experienced concrete. *Proceedings of 7th International Symposium on Fiber Reinforced Polymer Reinforcement for Reinforced Concrete Structures, Kansas City*, MO, American Concrete Institute (ACI), Farmington Hills, MI, 15151530.

Diab, H. M., et al. (2009). Theoretical solution for fatigue debonding growth and fatigue life prediction of FRP-concrete interfaces. *Advanced Structural Engineering, 12*(6), 781–792.

Ferrier, E., et al. (2005). Fatigue of CFRPs externally bonded to concrete. *Materials and Structures, 38*, 39–46.

fib Task Group 9.3 FRP, bulletin 14. (2001). Externally bonded FRP reinforcement for RC structures.

DAfStb-guideline. (2012). *DAfStb Guideline On the Strengthening of Concrete Members with Adhesively Bonded Reinforcement*. German Committee for Structural Concrete, Berlin: Beuth Verlag.

German Committee for Structural Concrete. (2013). *Commentary on the DAfStb Guideline Strengthening of Concrete Members with Adhesively Bonded Reinforcement with Examples, Report 595*. Berlin: Beuth Verlag.

Hankers, C. (1996). Zum Verbundtragverhalten laschenverstärkter Betonbauteile unter nicht vorwiegend ruhender Beanspruchung, Dissertation, Institut für Baustoffe, Massivbau und Brandschutz, TU Braunschweig.

Heffernan, P. J., et al. (2004). Fatigue behavior of reinforced concrete beams strengthened with carbon fiber reinforced plastic laminates. *Journal of Composites in Construction, 8*, 132–140.

Kim, Y. J., Heffernan, P. J. (2008). Fatigue behavior of externally strengthened concrete beams with fiber-reinforced polymers: state of the art. *Journal of Composites in Construction, 12*, 246–256.

ISIS Canada. (2001), Strengthening reinforced concrete structures with externally-bonded fibre reinforced polymers: Design manual No. 4.

Ko, H. (2007). Bond stress slip relationship between frp sheet and concrete under cyclic load. *Journal of Composites for Construction, 11*(4), 419–426.

Sutton, M. A. et al. (1983). Determination of displacements using an improved digital correlation method. *Image Vision Computing, 1*(3), 133–139.

Sutton, M. A., et al. (2009). *Image correlation for shape, motion and deformation measurements*. New York: Springer.

Wu, Z., et al. (2010). Tensile fatigue behavior of FRP and hybrid FRP sheets. *Composites: Part B, 41*, 396402.

Yao, J., et al. (2005). Experimental study on FRP-to-concrete bonded joints. *Composites Part B Engineering, 36*(2), 99–113.

Effects of Fire and High Temperature

ACI 318-11. (2013). Building Code Requirements for Structural Concrete and Commentary, ISBN-13 978-0870317446.

Bai, Y., Vallée, T., & Keller, T. (2007). Modeling of thermo-physical properties for FRP composites under elevated and high temperature. *Composites Science and Technology, 67*, 3098–3109.

Bisby, L. A., Green, M. F., Kodur, V. K. R. (2005). Response to fire of concrete structures that incorporate FRP. *Progress in Structural Engineering and Materials*, *7*(3), 136–149.

Blontrock, H., Taerwe, L., & Vandevelde, P. (2001). Fire testing of concrete slabs strengthened with fibre composite laminates. *Proceedings of 5th International Conference on Fibre Reinforced Plastics for Reinforced Concrete Structures,* Telford, Cambridge, U.K., pp. 547–556.

Cleary, D. B., Cassino, C. D., & Tortorice, R. (2003). Effect of elevated temperature on a fiber composite used to strengthen concrete columns. *Journal of Reinforced Plastics and Composites, 2210*, 881–895.

Dai, J. G., Sato, Y., Ueda, T., & Sato, Y. (2005). Static and fatigue bond characteristics of interfaces between CFRP sheets and frost damage experienced concrete. *Proceedings of FRPRCS-7, ACI-SP-230-86*, pp. 1515–1530.

Dai, J. G., Gao, W. Y., Teng, J. G. (2013). Bond-slip model for FRP laminates externally bonded to concrete at elevated temperature. *Journal of Composites for Construction, 17*(2). © ASCE, ISSN 1090-0268/2013/2-217-228.

Deuring, M. (1993). Brandversuche an nachtraglich verstarkten tragern aus beton. Research Rep. No. 148'795, EMPA, Dubendorf, Switzerland.

EN 1992-1-2. Eurocode 2. Design of concrete structures – Part 1-2: General Rules – Structural Fire Design, European committee for standardization, March, 2004.

Gamage, J. C. P. H., Wong, M. B., & Al-Mahaidi, R. (2005). Performance of CFRP strengthened concrete members under elevated temperatures. *Proceedings of International Symposium on Bond Behaviour of FRP in Structures*, IIFC, Hong Kong, pp. 113–118.

Jia, J., Boothby, T. E., Bakis, C. E., & Brown T. L. (2005). Durability evaluation of glass fiber reinforced-polymer-concrete bonded interfaces. *Journal of Composites for Construction*, *9*(4). ©ASCE, ISSN 1090-0268/2005/4-348–359.

Kodur, V. K. R., Bisby, L. A., & Green, M. F. (2007). Guidance for the design of FRP-strengthened concrete members exposed to fire. *Journal of Fire Protection Engineering, 175*, 5–26.

Mouritz, A. P., & Gibson, A. G. (2006). *Fire properties of polymer composite materials*. New York: Springer.

Mouritz, A. P., & Mathys, Z. (1999). Postfire mechanical properties of marine polymer composites. *Composite Structures, 471*, 643–653.

NTC. (2008). D.M.14/1/2008 "Norme Tecniche per le Costruzioni" (Testo integrato con la Circolare n°617/C.S.LL.PP. del 2/2/2009).

Nigro, E., Cefarelli, G., Bilotta, A., Manfredi, G., & Cosenza, E. (2011a). Fire resistance of concrete slabs reinforced with FRP bars. Part I: experimental investigations on the mechanical behaviour. *Composites: Part B*, *Engineering*, *42*, 1739–1750. doi:10.1016/j.compositesb.2011.02.025. ISSN 1359-8368.

Nigro, E., Cefarelli, G., Bilotta, A., Manfredi, G., & Cosenza, E. (2011b). Fire resistance of concrete slabs reinforced with FRP bars. Part II: experimental results and numerical simulations on the thermal field. *Composites: Part B*, *Engineering*, *42*, 1751–1763. doi:10.1016/j.compositesb.2011.02.026. ISSN 1359-8368.

Nigro, E., Cefarelli, G., Bilotta, A., Manfredi, G., & Cosenza, E. (2013). Adhesion at high temperature of FRP bars straight or bent at the end of concrete slabs. *Journal of Structural Fire Engineering*, pp. 71–86. doi:10.1260/2040-2317.4.2.71. ISSN 2040-2317.

Nigro, E., Bilotta, A., & Del Prete, I. (2013a). Flexural check at high temperatures of reinforced concrete bridge decks strengthened with EBR-FRP. Applications of Structural Fire Engineering, 19–20 April 2013, Prague, Czech Republic.

Nigro, E., Bilotta, A., & Del Prete, I. (2013b). High temperature effects on flexural performances of RC bridge decks strengthened with EBR-FRP. ACI Italy Chapter. Bergamo 3-4 October 2013.

Porter, M. L., & Harries, K. A. (2005). *Workshop on research in FRP composites in concrete construction*. Arlington: National Science Foundation.

Saafi, M., & Romine, P. (2002). Effect of fire on concrete cylinders confined with GFRP. *Proceedings of 2nd International Conference on Durability of Fibre Reinforced Polymer (FRP) Composites for Construction*, Université de Sherbrooke, pp. 512–521.

Long Term Properties of FRP Systems

ASTM D 638. (2003). Standard Test Methods for Tensile Properties of Plastics. *American Society for Testing and Materials* (ASTM), Pennsylvania, US.

ASTM 2990. (2001). Standard Test Methods for Tensile, Compressive, and Flexural Creep and Creep-Rupture of Plastics. *American Society for Testing and Materials* (ASTM), Pennsylvania, US, 20 pp.

Brinson, H. F., & Brinson, L. C. (2008). *Polymer engineering science and viscoelasticity: an introduction*. New York: Springer Science + Business Media.

Choi, K.-K., Meshgin, P., & Taha, M. M. R. (2007). Shear creep of epoxy as the concrete-FRP interfaces. *Composites Part B: Engineering, 38*(5–6), 772–780.

Costa, I. G. (2014). Prestressed carbon fibre laminates applied according to near surface mounted technique to increase the flexural resistance of reinforced concrete beams. Ph.D. Thesis, University of Minho.

Costa, I. G., & Barros, J. A. O. (2013). Assessment of the long term behaviour of structural adhesives in the context of NSM flexural strengthening technique with prestressed CFRP laminates. FRPRCS11, Guimarães, Portugal.

Costa, I. G., & Barros, J. A. O. (2011). Creep of Adhesives: Review. *Report 11-DEC/E-03*, University of Minho, Guimarães, Portugal, 39 pp.

Diab, H., & Wu, Z. (2007). A linear viscoelastic model for interfacial long-term behavior of FRP–concrete interface. *Composites Part B: Engineering, 39*(4), 722–730.

Dolan, C. W., Hamilton, H. R., Bakis, C. E., & Nanni, A. (2001). Design recommendations for concrete structures prestressed with FRP tendons. *FHWA-DTFH61-96-C-00019*, Federal Highway Administration, Washington, DC.

Feng, C. W., Keong, C. W., Hsueh, Y. P., Wang, Y. Y., & Sue, H. J. (2005). Modeling of long-term creep behavior of structural epoxy adhesives. *International Journal of Adhesion and Adhesives, 25*(5), 427–436.

ISO 527-2. (1993). Plastics: Determination of tensile properties—Part 2: Test conditions for moulding and extrusion plastics. *International Organization for Standardization* (ISO), Geneva, SZ, 8 pp.

ISO 899-1. (2003). Plastics: Determination of creep behaviour—Part 1: Tensile creep. *International Organization for Standardization* (ISO), Geneva, SZ.

Khoun, L., & Hubert, P. (2010). Cure shrinkage characterization of an epoxy resin system by two in situ measurement methods. *Polymer Composites*, *31*(9), 1603–1610.

Li, C., Potter, K., Wisnom, M. R., & Stringer, G. (2004). In-situ measurement of chemical shrinkage of MY750 epoxy resin by a novel gravimetric method. *Composites Science and Technology, 64*(1), 55–64.

Lopez-Anido, R. A., & Naik, T. R. (2000). *Emerging Materials for Civil Engineering Infrastructure: State of the Art*. Reston: American Society of Civil Engineers.

Majda, P., & Skrodzewicz, J. (2009). A modified creep model of epoxy adhesive at ambient temperature. *International Journal of Adhesion and Adhesives, 29*(4), 396–404.

Meaud, C., Jurkiewiez, B., & Ferrier, E. (2011). Investigation of creep effects in strengthened RC structures through double lap shear testing. *Composites Part B: Engineering, 42*(3), 359–366.

Sayed-Ahmed, E. Y. (2002). Single and Multi-Strand Steel Anchorage Systems for CFRP Tendons/Stays. *4th Structural Speciality Conference of the Canadian Society for Civil Engineering*, Montréal, Québec, 5–8 June, 10 pp.

Wang, W.-W., Dai, J.-G., Harries, K. A., & Bao, Q.-H. (2012). Prestress losses and flexural behavior of reinforced concrete beams strengthened with posttensioned CFRP sheets. *Journal of Composites for Construction, ASCE, 16*(2), 207–216.
Wu, Z., & Diab, H. (2007). A linear viscoelastic model for interfacial long-term behavior of FRP-concrete interface. *Journal of Composites for Construction, ASCE, 11*(5), 477–486.
Yu, H., Mhaisalkar, S., & Wong, E. (2005). Cure shrinkage measurement of nonconductive adhesives by means of a thermomechanical analyzer. *Journal of Electronic Materials, 38*(4), 1177–1182.

Anchorage Systems for External Strengthening with FRP

Ascione, L., & Berardi, V. P. (2011). Anchorage device for FRP laminates in the strengthening of concrete structures close to beam-column joints. *Composites Part B: Engineering*, *42*(7), 1840–1850.
Al-Mahaidi, R., & Kalfat, R. (2011). Investigation into CFRP plate end anchorage utilizing uni-directional fabric wrap. *Composite Structures, 93*(2), 821–830.
Antoniades, K. K., Salonikios, T. N., & Kappos, A. J. (2003). Cyclic tests on seismically damaged R/C walls strengthened using fiber reinforced polymer reinforcement. *ACI Structural Journal, 100*(4), 510–518.
Antoniades, K. K., Salonikios, T. N., & Kappos, A. J. (2005). Tests on seismically damaged reinforced concrete walls repaired and strengthened using fiber-reinforced polymers. *Journal of Composites for Construction, 9*(3), 236–246.
Bank, L. C. (2004). Mechanically-fastened FRP (MF-FRP): a viable alternative for strengthening RC members. In Seracino (Ed.), *Proceedings of the International Conference FRP Composites in Civil engineering, CICE 2004, Adelaide*, Australia, 8–10 December 2004, pp. 3–15. Taylor & Francis Group, London. ISBN 90 5809 638 6.
Blaschko. M. (2001). Anchorage device for FRP strips. In T. Telford (Ed.), *Proceedings of the Fifth Conference on Non-Metallic Reinforcement for Concrete Structures*, *Proceedings of FRPRCS5 International symposium*, Cambridge, UK, Vol. 2, pp. 1255–1264.
Bonaldo, E., Barros, J. A. O., & Lourenço, P. J. B. (2008). Efficient strengthening technique to increase the flexural resistance of existing RC slabs. *Journal of Composites for Construction, 12*(2), 149–159.
Ceroni, F. (2010). Experimental performances of RC beams strengthened with FRP materials. *Construction and Building Material, 24*, 1547–1559.
Ceroni, F., Pecce, M., Matthys, S., & Taerwe, L. (2008). Bond tests on concrete elements with CFRP and anchorage systems. *Composites: Part B, 39*, 429–441.
Ceroni, F., & Pecce, M. (2010). Evaluation of bond strength and anchorage systems in concrete elements strengthened with CFRP sheets. *Journal of Composites in Construction, ASCE, 14*(5), 521–530.
Eshwar, N., Nanni, A., & Ibell, T. J. (2008). Performance of two anchor systems of externally bonded fiber-reinforced polymer laminates. *ACI Materials Journal.*
Jinno, Y., Tsukagoshi, H., & Yabe, Y. (2001). RC beams with slabs strengthened by CF sheets and bundles of CF strands. In T. Telford (Ed.), *Proceedings of FRPRCS-5 International symposium,* July 2001, Cambridge, UK, pp. 981–987.
Karantzikis, M., Papanicolaou, C. G., Antonopoulos, C. P., & Triantafillou, T. C. (2005). Experimental investigation of nonconventional confinement for concrete using FRP. *ASCE Journal of Composite for Construction.*
Khalifa, A., Belarbi, A., & Nanni, A. (2000). Shear performance of RC members strengthened with externally bonded FRP wraps. *Proceedings of 12WCEE Conference.*
Khalifa, A., Alkhrdaji, A., Nanni, A., & Lansburg, S. (1999). Anchorage of surface mounted FRP reinforcement. *Concrete International: Design and Construction, 21*(10), 49–54.

Kim, Y. J., Wight, R. G., & Green, M. F. (2008). Flexural strengthening of RC beams with prestressed CFRP sheets: using nonmetallic anchor systems. *ASCE Journal of Composites for Construction, 12*(1), 44–52.

Kobayshi, K., Fujii, S., Yabe, Y., Tukagoshi, H., & Sugiyama, T. (2001). Advanced wrapping system with CF-anchor: stress transfer mechanism of CF-anchor. In T. Telford (Ed.), *Proceedings of FRPRCS5 International symposium,* July 2001, Cambridge, UK.

Mukhopadhyaya, P., Swamy, N., & Lynsdale, C. (1998). Optimizing structural response of beams strengthened with GFRP plates. *Journal of composites for construction.*

Nagy-Gyorgy, T., Mosoarca, M., Stoian, V., Gergely, J., & Dan, D. (2005). Retrofit of reinforced concrete shear walls with CFRP composites. *Proceedings of fib Symposium "Keep concrete Attractive,* Budapest, Hungary, 23–25 May 2005, pp. 897–902.

Niemitz, C. W., James, R., & Breña, S. F. (2010). Experimental behavior of carbon fiber-reinforced polymer CFRP sheets attached to concrete surfaces using CFRP anchors. *Journal of Composites for Construction, 14*(2), 185–194.

Orton, S. L., Jirsa, J. O., & Bayrak, O. (2008). Design considerations of carbon fiber anchors. *ASCE Journal of Composites for Construction, 12*(6), 608–616.

Özdemir, G. (2005). Mechanical Properties of CFRP Anchorages, Master of Science Thesis, Middle East Technical University, Istanbul Turkey.

Prota, A., Nanni, A., Manfredi, G., & Cosenza, E. (2004). Selective upgrade of undersigned reinforced concrete beam-column joints using carbon fiber reinforced polymers. *ACI Structural Journal, 101*(5), 699–707.

Prota, A., Tan, K., Nanni, A., Pecce, M., & Manfredi, G. (2006). Performance of RC shallow beams externally bonded with steel reinforced polymer. ACI Structural Journal.

Prota, A., Manfredi, G., Balsamo, A., Nanni, A., & Cosenza, E. (2005). Innovative technique for seismic upgrade of RC square columns. *Proceedings of FRPRCS7.*

Smith, S., Zhang, H., & Wang, Z. (2013). Influence of FRP anchors on the strength and ductility of FRP-strengthened RC slabs. *Construction and Building Materials, 49*, 998–1012.

Tan, K. H. (2001). Details of FRP reinforcement: an overview. In J. G. Teng (Ed.), *Proceedings of FRP Composite in Civil Engineering*, 12–15 December 2001, Hong Kong, China, Vol. 2, pp. 1247–1254.

Tan, K. H., Patoary, M. K. H., & Roger, C. S. K. (2003). Anchorage system for masonry walls strengthened with FRP composite laminates. *Journal of Reinforced Plastics and Composites*, *22*(15), 1353–1371.

Tan, K. H. (2002). Strength enhancement of rectangular reinforced concrete columns using fiber-reinforced polymer. *ASCE Journal Composite for Construction, 6*(3), 175–183.

Tan, K. H. (2003). Effect of cyclic loading on FRP-concrete interfacial bond strength. In *Proceedings of International Symposium on Latest Achievement of Technology and Research on Retrofitting Concrete structures*, JCI, Kyoto, July 2003, pp. 1–8.

Zhang, H., & Smith, S. (2012a). Influence of FRP anchor fan configuration and dowel angle on anchoring FRP plates. *Composites Part B Engineering, 43*(8), 3516–3527.

Zhang, H., & Smith, S. (2012b). FRP-to-concrete joint assemblages anchored with multiple FRP anchors. *Composite Structures*, *94*(2), 403–414.

Zehetmaier, Z. (2000). Entwicklung mechanischer endverankerungen für aufgeklebte CFK-lamellen. (Development of mechanical anchorages for externally bonded CFRP-strips). In K. Zilch (Ed.), Massivbau 2000, Forschung, Entwicklungen und Anwendungen, Münchner Massivbau- Seminar 2000, pp. 217–232. Springer VDI Verlag, Düsseldorf.

Mechanically Fastened Systems

Arora, D. (2003). Rapid Strengthening of Reinforced Concrete Bridge with Mechanically Fastened Fiber-Reinforced Polymer Strips. M.Sc. Thesis, University of Wisconsin, Madison, 353 pp.

Bank, L. C., Lamanna, A. J., Ray, J. C. & Velazquez, G. I. (2002). Rapid Strengthening of Reinforced Concrete Beams with Mechanically Fastened, Fiber Reinforced Polymeric Composites Materials. Report ERDC/GSL TR-02-4, US Army Corps of Engineers, 99 pp.

Bank, L. C. (2004). Mechanically-Fastened FRP (MF-FRP): A Viable Alternative for Strengthening RC Members. *Proceedings of the FRP Composites in Civil Engineering (CICE 2004), Adelaide, Australia*. December 8–10.

Bank, L. C., Arora, D. (2007). Analysis of RC beam strengthened with mechanically fastened FRP (MF-FRP) strips. *Composite Structures, 79*, 180–191.

Brown, V. L., Bank, L. C., Arora, D., Borowicz, D. T., Godat, A., Lamanna, A. J., Lee, J., Matta, F., Napoli, A. & Tan, K. H. Experimental Studies of Mechanically-Fastened FRP Systems: State-of-the-Art. ACI Special Publication-Proceedings of the FRPRCS-10. Tampa Bay, FL, USA, April 2011.

Dempsey, D. D., & Scott, D. W. (2006). Wood members strengthened with mechanically fastened FRP strips. *Journal of Composites for Construction, 10*(5), 392–398.

Ebead, U. (2011). Hybrid externally bonded/mechanically fastened fiber-reinforced polymer for RC beam strengthening. *ACI Structural Journal, 108*(6), 669–678.

Ekenel, M., Rizzo, A., Myers, J. J. & Nanni, A. (2005). Effect of fatigue loading on flexural performance of reinforced concrete beams strengthened with FRP fabric and pre-cured laminate systems. *Proceedings of 3rd International Conference of Composites in Construction*, Lyon, France.

Ekenel, M., Rizzo, A., Myers, J. J., & Nanni, A. (2006). Flexural fatigue behavior of reinforced concrete beams strengthened with FRP fabric and precured laminate systems. *Journal of Composites for Construction, 10*(5), 433–442.

Elsayed, W. E., Ebead, U. A., & Neale, K. W. (2009a). Studies of mechanically fastened fiber-reinforced polymer strengthening systems. *ACI Structural Journal, 106*, 49–59.

Elsayed, W. E., Ebead, U. A., & Neale, K. W. (2009b). Mechanically fastened FRP-strengthened two-way concrete slabs with and without cutouts. *Journal of Composites for Construction, 13* (1), 198–207.

El-Maaddawy, T. (2013). Mechanically fastened composites for retrofitting corrosion-damaged reinforced-concrete beams: experimental investigation. *Journal of Composites for Construction*, 1–9. ISSN 1090-0268/04013041(9).

Faella, C., Martinelli, E., & Nigro, E. (2008). Analysis of steel-concrete composite PR-frames in partial shear interaction: a numerical model and some applications. *Engineering Structures, 30* (4), 1178–1186.

Galati, D., Rizzo, A. & Micelli, F. (2007). Comparison of reinforced concrete beams strengthened with FRP pre-cured laminate systems and tested under flexural loading. *Proceedings of the 8th International Symposium on Fiber-Reinforced Polymer Reinforcement for Concrete Structures, FRPRCS-8*, Patras, Greece.

Johnson, D. (2011). An Investigation for Strengthening Existing Reinforced Concrete Beams in Shear Using a MF-FRP Retrofit System. Master Thesis 2011, University of Wisconsin-Madison, USA. pp. 1–202.

Lamanna, A. J., Bank, L. C., & Scott, D. W. (2001). Flexural strengthening of reinforced concrete beams using fasteners and fiber reinforced polymer strips. *ACI Structural Journal, 98*(3), 368–376.

Lamanna, A. J. (2002). Flexural Strengthening of Reinforced Concrete Beams with Mechanically Fastened Fiber Reinforced Polymer Strips. Ph.D. Dissertation, University of Wisconsin-Madison.

Lamanna, A. J., Bank, L. C., & Scott, D. W. (2004a). Flexural strengthening of reinforced concrete beams by mechanically attaching fiber-reinforced polymer strips. *Journal of Composites for Construction, 8*(3), 203–210.

Lamanna, A. J., Bank, L. C., & Borowicz, D. T. (2004b). Mechanically fastened FRP strengthening of large scale RC bridge T beams. *Advances in Structural Engineering, 7*(6), 525–538.

Lee, H. L., Lopez, M. M., Bakis, C. E. Flexural behavior of reinforced concrete beams strengthened with mechanically fastened FRP strip. *Proceedings of FRPRCS-8. Patras, Greece*, pp. 1–9.

Lee, J. H., Lopez, M. M., & Bakis, C. E. (2009). Slip effects in reinforced concrete members with mechanically fastened FRP strip. *Cement and Concrete Composites, 31*, 496–504.

Martin, J. A., & Lamanna, A. J. (2008). Performance of mechanically fastened FRP strengthened concrete beams in flexure. *Journal of Composites for Construction, 12*(3), 257–265.

Martinelli, E., Napoli, A., Nunziata, B. & Realfonzo, R. (2013). Flexural Response of RC Beams Strengthened by MF-FRP Laminates: Numerical Modeling. *Proceedings of SMAR 2013, Istanbul*, September 9–11.

Martinelli, E., Napoli, A., Nunziata, B., & Realfonzo, R. (2014). A 1D finite element model for the flexural behaviour of RC beams strengthened with MF-FRP strips. *Composite Structures, 107*, 190–204.

Napoli, A., Matta, F., Martinelli, E., Nanni, A., & Realfonzo, R. (2010). Modeling and verification of response of RC slabs strengthened in flexure with mechanically fastened FRP laminates. *Magazine of Concrete Research, 62*(8), 593–605.

Napoli, A., Bank, L. C., Brown, V. L., Martinelli, E., Matta, F., & Realfonzo, R. (2013). Analysis and design of RC structures strengthened with mechanically fastened FRP laminates: a review. *Composites Part B: Eng, 55*, 386–399.

Nardone, F., Lignola, G. P., Prota, A., Manfredi, G. & Nanni, A. (2012) Ultimate limit state of MF-FRP beams. In Biondini & Frangopol (Eds.), *Proceedings "6th International Conference on Bridge Maintenance, Safety and Management (IABMAS 2012). Stresa, Lake Maggiore*, Italy, July 8–12, pp. 1235–1242. London: Taylor & Francis Group. ISBN 978-0-415-62124-3.

Nardone, F., Lignola, G. P., Prota, A., Manfredi, G., & Nanni, A. (2011). Modeling of flexural behavior of RC beams strengthened with mechanically fastened FRP strips. *Elsevier Composite Structures, 93*(8), 1973–1985.

Newmark, N. M., Siess, C. P. & Viest, I. M. (1951). Tests and Analysis of Composite Beams with Incomplete Interaction. *Proceedings of the Society of Experimental Stress Analysis* (Vol. 9, no. 1, pp. 75–92).

Oliva, M., Bank. L. C., Borowicz, D. T., Arora, D. (2003). Rapid strengthening of reinforced concrete bridges. Technical Report 2003, University of Wisconsin Madison, Wisconsin Highway Research Program, USA. pp. 1–167.

Quattlebaum, J. B., Harries, K. A., & Petrou, M. F. (2005). Comparison of three flexural retrofit systems under monotonic and fatigue loads. *Journal of Bridge Engineering, 10*(6), 731–740.

Realfonzo, R., Martinelli, E., Napoli, A., & Nunziata, B. (2013). Experimental investigation of the mechanical connection between FRP laminates and concrete. *Composites Part B Engineering, 45*(1), 341–355.

Rizzo, A. (2005). Application of Mechanically Fastened FRP (MF-FRP) Pre-cured Laminates in Off- System Bridges. MSc thesis, University of Missouri-Rolla, USA. pp. 1–285.

Rizzo, A., Galati, N., Nanni, A. & Dharani, L. R. (2005a). Material Characterization of FRP Pre-Cured Laminates Used in the Mechanically Fastened FRP Strengthening of RC Structures. SP-230. *7th International Symposium on Fiber-Reinforced (FRP) Polymer Reinforcement for Concrete Structures, American Concrete Institute* (Vol. 230, pp. 135-152).

Rizzo, A., Galati, N., Nanni, A. & Bank, L. C. (2005b). Strengthening Concrete Structures with Mechanically Fastened Pultruded Strips. *Proceedings of Composites*. Columbus, Ohio, USA, September 28–30.

Rosner, C. N., & Rizkalla, S. H. (1995). Bolted connections for fiber reinforced composite structural members: experimental program. *Journal of Materials in Civil Engineering, 7*(4), 223–231.

Schorer, A. E., Bank, L. C., Oliva, M. G., Wacker, J. P. & Rammer, D. C. (2008). Feasibility of Rehabilitating Timber Bridges Using Mechanically Fastened FRP Strips. *Proceedings of the ASCE SEI Structures 2008 Conference: Crossing Borders, Vancouver, Canada*, April 24–26, 2008.

Sena-Cruz, J. M., Barros, J. A. O., Coelho, M. R. F. & Silva, L. F. F. T. (2011). Efficiency of different techniques in flexural strengthening of RC beams under monotonic and fatigue loading. *Construction and Building Materials*, *29*, 175–182.

Chapter 7
Prestressed FRP Systems

Julien Michels, Joaquim Barros, Inês Costa, José Sena-Cruz, Christoph Czaderski, Giorgio Giacomin, Renata Kotynia, Janet Lees, Carlo Pellegrino and Edmunds Zile

Abstract This chapter provides an overview on the state-of-the-art in prestressing systems for the structural retrofitting of reinforced concrete (RC) structures using Externally Bonded (EB) Fibre Reinforced Polymers (FRP). Focus is put on flexural strengthening, which currently is the most common application field for composite materials in structural engineering. The manuscript provides information regarding commercially available prestressing systems and their anchorage procedures. In addition to conventional mechanical anchorages, the innovative 'gradient anchorage' that lacks any remaining plates or bolts is also presented. Additionally, the authors mention various current prototypes at the laboratory-scale level. Performed experimental investigations, results, and conclusions represent the core content of this chapter. Several studies from various universities and research institutes worldwide are presented and explained. In these research projects, the previously mentioned systems are applied to specific reinforced or prestressed reinforced concrete members for strengthening purposes. Static and/or dynamic loading indicate the efficiency of the retrofitting concept compared to the reference structure. Generally, prestressed FRP will be demonstrated to follow the principle of conventional prestressed concrete by resulting in higher cracking, yielding, and bearing loads. Especially under service loads, the structural behaviour is improved.

J. Michels (✉) · C. Czaderski · E. Zile
Empa, Swiss Federal Laboratories for Materials Science and Technology, Dübendorf, Switzerland
e-mail: Julien.Michels@empa.ch

J. Barros · I. Costa · J. Sena-Cruz
University of Minho, Guimarães, Portugal

G. Giacomin
G&P Intech, Altavilla Vicentina, Italy

R. Kotynia
Lodz Technical University, Łódź, Poland

J. Lees
University of Cambridge, Cambridge, UK

C. Pellegrino
University of Padua, Padua, Italy

C. Pellegrino and J. Sena-Cruz (eds.), *Design Procedures for the Use of Composites in Strengthening of Reinforced Concrete Structures*, RILEM State-of-the-Art Reports 19, DOI 10.1007/978-94-017-7336-2_7

A special section is dedicated to prestressed near-surface-mounted (NSM) systems. In addition to the experiments section, calculation techniques for designing prestressed FRP for flexural strengthening are also handled. In shear strengthening and column confinement, prestressed FRP has been limited to notably few research applications to date. Nonetheless, an overview is given and future possible employment is discussed. Eventually, examples from real structural retrofitting projects should provide the practitioners with some background to better disseminate the retrofitting technique in question. The concluding section summarizes the actual situation and identifies needs for future research.

Keywords FRP · Strengthening · Prestressing · Reinforced concrete

Introduction

After pioneering works in the field of unstressed Fibre-Reinforced-Polymer (FRP) strips for structural retrofitting (Meier 1987, 1995; Kaiser 1989) these techniques have been studied for several years. Their application with an initial prestress is more recent and still in development. In structural engineering, the application of externally bonded FRP is currently somewhat limited to carbon. However, the literature also presents applications with Glass Fibre Reinforced Polymer (GFRP) sheets (e.g. Huang et al. 2005). An experimental study on prestressed Aramid tendons is presented in Lees and Burgoyne (1999). The material is in practice mainly used in the form of strips/laminates (flexure), straps (shear), and wraps (confinement). The information given in the present chapter will be limited to the three types of carbon reinforcements. A special section is dedicated to prestressed Near-Surface-Mounted (NSM) reinforcements.

The mounting of an unstressed outer reinforcement (EBR) has the disadvantage of to provide a very limited additional stiffness to the structure under service loads. In addition, a very large number of research activities have clearly demonstrated that ultimate failure at a structural level occurs after the debonding of the composite reinforcement from its concrete substrate. Generally, the FRP is far from reaching its ultimate capacity in tension. Hence, the composite high tensile strength is not exploited; often, only 20–30 % of the material's capacity is effectively used (Motavalli et al. 2011).

Like externally prestressed steel cables, pretensioned composite reinforcements offer the possibility to act against present dead loads of the respective construction and thus can reduce existing deflections and close existing cracks. Furthermore, prestressing more efficiently exploits the material's high tensile strength.

The present chapter provides an overview of prestressed FRP for enhancing the structural efficiency in flexure, shear, and column confinement. Special attention is paid to the following topics: prestressing technique, anchorage method, and current knowledge about the system's performance. Short subsections about shear strengthening and confinement provide information on recent developments in the field. Finally, practical cases complete the overview.

Flexural Strengthening

General Information

Several investigations have documented the structural advantages of prestressed FRP reinforcement in retrofitting (Deuring 1993). A first positive aspect is the possibility to actively act against dead loads and thus reduce the existing deflections and cracks in the structure. With a sufficient amount of initial prestrain in the laminate, the cracking load is considerably increased compared to a reference (un-strengthened) beam. The same is valid for the load at which the inner steel reinforcement begins to yield. The ultimate load carrying capacity is also generally enhanced. However, a decrease of ductility, resulting in a lower deflection when reaching the ultimate load, can be observed when an initial prestress is applied to the laminate. While a structure that is retrofitted with an unstressed strip always exhibits debonding failure at the peak load, tensile failure of the external reinforcement can be obtained if the initial prestrain ε_{fp} is sufficiently high. Figure 7.1 shows a schematic representation of the moment-curvature $(M - \chi)$ relationship for the three mentioned situations. An enhanced crack, yield, and ultimate load is shown in terms of an increase in the respective bearing moments ΔM_{cr}, ΔM_y and ΔM_u.

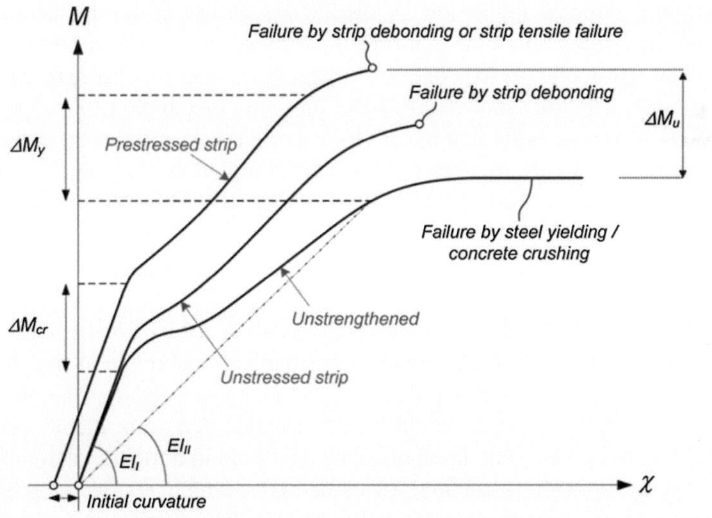

Fig. 7.1 Schematic Moment-Curvature relationships for an unstrengthened RC element, a strengthened RC element with an unstressed laminate and a strengthened RC element with a prestressed laminate

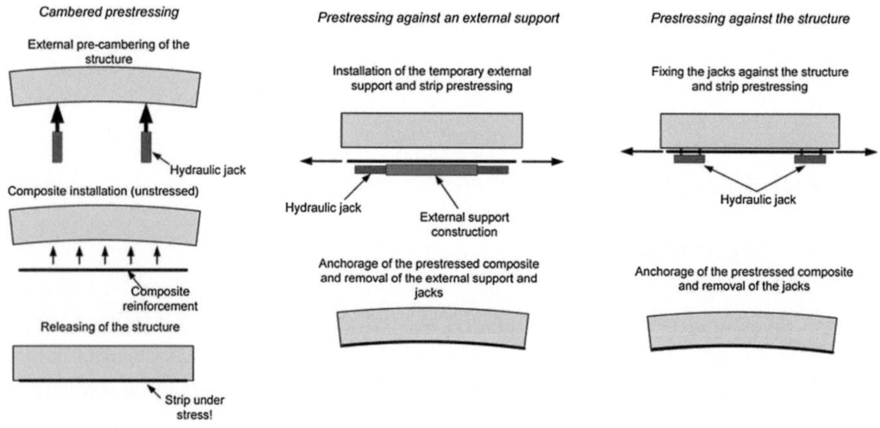

Fig. 7.2 Different types of prestressing of an existing RC construction (El-Hacha et al. 2001)

Prestressing Systems and Anchorage Techniques

Types of Prestressing

An existing structure can generally be prestressed via three methods, which are summarised in El-Hacha et al. (2001) and presented in Fig. 7.2. The first technique, known as the *cambered beam system*, requires an initial counter-deflection against the dead-loads by means of hydraulic jacks (El-Hacha et al. 2001). Subsequently, the FRP strip is applied and the structure is prestressed due to subsequent releasing of the initially inflicted deflection. A second possibility is the use of an *external support construction*, which the equipment for prestressing application is supported against. The third and most common category is the *prestressing against the structure* itself. This measure requires the previous installation of supporting elements, such as anchor bolts that are used to fixate a hydraulic jack. In most cases, these temporary elements are removed after the completion of the retrofitting action.

Commercially Available Prestressing and Anchorage Systems

Currently, several prestressing and anchorage systems for CFRP strips are available on the market. In general, they foresee a mechanical anchorage at the strip ends.

In most cases, the external reinforcement is prestressed against the existing concrete structure. To do so, a hydraulic jack installed in a frame element is fixed by means of several dowels. The laminate is usually held in a mobile clamping system and pushed towards the structural element ends, which results in prestressing. Figure 7.3 provides an example of such a device. The presented system by S&P Clever Reinforcement Company fixes an aluminium plate that applies compression to the strip. The anchorage system requires a minimum of adhesive curing, which

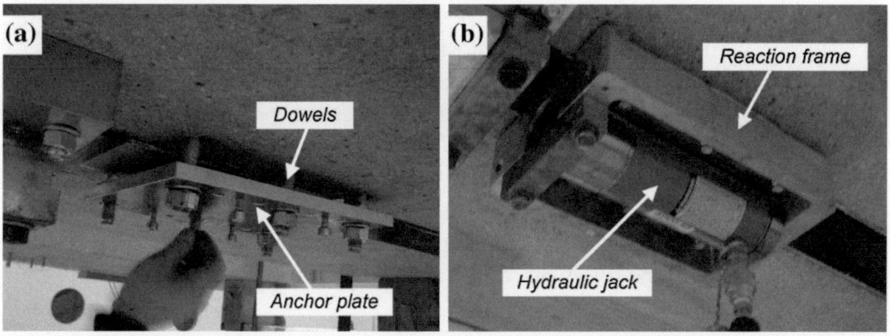

Fig. 7.3 **a** Anchor plate during installation and **b** reaction frame with hydraulic jack

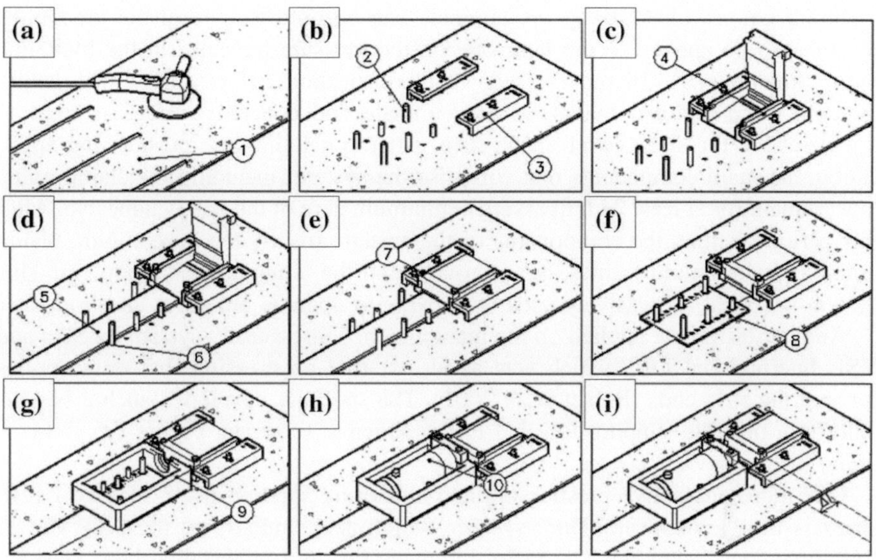

Fig. 7.4 Main steps for applying the prestress: **a** Surface preparation; **b** drilling the holes for the bolts of the anchors and base angles; **c** placing the clamp unit; **d**, **e** introducing the laminate in anchor and fixing it; **f** placing the aluminium plate; **g** placing the aluminium frame; **h** placing the hydraulic jack; **i** prestressing the laminate and controlling the deformation

implicates that the prestress force at the ends is generally fully released 1 day after the installation.

The system is relatively simple and light, and can be used with CFRP laminates that are up to 100 mm wide. The laminates used with this system are typically between 1.2 and 1.4 mm thick.

Figure 7.4 presents the main steps to be followed in order to prestress CFRP laminates with this system, which include the following:

(1) The surface is prepared to remove the concrete laitance and obtain a rough surface while minimising the exposure of aggregates (Fig. 7.4a).
(2) Holes are drilled to accommodate the bolts to fix the aluminium anchor plates and base angles. The locations of these holes are previously marked on the element to be prestressed (Fig. 7.4b).
(3) The clamp units are introduced between the base angles and cutting CFRP laminate with the desired length. The epoxy adhesive to bond the laminate to the concrete is then prepared and applied to the laminate. Finally, the laminate is inserted on the element to be strengthened between the clamp units (Fig. 7.4c–e).
(4) The steel anchor, aluminium frame, and hydraulic jack are successively placed at both extremities (Fig. 7.4f–h).
(5) The prestress is applied at the predefined load level (Fig. 7.4i).

Strain gauges glued along the laminate can be used to control the predefined prestress to be applied to the laminate via the pressure indicator in the hydraulic pump (manometer). In most cases, visual inspection by measuring the length increase on predefined marks on the laminate and concrete surface is used to double-check the stress level of the strip. After concluding the prestressing application, the main components that comprise the system (including the clamp units) must remain for at least 24 h to assure a minimum cure of the epoxy adhesive. After this period of time, the components of the system, mainly the base angles, clamp units, and aluminium frames, are removed, and the temporary bolts are cut. This system has been successively used worldwide to upgrade buildings and bridges.

Another technique applied on the market is the “Stresshead”-system by Sika and VSL International Ltd., which acts as an external prestressing that only applies force at the strip ends (Berset et al. 2002). The specially developed anchor head is held by a metallic support doweled to the concrete substrate. Figure 7.5 shows a photo of this device.

One more alternative prestressing and anchoring solution from Sika for CFRP strips is the “Leoba-CarboDur” system, which acts similarly to the S&P Clever Reinforcement technique. The main difference lies in a preinstalled base plate (in

Fig. 7.5 Stress-Head anchorage for prestressed CFRP strips

the concrete substrate), which acts as a support for the hydraulic jack during the prestressing and curing phase of the epoxy adhesive. Subsequently, another anchor plate is mounted and pressed against the laminate from the other side. The technique is documented in detail in Andrä et al. (2001) and Andrä and Maier (2005).

Professor Urs Meier developed the first and so far only non-mechanical anchorage system for prestressed CFRP strips at Empa. The "gradient-anchorage" foresees a gradual prestress force decrease at both strip ends over a defined length, Δl_i, to limit the shear stresses that appear in the anchorage zone and avoid premature debonding (Triantafillou et al. 1992). Debonding issues appear when too high prestress forces are transferred from the laminate to the concrete substrate at the ends. The application technique is based on the adhesive's ability to undergo accelerated curing at high temperatures (Czaderski et al. 2012; Michels et al. 2012). Figure 7.6 schematically presents the force evolution in the gradient anchorage.

Fig. 7.6 Schematic representation of a gradient anchorage (Michels et al. 2013)

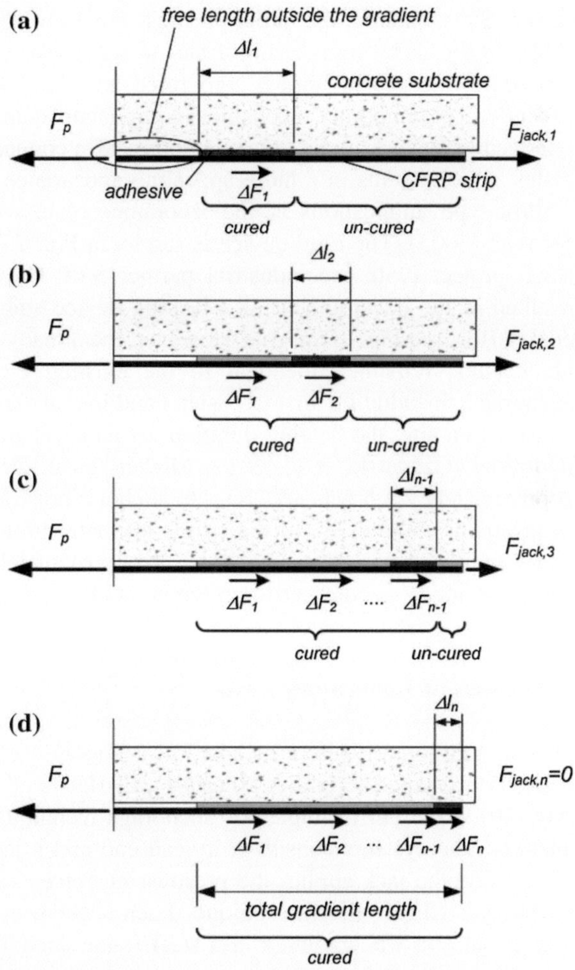

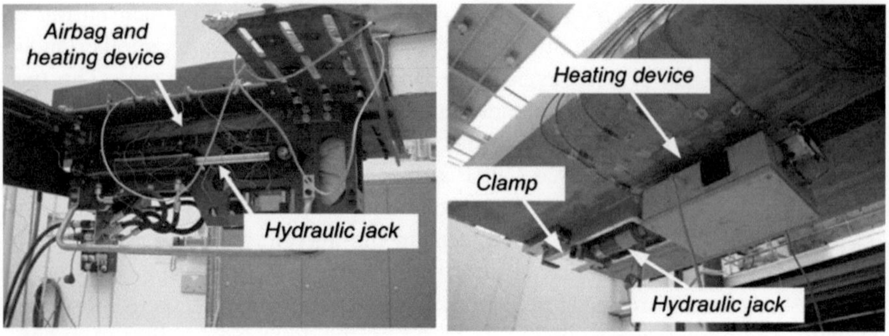

Fig. 7.7 First Empa prototype and newest heating device for application on site

An intermediate sector with bond length Δl_i is chosen for the accelerated adhesive curing. Subsequently, the initial prestress force level F_p in the strip is decreased by ΔF_i in the jack by transferring a part to the sector with the cured adhesive. This procedure is repeated until a zero force level is reached in the jacks. After the execution, all temporary mechanical components (support bolts, frame, etc.) can be removed to leave a purely concrete-epoxy-strip connection without any dowels and plates, which results in a more appealing appearance of the retrofitting.

Prototype applications at the laboratory scale were presented by Meier and Stöcklin (2005). The used device is shown in Fig. 7.7 (left). A recently completed R&D project (with the industrial partner S&P Clever Reinforcement) at Empa resulted in the development of a heating device suitable for on-site use (Michels et al. 2013), see Fig. 7.7 (right). However, the already existing prestressing devices, i.e. clamps, remained identical to the existing elements used for mechanical anchorage. In addition to the easier handling of the new heating device on the construction site, the heating duration for an accelerated adhesive curing was also optimised (Czaderski et al. 2012; Michels et al. 2013). A total heating time of approximately 2.5 h was necessary to anchor a prestress force F_p of 120 kN (0.6 % of prestrain, about 40 % of the strip's ultimate strain). Including preparation, one strip can be prestressed and anchored in approximately 4 h. Outside the anchorage zone, the epoxy is cured at room temperature.

Prototypes at Laboratory Level

Several additional types of mechanical anchorages can be found at the laboratory scale. For instance, Wight et al. (2001), El-Hacha et al. (2003), and El-Hacha and Aly (2013) present examples of anchorage techniques that include bar and plate anchors. The systems consist of a dead end and a jacking end anchor at which a fixed hydraulic jack applies the prestressing force. Wight et al. (2001) proposed a multi-layer CFRP sheets technique. Each sheet was separately prestressed using a steel bar and was mounted in a steel frame anchored on the RC member. This solution was only verified in experimental tests and has not been used in practical applications.

Non-metallic, mechanically anchored, and CFRP anchored U-wraps sheets were investigated in Kim et al. (2008d, e). The general idea of this system was an anchorage system based solely on composite materials without any steel or aluminium elements.

A general and very detailed overview of anchorage systems for CFRP reinforcements of all types was presented by Schlaich et al. (2012).

Beam and Girder Strengthening

The following paragraphs present a few literature examples that experimentally investigated the strengthening of reinforced concrete (RC) or prestressed reinforced concrete (PRC) elements. In general, all studies showed a general increase in the cracking load, yielding load, and ultimate load, as well as lower deflections and crack openings than the reference structure at a specific load level when a prestressed CFRP strengthening was applied.

Suter and Jungo (2001) conducted bending tests on beam elements with 6 m span. A mechanical anchorage with a bonded plate as shown in Fig. 7.3 was used. The study suggested prestrain values from 0.6 to 0.8 % to optimise the structural behaviour. The authors also mention an increasing number of problems induced by increasing the initial force in the laminate. A generally known phenomenon is the slipping out of the strip from the clamp during the prestressing procedure.

Similar test procedure was followed by Meier and Stoecklin (2005), Kotynia et al. (2011), Michels et al. (2013, 2014a), which all used a gradient anchorage system with an initial prestrain ε_{fp} of 0.6 %. For instance, the first research group obtained tensile failure in the strip when using 4 laminates with a cross-section of $50 \times 0.6\ \mathrm{mm}^2$ instead of 2 laminates with a cross-section of $50 \times 1.2\ \mathrm{mm}^2$. This effect was partially due to a reduction in the interfacial shear stress when using a higher number of strips with a smaller cross-section. Strip failure and the force-midspan deflection diagram are shown in Fig. 7.8. Investigations regarding the application of the gradient anchorage on a shotcrete substrate are documented in Michels et al. (2014b).

In the field of prestressed concrete girder retrofitting Fernandes et al. (2013) tested four prestressed high-strength concrete (HSC) girders with a span of 20 m under four point bending. Two girders were used as reference, while one was externally strengthened with unstressed CFRP laminates and one was externally strengthened with prestressed CFRP laminates. The cross-sectional geometry of the girders is depicted in Fig. 7.9. Two laminates with a rectangular cross-section of $100 \times 1.4\ \mathrm{mm}^2$ per girder were adopted to strengthen the girders. The previously presented commercial system proposed by S&P Clever Reinforcement Company for prestressing and anchoring the laminates was used. A prestrain level of 0.4 % was chosen, which corresponded to approximately 30–35 % of the strip's ultimate

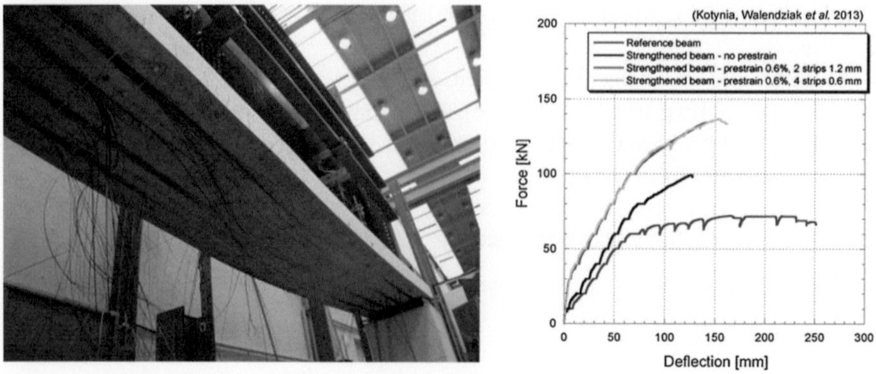

Fig. 7.8 Tensile failure of a prestressed CFRP laminate (*left*) and force-deflection diagram of the test series by Meier and Stoecklin (2005) and Kotynia et al. (2011)

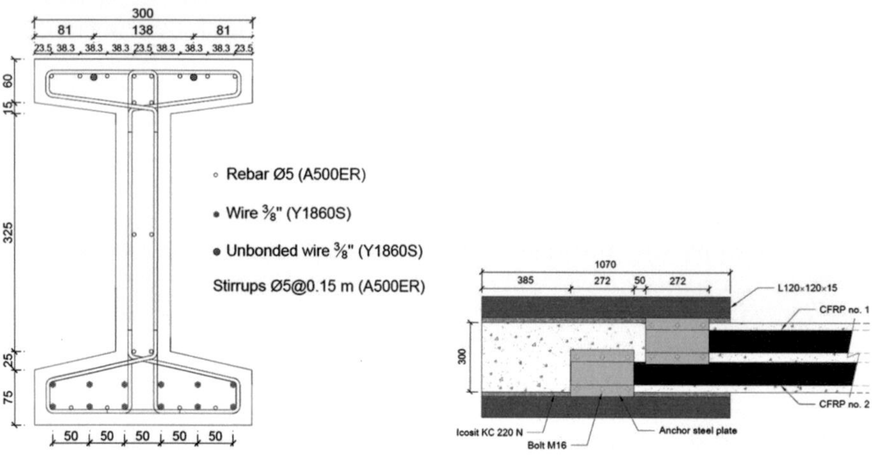

Fig. 7.9 Cross section (*left*) and bottom view of the anchorage zone (*right*) of a prestressed concrete girder retrofitted with two prestressed CFRP laminates with mechanical anchorage (Fernandes et al. 2013)

tensile strength. End-anchorage plates were used (see Figs. 7.3 and 7.9 (right)). The strengthening procedure is also interesting from a practical installation point of view, in which lateral steel L-profiles were used to temporarily increase the available surface for the strip clamping.

Figure 7.10 (left) shows the force versus the deflection at the mid span. Based on the presented experimental program, the following main conclusions can be drawn: (a) the load carrying capacity of unstressed and prestressed laminates increased by 22 and 35 %, respectively, compared to the reference girders; (b) the bending stiffness prior to the onset of yielding of the longitudinal steel bars of both retrofitted girders only slightly changed; (c) the benefits of prestressing the CFRP

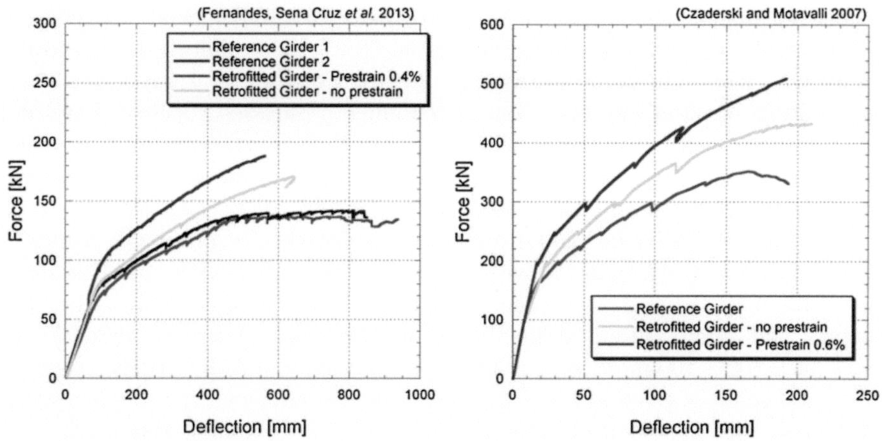

Fig. 7.10 Force-deflection diagram of the experimental program by Fernandes et al. (2013) and Czaderski and Motavalli (2007)

Fig. 7.11 Large-scale prestressed concrete bridge girder testing at Empa (Czaderski and Motavalli 2007)

laminates were materialised not only in terms of strength but also in terms of stiffness at different levels: crack initiation, yield initiation of the longitudinal steel bars, and ultimate load; (d) the initially unstressed CFRP laminate reached a strain of 0.65 %, whereas a value of 0.95 % was obtained with an initial prestrain of 0.4 %; (e) the girders failed via concrete crushing at the top flange.

Large-scale prestressed concrete girders taken from an existing bridge in Ticino (CH) were strengthened with four prestressed laminates with a gradient anchorage and subsequently tested under static loading (see Figs. 7.10 (right) and 7.11). The results are documented in Czaderski and Motavalli (2007). Both strengthened girders (with an unstressed and a prestressed CFRP strip reinforcement) showed significant increase in the load-carrying capacity. Conversely, Aram et al. (2008) reported that the structural behaviour of retrofitted prestressed concrete beams did not show any improvement. This lack of improvement was attributed to a premature debonding of the strip due to very high shear stresses in the gradient anchorage zone because of the short beam span.

Pellegrino and Modena (2009) presented another case of prestressed concrete elements that were strengthened with external prestressed CFRP strips. This study also utilised mechanical anchorage. Five 10 m long real-scale beams (four RC beams and one PRC beam with pretensioned internal strands) of cross-Sect. 300 × 500 mm^2, were tested at the University of Padova (Italy). A photo of the test setup is given in Fig. 7.12. Regarding the strengthening, unidirectional CFRP pultruded laminates of 1.2 × 100 mm^2 and 1.2 × 80 mm^2 were used.

The RC-C (reference) diagram presented in Fig. 7.12 showed the typical flexural behaviour of RC beams with pre-cracked, post-cracked, and post-yielded stages. The RC-N (with EBR—no end anchorage) diagram shows a brittle behaviour due to the sudden strip delamination. The RC-EA (with EBR and end-anchorage) diagram shows similar behaviour but with a higher value of the ultimate load. This was due to the presence of end-anchorage devices. Intermediate delamination of the CFRP occurred in this case, with failure of the end-anchorage device.

Failure of beams RC-PrEA (RC beam with prestressed EBR) and PRC-PrEA (PRC beam with prestressed EBR) was again due to delamination of the CFRP, but prestressing and end-anchorage action delayed complete failure. A relevant increment in the ultimate load and an increment of the load at which the first crack appeared occurred for beams with pretensioned laminates (RC-PrEA, PRC-PrEA) with respect to the control beam (RC-C) due to the compressive axial force transferred by pretensioning and, for beam PRC-PrEA, also by internal strands.

In general, mechanical anchor devices increase the ultimate capacity of the structural element, delaying the end and/or intermediate delamination. The CFRP

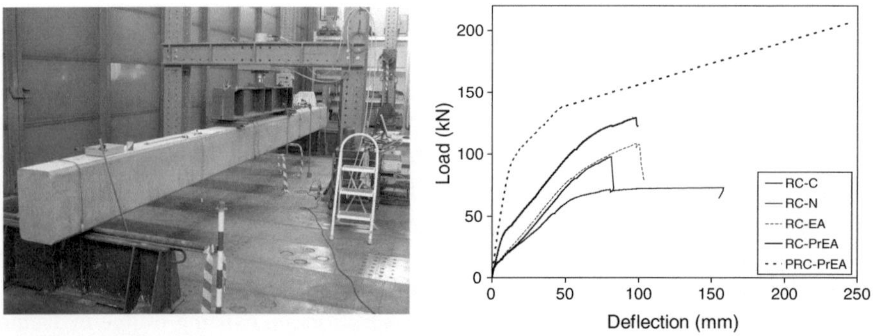

Fig. 7.12 Test setup and force-deflection curve of the beam tests performed by Pellegrino and Modena (2009)

strengthening system was capable of significantly increasing the load carrying capacity of the structural element at which first cracking occurs. Furthermore, it allows reduction of crack amplitudes, a more uniform distribution of cracks and a better utilization of CFRP material characteristics with strain values near the ultimate.

For additional literature on experimental investigations about prestressed CFRP laminates in bending, the reader is invited to also consult (França and Costa 2007; Kim et al. 2007, 2008a, 2010a, b; Quantrill and Hollaway 1998; Wu et al. 2003; Neubauer et al. 2007).

Influential Parameters

As mentioned, experimental research proved that unstressed strengthening precludes the full utilisation of the CFRP tensile strength. The limited tensile strength of concrete results in the debonding of CFRPs from the concrete surface. A review of the available literature on the strengthening of RC members with unstressed and prestressed laminates shows that the strengthening effect significantly depends on a number of factors, including the type of laminate, its stiffness, the number of layers, and the existing longitudinal and shear reinforcement ratios (Garden and Hollaway 1998; Teng et al. 2002). In addition to previously presented publications, the reader is invited to consult among other the following literature regarding strengthening efficiency of externally bonded unstressed and prestressed CFRP strips: You et al. (2012), Kim et al. (2008d), Pellegrino and Modena (2009), Yu et al. (2008), Wight et al. (2001), Kotynia and Kaminska (2003), Meier and Stoecklin (2005), Kotynia et al. (2011).

The following paragraphs present a study on influence parameters based on database presented in Kotynia et al. (2013b).

The efficiency of strengthening RC members with prestressed CFRP composites has been analysed based on experimental test results. The gathered data include all variable parameters that influence the test results: cross-sectional dimensions (*b, h*), compressive concrete strength (f_{ck}), tensile (yield) strength and elastic modulus of the steel (f_{yk}, E_s) and composite (f_{fu}, E_f), cross-sectional area of longitudinal steel (A_s) and composite reinforcement (A_f) with corresponding reinforcement ratios (ρ_s, ρ_f), equivalent composite reinforcement ratio defined by $\rho_{f,eq} = \rho_f \cdot (E_f/E_s)$, initial prestressing strain of the composite (ε_{fp}), maximal strain of the composite registered during the test $(\varepsilon_{f,max})$, failure mode, strengthening ratio in terms of load carrying capacity $(\eta_u = (M_u - M_{u0})/M_{u0})$ and in terms of cracking load $(\eta_{cr} = (M_{cr} - M_{cr0})/M_{cr0}$, where: M_u, M_{u0}, M_{cr}, M_{cr0} stand for ultimate bending moment of the strengthened and reference member as well as cracking moment of such members, respectively). All elements were strengthened with CFRP. The main differences between the used materials were the type of the composite (*L*—laminate, or *S*—sheet) and the modulus of elasticity (E_f). To reduce the influence of the CFRP elastic modulus, an equivalent composite reinforcement ratio $(\rho_{f,eq})$ was introduced, for which the FRP ratio is defined by $\rho_f = A_f / (b \cdot d_f)$ and d_f is the effective composite reinforcement depth.

The primary classification of the analysed members was based on the failure mode obtained in the test. Three groups were distinguished. The first two include elements that failed due to intermediate crack debonding (IC) of the composite or due to the rupture of fibres (R). In the last group, the composites were not properly applied and utilised, which resulted in the following failure modes: concrete crushing (CC), debonding of the composite's ends (ED), concrete cover separation (CCS), or failure of the anchorage system (A). Due to the very low efficiency of the strengthening of the members from the third group, these cases were not considered in the analysis (Kotynia et al. 2013b).

The longitudinal steel reinforcement ratio (ρ_s) significantly affects the strengthening gain (ratio) defined as $\eta_u = (M_u - M_{u0})/M_{u0}$, which was analysed in terms of ultimate loads and is shown in Fig. 7.13 (Kotynia et al. 2013b). The experimental test results were divided into two groups of different equivalent composite reinforcement ratios $(\rho_{f,eq})$, which were equal to 0.05 and 0.10 % (Fig. 7.13). The comparison proved that the strengthening efficiency of members with a higher steel reinforcement ratio was lower. The strengthening efficiency of beams with less steel reinforcement ($\rho_s = 0.44\,\%$) was higher ($\eta_u = 1.55$) than that of beams with higher reinforcement ($\rho_s = 0.50\,\%$ and $\rho_s = 0.83\,\%$, corresponding efficiency of $\eta_u = 0.86$ *and* $\eta_u = 0.59$, respectively) for lower composite reinforcement ratios.

The same observation was made in a group of members with higher composite reinforcement ratio ($\rho_{f,eq} = 0.10\,\%$, see Fig. 7.13) both for strengthening with mechanically anchored and non-anchored CFRP laminates/sheets. In general, when the steel reinforcement ratio increased twofold (from $\rho_s = 0.44\,\%$ to $\rho_s = 0.89\,\%$), the strengthening efficiency value decreased more than twofold (from $\eta_u = 1.55$ to $\eta_u = 0.59$).

The different prestressing strains of the composites (ε_{fp}) can explain the difference in the achieved strengthening efficiency for the members of the same steel and

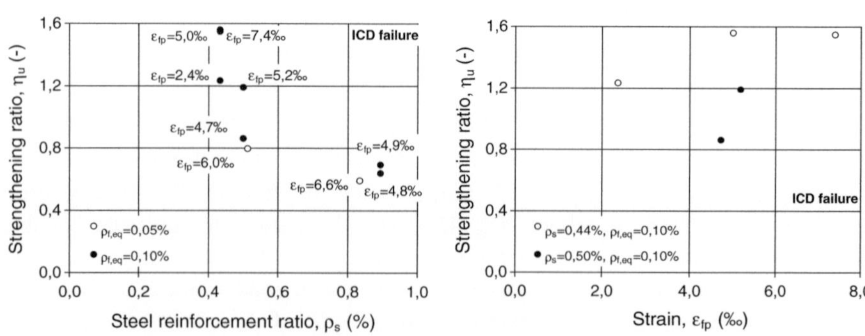

Fig. 7.13 Influence of steel reinforcement ratio ρ_s on the strengthening efficiency η_u (*left*), and of composite prestrain ε_{fp} on the strengthening efficiency η_u (*right*)

composite reinforcement ratio (Fig. 7.13). The same members shown on the graph of the strengthening efficiency as a function of the prestrain (Fig. 7.13) demonstrate the positive influence of the prestrain. In both groups ($\rho_s = 0.44\%$ and $\rho_s = 0.50\%$) an increase in the prestrain ε_{fp} significantly increased the strengthening efficiency. The beneficial effect of the higher CFRP prestressing strain is noticeable for members that failed due to intermediate crack debonding (IC). Otherwise, the prestressing strain did not affect the ultimate loads of the members with anchored CFRP composites, which failed by CFRP fracture.

The relationship between the equivalent composite reinforcement ratio ($\rho_{f,eq}$) and the strengthening efficiency (η_u) for the members that failed due to composite rupture (*R*) is presented in Fig. 7.14. An increase in the equivalent strengthening ratio resulted in an almost linear increase of the strengthening efficiency.

The graphic interpretation of the strengthening efficiency is presented in Fig. 7.15 in terms of the ultimate loads (η_u—solid lines on the graph) and cracking loads (η_{cr}—dashed lines). The FRP prestressing strain significantly affected the slopes of the curves, and the influence of the FRP prestressing strain was significantly more beneficial to the increase of the cracking loads than the load capacity. The increase in the strengthening efficiency in terms of the serviceability conditions ($\Delta\eta_{cr}$) was up to 4 times higher than in terms of the ultimate limit state ($\Delta\eta_u$) (see Fig. 7.15).

A CFRP prestressing strain equal to $\varepsilon_{fp} = \varepsilon_{fu} - \varepsilon_{f,max}$ is recommended (where ε_{fu} is the ultimate CFRP strain and $\varepsilon_{f,max}$ should be presumed of 0.7 %, as suggested by the authors) to maximise the strengthening efficiency (with simultaneous CFRP rupture and debonding). However, concrete crushing may occur if the concrete compressive strength is lower.

A limited number of research programs have considered the preloading effect of the RC member prior to the strengthening on the strengthening efficiency. Tests by Kotynia et al. (2013a) proved the high efficiency of flexural strengthening with prestressed CFRP laminates, even for highly loaded structures prior to

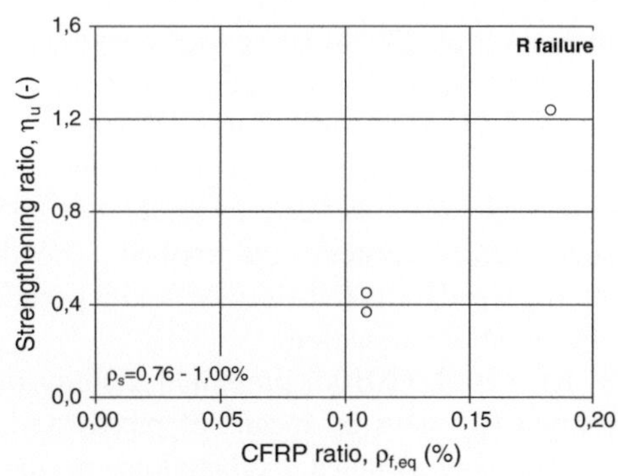

Fig. 7.14 Influence of composite equivalent reinforcement ratio $\rho_{f,eq}$ on the strengthening efficiency η_u

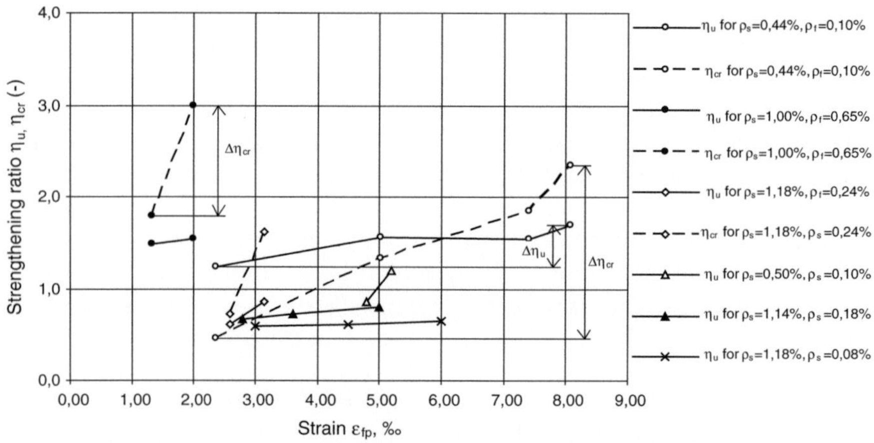

Fig. 7.15 Influence of the prestressing strain ε_{fp} on the strengthening efficiency η_u and η_{cr}

strengthening. The RC beams exhausted to 25 and 75 % of their ultimate loads and strengthened with the prestressed CFRP laminates indicated optimistic results.

Results indicated minor differences caused by the effect of preloading. The strengthening ratio was shown to be inversely proportional to the steel reinforcement ratio. The preloading level did not affect the ultimate concrete tensile strains.

Adhesion between the CFRP laminate and the concrete significantly affects slab deformation after the steel reinforcement yields. The load-induced strain in the unbonded laminates ($\varepsilon_{f,max}$) ranged from 0.0050 to 0.0069, while the bonded laminates reached strains of 0.0093–0.0069 (Kotynia et al. 2013a). Similarly, the CFRP strain efficiency ($\eta_{\varepsilon f}$) ranged from 0.68 to 0.87 for slabs strengthened with bonded laminates and from 0.56 to 0.68 for the slabs with unbonded laminates.

Two-Way Slab Strengthening

In Kim et al. (2008b) large-scale two-way RC slabs were strengthened with prestressed composite strips anchored with a mechanical anchorage system composed of a bolted plate. Similar to applications with unidirectional flexural elements, an enhancement in the cracking, yielding, and ultimate load was noticed. Important reductions of the deflections under service load were observed when a prestressed strengthening system was applied. The ductile failure of the control slab (unstrengthened) was transformed into a pseudo-ductile mode in the post-peak domain after strip delamination.

Kim et al. (2010c) presented a study on the punching shear behaviour of retrofitted RC slabs by means of prestressed CFRP sheets. For additional considerations, more detailed explanations of the related phenomenon are provided in section “Shear strengthening”.

NSM-FRP Prestressed Systems

Introduction

According to the literature review, prestressed FRP systems for the flexural strengthening of reinforced concrete (RC) elements have already been applied successfully using the externally bonded reinforcing technique (EBR). In the context of prestressed EBR, significant improvements are reported in RC elements in serviceability and ultimate limit state conditions, such as increase of the load carrying capacity, durability, and structural integrity. The NSM technique is, however, more effective for the flexural strengthening of RC elements than EBR (Barros et al. 2007). Therefore, some efforts are being done in order to combine the intrinsic benefits of using NSM-CFRP with those derived from the application of prestressed EBR-CFRP. This section resumes the fundamental research carried out in this context.

Currently, there is few work reported in the literature on NSM-based prestress technique that can actually be applied on job site. Most of the tested specimens were strengthened in the sagging region, but the strengthening tasks were performed as if it was a hogging region (Nordin and Täljsten 2006; Rezazadeh et al. 2013), i.e., the elements are initially turned over, strengthened, and finally turned over again to its original position in order to be tested. The scheme in which the hydraulic jacks are being placed (in line with the FRP and beyond the boundaries of the element) is impracticable in real cases. Only Gaafar and El-Hacha (2008) claim to have a system that allows this technique to be applied in job-site, and other author is currently refining the design of a system for the application of prestress to Carbon FRP (CFRP) laminates (Barros 2009). Barros (2009) has proposed an innovative approach for applying prestressed NSM-CFRP laminates in real practice, but the mechanical system subjacent to this technique was not yet available.

Flexurally Strengthened Beams and Plates

Several experimental investigations on beam strengthening can be found in the literature. Nordin and Täljsten (2006) using 10×10 mm^2 prestressed CFRP rods applied according to the NSM technique to strengthen $200 \times 300 \times 4000$ mm^3 reinforced concrete beams, realized that the applied stress was efficiently transferred to the surrounding concrete even without the use of any special device to anchor the CFRP. Analysing their results, the loss of ductility in relation to the non-prestressed beams was remarkable. While in the series strengthened with a CFRP of 160 GPa elastic modulus the final deflection of the non-prestressed beam was about 50–55 mm, after the application of 20 % of prestress it was reduced by 40 %, i.e. 33 mm. The same observation can me made based on the results using a CFRP laminate of 250 GPa elastic modulus where the deformation of the 0 % prestress beams ranged between 37 and 41 mm and the reduction observed by the application of 19 % of prestress was nearly 30 %.

Gaafar and El-Hacha (2008) reported the tests performed on $200 \times 400 \times 5150$ mm^3 reinforced concrete beams prestressed with two NSM strips of 2×16 mm^2, and verified a considerable increase of the load at cracking and yielding initiation. This increase was, however, followed by a significant reduction of the ductility, since the deflection at failure was dramatically decreased (60 % of prestress conducted to failure at a deflection level of approximately 50 % of the deflection observed in the non-prestressed beam).

Rezazadeh et al. (2013) carried out an experimental program on RC beams flexurally strengthened with passive and prestressed CFRP laminates and verified that a CFRP reinforcement ratio of $\rho_f = A_f/(b \cdot d_f) = 0.06\,\%$ conducted to an increase of approximately 63 % in the ultimate load carrying capacity of beams with a steel reinforcement ratio of $\rho_s = A_s/(b \cdot d_s) = 0.39\,\%$, regardless the fact the CFRP laminate is passive or applied with a prestress level of 20 and 40 %. A prestress level of 20 and 40 % conducted to an increase of 47 and 55 % in terms of load carrying capacity at deflection corresponding to the serviceability limit state, while passive CFRP laminate provided an increase of 32 %. Such in the previous experimental programs, they also verified that the load at cracking and steel yielding initiation increased with the prestress level, but the deflection at failure of the RC beams decreased with the increase of the prestress level, since all the strengthened RC beams failed by the rupture of the CFRP laminate.

In the experimental program carried out by Costa (2014), composed of ten RC beams flexurally strengthened with passive and prestressed CFRP laminates (grouped in three series of RC beams), it was verified that the long term losses of prestress have occurred between a couple of days to a couple of weeks to become stabilized. In this experimental program, Costa (2014) has also verified that the load at crack initiation and at steel yield initiation has increased with the prestress level, but the corresponding deflection was not significantly affected. The force-deflection of the series corresponding to the quasi real-scale beams is represented in Fig. 7.16. The prestressing level (percentage of the tensile strength of the CFRP provided by the supplier) is indicated in the designation attributed to each strengthened beam (CFRP laminate of 1.4×20 mm^2 cross-sectional area was applied in each beam, $\rho_l = 28/(150 \times 290) = 0.064\,\%$). In all the series of beams it was verified that the prestress level applied to the CFRP laminates has no influence on the ultimate load carrying capacity of the strengthened beams, since failure was in all cases dominated by the CFRP rupture. However, the deflection at failure has significantly decreased with the increase of the prestress level. The total cracked length of the beams has also decreased with the increase of the prestress level. In terms of average crack spacing, it was similar in all the tested beams and equal to the spacing of the steel stirrups.

Based on the bibliographic survey on the topic, it can be concluded that, in general, the deflections measured in the beams decreased with the increase of the applied prestress level and thus, the behaviour of these elements under

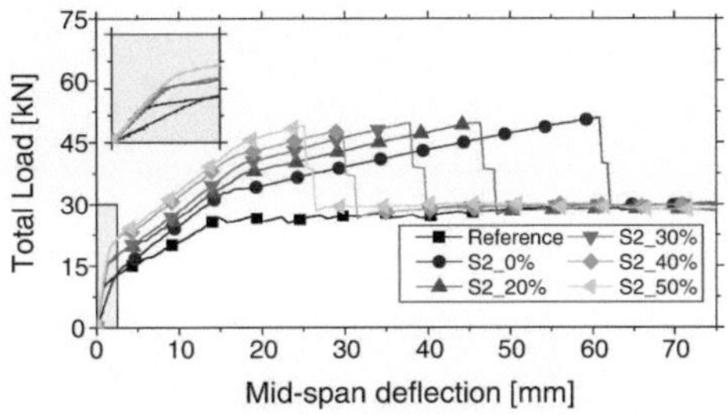

Fig. 7.16 Load-midspan deflection curve by Costa (2014) for various prestress NSM prestress levels

serviceability limit states significantly improved, although this evidence has as immediate consequence in the reduction of the ductility of the reinforced concrete element.

Unlike the experimental programs previously presented in which the beams were monotonically loaded up to failure, (Badawi and Soudki 2008) performed cyclic tests on beams strengthened with prestressed NSM-FRPs up to 40 and 60 % of their ultimate capacity. According to the obtained results, the application of prestress increased the fatigue capacity of the original reinforced concrete beams. Failure was essentially dominated by the rupture of the tensile reinforcement, mainly due to the accumulation of slippage between CFRP and adhesive that caused an increase of the average stress installed on the steel bars.

Regarding plates, the efficacy of the NSM strengthening technique with passive FRP reinforcements to increase the flexural resistance was assessed by Bonaldo et al. (2008), Dalfré and Barros (2011), and Breveglieri et al. (2012). In fact, NSM CFRP laminates without any prestress level can increase significantly the ultimate load carrying capacity of RC structural elements and high mobilization of the tensile properties of the CFRP can be assured. However, for deflection levels corresponding to the serviceability limit states the benefits of the CFRP are, in general, of small relevance. By prestressing the CFRP, its high tensile capacity is more effectively used, contributing to increase significantly the load carrying capacity of the strengthened elements under both service and ultimate conditions. The prestress can also contribute to close eventual existing cracks, to decrease the tensile stress installed in the existing flexural reinforcement, and to increase the shear capacity of these elements. Thus, prestressing the CFRP seems to be a cost-effective solution to increase both the structural performance and the durability of the strengthened RC structure. The use of FRP prestressed systems for the flexural strengthening of RC slabs is however quite limited.

Hosseini et al. (2014) have carried out an experimental program composed of four RC slabs with the purpose of evaluating the influence of the prestress level in the behaviour of this kind of elements in terms of serviceability and ultimate limit states. The adopted reinforcement systems were designed to assure flexurally failure mode for all the tested slabs (reinforcement yielding). The SREF is the reference

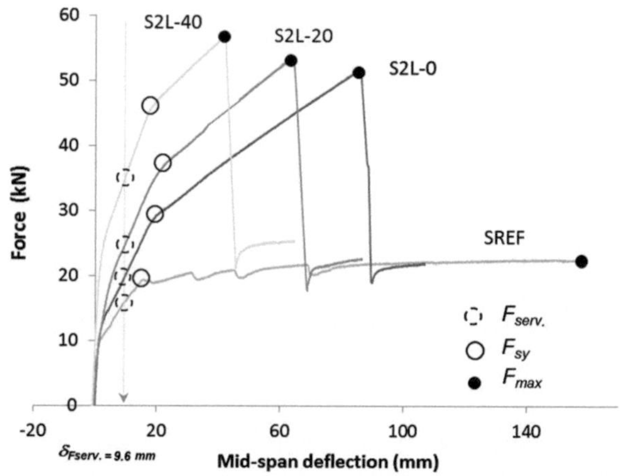

Fig. 7.17 Force-midspan deflection of the tested RC slabs (Hosseini et al. 2014)

slab without CFRP, and the S2L-0, S2L-20 and S2L-40 slabs are those flexurally strengthened using two NSM CFRP laminates with different prestress level: 0 % (S2L-0), 20 % (S2L-20) and 40 % (S2L-40) of the ultimate tensile strength of the CFRP laminates. The CFRP laminates used in the present experimental program have a cross section of 1.4 × 20 mm^2. The tested slabs had a percentage of longitudinal tensile steel bars (ρ_{sl}) of approximately 0.35 %, while the CFRP strengthening percentage (ρ_f) is approximately 0.08 %.

Figure 7.17 shows the relationship between the applied force and the deflection at mid-span $(F - \delta)$, for the tested RC slabs. It has been verified that:

- Strengthening RC slabs with prestressed NSM CFRP laminates resulted in a significant increase of the load carrying capacity at serviceability and ultimate limit states. By applying 20 % of prestress in the NSM CFRP laminates, the service and ultimate loads have increased, respectively, 55 and 136 % when compared to the corresponding values of the reference slab, while 40 % of prestress has guaranteed an increase of 119 and 152 %.
- By increasing the prestress level in the NSM CFRP laminates the overall flexural behaviour of the slabs at service and ultimate states has improved, but the deflection at the maximum load and at yield initiation of the steel reinforcement of the slabs has decreased with the increase of the prestress level. However, the deflection at maximum load was more than two times the deflection at yield initiation, with a significant plastic incursion on the steel reinforcement, which assures the required level of deflection ductility for this type of RC structures.
- Regardless the prestress level applied to the CFRP laminates, all the strengthened slabs failed by rupture of the laminates after yielding of the tension steel reinforcement. This failure mode proved the high effectiveness of the NSM technique for the flexural strengthening of RC slabs.

Analytical Models

By using FEM-based advanced constitutive models for the material nonlinear analysis of RC beams flexurally strengthened with passive and prestressed CFRP laminates Rezazadeh et al. (2013) have demonstrated that existing commercial FEM-based software can be used to predict with good accuracy the behaviour of this type of structures, as long as the data for the model parameters is correctly estimated.

Barros et al. (2012) developed a closed form formulation to determine the moment-curvature response of rectangular cross section of RC elements failing in bending that can be strengthened by prestressed FRP systems. Using the moment-curvature relationship predicted by the model and implementing an algorithm based on the virtual work method, a numerical strategy was developed for the prediction of the force-deflection response of statically determinate beams. This approach was extended to statically indeterminate RC elements failing in bending (Barros and Dalfré 2013).

Costa (2014) developed a spreadsheet for the determination of the most significant points of the moment-curvature of reinforced concrete beams using a closed form formulation. This spreadsheet allows the calculation of the cracking, yielding and ultimate curvature, as well as the corresponding bending moment of rectangular reinforced concrete sections with one layer of conventional tensile reinforcement, one layer of conventional compressive reinforcement, and one layer of composite strengthening to which a certain amount of prestress can be applied. The formulation used in this spreadsheet is based on conventional sectional analysis theory according to which the distribution of strain is assumed to be linear along the height of the beam. In this spreadsheet, the behaviour of concrete and steel are assumed to be in accordance with Eurocode 2 (CEN 2004) while the FRP was assumed as having linear elastic behaviour up to failure. This spreadsheet also includes a formulation capable of using the moment-curvature for the evaluation of the mid-span deflection of a simply supported RC element subjected to four-point loading configuration with notable accuracy.

Design Issues

In general, reinforced or prestressed reinforced concrete members strengthened with prestressed CFRP exhibit strip debonding when they reach their ultimate load carrying capacity. In some cases, tensile failure was reported (Meier and Stoecklin 2005). Currently, design codes have not yet been elaborated for prestressed composites.

A semester project at ETH Zürich (Harmanci 2012) revealed that the application of the conventional design rules for debonding in the free length due to excessive interfacial bond shear stress and/or CFRP tensile strain, in this case according to the

Swiss SIA 166 (SIA 2004) for externally bonded strip reinforcement, can also be applied to the prestressed case with reasonable precision. Several design codes include a plate-end debonding criterion when unstressed laminates are loaded. For a prestressed system, the end-anchorage needs to be defined in terms of the ultimate load at which the strip would be eventually pulled out of the anchorage. Theoretical considerations regarding the prestressed laminates bonded to a concrete substrate without a mechanical anchorage are presented in Triantafillou and Deskovic (1991). Gradient anchorage design, with different ultimate crack locations, was presented by Czaderski (2012). Harmanci (2013) validated the results by implementing the failure criteria in a numerical code to calculate bonded and unbonded (on the free length) prestressed CFRP strips with gradient anchorage by comparing them to experimental data taken from Czaderski (2012).

The overall ductility of the strengthened system is decreased when the prestrain applied to the laminate increases (Michels et al. 2011). A structural designer should respect common practice and leave sufficient ductility to the structure. For instance, such ductility can be ensured by guaranteeing a significant difference between the curvature at steel yielding and ultimate strip debonding. Detailed investigations that also include energy dissipation concepts are presented in Kim et al. (2008c), Oudah and El-Hacha (2011, 2012).

Shear Strengthening

Prestressed Shear Reinforcement for Concrete

The governing aim for prestressed shear reinforcement is to enhance the shear capacity of concrete. In terms of the concrete web shear performance, the addition of transverse prestress will result in higher cracking loads and steeper crack angles. A prime motivation for early examples of prestressed steel shear reinforcement was to enable the use of deep, thin webbed members, e.g., Freyssinet's post-war reconstruction of the Esbly bridge in France (Freyssinet 1950).

Prestressed shear reinforcement is not generally used in current practice for new construction. However, this technique has received growing interest for the strengthening and repair of existing reinforced concrete structures. Strengthening options that are external to the structure are likely to be more practical and less disruptive. Therefore, adequate protection against corrosion is a main concern for external steel reinforcements. Meier et al. (1993) patented a method to apply prestressed FRP shear reinforcement for strengthening applications. FRPs are expected to be durable in external environments and have a high strength-to-weight ratio. Thus, FRP cross-sections can be used to deliver strength enhancements similar to those of steel while being thinner and lighter. The strain capacity of FRPs can be relatively high and using prestressed systems represents a more efficient use of the materials (Burgoyne 2001). FRPs with a high strength and stiffness and the ability to sustain stress over the long-term are required for prestressing applications.

FRP Prestressed Shear Strengthening

The force in prestressed shear reinforcement consists of the initial prestress (minus losses), ε_{fp}, plus additional strains $\Delta\varepsilon_f$ generated due to the applied load. The compressive prestress relieves the strain in the existing internal transverse steel and increases the crack bridging force for a given shear crack width. The additional strains, $\Delta\varepsilon_f$, are small when the concrete is uncracked but will increase as shear cracks open. The FRP strengthening material properties, e.g. stiffness and strength, play a significant role after cracking. Aggregate interlock mechanisms are enhanced because a higher crack bridging force leads to smaller crack widths. In addition, FRPs are elastic and, therefore, they can continue to sustain load after yielding of any internal transverse steel.

While either bonded or unbonded FRP prestressed systems could be used in principle, more research has focused on unbonded systems so far. A non-laminated strap system developed by Winistörfer (1999) has been used for prestressed shear elements. The system uses a thin thermoplastic CFRP tape, and the tape layers can be fusion bonded together to make a closed strap (Lees and Winistörfer 2011). The flexible, self-anchored straps avoid some of the difficulties associated with gripping FRPs and overcome the difficulties of strengthening reinforced concrete with complex geometries. Figure 7.18 also presents a concept for a prestressed shear reinforcement with CFRP straps (Motavalli et al. 2011).

In an unbonded system, the crack opening displacements are averaged over the unbonded length of the reinforcement, which avoids local crack stress concentrations in the FRP. However, unbonded systems experience size effects because the induced strap strain reduces as the beam depth increases for a given crack opening. Stenger (2000) found that the initial strap prestress is an important parameter for the shear resistance in deep beams. Experiments on small scale rectangular shear critical beams strengthened with CFRP straps (Kesse and Lees 2007) identified five different failure modes. The three shear modes were: (1) Strap failure, which eventually leads to global beam shear failure; (2) Shear failure occurs in an unstrengthened concrete region adjacent to the straps, which do not fail; and

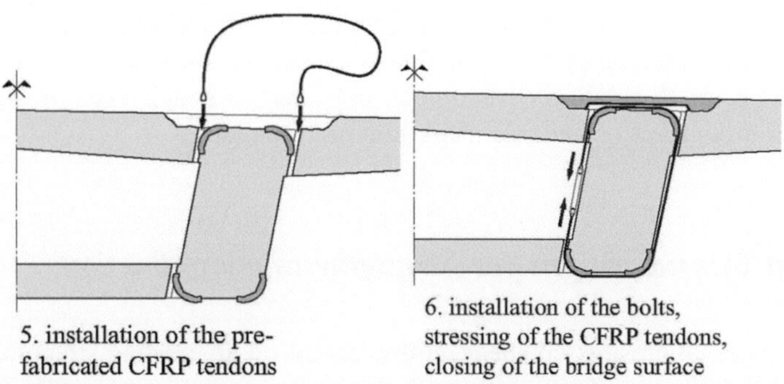

Fig. 7.18 Concept for a prestressed shear reinforcement (Motavalli et al. 2011)

(3) Extensive concrete crack opening and damage followed by strap failure. The two flexural modes were distinguished by whether the flexural failure was followed by a strap failure with little ductility or whether concrete crushing eventually followed adequate ductility characterised by the yielding of the longitudinal steel. The propensity for a given failure mode depended on the initial prestress, the spacing and the strap stiffness. This system also featured trade-offs, e.g. a certain level of strap prestress was necessary to ensure effective crack bridging. However, the reserve strain capacity to accommodate crack opening prior to strap failure was limited if the prestress was excessive.

Analysis and Design of Prestressed FRP Shear Strengthening Systems

Because FRP materials are linearly elastic, one difficulty in the analysis of prestressed FRP shear reinforcement is the prediction of the FRP force at failure. The force in the FRP is compatible with the transverse expansion of the base structure. Therefore, analysis techniques that consider strain compatibility/crack opening can allow for this relationship. A general consensus about the shear resistance of reinforced concrete is lacking; thus, the inclusion of additional FRP prestressed reinforcement is a further complication. One approach is to add the contribution of the FRP (V_{frp}) to concrete, V_c, and steel, V_s, contributions. Chen and Teng (2001) have proposed an expression for V_{frp} for prestressed FRP straps based on an assumed crack profile, whereas Hoult and Lees (2009) used a shear friction approach and a compatibility relationship between the shear friction and crack opening to determine the force in the FRP. Another approach consists of using a model that considers equilibrium, compatibility and material laws of reinforced concrete cracked in response to shear, such as the modified compression field theory (MCFT) developed by Vecchio and Collins (1988). Unbonded prestressed CFRP straps have been incorporated into the MCFT in an average sense (Lees et al. 2002) and the inclusion of straps with non-uniform spacing has also been investigated (Yapa and Lees 2013). In a two-dimensional finite element analyses of CFRP strap-strengthened T-beams (Dirar et al. 2013) the strap strains were generally underpredicted and the results depended on the shear models and concrete input parameters. A rotating crack model appeared to match experimental results better than a constant or variable shear retention model.

Practical Considerations for Strengthening Applications

Installing external prestressed shear reinforcement on an existing structure presents a number of challenges. Drilling through the structure may be necessary to connect the compression and tension chords to allow for the insertion of the shear

Fig. 7.19 CFRP strap strengthening of a T-beam (©University of Cambridge) and a flat slab (courtesy of Dr. A.U. Winistörfer)

reinforcement, e.g. in slab-on-beam or flat slab (see Fig. 7.19) structures. Prestress must also be imparted on the FRP shear reinforcement. Several different installation methods have been investigated for unbonded CFRP strap systems. For a strap installed in situ, a support pad can be inserted under the strap (see Fig. 7.1a) and lifted (Lees et al. 2002) while shims are placed under the pad to lock in the prestress. Another option is to pre-fabricate a flexible CFRP strap with loops at either end (Czaderski et al. 2008; Koppitz et al. 2013; Keller et al. 2013). The strap is placed around the region to be strengthened, supported on saddles and stressed by joining the two ends of the strap together using a turn-buckle or a threaded rod system. Frictional losses around the saddles can reduce the effective prestress (Czaderski et al. 2008), which would need to be considered in the design. Constraints for FRP strap systems include the need for a smooth bearing surface and a minimum radius to avoid failure (Lees and Winistörfer 2011). As with any external system, adequate protection against vandalism must be ensured.

Confinement

Introduction

FRPs can be used to confine the lateral deformation of concrete columns subjected to axial compressive loadings. In a confined concrete column, the confinement is loaded in the hoop direction, while the concrete is loaded in tri-axial compression such that both materials are used to their best advantages. The confinement can greatly enhance both the strength and the ultimate strain of the concrete. Therefore, the ductility of confined concrete is greatly enhanced. However, the advantage of prestressed confinement compared to unstressed confinement is not as obvious. In this section, the benefits of confinement prestressing, namely a smaller decrease in the residual strength and lower deformations, will be presented.

Effect of Prestress on Confined Column Response

Janke et al. (2009) presented an experimental study of concrete cylinders confined with unstressed and prestressed steel and CFRP bands. The steel or CFRP bands were spirally wounded around the cylinders. Neither of the band types was bonded to the concrete surface. To maintain the pretension, steel anchoring clamps were mounted at the end of the concrete cylinders. The prestress and tensile stiffness of the tangential confinement reinforcement were varied. A concentric load was applied to the confined cylinders until failure to obtain data on the ductility and ultimate load of the different specimens. Typical axial load-compressive strain curves are shown in Fig. 7.20.

The slope of the load-deformation curve changed at higher loads when the confinement was prestressed compared to the unstressed confinement for both confinement materials. After this slope change, the curves for the unstressed and prestressed specimens continued in an almost parallel formation. The peak load and axial strain of the prestressed CFRP confined cylinder was lower than that of the unstressed specimen. This difference was likely due to the initial prestrain of the CFRP band, which reduced the usable load strain. The effect of prestressed confinement on the slope change and peak load was the same in experimental studies performed by Zile et al. (2009) and Ciniña et al. (2012). In both campaigns, the columns were confined by unstressed and prestressed basalt (Ciniña et al. 2012) and

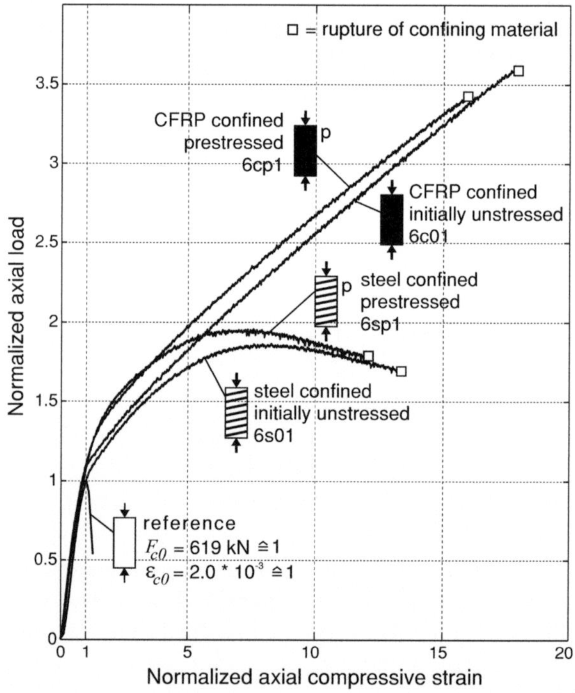

Fig. 7.20 Typical axial load compressive strain curves of confined columns (Janke et al. 2009)

CFRP (Zile et al. 2009) bands. However, contrary to Janke et al. (2009) the bands were impregnated with an epoxy resin prior to placing them around the cylinders. Therefore, one might argue that prestressed confinement resulted in a small benefit. However, Janke et al. (2009) presented other arguments showing that prestressed confinement is advisable.

Residual Strength of Concrete

External confinement can become ineffective or may need to be removed as a result of fire damage, vandalism, or repair measures. Fires occur frequently after earthquakes. Vandalism, i.e. deliberate destruction, can also be a problem especially for freestanding concrete columns that are externally confined. Corrosion protection measures are a possible reason for the removal and exchange of confinement long after they have been applied. The term 'confined concrete strength' loses its meaning for the above mentioned cases, and the residual strength determines the capability of the column to resist the existing loads. Confined concrete strength is a variable of the confined system. In contrast, the residual strength of unconfined concrete, which has previously been part of a confined system, can be considered a material parameter. This parameter is assumed to depend on the loading history in the confined state.

To elaborate, when the confinement is removed after the confined system has been subjected to loads above the unconfined concrete strength (i.e. to a certain overload) the residual strength of the concrete component is significantly diminished due to concrete damage (i.e. microcracking). This effect is especially pronounced for unstressed confinement because the activation of confinement pressure requires considerable lateral strain, which significantly increases the microcrack density and crack width.

Janke et al. (2009) found that prestressing the confinement reinforcement significantly affects the residual strength of columns after an overload. To estimate the residual strength, the concentric load on the confined cylinders was increased only up to a predefined overload above the unconfined load capacity (Fig. 7.21a). The confinement reinforcement was then removed and the residual capacity was determined in another compression test (Fig. 7.21b).

In the experiments the axial compressive strain and plastic deformation of the unstressed specimens was significantly higher than that of the prestressed confinement in response to the same overload (Fig. 7.21a). In the example shown in Fig. 7.21 the axial compressive strain of the initially unstressed confined specimen and the prestressed confined specimen were 3.5 times and 1.7 times higher than that of the unconfined reference specimen at peak load, respectively. The residual strength was 56 and 95 %, respectively. Therefore, prestressing the confinement reinforcement significantly reduces damage to the confined concrete under overloads compared to unstressed confinement reinforcement, which was demonstrated by a higher residual capacity (Fig. 7.21b) and lower axial strain (Fig. 7.21a). The

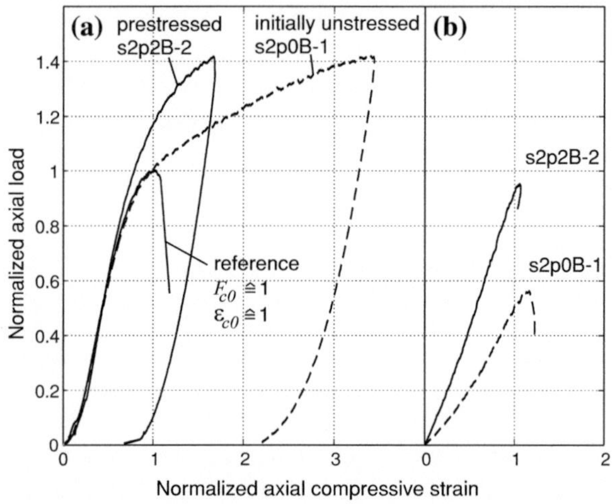

Fig. 7.21 **a** Load cycle to 140 % of unconfined compressive strength with confinement, **b** breaking test after removal of confinement (Janke et al. 2009)

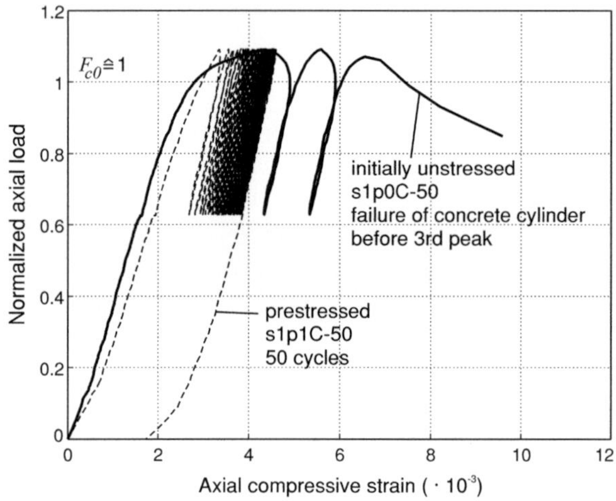

Fig. 7.22 Cyclic loading of specimen with low confinement modulus (Janke et al. 2009)

latter could be beneficial to serviceability limit states (lower displacements), while the former significantly improves the safety of construction under special circumstances.

Residual Strength Under Cyclic Loading

Prestressed confinement reinforcement was found to maintain the residual strength under cyclic loading compared to the unstressed variant (Janke et al. 2009). Figure 7.22 shows an example of a cyclically loaded specimen with prestressed and unstressed confinement. The unstressed specimen with low confinement stiffness

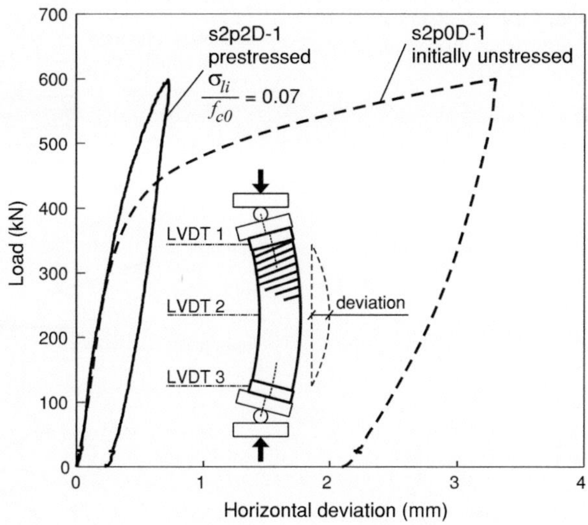

Fig. 7.23 Horizontal deviation of confined specimen at combined bending and axial load (Janke et al. 2009)

broke prior to the third peak of the cyclic loading process. The peak of the overload was 109 % of the unconfined compressive strength. In contrast, the residual capacity of the prestressed reference cylinder was 95 % after 50 cycles, which demonstrated the effectiveness of prestressed confinement. The residual capacity values of the prestressed specimens were nearly independent of the number of load cycles for the defined level of overload. Therefore, prestressing the confinement is also beneficial for cyclic loading.

Eccentric Loading of Confined Columns

Unstressed confinement only becomes effective for lateral expansion. Therefore, the effect of such confinement is small under bending stress, because pure bending creates only a slight overall lateral expansion. The typical response of confined specimen subjected to combined bending and axial load is shown in Fig. 7.23. The results show that even moderate confinement prestress is effective in members under bending stress. A significantly smaller residual deviation compared to unstressed confined specimen was found. Janke et al. (2009) related this phenomenon to a lower degree of damage to the concrete. Once more, the beneficial effect of prestressed confinement was shown.

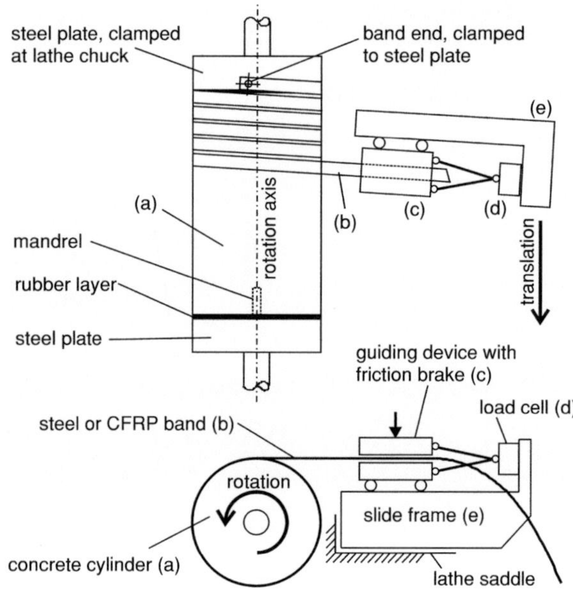

Fig. 7.24 Mechanical prestressing procedure (Janke et al. 2009)

Technical Aspects of Confinement Prestressing

In Janke et al. (2009) the steel or CFRP bands were spirally wound around the cylinders using a stationary lathe. This procedure is illustrated in Fig. 7.24. During winding, the slowly rotating cylinder pulled the band from the coil through the guiding device, which served as a friction brake to create axial tension in the band. However, prestressing FRP on site remains technically complex.

Practical Applications

The present section shows several examples of practical applications of flexural strengthening by means of prestressed CFRP strips (Figs. 7.25, 7.26, 7.27, 7.28, 7.29, 7.30 and 7.31).

Durability

Durability is one key aspect in structural engineering. In the field of prestressed FRP reinforcements only few investigations have been presented. El-Hacha et al. (2004a, b) present studies on strengthened reinforced concrete beams with prestressed CFRP sheets at room and low temperatures. However, these investigations

Fig. 7.25 Bridge box girder strengthening, Rijeka (Croatia), courtesy of S&P Clever Reinforcement AG

Fig. 7.26 Bridge girder strengthening, Bangok (Thailand), courtesy of S&P Clever Reinforcement AG

have to be qualified as short-term tests. Since 2000, a prestressed CFRP strip (prestrain 0.55 %) with a gradient anchorage is regularly monitored in terms of strain evolution in time (Michels et al. 2013). Up to now, only a slight decrease (0.04 %) with a stable tendency could be noticed. In general, it is presumed that, in addition to concrete degradation phenomena, creep behaviour of both concrete in compression as well as of the epoxy resin play a major role in long-term efficiency

Fig. 7.27 Office building, Zurich Altstetten (Switzerland), courtesy of S&P Clever Reinforcement AG

of a CFRP prestressing system (Diab et al. 2009). At Empa investigations in the field of durability for bridge retrofitting applications are currently going on. Of special interest is, next to long-term durability, the temperature stability of prestressed CFRP strips with a gradient anchorage when being used as a negative bending moment reinforcement on top of the structure at the moment of the asphalt installation and the related high temperatures.

Fatigue tests are documents in Kotynia et al. (2011), Oudah and El-Hacha (2013a, b).

Concluding Remarks

This chapter presents several aspects of structural concrete strengthening with prestressed FRPs in flexure, shear, and confinement modes. Prestressed FRP systems have already been widely used in the academic community for several years, whereas practical applications in the field are rather limited. However, the presented investigations show a large potential of a much wider use in the future.

Fig. 7.28 Punching shear strengthening in an office building, Nyon (Switzerland), courtesy of S&P Clever Reinforcement AG

Fig. 7.29 Carbo Stress prestressing system, courtesy of Sika Switzerland

Fig. 7.30 Bridge girder strengthening, Winnepeg (Canada), courtesy of Prof. Dr. Y. Kim

Fig. 7.31 Strengthening of the girders of the Battiferro-Navile viaduct (A14 Highway Bologna-Taranto, Italy), courtesy of Prof. Dr. Carlo Pellegrino and Giorgio Giacomin

The most obvious advantage of additional flexural reinforcement is a reduction of existing deflections and crack openings for use under service loads. Monotonic loading tests have also shown an enhancement of the yielding and ultimate load. In most cases, the debonding of the additional strip reinforcement was the governing failure mode, while a tensile failure of the composite could be observed in isolated cases. The anchorage system is important in terms of practicability: similar to conventional prestressed concrete, adequate anchorage is necessary for load transfer to the structure during and after the installation phase. Currently, several mechanical systems with anchor bolts and plates are available on the market. These systems require drilling into the existing concrete substrate. The gradient anchorage method, as presented in previous sections, leaves a pure strip-epoxy-concrete connection without any remaining mechanical devices. This technique is promising both in terms of durability and aesthetics.

Prestressed carbon composites in shear and confinement modes are currently limited to academic research. Practical applications have been seldom implemented. Such a procedure seems to be highly beneficial and would be worthwhile to investigate and install.

The anchorage system remains a crucial factor. A wider application range will eventually also depend on the practicability of the prestressing system. Flexural strengthening usually features a better access to the structure, while shear strengthening might often be more complicated due to geometric restrictions.

Future research and development should primarily focus on three aspects: (a) the elaboration of design codes to allow structural designers to dispose of a calculation tool, (b) the assessment of durability of the different systems and (c) the development of practical prestressing and anchoring systems.

Acknowledgments The authors from the Structural Engineering Research Laboratory at Empa express their gratitude to the Swiss innovation promotion agency for their support in the framework of the project CTI No. 10493.2 PFIW-IW. The authors from University of Minho would like to thank FEDER for supporting this work via funds through the Operational Program for Competitiveness Factors—COMPETE and national funds through FCT—Portuguese Foundation for Science and Technology under the project FRPreDur—PTDC/ECM-EST/2424/2012. All industrial partners involved in the respective research activities are kindly acknowledged.

References

Andrä, H.-P., König, G., & Maier, M. (2001). Einsatz vorgespannter Kohlefaser-Lamellen als Oberflächenspannglieder. *Beton- und Stahlbetonbau, 96*(12), 737–747.

Andrä, H.-P., & Maier, M. (2005). Instandsetzung von Brücken mit einer neuen Generation von Spanngliedern auf Basis von CFK-Bändern. *Bauingenieur, 80*, 7–16.

Aram, M. R., Czaderski, C., & Motavalli, M. (2008). Effects of gradually anchored prestressed CFRP strips bonded on prestressed concrete beams. *Journal of Composites for Construction, 12*(1), 25–34.

Badawi, M., & Soudki, K. A. (2008). *Fatigue of RC beams strengthened with prestressed NSM CFRP rods*. Paper presented at the CICE2008, Zurich, Switzerland.

Barros, J. A. O. (2009). *Pre-stress technique for the flexural strengthening with NSM-CFRP strips*. Paper presented at the FRPRCS-9, Sydney, Australia.

Barros, J. A. O., & Dalfré, G. M. (2013). *A model for the prediction of the behavior of continuous RC slabs flexurally strengthened with CFRP systems*. Paper presented at the FRPRCS-11, Guimaraes, Portugal.

Barros, J. A. O., Dias, S. J. E., & Lima, J. L. T. (2007). Efficacy of CFRP-based techniques for the flexural and shear strengthening of concrete beams. *Cement and Concrete Composites, 29*(3), 203–217.

Barros, J. A. O., Taheri, M., Salehian, H., & Mendes, P. J. D. (2012). A design model for fibre reinforced concrete beams pre-stressed with steel and FRP bars. *Composite Structures, 94*(8), 2494–2512.

Berset, T., Schwegler, G., & Trausch, L. (2002). Verstärkung einer Autobahnbrücke mit vorgespannten CFK-Lamellen. *tec21, 128*, 27–29.

Bonaldo, E., De Barros, J. A. O., & Loureņo, P. B. (2008). Efficient strengthening technique to increase the flexural resistance of existing RC slabs. *Journal of Composites for Construction, 12*(2), 149–159.

Breveglieri, M., Barros, J. A. O., Dalfré, G. M., & Aprile, A. (2012). A parametric study on the effectiveness of the NSM technique for the flexural strengthening of continuous RC slabs. *Composites: Part B—Engineering, 43*(4), 1970–1987.

Burgoyne, C. J. (2001). Rational use of advanced composites in concrete. *Proceedings of the Institution of Civil Engineers: Structures and Buildings, 146*(3), 253–262.

CEN. (2004). *Eurocode 2: Design of concrete structures—Part 1-1: General rules and rules for buildings, EN 1992-1-1 2004*. Brussels: Comité Européen de Normalisation.

Chen, J. F., & Teng, J. G. (2001). *On the strength of RC beams shear strengthened with prestressed FRP straps*. Paper presented at the Proceedings of the International Conference on FRP Composites in Civil Engineering.

Ciniña, I., Zile, E., & Zile, O. (2012). Mechanical behavior of concrete columns confined by basalt FRP windings. *Mechanics of Composite Materials, 48*(5), 539–546.

Costa, I. G. (2014). *Prestressed carbon fibre laminates applied according to near surface mounted technique to increase the flexural resistance of reinforced concrete beams*. Ph.D. thesis, University of Minho.

Czaderski, C. (2012). *Strengthening of reinforced concrete members by prestressed, externally bonded reinforcement with gradient anchorage*. Ph.D. thesis, ETH Zürich.

Czaderski, C., Martinelli, E., Michels, J., & Motavalli, M. (2012). Effect of curing conditions on strength development in an epoxy resin for structural strengthening. *Composites: Part B—Engineering, 43*(2), 398–410.

Czaderski, C., & Motavalli, M. (2007). 40-Year-old full-scale concrete bridge girder strengthened with prestressed CFRP plates anchored using gradient method. *Composites: Part B—Engineering, 38*(7–8), 878–886.

Czaderski, C., Motavalli, M., & Winistörfer, A. (2008). *Prestressed shear strengthening of a box girder bridge with non-laminated CFRP straps*. Paper presented at the CICE2008, Zürich, Switzerland.

Dalfré, G. M., & Barros, J. A. O. (2011). Flexural strengthening of RC continuous slab strips using NSM CFRP laminates. *Advances in Structural Engineering, 14*(6), 1223–1245.

Deuring, M. (1993). *Verstärken von Stahlbeton mit gespannten Faserverbundwerkstoffen*. Ph.D. thesis, ETH Zürich.

Diab, H., Wu, Z., & Iwashita, K. (2009). Short and long-term bond performance of prestressed FRP sheet anchorages. *Engineering Structures, 31*(5), 1241–1249.

Dirar, S., Lees, J. M., & Morley, C. T. (2013). Precracked reinforced concrete t-beams repaired in shear with prestressed carbon fiber-reinforced polymer straps. *ACI Structural Journal, 110*(5), 855–865.

El-Hacha, R., & Aly, M. Y. E. (2013). Anchorage system to prestress FRP laminates for flexural strengthening of steel-concrete composite girders. *Journal of Composites for Construction, 17* (3), 324–335.

El-Hacha, R., Green, M. F., & Wight, R. G. (2004a). Flexural behaviour of concrete beams strengthened with prestressed carbon fibre reinforced polymer sheets subjected to sustained loading and low temperature. *Canadian Journal of Civil Engineering, 31*(2), 239–252.

El-Hacha, R., Wight, R. G., & Green, M. F. (2001). Prestressed fibre-reinforced polymer laminates for strenghtening structures. *Progress in Structural Engineering and Materials, 3*, 111–121.

El-Hacha, R., Wight, R. G., & Green, M. F. (2003). Innovative system for prestressing fiber-reinforced polymer sheets. *ACI Structural Journal, 100*(3), 305–313.

El-Hacha, R., Wight, R. G., & Green, M. F. (2004b). Prestressed carbon fiber reinforced polymer sheets for strengthening concrete beams at room and low temperatures. *Journal of Composites for Construction, 8*(1), 3–13.

Fernandes, P., Sena Cruz, J., Fernandes, C., Silva, P., & Dias da Costa, J. (2013). *Flexural response of HSC girders strengthened with non- and prestressed CFRP laminates*. Paper presented at the FRPRCS-11, Guimaraes, Portugal.

França, P., & Costa, A. (2007). *Behaviour of flexural strengthened beams with prestressed CFRP laminates*. Paper presented at the FRPRCS-8, Patras, Greece.

Freyssinet, E. (1950). Prestressed concrete: Principles and applications. *Journal of the Institution of Civil Engineers, 33*(4), 331–380.

Gaafar, M. A., & El-Hacha, R. (2008). *Strengthening reinforced concrete beams with prestressed FRP nea surface mounted technique*. Paper presented at the CICE2008, Zurich, Switzerland.

Garden, H. N., & Hollaway, L. C. (1998). An experimental study of the failure modes of reinforced concrete beams strengthened with prestressed carbon composite plates. *Composites: Part B—Engineering, 29*(4), 411–424.

Harmanci, Y. E. (2012). *Externally bonded reinforcement (EBR) for flexural strengthening of reinforced concrete beams with regular cross sections*. Semester work, ETH Zürich.

Harmanci, Y. E. (2013). *Prestressed CFRP for structural retrofitting—experimental and analytical investigation*. M.Sc. thesis, ETH Zürich.

Hosseini, M. R. M., Dias, S. J. E., & Barros, J. A. O. (2014). Effectiveness of prestressed NSM CFRP laminates for the flexural strengthening of RC slabs. *Composite Structures, 111* (1), 249–258.

Hoult, N. A., & Lees, J. M. (2009). Modeling of an unbonded CFRP strap shear retrofitting system for reinforced concrete beams. *Journal of Composites for Construction, 13*(4), 292–301.

Huang, Y.-L., Hung, C.-H., Yen, T., Wu, J.-H., & Lin, Y.-C. (2005). Strengthening reinforced concrete beams using prestressed glass fiber-reinforced polymer—Part I: Experimental study. *Journal of Zhejiang University: Science, 6*(3), 166–174.

Janke, L., Czaderski, C., Ruth, J., & Motavalli, M. (2009). Experiments on the residual load-bearing capacity of prestressed confined concrete columns. *Engineering Structures, 31* (10), 2247–2256.

Kaiser, H. (1989). *Bewehren von Stahlbeton mit kohlenstofffaserverstärkten Epoxidharzen*. Ph.D. thesis, ETH Zürich.

Keller, T., Kenel, A., & Koppitz, R. (2013). Carbon fiber-reinforced polymer punching reinforcement and strengthening of concrete flat slabs. *ACI Structural Journal, 110*(6), 919–927.

Kesse, G., & Lees, J. M. (2007). Experimental behavior of reinforced concrete beams strengthened with prestressed CFRP shear straps. *Journal of Composites for Construction, 11*(4), 375–383.

Kim, Y. J., Green, M. F., & Gordon Wight, R. (2010a). Bond and short-term prestress losses of prestressed composites for strengthening PC beams with integrated anchorage. *Journal of Reinforced Plastics and Composites, 29*(9), 1277–1294.

Kim, Y. J., Green, M. F., & Wight, R. G. (2007). Flexural behaviour of reinforced or prestressed concrete beams including strengthening with prestressed carbon fibre reinforced polymer sheets: Application of a fracture mechanics approach. *Canadian Journal of Civil Engineering, 34*(5), 664–677.

Kim, Y. J., Green, M. F., & Wight, R. G. (2008a). Live load distributions on an impact-damaged prestressed concrete girder bridge repaired using prestressed CFRP sheets. *Journal of Bridge Engineering, 13*(2), 202–210.

Kim, Y. J., Green, M. F., & Wight, R. G. (2010b). Effect of prestress levels in prestressed CFRP laminates for strengthening prestressed concrete beams: A numerical parametric study. *PCI Journal, 55*(2), 96–108.

Kim, Y. J., Longworth, J. M., Wight, R. G., & Green, M. F. (2008b). Flexure of two-way slabs strengthened with prestressed or nonprestressed CFRP sheets. *Journal of Composites for Construction, 12*(4), 366–374.

Kim, Y. J., Longworth, J. M., Wight, R. G., & Green, M. F. (2010c). Punching shear of two-way slabs retrofitted with prestressed or non-prestressed CFRP sheets. *Journal of Reinforced Plastics and Composites, 29*(8), 1206–1223.

Kim, Y. J., Shi, C., & Green, M. F. (2008c). Ductility and cracking behavior of prestressed concrete beams strengthened with prestressed CFRP sheets. *Journal of Composites for Construction, 12*(3), 274–283.

Kim, Y. J., Wight, G. R., & Green, M. F. (2008d). Flexural strengthening of RC beams with prestressed CFRP sheets: Using nonmetallic anchor systems. *Journal of Composites for Construction, 12*(1), 44–52.

Kim, Y. J., Wight, R. G., & Green, M. F. (2008e). Flexural strengthening of RC beams with prestressed CFRP sheets: Development of nonmetallic anchor systems. *Journal of Composites for Construction, 12*(1), 35–43.

Koppitz, R., Keller, T., & Kenel, A. (2013). *Punching shear strengthening of reinforced concrete flat slabs with carbon-fiber rienforced polymer reinforcement*. Paper presented at the FRPRCS-11, Guimaraes, Portugal.

Kotynia, R., & Kaminska, M. E. (2003). *Ductility and failure mode of RC beam strengthened for flexure with CFRP*. Report No. 13, Department of Concrete Structures, Faculty of Civil Engineering, Architecture and Environmental Engineering, Technical University of Lodz.

Kotynia, R., Lasek, K., & Staskiewicz, M. (2013a). Flexural behaviour of preloaded RC slabs strengthened with prestressed CFRP laminates. *Journal of Composites for Construction, 18*(3).

Kotynia, R., Staskiewicz, M., & Lasek, K. (2013b). *Efficiency analysis of strengthening of RC structures with prestressed CFRP composites*. Paper presented at the APFIS 2013, Melbourne, Australia.

Kotynia, R., Walendziak, R., Stoecklin, I., & Meier, U. (2011). RC slabs strengthened with prestressed and gradually anchored CFRP strips under monotic and cyclic loading. *Journal of Composites for Construction, 15*(2), 168–180.

Lees, J. M., & Burgoyne, C. J. (1999). Experimental study of influence of bond on flexural behavior of concrete beams pretensioned with aramid fiber reinforced plastics. *ACI Structural Journal, 96*(3), 377–385.

Lees, J. M., & Winistörfer, A. U. (2011). Nonlaminated FRP strap elements for reinforced concrete, timber, and masonry applications. *Journal of Composites for Construction, 15*(2), 146–155.

Lees, J. M., Winistörfer, A. U., & Meier, U. (2002). External prestressed carbon fiber-reinforced polymer straps for shear enhancement of concrete. *Journal of Composites for Construction, 6* (4), 249–256.

Meier, U. (1987). Brückensanierungen mit Hochleistungs-Faserverbundwerkstoffen. *Material und Technik, 15*(4), 125–128.

Meier, U. (1995). Strengthening of structures using carbon fibre/epoxy composites. *Construction and Building Materials, 9*(6), 341–351.

Meier, U., Deuring, M., & Meier, H. (1993). *Method and apparatus for increasing the shear strength of a construction structure*. US Patent No. 5.617.685.

Meier, U., & Stoecklin, I. (2005). *A novel carbon fiber reinforced polymer (CFRP) system for post-strengthening*. Paper presented at the ICCRRR, Cape Town, South Africa.

Michels, J., Czaderski, C., El-Hacha, R., Brönnimann, R., & Motavalli, M. (2012). Temporary bond strength of partly cured epoxy adhesive for anchoring prestressed CFRP strips on concrete. *Composite Structures, 94*(9), 2667–2676.

Michels J, Czaderski C, Motavalli M (2011) Prestressed CFRP for structural strenghtening. Paper presented at the 1st SMAR conference, Dubai, UAE.

Michels, J., Martinelli, E., Czaderski, C., & Motavalli, M. (2014a). Prestressed CFRP strips with gradient anchorage for structural concrete retrofitting: Experiments and numerical modeling. *Polymers, 6*, 114–131.

Michels, J., Sena Cruz, J., Czaderski, C., & Motavalli, M. (2013). Structural strengthening with prestressed CFRP strip with gradient anchorage. *Journal of Composites for Construction, 17* (5), 651–661.

Michels, J., Staskiewicz, M., Czaderski, C., Lasek, K., Kotynia, R., & Motavalli, M. (2014b). Anchorage resistance of CFRP strips externally bonded to various cementitious substrates. *Composites: Part B—Engineering, 63*, 50–60.

Motavalli, M., Czaderski, C., & Pfyl-Lang, K. (2011). Prestressed CFRP sor strengthening of reinforced concrete structures: Recent developments at Empa, Switzerland. *Journal of Composites for Construction, 15*(2), 194–204.

Neubauer, U., Vom Berg, W., & Onken, P. (2007). *Structural strengthening with a new system of prestressed CFRP strips*. Paper presented at the FRPRCS-8, Patras, Greece.

Nordin, H., & Täljsten, B. (2006). Concrete beams strengthened with prestressed near surface mounted CFRP. *Journal of Composites for Construction, 10*(1), 60–68.

Oudah, F., & El-Hacha, R. (2011). *Ductility of reinforced concrete beams strengthened with prestressed NSM CFRP strips/rebars—analytical study*. Paper presented at the 1st SMAR conference, Dubai, UAE.

Oudah, F., & El-Hacha, R. (2012). A new ductility model of reinforced concrete beams strengthened using Fiber Reinforced Polymer reinforcement. *Composites Part B Engineering, 43*(8), 3338–3347.

Oudah, F., & El-Hacha, R. (2013a). Analytical fatigue prediction model of RC beams strengthened in flexure using prestressed FRP reinforcement. *Engineering Structures, 46*, 173–183.

Oudah, F., & El-Hacha, R. (2013b). Research progress on the fatigue performance of RC beams strengthened in flexure using Fiber Reinforced Polymers. *Composites Part B Engineering, 47*, 82–95.

Pellegrino, C., & Modena, C. (2009). Flexural strengthening of real-scale RC and PRC beams with end-anchored pretensioned FRP laminates. *ACI Structural Journal, 106*(3), 319–328.

Quantrill, R. J., & Hollaway, L. C. (1998). The flexural rehabilitation of reinforced concrete beams by the use of prestressed advanced composite plates. *Composites Science and Technology, 58* (8), 1259–1275.

Rezazadeh, M., Costa, I.G., & Barros, J. A. O. (2013). *Assessment of the effectiveness of prestressed NSM CFRP laminates for the flexural strengthening of RC beams*. Paper presented at the FRPRCS-11, Guimaraes, Portugal.

Schlaich, M., Zwingmann, B., Liu, Y., & Goller, R. (2012). Zugelemente aus CFK und ihre Verankerungen. *Bautechnik, 89*(12), 841–850.

SIA. (2004). *SIA 166 - Klebebewehrungen*. Zürich, Switzerland.

Stenger, F. (2000). *Tragverhalten von Stahlbetonscheiben mit vorgespannter externer Kohlenstoffaser-Schubbewehrung*. Ph.D. thesis, ETH Zürich.

Suter, R., & Jungo, D. (2001). Vorgespannte CFK-Lamellen zur Verstärkung von Bauwerken. *Beton- und Stahlbetonbau, 96*(5), 350–358.

Teng, J. G., Chen, J. F., Smith, S. T., & Lam, L. (2002). *FRP strengthened RC structures*. Chichester: Wiley.

Triantafillou, T. C., & Deskovic, N. (1991). Innovative prestressing with FRP sheets. Mechanics of short-term behavior. *Journal of Engineering Mechanics, 117*(7), 1652–1672.

Triantafillou, T. C., Deskovic, N., & Deuring, M. (1992). Strengthening of concrete structures with prestressed fiber reinforced plastic sheets. *ACI Structural Journal, 89*(3), 235–244.

Vecchio, F. J., & Collins, M. P. (1988). Predicting the response of reinforced concrete beams subjected to shear using modified compression field theory. *ACI Structural Journal, 85*(3), 258–268.

Wight, R. G., Green, M. F., & Erki, M. A. (2001). Prestressed FRP sheets for poststrengthening reinforced concrete beams. *Journal of Composites for Construction, 5*(4), 214–220.

Winistörfer, A. (1999). *Development of non-laminated advanced composite straps for civil engineering applications*. Ph.D. thesis, University of Warwick.

Wu, Z. S., Iwashita, K., Hayashi, T., Higuchi, T., Murakami, S., & Koseki, Y. (2003). Strengthening prestressed-concrete girders with externally prestressed PBO fiber reinforced polymer sheets. *Journal of Reinforced Plastics and Composites, 22*(14), 1269–1286.

Yapa, H. D., & Lees, J. M. (2013). Rectangular reinforced concrete beams strengthened with CFPR straps. *Journal of Composites for Construction (ASCE)*. doi:10.1061/(ASCE)CC.1943-5614.0000416.

You, Y. C., Choi, K. S., & Kim, J. (2012). An experimental investigation on flexural behavior of RC beams strengthened with prestressed CFRP strips using a durable anchorage system. *Composites: Part B—Engineering, 43*(8), 3026–3036.

Yu, P., Silva, P. F., & Nanni, A. (2008). Flexural strength of reinforced concrete beams strengthened with prestressed carbon fiber-reinforced polymer sheets—Part II. *ACI Structural Journal, 105*(1), 11–20.

Zile, E., Daugevičius, M., & Tamužs, V. (2009). The effect of pretensioned FRP windings on the behavior of concrete columns in axial compression. *Mechanics of Composite Materials, 45*(5), 457–466.

Chapter 8
NSM Systems

José Sena-Cruz, Joaquim Barros, Vincenzo Bianco, Antonio Bilotta, Dionysios Bournas, Francesca Ceroni, Glaucia Dalfré, Renata Kotynia, Giorgio Monti, Emilio Nigro and Thanasis Triantafillou

Abstract This chapter gives an overview on the state-of-the-art of the Near-Surface Mounted (NSM) technique for structural retrofitting of reinforced concrete structures using Fibre Reinforced Polymers (FRP). The chapter is divided in 5 sections. Firstly, general aspects of the technique are introduced, including main advantages, nomenclature adopted, type of FRP reinforcement and adhesive used, groove geometry/location, and constructional aspects. Then, the existing

---

Final Remarks This chapter gave an overview about the current state of the near surface-mounted (NSM) strengthening technique. This technique presents high levels of efficiency for upgrading the flexural and shear carrying capacities of reinforced concrete (RC) structures. In the last decade several efforts were done to increase the knowledge in terms of bond, flexural, and shear responses of NSM systems. Some analytical formulations were proposed to predict its behaviour. Design rules were also presented. However, due to the limited number of experimental results, the existing formula have a limited application. In addition to that, there are several areas that need a lot of effort in order to appraise the response of NSM systems, such as durability and long-term behaviour.

J. Sena-Cruz (✉) · J. Barros
University of Minho, Guimaraes, Portugal
e-mail: jsena@civil.uminho.pt

V. Bianco · G. Monti
Sapienza University of Rome, Rome, Italy

A. Bilotta · E. Nigro
University of Naples Federico II, Naples, Italy

D. Bournas
The University of Nottingham, Nottingham, UK

F. Ceroni
University of Sannio, Benevento, Italy

G. Dalfré
UNILA, Foz do Iguaçu, Brazil

R. Kotynia
Lodz Technical University, Łódź, Poland

T. Triantafillou
University of Patras, Patras, Greece

C. Pellegrino and J. Sena-Cruz (eds.), *Design Procedures for the Use of Composites in Strengthening of Reinforced Concrete Structures*, RILEM State-of-the-Art Reports 19, DOI 10.1007/978-94-017-7336-2_8

knowledge on the bond behaviour is addressed focusing on the performed experimental investigations, results, and conclusions in terms of test setups, failure modes, and influence of different parameters on the bond performance. The performance of the existing local bond-slip behaviour laws, both proposed by codes and researches, is also addressed. Similar analysis is also followed for the case of flexural strengthening. The current state on shear strengthening includes two formulations for predicting the NSM shear carrying capacity. The chapter ends with a concluding section summarizing the current state and identifying the needs for future research.

Keywords FRP · Strengthening · NSM · Reinforced concrete

Introduction

In early 2000s the near surface-mounted (NSM) strengthening technique was proposed and used as an alternative system to the externally bonded reinforcement (EBR). In the NSM the fibre reinforced polymer (FRP) reinforcements are inserted into pre-cut grooves opened in the concrete cover. The FRP is then fixed to concrete with an adhesive. The NSM concept is not new. In fact, in the 1940s it started to be used in Europe for the strengthening of reinforced concrete structures. This pioneering technique consisted on placing rebar into grooves located at the concrete cover. These grooves were then filled with cement mortar (Asplund 1949).

After more than 10 years of research and applications, the NSM strengthening technique has prevailed and, when compared with the EBR, presents the following main advantages (De Lorenzis and Teng 2007):

- The surface preparation is reduced and in some cases the amount of site installation work may be reduced;
- The removal of the week concrete laitance layer is no longer needed;
- The levels of efficiency are higher since the technique is less prone to FRP debonding from the concrete substrate;
- The FRP reinforcements can be more easily anchored into adjacent members (preventing debonding failures);
- The FRP reinforcements are protected by the concrete cover and, consequently, are less exposed to mechanical damage, impact loading, fire and vandalism;
- The visual aspect of the strengthened structure is virtually unchanged.

Typically, glass FRP (GFRP) or carbon FRP (CFRP) are used as reinforcement materials. Depending on the cross-section geometry the FRP can be named by bar or strip (see Figs. 8.1 and 8.2). A FRP bar has square or round cross-section, whereas the FRP strip is characterized by a rectangular cross-section where the width is significantly higher than the thickness. Ribbed, sand-blasted, sand-coated, smooth or spirally wounded are the most common external-surfaces of the bars, while smooth surface usually characterize the FRP strips.

Figure 8.1 shows the two most commonly tested and used solutions when the NSM strengthening technique is applied. In the FRP bars, the magnitude of groove width and depth, b_g and h_g, respectively, are similar. In general, to attain this type of grooves, two saw cuts and the removal of concrete between them with a chisel is required. In the case of FRP strips, narrow grooves are open in the concrete cover yielding to distinct magnitudes for the groove width and depth. Typically only one saw cut is required. The minimum values of groove dimensions recommended by ACI document (2008) are $1.5d_b$ for both, in the case of bars, and $3t_f$ and $1.5w_f$, for thickness and width, respectively, in the case of strips (see also Fig. 8.1).

Due to the final cross-section geometry of the grooves, square and rectangular FRP reinforcements explore it better than round FRP ones, since a uniform adhesive thickness is achieved. On the other hand, in round bars the normal stresses (perpendicular to the FRP) accompanying the tangential bond stresses (parallel to the FRP) tend to split the epoxy cover while in square and rectangular bars they act mainly towards the groove lateral concrete. Between these two, rectangular bars have been shown to be more efficient since they maximize the ratio of surface to cross-sectional area, which minimizes the bond stresses associated with a given tensile force in the FRP.

In the NSM system, the adhesive is responsible for stresses transfer between the FRP reinforcement and the concrete substrate. The main properties that affect the

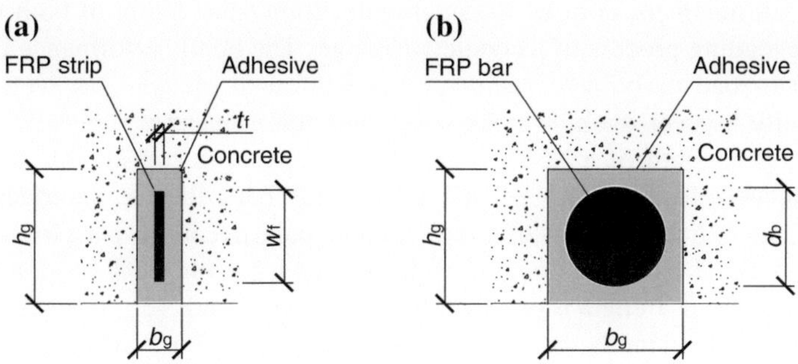

Fig. 8.1 NSM system using FRP **a** strip and **b** bar

Fig. 8.2 FRP systems

performance of the adhesive are its tensile and shear strength, modulus of elasticity and adhesion to the FRP. Up to now epoxy resins are the most frequently used adhesive; however, a few studies using cement mortar can be found.

The NSM application involves the following steps:

a. Open grooves in the concrete cover using a saw cut machine;
b. Clean the grooves with compressed air;
c. Clean the FRP with an appropriate cleaner (e.g., acetone);
d. Prepare the adhesive according to the supplier's recommendations;
e. Fill the grooves and, if possible, cover the lateral faces of the FRP with the adhesive;
f. Insert the FRP into the grooves, and slightly press it to force the adhesive to flow between the FRP and the grooves' borders. This phase requires a special care in order to assure that the grooves are completely filled with adhesive. When this is not the case the formation of voids might occur;
g. Finally, the adhesive in excess is removed and the surface is levelled;
h. The time of adhesive cure indicated by the supplier must be respected before its expected performance becomes fully available.

Bond

In the present context, bond means the transfer of stresses between the concrete and the FRP reinforcement in order to develop the composite action of both materials, during the loading process of a concrete element. The bond performance influences the ultimate load carrying capacity of the reinforced element, as well as some serviceability aspects, such as crack width and crack spacing.

The bond strength of a NSM system is the maximum transferable load and is directly related to the type of failure (at the bar-adhesive interface, at the adhesive-concrete interface, within the concrete, cohesive at the adhesive and in the FRP material). The overall bond strength is dependent on local bond strength. In general, local bond strength is obtained from specimens with very short bond lengths or with large bond lengths where the strains (and/or slip) are measured. The local bond-slip behaviour is affected by the following main parameters: materials' mechanical properties, FRP reinforcement and grooves surface treatment, geometry of the strengthening system (bars or strips), grooves' dimensions and depth of the FRP reinforcement in the groove. The shape ratio, k, namely the ratio between groove and FRP dimensions, also affects the failure mode of the strengthening system (Sena Cruz and Barros 2004; De Lorenzis and Teng 2007; Seracino et al. 2007).

Experimental bond tests on concrete elements strengthened with NSM FRP bars or strips have been performed by several researchers during the last decade. In (De Lorenzis et al. 2004), GFRP and CFRP ribbed bars with 9.5 mm diameter and spirally wounded CFRP bars with 7.5 mm diameter were tested in concrete elements with mean compressive strength of 22 MPa. Most of the specimens failed by adhesive splitting. In De Lorenzis and Nanni (2002), GFRP deformed bars (13 mm diameter) and CFRP deformed bars (9.5 mm diameter) were tested in concrete elements (27.5 MPa). All specimens failed by adhesive splitting. Moreover, sand blasted CFRP bars (9.5 and 13 mm diameter) were tested too and a FRP-adhesive interface failure occurred. In De Lorenzis et al. (2002) spirally wounded CFRP bars (7.5 mm diameter) were tested in concrete elements (22 MPa). All specimens attained an adhesive-concrete interface failure. In Sena Cruz and Barros (2004) and (Sena Cruz et al. 2006) smooth carbon strips with dimensions 10 × 1.4 mm were tested in specimens characterized by three values of concrete strength (35, 45, and 70 MPa). Failure always occurred at the FRP-adhesive interface. In Seracino et al. (2007) smooth carbon strips with thickness of 1.2 mm and width of 10, 15, and 20 mm were tested. Concrete strength varied in the range 30–60 MPa. For the specimens with concrete strength equal or higher than 50 MPa a tensile failure of the FRP occurred, except for two cases which showed a FRP-adhesive interface failure. For lower values of strength, an adhesive-concrete interface failure was observed. In Teng et al. (2006) smooth carbon strips (16 × 2 mm) were tested in concrete specimens (44 MPa) and a FRP-adhesive interface failure occurred. In Novidis and Pantazopoulou (2008) sand blasted deformed CFRP bars (12 mm diameter) were tested in concrete specimens (30 MPa) and an adhesive-concrete interface failure was observed.

Regarding the setup, both numerical and experimental studies (Novidis and Pantazopoulou 2008) showed that different test setups can significantly change the experimental results. Nevertheless, at present, no consensus on a standard test procedure has been still reached. For this reason, a Round Robin initiative was recently carried out involving several research laboratories (En-Core and fib TG 9.3 2010; Palmieri et al. 2012) and aimed at testing the bond strength of the same FRP materials according to different setups in different laboratories. In particular, the laboratories of the Universities of Naples, Ghent, Minho and Budapest tested various NSM systems. All laboratories adopted a double shear test (DST), where two concrete blocks were connected by the NSM reinforcements on two opposite sides. The only exception was the laboratory of Naples (Bilotta et al. 2011), which used a single shear test (SST) setup. Failure modes were quite different; in particular in the tests carried out by the University of Minho an adhesive-concrete interface failure occurred in all cases, unless in the CFRP bars (6 mm diameter), which failed at the adhesive-FRP interface. In the tests performed at the University of Ghent, the failure happened at the FRP-adhesive interface for the smooth carbon bars (8 mm diameter) and for the thicker carbon strips (2.5 × 15 mm). Adhesive splitting occurred for both types of sand coated basalt bars and for the ribbed glass bars. In the other cases, an adhesive-concrete interface failure occurred. In most of the tests carried out by the University of Budapest failure occurred in the concrete block and only in some cases it was due to adhesive splitting. Finally, a concrete-adhesive interface failure occurred in most specimens strengthened with NSM systems at the University of Naples. Even if the DST setup led, in some cases, to the failure of the concrete due to an incorrect alignment of the two blocks, it is worth noting that the concrete strength could affect the failure mode. Indeed concrete strength varied in the range 27–32 MPa for the tests at University of Ghent, was 43 MPa at University of Budapest and 35 MPa at University of Minho. At the University of Naples, a lower concrete strength (about 19 MPa) was used in order to replicate the conditions of existing RC buildings. For low strength, a concrete cohesive failure is more probable.

Based on the collected experimental results, the possible failure modes can be defined as:

i. Debonding at FRP bar/strip—adhesive interface
A pure interfacial mode can be critical for bars with a smooth or lightly sand-blasted surface (De Lorenzis and Nanni 2002), i.e. when the bond resistance relies primarily on the adhesion between the bar and the adhesive. This type of failure is identified by the virtual absence of adhesive attached to the bar surface after failure. For round bars, longitudinal cracking of the adhesive cover produced by the radial components of the bond stresses may accelerate the occurrence of an interfacial failure.

ii. Cohesive shear failure within the adhesive
The cohesive shear failure of the adhesive was observed for strips with a roughened surface (Blaschko 2003). Such a failure is identified by the presence of adhesive on both strip and concrete after failure and occurs when the tensile strength of the adhesive is exceeded. Since surrounding concrete is much stiffer than the adhesives, it introduces some confinement at the concrete-adhesive interface. Therefore, the Mohr-Coulomb principles contribute to avoid a pure sliding failure mode at the concrete-adhesive interface and induce the failure within the adhesive.

iii. Debonding at adhesive—concrete interface
Bond failure at the adhesive-concrete interface may occur as pure interfacial failure or as cohesive shear failure in the concrete. The pure interfacial failure mode was found to be critical for pre-cast grooves (De Lorenzis et al. 2002) and, in general, for grooves with smooth surfaces. Indeed, in this case the cohesion/adhesion phenomena are very small, as well as the internal friction angle of the materials in contact, therefore the shear strength at the interface due to the Mohr-Coulomb effect is small, being the weakest link of the all system. Cohesive shear failure within the concrete is the most common failure mode since concrete is the weakest material. Indeed, the surrounding concrete at the loaded end is subjected to tensile stresses that exceeds the tensile strength of the concrete, as long as the bond length is long enough (Ceroni et al. 2012). It has been observed also in bond tests on concrete specimens strengthened with smooth NSM strips (Seracino et al. 2007). In general, the failure at the adhesive-concrete interface was experimentally observed for values of the shape factor $k > 1.5$–2 (De Lorenzis and Teng 2007); these limits, indeed, warrant a sufficient adhesive thickness around the FRP bar/strip in order to avoid the adhesive cover splitting.

iv. Adhesive cover splitting
Longitudinal cracking of the adhesive is generally identified as cover splitting. This was observed to be the critical failure mode for deformed (i.e. ribbed and spirally wound) round bars in moderate strength concrete (De Lorenzis and Nanni 2002; De

Lorenzis et al. 2004). Specimens with low values of groove size (k ≈ 1) and very brittle adhesive can show its splitting without significant damage in the surrounding concrete. For increasing groove depth and adhesive thickness, the resistance of the adhesive cover to splitting increases and failure is controlled by cracking of the surrounding concrete: in these cases the bond strength can also increase with the groove size until the failure load corresponding to the adhesive-concrete interface failure is attained. Moreover, as deeper the FRP is installed into the groove as larger confinement the surrounding concrete introduces to the FRP, resulting in beneficial effects in terms of bond strength (Costa and Barros 2011). A minimum value of k = 1.5 is suggested to avoid splitting (De Lorenzis and Nanni 2002). Splitting failure is also typical of specimens strengthened with cement filled grooves for low values of the k factor, due to the lower tensile strength of the cement fillers.

v. Concrete splitting

When an NSM bar/strip is close to the edge of a concrete member, failure can involve the splitting of the edge concrete (Blaschko 2001; Galati and De Lorenzis 2009). This kind of failure mode can be easily eliminated by keeping a minimum distance from the edge. Moreover this kind of failure can be quite common mainly in elements of relatively low strength class.

Evaluation of the Existing NSM ACI Formulation

The ACI 440.2R-08 (2008) guide includes a simple formulation for predicting the maximum pullout strength. In this formulation the key parameter is the maximum bond strength for the entire NSM system. If the bonded length (L_b) is at least equal to a development length (L_d), then the experimental bilinear shear stress versus slip law can be assumed to be constant along the bond length and equal to the average bond strength (τ_{avg}). ACI assumes τ_{avg} equal to 6.9 MPa for all FRP NSM systems. Hence, by imposing this limit τ_{avg} to the connection's maximum capacity, L_d and the maximum pullout force installed in the FRP ($F_{f,max}$) can be estimated. Then, if $L_b \geq L_d$ the failure will occur by FRP rupture. Otherwise it will occur by one of the five failure modes described in the previous section.

It is interesting to notice that only four parameters need to be known *à priori* in order to predict $F_{f,max}$ using ACI formulation, namely: FRP perimeter (p_f), cross-section area (A_f), tensile strength (f_{fu}), and bonded length (L_b).

$$L_d = \frac{A_f f_{fu}}{p_f \tau_{avg}} \tag{8.1}$$

$$F_{f,\max} = \begin{cases} A_f f_{fu} & \text{if } L_b \geq L_d \\ A_f f_{fu} \frac{L_b}{L_d} & \text{if } L_b < L_d \end{cases} \tag{8.2}$$

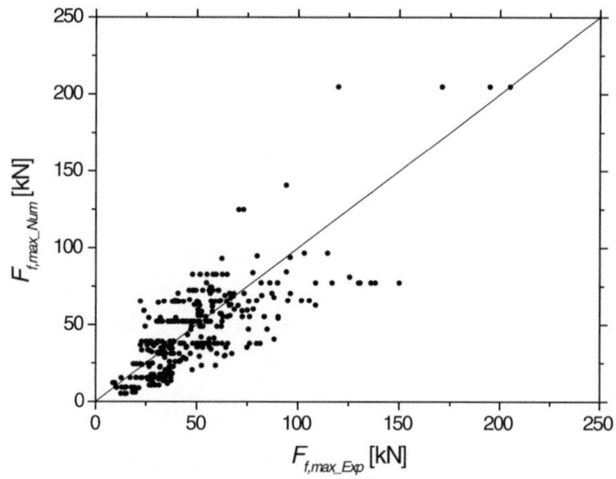

Fig. 8.3 Comparison between the experimental (F_{f,max_Exp}) and ACI numerical (F_{f,max_Num}) maximum pullout force prediction

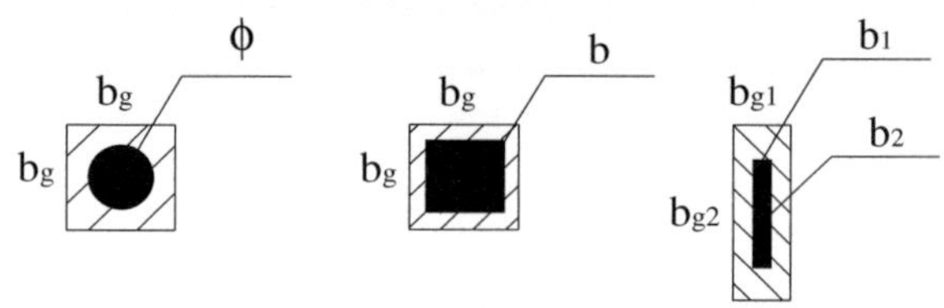

$k=b_g/2\phi$ $k=b_g/2b$ $k=b_{g2}/2b_2$

$p_{f,NSM}=\pi\phi$ $p_{f,NSM}=4b$ $p_{f,NSM}=2b_2+b_1$

$p_{f,g}=3b_g$ $p_{f,g}=3b_g$ $p_{f,g}=2b_{g2}+b_{g1}$

Fig. 8.4 Definition of geometrical parameters of the grooves

An ongoing analytical work is being performed in order to access the accuracy of ACI formulation (Coelho et al. 2013, 2014). Preliminary results from a database with 363 direct pullout tests using different setups and materials are shown in Fig. 8.3.

As can be seen, a large scatter exists which is also verified by computing the error measures associated to the ratio between numerical and experimental values of the maximum pullout force. A standard variation of 0.4 was found, corresponding to a coefficient of variation of almost 42 %. This reveals that further improvements are needed for the ACI NSM formulation.

Predicting Formulae for NSM

A reference database of 167 results from bond tests available in the literature (93 tests on round bars, 12 on square bars, and 62 on rectangular strips) were collected. Figure 8.4 and Table 8.1 show the ranges of the main geometrical and mechanical parameters of the FRP NSM systems used in the tests. Note that the groove dimensions are compatible with the usual concrete cover values.

Table 8.1 Ranges of geometrical and mechanical parameters of the experimental database

Geometry	FRP	Groove
93 round bars (mm)	Ø = 6÷10	b_g = 10–25
12 square bars (mm)	b = 10	b_g = 15
62 strips (mm)	b_1 = 1.2–2.5 b_2 = 10–20	b_{g1} = 3.3–8; b_{g2} = 12–25
Concrete	Strength f_{cm} (MPa)	19–70
FRP	Youngs' modulus (GPa)	37–182

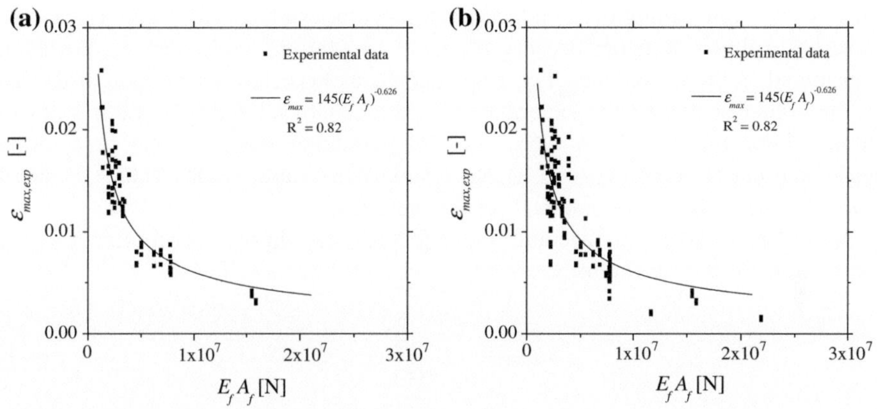

Fig. 8.5 Experimental maximum strain for bond failure versus $E_f \cdot A_f$: **a** experimental results in the case of A/C failure; **b** extended database for different types of bond failure modes

An approach to have a design formulation for the maximum strain at failure in the FRP reinforcement consists in extrapolating a trend based on experimental measures of strain at failure. Thus, the maximum experimental strain calculated dividing the failure load to the reinforcement axial stiffness, $E_f \cdot A_f$, is plotted in Fig. 8.5a versus the axial stiffness itself for the experimental results of the bond tests of (De Lorenzis et al. 2002, 2004; Seracino et al. 2007; En-Core and fib TG 9.3 2010; Bilotta et al. 2011, 2012), where an adhesive-concrete interface failure (A/C) occurred (totally 78 data). Figure 8.5a shows that the maximum strain decreases as the axial stiffness increases, as well as usually observed for the EBR systems (Triantafillou and Antonopoulos 2000). The following law is able to fit the experimental results ($R^2 = 0.82$):

$$\varepsilon_{\max,\mathrm{th}} = a \cdot \frac{1}{\left(E_f \cdot A_f\right)^b} \tag{8.3}$$

where $a = 145$ and $b = 0.625$ clearly depend on the database used for the regression.

Indeed, in Fig. 8.5b further experimental data (De Lorenzis and Nanni 2002; Sena Cruz and Barros 2004; Sena Cruz et al. 2006; Teng et al. 2006) were added to

Table 8.2 Regression coefficients of Eqs. (8.3) and (8.4) for different sets of experimental results

Eq.	Database	Failure mode	a	b	c	R^2
(8.3)	78 data	A/C	145	0.625	–	0.82
	137 data	A/C, A/F, S	112	0.61	–	0.66
(8.4)	78 data	A/C	145	0.74	0.45	0.85
	137 data	A/C, A/F, S	272	0.85	0.71	0.76

take into account also the results of specimens where adhesive-FRP interface failure (A/F) or splitting failure of the adhesive (S) occurred (totally 137 data). These data have been plotted together because the latter two failure modes can be, however, considered as 'bond' failures. The graph shows, as expected, a reduction of R^2 due to the higher scatter of the experimental results characterized by several types of bond failure modes ($R^2 \approx 0.66$). For the extended database, the best fitting regression coefficients are $a = 112$ and $b = 0.61$. In Table 8.2 the values of a and b for the two sets of experimental results are listed.

In order to take into account the effect of the groove perimeter, $p_{f,g}$, the expression for the maximum strain has been modified as follows:

$$\varepsilon_{\max,\text{th}} = a \cdot \frac{\left(p_{f,g}\right)^{c}}{\left(E_f \cdot A_f\right)^{b}} \tag{8.4}$$

The regression coefficients a, b and c have been evaluated according to a least-square best fitting criterion for the same sets of data used for assessing the coefficients of Eq. (8.3). In Table 8.2 the values of these coefficients are listed together with the corresponding value of R^2.

When only the A/C interface failure modes are considered, a slight improvement of the experimental-theoretical fitting is achieved as evidenced by the R^2 values (0.82 for Eq. (8.3) and 0.85 for Eq. (8.4)). By contrast, when the regression is extended to the whole database (137 data) considering different failure modes, a sensible improvement is obtained using Eq. (8.4): $R^2 = 0.76$ for Eq. (8.4) versus $R^2 = 0.66$ for Eq. (8.3). In Fig. 8.6 the experimental strain at failure is compared with the provisions given by Eq. (8.4) where the best fitting values of coefficients calibrated on the extended database are used (137 data for different types of bond failure modes).

Another parameter that might influence the maximum strain is the concrete compressive strength. This effect has been investigated too for the available data, but no clear influence was evidenced for now. Therefore, using Eq. (8.4) is preferable with respect to Eq. (8.3) due to the higher R^2 of the regression on all sets of experimental data considered.

In Fig. 8.7 are shown the predictions given by Eq. (8.4) when coefficients calibrated on 78 and 137 data are used. To make the comparison, a value of the perimeter $p_{f,g} = 40$ mm has been fixed, by taking into account that it generally varies between 20–60 mm, with an upper bound due to the ordinary values of

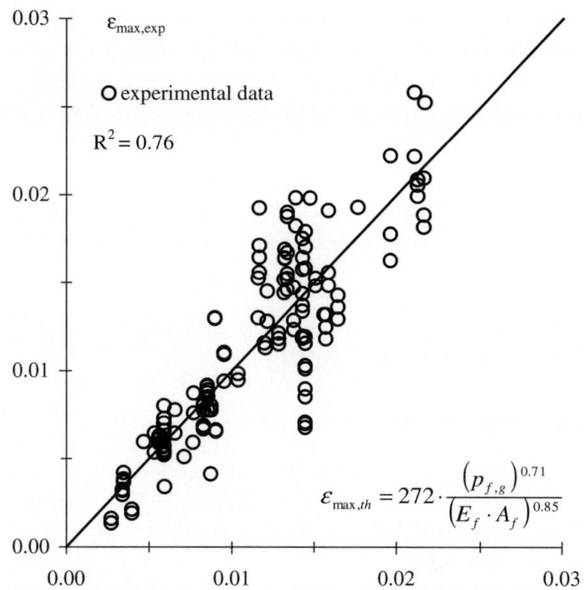

Fig. 8.6 Experimental maximum strain versus theoretical values given by Eq. (8.4) calibrated on the extended database (137 data)

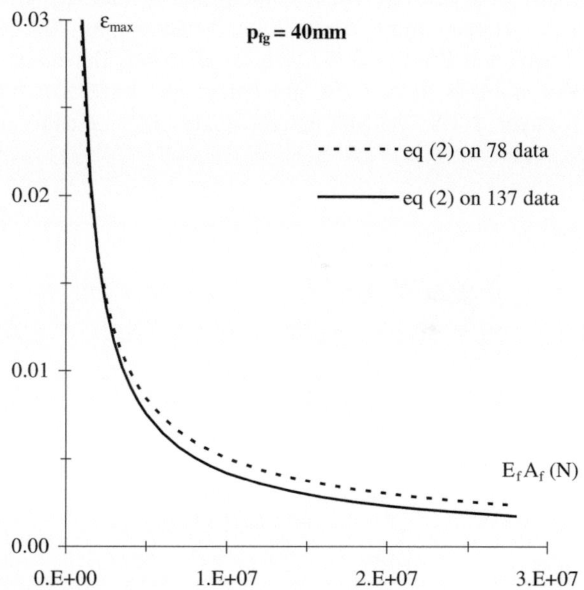

Fig. 8.7 Theoretical prediction by Eq. (8.4) for $p_{f,g}$ = 40 mm

concrete cover (20–30 mm). It is clear that the Eq. (8.4) with coefficient calibrated on 78 data provides predictions higher than those obtained on 137 data (i.e. +10–20 %). Similar results are obtained for different perimeters: as lower the value of $p_{f,g}$ as larger the differences between the predictions. However, the less conservative formula can be used only if a failure at the interface A/C is expected. Hence, aimed at giving design provisions, using the most conservative formula is preferable, because not sure indications are available to avoid failure modes in the adhesive (splitting or debonding at FRP-adhesive interface). Note that in the experimental database the parameter $p_{f,g}/p_{f,NSM}$ varies between 1.2 and 2.5. So the formula given by Eq. (8.4) should be used for values of $p_{f,g}/p_{f,NSM}$ ranging in this interval.

To finalize, it should be stressed that Eqs. (8.3) and (8.4) give average values of the maximum strain in the FRP reinforcement, whereas a characteristic value has to be estimated to give design provisions.

Flexural Strengthening

Beams and Slabs

When the NSM is used for flexural strengthening, the most common FRP reinforcement used consists of bars (circular and square cross-section) or rectangular strips (Fig. 8.8).

The existing experimental tests on RC members strengthened in flexure with NSM FRP reinforcement contain RC beams (De Lorenzis et al. 2000; Blaschko 2001; Hassan and Rizkalla 2003; Täljsten et al. 2003; EI-Hacha and Rizkalla 2004; Barros and Fortes 2005; Teng et al. 2006; Barros et al. 2007; Castro et al. 2007; Kotynia 2007a; Novidis and Pantazopoulou 2007; Yost et al. 2007; Burke 2008; Kalayci 2008; Al-Mahmoud et al. 2009; Costa and Barros 2010) and slabs

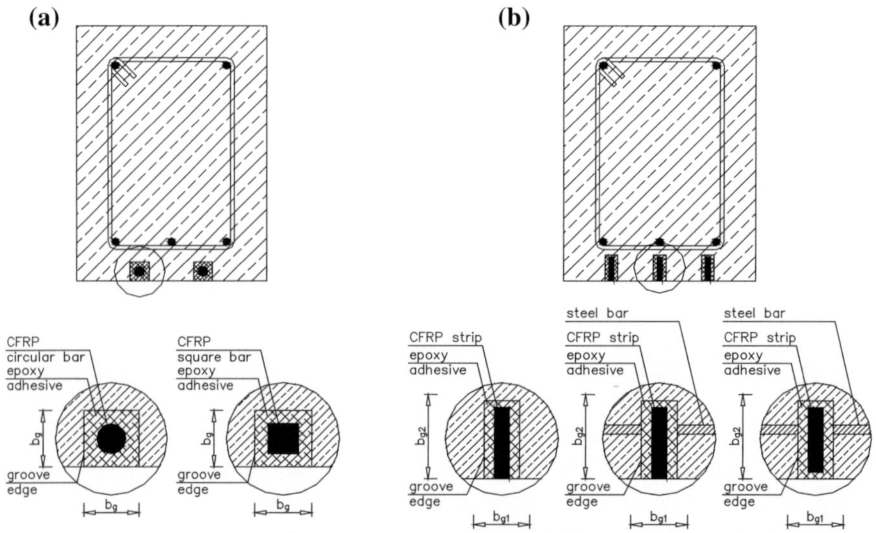

Fig. 8.8 Flexural strengthening of RC members with: **a** NSM bars and **b** strips

(Asplund 1949; Liu et al. 2006; Bonaldo et al. 2008; Dalfré and Barros 2011). The test database collected for analysis refers almost 50 RC members strengthened with NSM FRP. More than 80 % of them contain beams and only less than 20 % slabs.

The most popular FRP reinforcement used for NSM strengthening is made of CFRP (with more than 70 % for strips and 20 % for bars), there are only 8 % of members strengthened with GFRP bars. The tested specimens differ in geometry, load and static schemes, internal steel reinforcement ratio, NSM FRP percentage, FRP shape, its cross-section, concrete and adhesive strengths, concrete cover thickness and size of RC members.

Based on published research in the field of flexural strengthening with NSM FRP bars/strips, the most interesting observations are described and discussed in the following paragraphs.

De Lorenzis et al. (2000) conducted one of the first tests on RC T-section beams strengthened with CFRP and GFRP bars. Test results indicated increase in the ultimate load of the strengthened specimens in comparison with the reference ones. Also, they showed that the efficacy of the NSM technique depends on the bond length of the NSM reinforcement.

Täljsten et al. (2003) tested rectangular beams in four point bending test configuration. Two different dimensions of square grooves with 10 mm for cement grout and 15 mm for epoxy grout were used. Rupture of the rods occurred in the beam with the epoxy adhesive while FRP-adhesive slipping occurred in the beam with the cement grout.

Hassan and Rizkalla (2003) carried out bond tests on nine T-section RC beams strengthened with NSM CFRP strips with variable embedment lengths. The maximum strain of the CFRP bars ranged from 0.7 to 0.8 % for embedment lengths below 800 mm. Results showed an increase in the CFRP strain during its debonding with the increase in the embedment length. Failure of the beams with ribbed NSM FRP round bars occurred by splitting of the adhesive which occurred due to CFRP bond length below 800 mm. Whereas, in the case of beams strengthened with NSM strips, rupture of the strips occurred when the embedment length was larger than 850 mm.

El-Hacha and Rizkalla (2004) tested T-section beams strengthened with CFRP strips or bars and thermoplastic GFRP strips in three point bending test. The use of NSM FRP reinforcements enhanced the flexural stiffness and significantly increased the ultimate load-carrying capacity of strengthened specimens. FRP-adhesive splitting was the dominant failure mode for the beams strengthened with NSM CFRP bars as a result of the high tensile stresses at that interface.

Barros and Fortes (2005) and Barros et al. (2007) investigated RC beams strengthened in flexure with variable number of NSM CFRP strips and different steel reinforcement ratios. Test results indicated an almost double increase in the load carrying capacity. Significant increases in the load at steel yielding and concrete cracking points for the strengthened beams, proved the higher efficiency of the NSM technique in comparison with EBR one.

Teng et al. (2006) investigated the influence of the embedment length of the strip. Test results of the beams strengthened with the shortest embedment length of

500 mm confirmed no effect of the strengthening on the ultimate load and on the beam's stiffness. The medium embedment length beams, ranging from 1200 to 1800 mm, indicated increases in the load bearing capacity. Those beams failed by concrete separation starting from the cut-off region towards the maximum moment region. Finally, the longest embedment length showed the propagation of debonding from the maximum moment region towards the cut-off section.

The results of 12 T-section RC beams by Castro et al. (2007) indicated failure due to intermediate crack debonding in the beams strengthened with CFRP strips and GFRP bars. Beams strengthened with CFRP bars failed by the bar-adhesive slipping.

Novidis and Pantazopoulou (2007) confirmed very promising results of the NSM technique in comparison to the EBR. The results indicated that the depth at which the FRP is bonded into the longitudinal grooves influences the strengthening gain.

Kotynia (2007a) tested three series of RC beams strengthened with NSM CFRP strips. The influence of the following parameters on the strengthening efficacy was investigated: CFRP depth, concrete cover thickness, longitudinal tensile steel, CFRP percentages, and concrete strength. Cutting of the steel stirrups in the tensile zone of the beam during the strips application did not influence the ultimate load capacity.

Based on 12 specimens, Kalayci (2008) investigated the influence of the groove size on the strengthening gain. The beams where tested with one type of strip/bar bonded into three different groove sizes. The ultimate loads reached for undersized groove specimens strengthened with CFRP strips was similar to the loads in the control specimens, even though the mid-span deflections were lower. Undersized and control groove beams had identical modes of failure: concrete and adhesive splitting. However, in the oversized specimens only concrete splitting occurred. Beams strengthened with CFRP bars reached similar deflections and ultimate loads in the control and undersized grooves specimens but, in the undersized specimens, adhesive splitting failure was observed. One of the oversized specimens failed by adhesive splitting while the other one by concrete splitting. Tests showed that smaller grooves led to adhesive splitting failures and bigger ones led to concrete splitting failures.

All the tests mentioned showed a significant effect of the FRP and steel reinforcement ratios likewise CFRP elasticity modulus on the ultimate loads and the CFRP strain utilization. The increase in the CFRP stiffness led to an ultimate load increase. However, it causes a decrease in the CFRP debonding strain.

Failure Modes

Fundamental division of the test specimens refers to the failure mode. The most expected failure mode is the intermediate crack (IC) debonding with the adjacent concrete cover separation and FRP rupture, which clearly show the full FRP strain utilisation. While the adhesive splitting (AS) or concrete crushing (CC) have been confirmed as premature failure modes, which do not indicate attractive strengthening efficiency. The AS failure with the FRP debonding from the adhesive indicates a low strength of the adhesive and strongly depends on the adhesive tensile strength (which, for this failure mode, is lower than the concrete strength) and the groove size (for higher groove widths adhesive splitting is observed more often).

The following failure modes appeared in the experimental tests of the RC members strengthened in flexure with NSM FRP reinforcement and are considered in the analysis of the strengthening efficiency:

Interfacial debonding

Interfacial debonding or adhesive cover splitting at the FRP-adhesive interface near the anchorage zone observed in the RC members NSM strengthened in flexure referred to similar failure modes observed in the bond tests.

Concrete cover separation

Concrete cover separation is more common for RC members strengthened with lower distances between the several grooves of the strengthening system since this can led to a undesirable group effect. This is also more frequent for decreasing tensile strengths of the concrete cover.

In many tests (Asplund 1949; De Lorenzis et al. 2000; Barros and Fortes 2005; Barros et al. 2007; Kotynia 2007a; Bonaldo et al. 2008) bond cracks inclined at approximately 45° to the beam axis formed on the soffit of the beam. Upon reaching the edges of the beam's soffit, these cracks may propagate upwards on the beam sides maintaining a 45° inclination within the cover thickness. Then, they can propagate horizontally at the level of the steel tension bars.

(a) Bar/strip end cover separation

Concrete cover separation is typical for the extremities of NSM FRP reinforcement at a significant distance from the supports. This failure starts from the cut-off section and propagates to the midspan of the RC member (Teng et al. 2006; Barros et al. 2007; Al-Mahmoud et al. 2009; Costa and Barros 2010) (Fig. 8.9).

(b) Localized cover separation

Bond cracks within or close to the maximum moment region together with pre-existing flexural and/or flexural-shear cracks may isolate triangular or

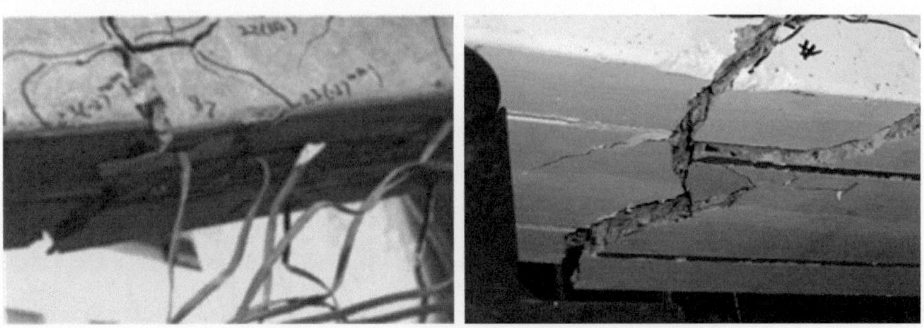

Fig. 8.9 Failure by strip end cover separation (Teng et al. 2006; Costa and Barros 2010)

trapezoidal concrete wedges. From those, one or more will eventually split off (Barros et al. 2007; Costa and Barros 2010).

(c) Flexural crack-induced cover separation

This failure mode is similar to the intermediate crack debonding in reinforced concrete members externally bonded with FRP materials. Concrete cover separation is followed by flexural concrete cracking propagating along the NSM reinforcement, involving one of the shear spans and the maximum bending moment region (De Lorenzis et al. 2000; Barros and Fortes 2005; Barros et al. 2007; Kotynia 2007a; Bonaldo et al. 2008; Costa and Barros 2010) (Fig. 8.10).

(d) Flexural-shear crack-induced cover separation

Similar to the EBR technique, when diagonal shear crack intersects the FRP, debonding initiates due to shear and normal interfacial stresses on the side of the crack and propagates towards the FRP reinforcement end. The failure generates in the concrete adjacent to the adhesive-concrete interface and promotes the concrete cover separation (Fig. 8.11a) (Costa and Barros 2010; Dalfré and Barros 2011).

When using NSM with the high depth strips, a longitudinal fracture along the strip can be formed due to the relatively high moment of inertia, which leads to the fracture along the FRP strip (Fig. 8.11b) (Costa and Barros 2010).

(e) Beam edge cover separation

Failure mode typical when the FRP NSM bar is located near the beam's edge. Detachment of the concrete cover appears along this edge (Fig. 8.12).

Fig. 8.10 Failure by intermediate crack debonding with adjacent cover concrete (Kotynia 2007a)

(a) (b)

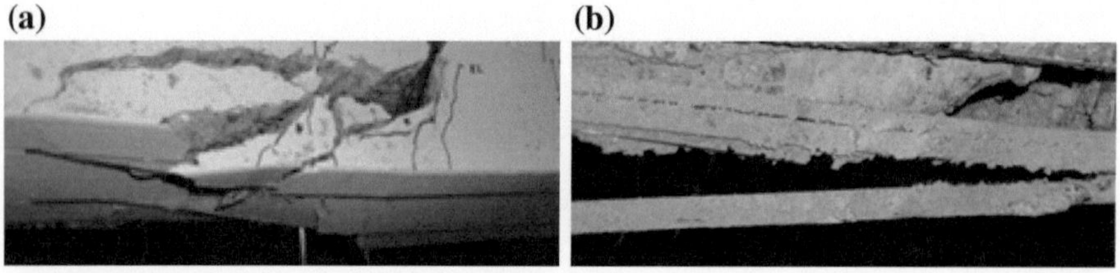

Fig. 8.11 Failure by concrete cover separation: **a** Followed by flexural shear failure crack propagation; **b** Fracture along the NSM strip (Costa and Barros 2010)

Fig. 8.12 Failure by the beam edge concrete cover separation (De Lorenzis and Teng 2007)

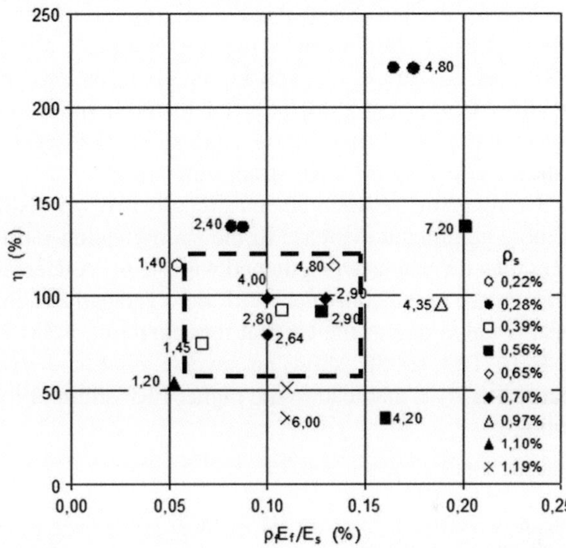

Fig. 8.13 Effect of the equivalent FRP percentage on the strengthening ratio (values of parameter A_f/b_f describe test result) (Cholostiakow et al. 2013)

Influence of Different Parameters on the Flexural Strengthening Performance

To unify the test results and to preserve the highest NSM efficiency in terms of the FRP tensile strain utilization and the gain of the ultimate load, only the specimens which failed due to concrete cover separation (CCS) were taken into analysis. The first parameter taken into consideration was the equivalent FRP reinforcement ratio expressed by $\rho_{f,eq} = \rho_f\, E_f/E_s$. The influence of the equivalent FRP ratio on the strengthening ratio (η_f) is shown on Fig. 8.13. The collected test data was divided into several series with longitudinal steel reinforcement ratios, ρ_s, ranging from 0.22 to 1.19 %. Moreover, the FRP cross section to its depth/width ratio, expressed by parameter A_f/b_f, identifies each test result (with values written next to the test results on Fig. 8.13). This figure indicates that specimens with similar steel ratio show increase in the strengthening ratio ($\eta_{f,}$) with increase in the strengthening ratio

($\rho_{f,eq}$). It should be noted that the strengthening efficiency increases with a decrease in the internal steel ratio (Cholostiakow et al. 2013).

The highest strengthening ratio of 221 % was obtained for the steel ratio $\rho_s = 0.28$ %, whereas the lowest strengthening ratio of 35 % was obtained for steel ratios of 1.19 and 0.56 %. It seems that the most effective NSM FRP strengthening was obtained in cases of RC members reinforced with steel ratios in the range of 0.38–0.71 % which are strengthened with the equivalent FRP reinforcement ratios in the range 0.07–0.15 %.

The data corresponding to the most effective NSM FRP strengthening cases are presented in the area confined with a dashed line (Fig. 8.13). For this region of effective strengthening combinations it can be said that a minimum of 2.6 mm for the ratio A_f/b_f is necessary.

The results indicate that with an increase of this parameter, the strengthening ratio increases. Moreover, A_f/b_f parameter has a crucial influence on the maximum FRP maximal strain, $\varepsilon_{f,test}$, during failure of the RC specimens by CCS. Test results confirm that $\varepsilon_{f,test}$ is not affected by both the concrete strength and steel reinforcement ratio. Whereas the equivalent FRP reinforcement ratio, $\rho_{f,eq}$, has significant effect on the FRP strain utilization.

Comparison between beams strengthened with NSM FRP strips and bars shows a more significant decrease in the strengthening ratio for beams strengthened with bars than for beams strengthened with strips. A decrease in the FRP strain $\varepsilon_{f,test}$ with an increase in the equivalent FRP ratio is clearly visible in Fig. 8.14. Most of tested beams failed due to the CCS at the maximum NSM FRP strains in a range of 1.2–1.4 %. Test results show an inverse relation between the FRP strain and the parameter A_f/b_f that lead to the higher FRP strain utilization for lower values of that ratio.

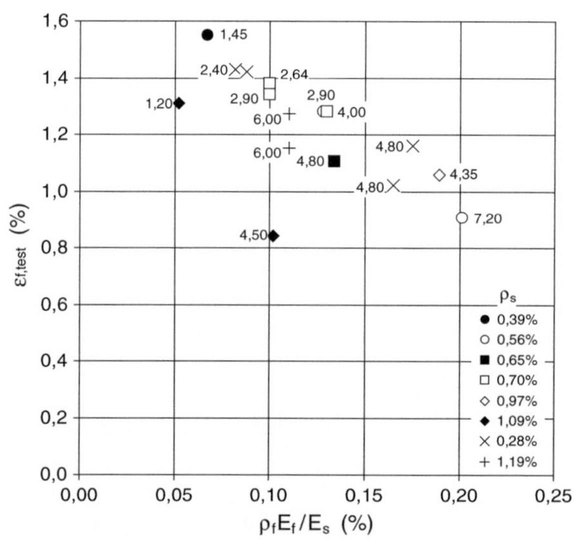

Fig. 8.14 Effect of equivalent FRP ratio on maximal strain in CFRP strips with values of a pramater A_f/b_f next to test results (Cholostiakow et al. 2013)

Basically, increase in the FRP stiffness causes decrease in the maximum FRP strain $\varepsilon_{f,test}$. Moreover this observation corresponds to a tendency of decreasing strain efficiency with increasing the equivalent FRP reinforcement ratio $\rho_f E_f / E_s$.

Also the shape of the FRP cross-section (rectangular, square, or circle) and its position (horizontal or vertical) strongly affect the FRP strain efficiency $\varepsilon_{f,test}$ registered in the tests (Cholostiakow et al. 2013).

Flexural Strengthening According to ACI 440.2R-08 versus Experimental Results

The ACI 440.2R-08 design guideline presents guidance on the calculation of the flexural strengthening effect of adding longitudinal NSM FRP reinforcement to the tension face of a reinforced concrete member. A specific illustration of the concepts applied in this section to the strengthening of existing rectangular or T-section RC members in the tension zone with non-prestressed steel is given on Fig. 8.15.

The following assumptions are made when calculating the flexural resistance of a section strengthened with an externally applied FRP system:

- Design calculations are based on the dimensions, internal reinforcing steel arrangement, and material properties of the existing member being strengthened;
- The strains in the steel reinforcement and concrete are directly proportional to the distance from the neutral axis (plane section principle);
- There is no relative slip between external FRP reinforcement and the concrete;
- The shear deformation within the adhesive layer is neglected because the adhesive layer is very thin with slight variations in its thickness;
- The maximum usable compressive strain in the concrete is 0.003;
- The tensile strength of concrete is neglected and the FRP reinforcement has a linear elastic stress-strain relationship until failure.

Unless all loads on a member, including self-weight and any prestressing forces, are removed before installation of FRP reinforcement, the substrate to which the FRP is applied will be strained. These strains should be considered as initial strains and should be excluded from the FRP strain.

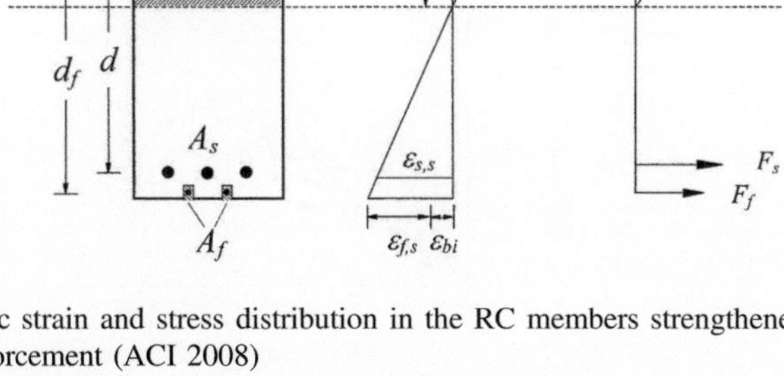

Fig. 8.15 Elastic strain and stress distribution in the RC members strengthened in flexure with NSM FRP reinforcement (ACI 2008)

The initial strain level on the bonded substrate, ε_{bi}, can be determined from an elastic analysis of the existing member, considering all loads that will be on the member during the installation of the FRP system. The elastic analysis of the existing member should be based on cracked section properties.

The maximum strain level that can be achieved in the FRP reinforcement will be governed by either the strain level developed in the FRP at the point at which concrete crushes, the point at which the FRP ruptures, or the point at which the FRP debonds from the substrate. The effective strain level in the FRP reinforcement at the ultimate limit state ε_{fe} can not exceed the effective design strain for FRP reinforcement at the ultimate limit state ε_{fd}.

For NSM FRP applications, the value of ε_{fd} may vary from 0.6 to 0.9 ε_{fu} depending on many factors such as member dimensions, steel and FRP reinforcement ratios and surface roughness of the FRP bar. Based on existing studies the committee recommends the use of $\varepsilon_{fd} = 0.7\ \varepsilon_{fu}$. To achieve the debonding design strain of NSM FRP bars ε_{fd}, the bonded length should be greater than the development length.

The design strategy explained above was applied to a database of RC beams under the following assumptions: ACI unit safety coefficients, mean material properties, steel in compression accounted, and parabola-rectangle concrete diagram with ε_{cu} = 3.0 ‰ (Dias et al.) (Fig. 8.16).

Continuous RC Slabs

The Externally Bonded Reinforcement (EBR) and the Near Surface Mounted (NSM) are the most used FRP-based techniques for the strengthening of RC elements. The efficiency of the NSM technique for the flexural and shear strengthening

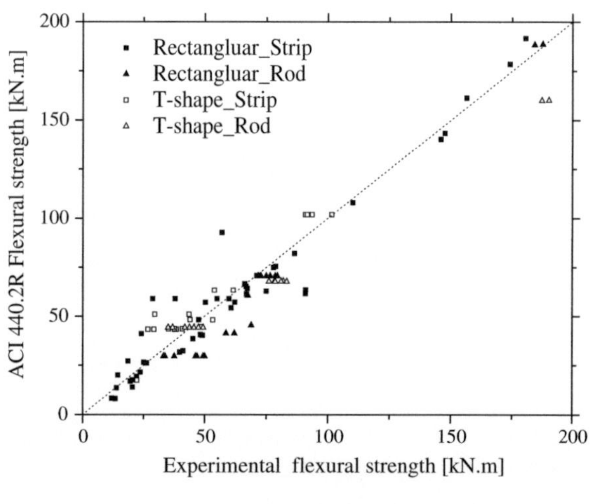

Fig. 8.16 Comparison between experimental and ACI prediction of the flexural strength (Dias et al.)

of RC members has already been assessed. However, most of the tests were carried out with NSM strengthened simply supported elements. Although many in situ RC strengthened elements are of continuous construction nature, there is a lack of experimental and theoretical studies in the behaviour of statically indeterminate RC members strengthened with FRP materials. The majority of research studies dedicated to the analysis of the behaviour of continuous elements reports the use of the EBR technique. Limited information is available in the literature dealing with the behaviour of continuous structures strengthened according to the NSM technique. Thus, to contribute for a better understanding of the influence of the strengthening arrangement (hogging, sagging, or both regions) and percentage of FRP in terms of load carrying capacity, moment redistribution capacity, and ductility performance, an experimental program formed by continuous slab strips strengthened in flexure with near surface mounted (NSM) Carbon Fiber Reinforced Polymer (CFRP) laminates was carried out at the University of Minho (Bonaldo 2008; Dalfré 2013). The experimental program was composed of seventeen $120 \times 375 \times 5875\ mm^3$ RC slab strips strengthened with NSM CFRP laminates, grouped in two series that are different in terms of strengthening configuration: H series, where H is the notation to identify the slabs strengthened with NSM CFRP laminates exclusively applied in the hogging region; HS series, where HS is the notation to identify the slabs strengthened with NSM CFRP laminates applied in both hogging and sagging regions. Figure 8.17 presents the test configuration adopted in the experimental program. The amount and disposition of the steel bars were designed to assure moment redistribution percentages (η) of 15, 30 and 45 %. The NSM CFRP systems applied in the flexural strengthened RC slabs were designed to increase in 25 and 50 % the load carrying capacity of the reference slab. From the obtained results, it was verified that the strengthening configurations composed by laminates only applied in the hogging region did not attain the target increase of the load carrying capacity. When applying CFRP laminates in both sagging and hogging regions (HS series), the target increase of the load carrying capacity was attained. Therefore, to increase significantly the load carrying capacity of the RC slabs, the sagging zones need also to be strengthened. A moment redistribution percentage lower than the predicted one was determined in the slabs strengthened with CFRP laminates in the hogging region (H). For this strengthening configuration, η has decreased with the increase of the CFRP percentage. However, adopting a flexural strengthening strategy composed of CFRP laminates applied in both hogging and sagging regions, the target values for the moment redistribution capacity was attained and the influence of the percentage of CFRP on η was marginal.

Simulation of RC slab strips strengthened with NSM CFRP laminates

Numerical analyses were carried out to simulate the load-deflection relationship of concrete elements reinforced with conventional steel bars and strengthened by NSM CFRP laminate strips. For assessing the predictive performance of a FEM-based computer program, the experimental tests were simulated by considering the nonlinear relevant aspects of the intervening materials. In general, the numerical simulations have reproduced with high accuracy the behaviour of the carried out tests. Later, a parametric study composed of 144 numerical simulations

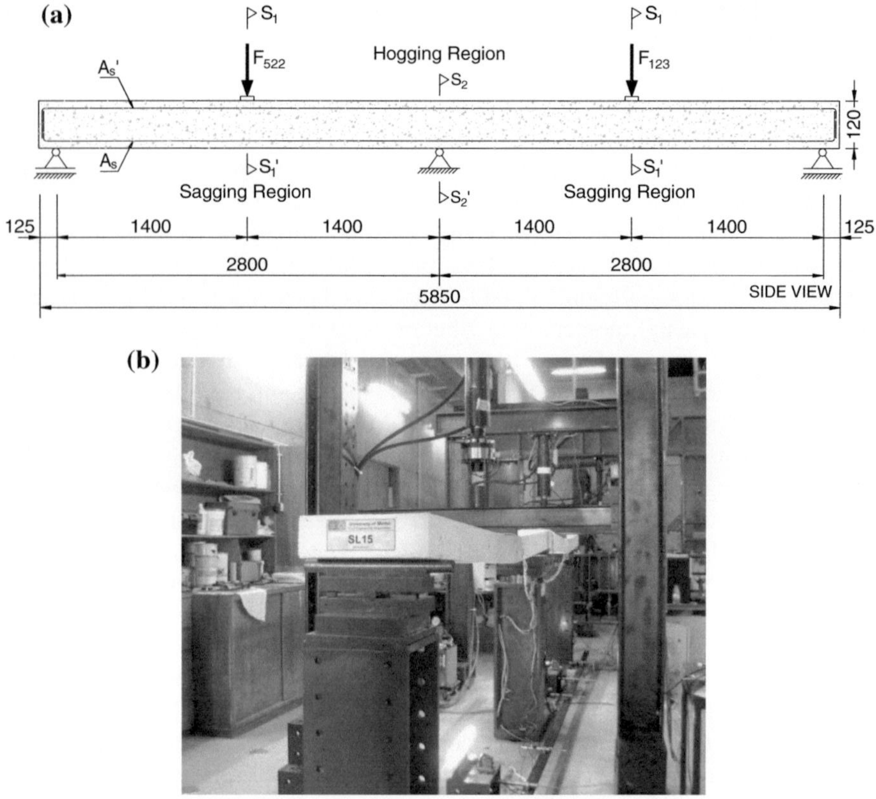

Fig. 8.17 Test configuration for continuous RC slabs tests: **a** Scheme; **b** foto. Note: specimens dimensions are in mm

was carried out to investigate the influence of the strengthening arrangement and CFRP percentage in terms of load carrying capacity and moment redistribution capacity of continuous RC slab strips flexural strengthened by the NSM technique.

According to the results, the load carrying and the moment redistribution capacities strongly depend on the flexural strengthening arrangement. The load carrying capacity of the strengthened slabs increases with the equivalent reinforcement ratio $\left[\rho_{s,eq} = A_s/(bd_s) + (A_f E_f/E_s)/(bd_f)\right]$ applied in the sagging and hogging regions ($\rho^S_{s,eq}$ and $\rho^H_{s,eq}$, respectively), but the increase is much more pronounced with $\rho^S_{s,eq}$, specially up to the formation of the plastic hinge in the hogging region (Fig. 8.18).

The moment redistribution (*MRI)* is defined as the ratio between the η of a strengthened slab, η_{streng}, and the η of its reference slab, η_{ref}, $\left(MRI = \eta_{streng}/\eta_{ref}\right)$, where η is the moment redistribution percentage at the formation of the second hinge (in the sagging region). According to the results, the moment redistribution

Fig. 8.18 Relationship between the moment redistribution index and $\rho^S_{s,eq}/\rho^H_{s,eq}$ for series: **a** SL15, **b** SL30, and **c** SL45

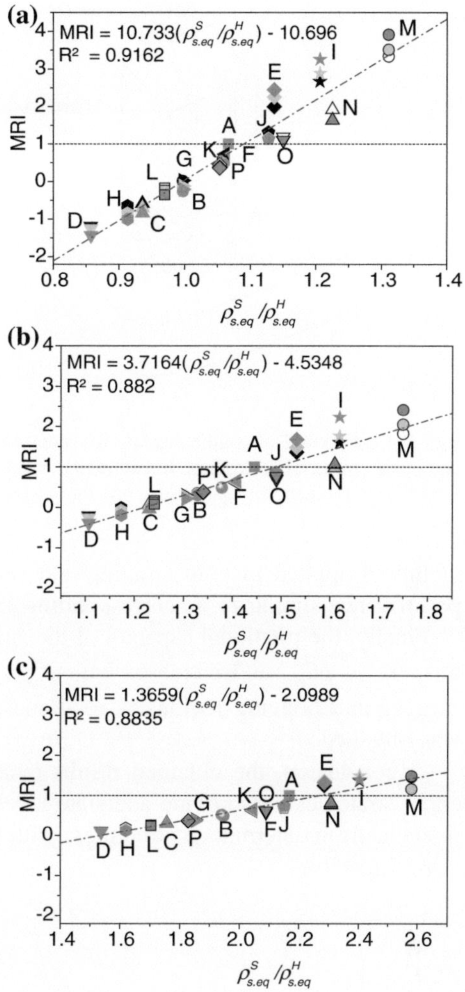

has increased with $\rho^S_{s,eq}/.\rho^H_{s,eq}$ and positive values ($MRI > 0$, which means that the moment redistribution of the strengthened slab was higher than its corresponding reference slab) were obtained when $\rho^S_{s,eq}/\rho^H_{s,eq}$ > 1.09, $\rho^S_{s,eq}/\rho^H_{s,eq}$ > 1.49 and $\rho^S_{s,eq}/\rho^H_{s,eq}$ > 2.27 for η equal to 15, 30 and 45 %, respectively. Thus, the moment redistribution percentage can be estimated if $\rho^S_{s,eq}/\rho^H_{s,eq}$ is known. Figure 8.19 presents the relationship between MRI and $\rho^S_{s,eq}/\rho^H_{s,eq}$ for series SL15, SL30 and SL45. The results evidenced that the use of efficient strengthening strategies can provide adequate levels of ductility and moment redistribution in statically indeterminate structures, with a considerable increase in the load carrying capacity. Also, according to the results, a flexural strengthening strategy composed of CFRP

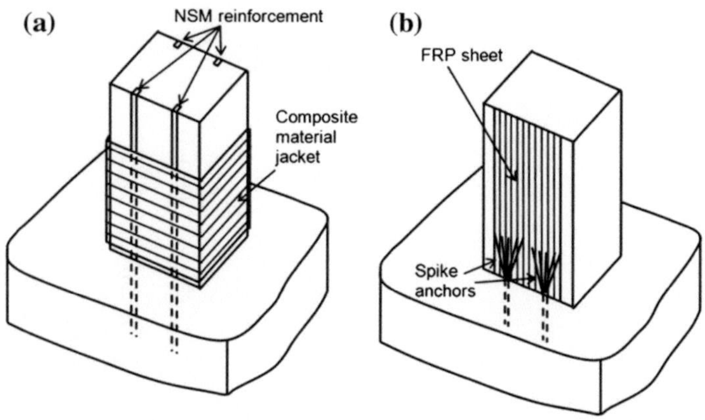

Fig. 8.19 Flexural strengthening of RC column with: **a** NSM reinforcement combined with composite material jacketing; **b** externally bonded FRP sheets combined with spike anchors (possibly combined with jacketing, not shown for the sake of clarity)

laminates applied in both hogging and sagging regions has a deflection ductility performance similar to its corresponding RC slab.

Finally, the rotational capacity of the strengthened slab strips decreases with the increase of $\rho^{H}_{s,eq}$, and increases with $\rho^{S}_{s,eq}$. In the slab strips strengthened in both sagging and hogging regions, a rotational capacity lower than its reference slabs was obtained.

In conclusion, the obtained results evidence that the use of efficient strengthening strategies can provide adequate level of ductility and moment redistribution in statically indeterminate structures, with a considerable increase in the load carrying capacity.

Columns

Flexural strengthening of RC columns is typically achieved today by using RC jackets or some forms of steel jackets, namely steel "cages", also followed by shotcreting. RC jackets or steel cages covered by shotcrete require intensive labour and artful detailing, they increase the dimensions and weight of columns and result in substantial obstruction of occupancy. On the other hand, fibre reinforced polymer (FRP) jacketing, which addresses all of the above mentioned difficulties, is not applicable, as effective strengthening of columns in flexure calls for the continuation of externally applied longitudinal reinforcement beyond the end cross-sections, where moments are typically higher. To overcome the aforementioned difficulties and problems associated with conventional techniques and FRP jacketing, recent research efforts have focused on the use of innovative strengthening techniques:

flexural strengthening of RC columns may be achieved through the use of near-surface mounted (NSM) FRP (Alkhrdaji et al. 2001; Bournas and Triantafillou 2009; Perrone et al. 2009; Maaddawy and Dieb 2011) or through a combination of externally bonded (EBR) FRP sheets (or laminates) and anchors (Prota et al. 2005; Realfonzo and Napoli 2009; Vrettos et al. 2013). This form of externally applied longitudinal reinforcement is prevented from local buckling in highly compressed areas through the use of confining jackets made of composite materials with polymer-based (FRP) or inorganic matrices (textile-reinforced mortars—TRM (e.g. Bournas et al. 2007; Bournas and Triantafillou 2009; see Chap. 9). These concepts are illustrated in Fig. 8.19.

Test Specimens and Tests Setup

A number of experimental programs studied the flexural strengthening of old-type RC columns with NSM reinforcement (Alkhrdaji et al. 2001; Bournas and Triantafillou 2009; Perrone et al. 2009). In Bournas and Triantafillou (2009) and Perrone et al. (2009) large-scale RC columns were tested under cyclic uniaxial flexure with constant axial load (Fig. 8.20a). The specimens were flexure-dominated cantilevers, with a height to the point of application of the load of 1.6 or 1.5 m in Bournas and Triantafillou (2009) and Perrone et al. (2009), respectively, and a cross-section of 250 × 250 mm^2. The geometry of a typical cross-section is shown in Fig. 8.20b. The specimens were designed such that the effect of a series of parameters on the flexural capacity of RC columns could be investigated. These parameters comprised: type of NSM reinforcement (CFRP strips, GFRP bars, stainless steel rebar); configuration of NSM reinforcement (CFRP strips placed with their large cross-section side perpendicular or parallel to the column sides, depending on whether a proper concrete cover is available or not); amount—that is geometrical reinforcing ratio—of NSM or internal reinforcement; type of bonding agent for the NSM reinforcement (epoxy resin vs. cement-based mortar); and NSM reinforcement with or without local jacketing at the member ends.

A short description of the specimens is given in Fig. 8.20c. Of crucial importance in the selection of NSM reinforcement was the requirement of equal tensile strength (not area or stiffness) for each of the reinforcing elements (CFRP strips, GFRP bars, stainless steel bars). The test setups adopted in Bournas and Triantafillou (2009) and Perrone et al. (2009) are identical with the columns fixed into a heavily reinforced base block, within which the longitudinal bars were anchored. The columns were subjected to lateral cyclic loading which consisted of successive cycles progressively increasing by 5 mm (or 2.5 mm in Perrone et al. 2009) of displacement amplitudes (0.31 % drift ratio) in each direction. At the same time a constant axial load was applied to the columns. The lateral load was applied using a horizontally positioned servo-hydraulic actuator. The axial load was exerted

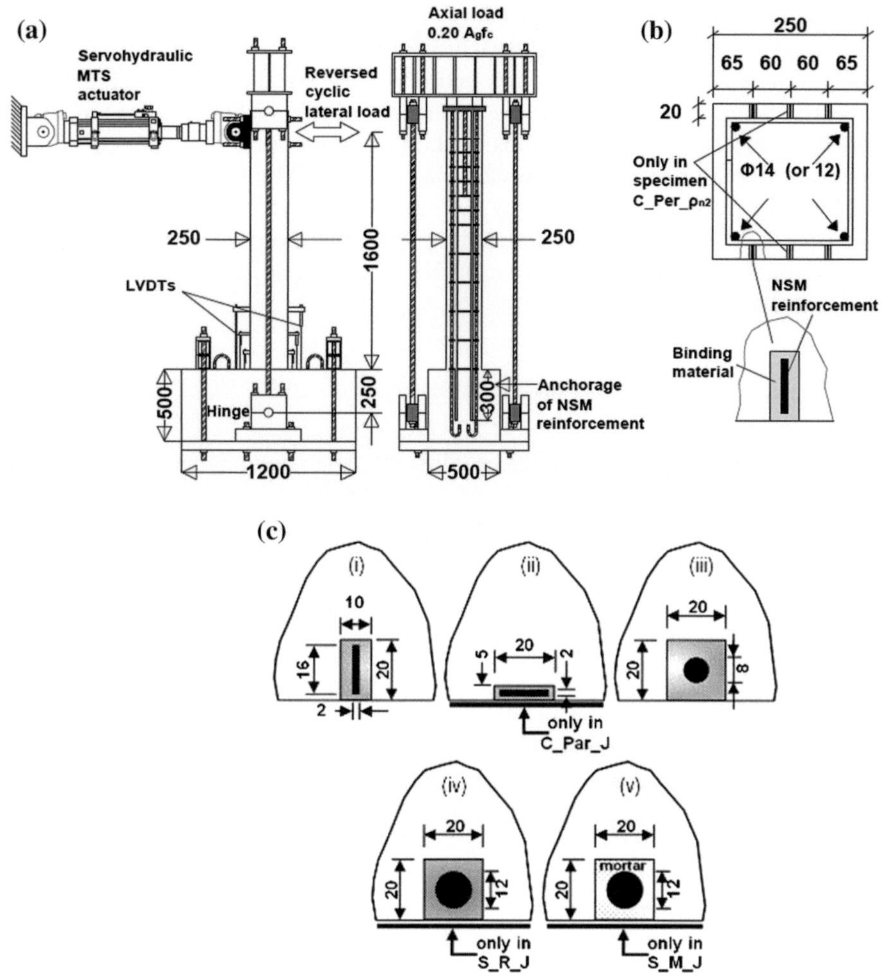

Fig. 8.20 **a** Schematic of test setup. **b** Cross-section of columns. **c** Detail of NSM reinforcement configuration in columns: (i) C_Per, C_Per_ρn2, C_Per_ρs2; (ii) C_Par and C_Par_J; (iii) G; (iv) S_R and S_R_J; and (v) S_M and S_M_J. (dimensions in mm)

by a set of four hydraulic cylinders, acting against two vertical rods connected to the strong floor of the testing frame through a hinge (Fig. 8.20a). Displacements and axial strains at the plastic hinge region were monitored using LVDTs fixed at the cross sections close to the column base (Fig. 8.20a). The instrumentation also comprised a number of strain gages which were mounted on the NSM reinforcing elements to calculate the NSM reinforcement strain at failure.

Failure Modes and Effective Strain

The flexural failure of RC columns strengthened in flexure with NSM reinforcement (FRP or stainless steel bars) occurs due to concrete crushing at the ultimate compressive strain, $\varepsilon_c = \varepsilon_{ccu}$ (or ε_{cu} for unconfined concrete), or due to failure of the NSM reinforcement at a limiting strain $\varepsilon_{n,lim}$, which generally develops after yielding of the internal tension steel bars. This strain depends on the failure mode of the external reinforcement, which may be debonding or tensile rupture. For design purposes, the simplest approach to account for debonding is to calculate the strain in the external reinforcement at debonding, $\varepsilon_{n,b}$, as the product of the ultimate strain $\varepsilon_{n,u}$ (the yield strain, in case of steel NSM reinforcement) and a bond reduction factor k_b. Debonding of the external reinforcement depends mainly on the anchorage length, configuration of rebar, nature of loading (monotonic or cyclic) and presence of external confinement. Experimental evidence (Bournas and Triantafillou 2009; Vrettos et al. 2013) suggests that even if failure of the external reinforcement is due to tensile rupture (in the case of FRP), the failure strain is, in general, less than the (monotonic) uniaxial ultimate strain derived from material testing. The effective ultimate tensile strain of FRP is calculated as the product of the ultimate tensile strain and an effectiveness reduction factor k_r, which depends mainly on the nature of loading (monotonic or cyclic) and the type of anchorage of the longitudinal reinforcement (e.g. configuration of spike anchors). Examples for the estimation of k_b and k_r may be found in Bournas and Triantafillou (2009) and Vrettos et al. (2013). As documented in these studies, if (a) flat FRP strips are used as NSM flexural reinforcement with their long side parallel to the column side, (b) the loading is cyclic, and (c) no confining jackets are used, then failure is due to debonding and k_b is about 0.25. If the same configuration is combined with jacketing, debonding is suppressed and failure is governed by rupture of the external reinforcement with k_r about 0.67. If the strips are used with their long side perpendicular to the column side (without confining jacket) failure is governed by rupture with k_r of about 0.5. If the NSM FRP is in the form of rebars with circular section, failure is controlled by debonding with k_b about 0.35. If FRP spike anchors are used with sufficient anchorage length so that failure is controlled by rupture (Vrettos et al. 2013), the effective strength of those anchors (in the case of cyclic loading) is approximately equal to 30 % of the tensile strength of straight fibres, hence $k_r = 0.30$. Finally, if the NSM reinforcement comprises properly anchored stainless steel rebar, those rebar yield prior to failure of the cross section due to concrete crushing, hence $k_r = 1$. All values for k_b and k_r given above should be considered as indicative, as they have been derived on the basis of relatively limited test results. The point to be made here is that, if the cross-section failure mode involves failure of the NSM flexural reinforcement, the cross-section analysis can be made with the strain in the external reinforcement equal to a (known) fraction of the ultimate uniaxial tension strain, to be determined through testing.

Load-Deflection Response

The response of the columns tested in Bournas and Triantafillou (2009) and Perrone et al. (2009) is given in Fig. 8.21 in the form of load-drift ratio and load-deflection loops, respectively. In Bournas and Triantafillou (2009) the control specimen attained a peak load of about 33 kN and a drift ratio at failure of 6.25 %. With only one exception (column C_Par) all strengthened specimens displayed considerably higher (from about 25 % up to about 100 %) flexural resistance compared to the control specimen.

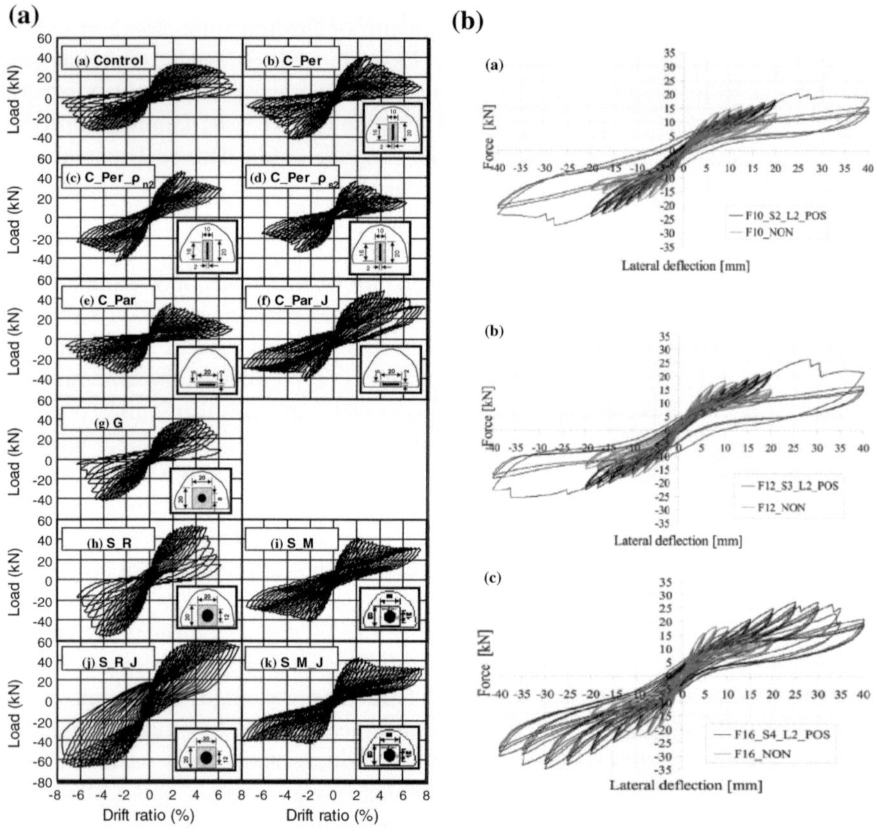

Fig. 8.21 **a** Load versus drift ratio curves for tested specimens in Bournas and Triantafillou (2009); **b** Load versus displacement curves for tested specimens in Perrone et al. (2009)

Influence of Different Parameters on the Flexural Strengthening Performance

Type of NSM reinforcement (C_Per vs. G vs. S_R). Despite the roughly equal (monotonic) uniaxial strength of CFRP, GFRP and stainless steel bars, the latter were more effective, resulting in a strength increase equal to 64 %. The respective values for FRPs were lower (26 % for CFRP and 22 % for GFRP), due to failure of the FRP reinforcing elements at strains less than those corresponding to peak stress, as a result of cyclic loading. In terms of deformation capacity, quantified here by the drift ratio at conventional failure, stainless steel and GFRP bars outperformed CFRP strips by approximately 25 %, due to the lower deformability of carbon fibres in comparison with the other two materials.

Geometrical reinforcing ratio of NSM reinforcement (C_Per vs. C_Per_ρ_{n2}). Increasing the NSM reinforcing ratio by 50 % (three vs. two strips in each side) resulted in a nearly proportional increase in column's strength, which is from 26 % in specimen C_Per to 35 % in specimen C_Per_ ρ_{n2}. This linearity may not apply in the case of large NSM reinforcing ratios.

Configuration of NSM strips (C_Per versus C_Par). In the absence of local jacketing, NSM strips placed with their larger cross-section side perpendicular to the column side were far more effective than those with their larger cross section side parallel to the column side, due to the more favourable bond conditions. The strength increase in the former case was 26 %, but only 4 %, that is marginal, in the latter case.

NSM reinforcement with or without local jacketing (C_Par versus C_Par_J, S_R versus S_R_J, S_M versus S_M_J). TRM jackets resulted in dramatic improvements of the retrofitted specimens' response, by increasing both strength and deformation capacity. Jacketing with TRM prevented buckling of the NSM reinforcement, thereby making the strength increase from 4 to 36 % in the case of CFRP and from 64 to 90 % in the case of stainless steel. In columns retrofitted with NSM bars placed inside mortar, jacketing offered a marginal increase in strength and a moderate increase in deformation capacity. Of all NSM systems examined, the one comprising stainless steel bars and TRM jacketing displayed the best response with stable post peak behaviour and minimal strength degradation up to large drift ratios. NSM FRP or stainless steel reinforcement is a viable solution towards enhancing the flexural resistance of reinforced concrete columns subjected to seismic loads. With proper design, which should combine compulsory NSM reinforcement with local jacketing at column ends, it seems that column strength enhancement does not develop at the expense of low deformation capacity.

Type of bonding agent (S_R versus S_M, S_R_J versus S_M_J). Epoxy resin was a much more effective bonding agent for NSM stainless steel. For the unjacketed specimens, when mortar was used (S_M) instead of resin (S_R) the increase in strength dropped from 64 to 24 %; the corresponding values for jacketed specimens were 90 and 29 %. Hence, the use of mortar instead of resin reduced the effectiveness of the strengthening scheme to about 1/3, due to pullout of the NSM bars.

Design of NSM Strengthened Columns Subjected to Biaxial Bending

Numerical and analytical modelling of columns strengthened with the NSM technique was implemented in Barros et al. (2008) and Bournas and Triantafillou (2013), respectively. Cyclic material constitutive laws were implemented in a finite element program and the tests with RC columns strengthened with the NSM technique were numerically simulated under cyclic loading in Barros et al. (2008). These numerical simulations reproduce the experimental load displacement diagrams satisfactorily. The modelling of the biaxial bending in columns strengthened with NSM reinforcement in combination with confinement proposed in Bournas and Triantafillou (2013) is presented. For any rectangular cross-section subjected to biaxial bending with axial force, the neutral axis is inclined, as shown in Fig. 8.22. The corresponding slope depends on the ratio of the bending moments in the two orthogonal directions and the mechanical properties of the cross-section. Some details of the cross-section analysis are given next.

The strains in the internal reinforcement, ε_{si} (i = 1, 2, 3, 4), and in the external reinforcement, ε_{nib} and ε_{nih} (i = 1, 2, 3, 4), may be found from similar triangle in terms of the strain at the extreme concrete fibre, ε_c, and the geometrical quantities x, h, b, d_s, b_s, d_b, b_b, d_h, b_h, and θ (angle of the neutral axis with respect to side h), all defined in Fig. 8.22. To account for internal steel yielding, failure of the external reinforcement (debonding or fracture in case of FRP, yielding in case of stainless steel) or concrete crushing, strain compatibility should be checked. For the case of external FRP the conditions to be checked are the following:

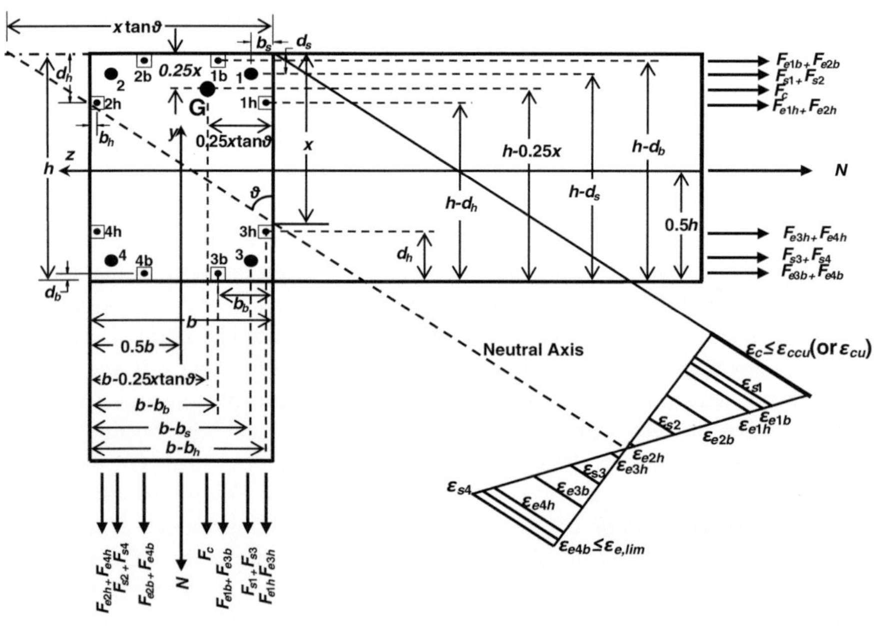

Fig. 8.22 Internal forces and strains at cross section subjected to biaxial bending with axial force

$$\text{If } \varepsilon_{si} \geq \frac{f_y}{E_s} \quad \text{then } f_{\text{si}} = f_{\text{y}} \tag{8.5a}$$

$$\text{If } \frac{f_y}{E_s} > \varepsilon_{si} > -\frac{f_y}{E_s} \quad \text{then } f_{\text{si}} = E_{\text{s}} \cdot \varepsilon_{\text{si}} \tag{8.5b}$$

$$\text{If } \varepsilon_{si} \leq -\frac{f_y}{E_s} \quad \text{then } f_{\text{si}} = -f_{\text{y}} \tag{8.5c}$$

$$\max(\varepsilon_{e4b}, \varepsilon_{e4h}) \leq \varepsilon_{e,\text{lim}} \tag{8.6}$$

$$\varepsilon_c \leq \varepsilon_{ccu} \text{ for confined concrete or } \varepsilon_{\text{cu}} \text{ for unconfined} \tag{8.7}$$

$$\text{If } \varepsilon_{si} \geq \frac{f_y}{E_s} \quad \text{then } f_{\text{si}} = f_{\text{y}} \tag{8.8a}$$

$$\text{If } \frac{f_y}{E_s} > \varepsilon_{si} > -\frac{f_y}{E_s} \quad \text{then } f_{\text{si}} = \text{E}_{\text{s}} \cdot \varepsilon_{\text{si}} \tag{8.8b}$$

$$\text{If } \varepsilon_{si} \leq -\frac{f_y}{E_s} \quad \text{then } f_{\text{si}} = -f_{\text{y}} \tag{8.8c}$$

$$\max(\varepsilon_{e4b}, \varepsilon_{e4h}) \leq \varepsilon_{e,\text{lim}} \tag{8.9}$$

$$\varepsilon_c \leq \varepsilon_{ccu} \text{ for confined concrete or } \varepsilon_{cu} \text{ for unconfined} \tag{8.10}$$

The forces of internal steel bars and external reinforcement can be computed as follows:

$$F_{si} = A_{si} \cdot f_{si} \tag{8.11}$$

$$F_{eib} = A_{eib} \cdot E_e \cdot \varepsilon_{eib} \tag{8.12}$$

$$F_{eih} = A_{eih} \cdot E_e \cdot \varepsilon_{eih} \tag{8.13}$$

where A_{si} = $A_s/4$ = area of internal steel reinforcement concentrated at section's corners, A_{nib} = $A_{nb}/4$ = area of NSM reinforcement lumped at each position on side b, A_{nih} = $A_{nh}/4$ = area of external reinforcement lumped at each position on side h and E_n = elastic modulus of NSM reinforcement. For concrete in the "elastic" range (stresses less than f_{co}) the magnitude of concrete's compressive stress resultant F_c is equal to the volume of a triangular pyramidal stress block (OABC in Fig. 8.23) with a height taken to be the maximum stress of concrete at a corner point. For concrete strains above ε_{co} the material behaves nonlinearly and the compression stress block (OABGDE in Fig. 8.23) becomes complex. The complexity of the analysis in this case can be reduced and the nonlinear stress calculations can be reduced to a sequence of linear ones, by computing the volume

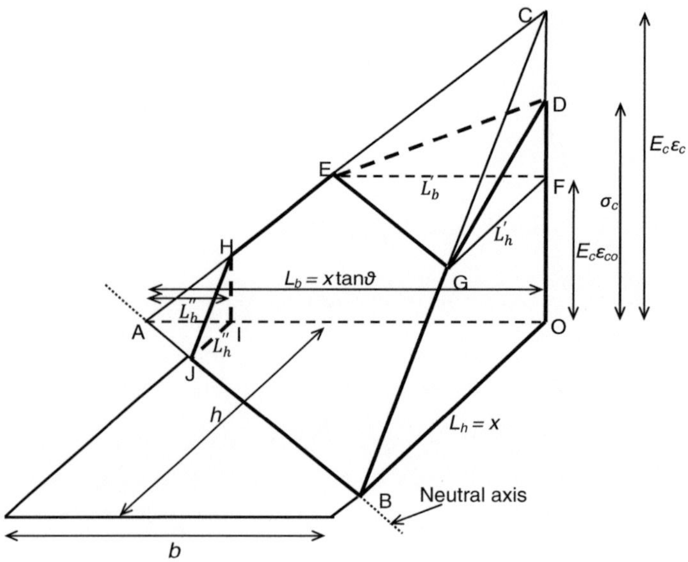

Fig. 8.23 Pyramidal compressive stress block in concrete

of the nonlinear stress block (that is the compressive stress resultant) as the algebraic sum of volumes of simple triangular pyramidal stress blocks as follows:

$$F_c = V_{OABC} - V_{FEGC} + V_{FEGD} - V_{AIJH} \quad (8.14)$$

Hence:

$$F_c = \frac{1}{6}L_h L_b E_c \varepsilon_c - \frac{1}{6}L'_h L'_b E_c(\varepsilon_c - \varepsilon_{co}) + \frac{1}{6}L'_h L' \frac{(f_{cc} - f_{co})}{(\varepsilon_{ccu} - \varepsilon_{co})} - \frac{1}{6}L''_h L''_b E_c\left(\varepsilon_c \frac{L''_b}{L_b}\right) \quad (8.15)$$

where $L_h = x$ = neutral axis depth parallel to side h (Fig. 8.23), $L_b = x\tan\theta$ = neutral axis depth parallel to side b, E_c = elastic modulus of concrete and L'_h, L'_b, L''_h, L''_b as given by the following relationships (Fig. 8.23):

$$L'_h = \frac{(\varepsilon_c - \varepsilon_{co})}{\varepsilon_c}L_x, \quad L'_h = \frac{(\varepsilon_c - \varepsilon_{co})}{\varepsilon_c}L_b, \quad L''_h = L_h - b\tan\theta, \quad L''_b = L_h - b \quad (8.16)$$

The equations presented above have three unknown quantities: x and $\tan\theta$, which define the position of the neutral axis, as well as ε_c that is the maximum compressive strain in the concrete. These unknowns can be determined through the use of axial force and moment equilibrium:

$$N = F_c + \sum_{i=1}^{4} F_{si} + \sum_{i=1}^{4} F_{eib} + \sum_{i=1}^{4} F_{eih} \tag{8.17}$$

$$\begin{aligned} M_y = F_c\left(\frac{b}{2} - 0.25 \cdot x \cdot tg\theta\right) + (F_{s1} - F_{s2} + F_{s3} - F_{s4})\left(\frac{b}{2} - b_s\right) \\ + (F_{e1b} - F_{e2b} + F_{e3b} - F_{e4b})\left(\frac{b}{2} - b_b\right) \\ + (F_{e1h} - F_{e2h} + F_{e3h} - F_{e4h})\left(\frac{b}{2} - b_h\right) \end{aligned} \tag{8.18}$$

$$\begin{aligned} M_z = F_c\left(\frac{b}{2} - 0.25 \cdot x\right) + (F_{s1} + F_{s2} - F_{s3} - F_{s4})\left(\frac{h}{2} - d_s\right) \\ + (F_{e1b} + F_{e2b} - F_{e3b} - F_{e4b})\left(\frac{h}{2} - d_b\right) \\ + (F_{e1h} + F_{e2h} - F_{e3h} - F_{e4h})\left(\frac{h}{2} - d_h\right) \end{aligned} \tag{8.19}$$

where N is the (compressive) axial force in the column and M_y and M_z are the bending moments with respect to the two centroidal axes y and z, respectively (Fig. 8.22). To account for the effect of possible confinement provided by composite material jacketing, the compressive stress–strain response of concrete is modelled as bilinear, in agreement with extensive experimental evidence. According to the typical approach toward modelling confinement of concrete by composite materials, the confined strength f_{cu} and ultimate strain ε_{ccu}, depend on the confining stress at failure (fracture of the jacket in the circumferential direction), σ_{lu}, as follows (Bournas and Triantafillou 2013):

$$\frac{f_{cc}}{f_{co}} = 1 + k_1\left(\frac{\sigma_{lu}}{f_{co}}\right)^m \tag{8.20}$$

$$\frac{\varepsilon_{ccu}}{\varepsilon_{co}} = 1 + k_2\left(\frac{\sigma_{lu}}{f_{co}}\right)^n \tag{8.21}$$

The confining stress σ_l is, in general, non-uniform, especially near the corners of rectangular cross sections. As an average for σ_l in a cross section with dimensions b and h one may write (Bournas and Triantafillou 2013):

$$\sigma_l = \frac{\sigma_{lh} + \sigma_{lb}}{2} = \frac{1}{2}\alpha_f\left(\frac{2t_j}{h}E_j\varepsilon_j + \frac{2t_j}{h}E_j\varepsilon_j\right) \tag{8.22}$$

where E_j and ε_j are the elastic modulus and strain, respectively, of the composite material jacket in the circumferential direction, t_j is the jacket thickness and α_f is a confinement effectiveness coefficient, which is defined as the ratio of effectively confined area A_e to the total cross-sectional area A_g as follows (fib 2001):

$$\alpha_f = 1 - \frac{(b - 2r_c)^2 + (h - 2r_c)^2}{3bh} \tag{8.23}$$

Hence, the confining stress at failure, σ_{lu}, is given by Eq. (8.24) with $E_j\varepsilon_j$ replaced by f_{je}, the effective jacket strength in the circumferential direction. Finally, the normalized confining stress at failure is written as:

$$\frac{\sigma_{lu}}{f_{co}} = \frac{\sigma_{luh} + \sigma_{lub}}{2f_{co}} = \frac{1}{2}\alpha_f\left(\frac{2t_j}{h}\frac{f_{je}}{f_{co}} + \frac{2t_j}{b}\frac{f_{je}}{f_{co}}\right) = \frac{1}{2}\alpha_f\left(\omega_{fh} + \omega_{fb}\right) \tag{8.24}$$

where $\omega_{fh} = (2t_j/h)(f_{je}/f_{co})$ and $\omega_{fb} = (2t_j/b)(f_{je}/f_{co})$ are the mechanical ratios of composite confining jacket in the direction perpendicular to side h and b, respectively. The literature on the precise form of confinement models for concrete, that is, on values for the empirical constants in Eqs. (8.20) and (8.21), is vast. In this study it was assumed that $k_1 = 2.6$, $m = 2/3$, $k_2 = 7.5$, and $n = 0.5$. However, the confinement model may be used with any other set of data for those empirical constants.

Comparison of the Analytical and Numerical Models with Test Results

The analytical procedure described was implemented in a computer program (Bournas and Triantafillou 2013). The program uses geometrical and material data as input to yield, through an iterative numerical procedure, bending moment—axial force interaction diagrams for different cross-sections. This computer program was used to facilitate the comparison of analytical predictions with test results identified in the literature on flexural strengthening of RC columns with NSM reinforcement. Such results are limited and apply mainly to the case of uniaxial or biaxial bending with axial force in columns strengthened with NSM rebar (Alkhrdaji et al. 2001; Bournas and Triantafillou 2009; Perrone et al. 2009; Maaddawy and Dieb 2011), whereas a few test results may be found for the case of uniaxial bending with axial force in columns with anchors or anchorage devices (fibre-based or metallic) at the critical cross-sections (Prota et al. 2005; Realfonzo and Napoli 2009; Vrettos et al. 2013). These results are summarized in Fig. 8.24a, where the predicted to experimental moment ratio is plotted for each column. The mean and the standard deviation of the ratio of the predicted-to-experimental value of normalized ultimate moments μ_{max} are equal to 0.985 and 0.111, respectively. The overall agreement between analysis and test results is quite satisfactory. Finally, it should be mentioned that the numerical model developed in Barros et al. (2008) reproduced with good agreement the load-defection response of NSM strengthened columns as illustrated in Fig. 8.24b.

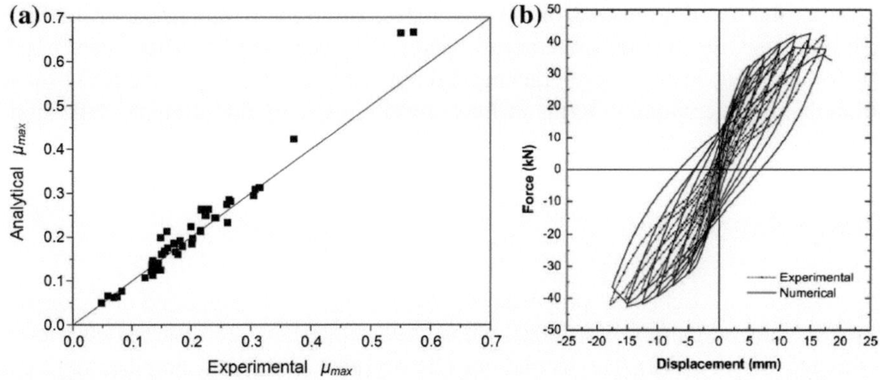

Fig. 8.24 **a** Comparison of experimental normalized bending moment capacity with value predicted by analytical model proposed in Bournas and Triantafillou (2013). **b** Comparison of experimental force-displacement response of an NSM strengthened column with the numerical model developed in Barros et al. (2008)

Shear Strengthening

Experimental research has demonstrated that the near surface mounted technique (NSM) is very effective for the shear strengthening of reinforced concrete (RC) beams (Barros and Dias 2006; Kotynia 2007b; El-Hacha and Wagner 2009). Available experimental results evidence that FRP reinforcements of rectangular cross section provide the greater shear strengthening effectiveness, due to larger ratio between the FRP-concrete bond perimeter and the cross sectional area of the FRP reinforcement, as well as larger confinement provided by the surrounding concrete to the FRP (Costa and Barros 2011). The FRP reinforcements are positioned orthogonally to the beam's axis, or as orthogonal as possible to the predicted direction of the shear failure crack, or to the already existing shear cracks. Carbon FRP (CFRP) laminates of rectangular cross section have been the most used in the NSM technique, so the design formulations herein proposed were mainly developed and calibrated by using experimental results from tests executed with RC beams shear strengthened with CFRP laminates, but their good predictive performance was also demonstrated when applied to other types of NSM FRP shear reinforcements.

The available experimental research also demonstrates that the NSM shear strengthening effectiveness is mainly dependent on the following parameters: percentage and orientation of the FRP reinforcements, concrete strength, and percentage of existing steel stirrups. The strengthening intervention often involves concrete elements already cracked. However, the experimental tests show that the main difference of the behaviour of NSM FRP beams with and without pre-cracks resides in an expected loss of initial stiffness in the pre-cracked beams (Dias and Barros 2012). In these beams the mobilization of the FRP reinforcements started just after the opening process of the pre-cracks, while the mobilization of the FRP reinforcements in the non-pre-cracked beams only occurred when the shear crack has formed. However, the pre-cracking did not affect the efficacy of the NSM shear strengthening technique in terms of load carrying capacity and ultimate deflection.

Design Approaches

Two design approaches are proposed in this chapter. One is supported on extensive experimental program (Dias 2008), whose relevant results are resumed elsewhere (Dias and Barros 2013). The second one (Bianco et al. 2013) is a simplification of a more general formulation based on an original interpretation of the NSM shear strengthening phenomena for RC beams that fulfils equilibrium, kinematic compatibility, and constitutive law of both the adhered materials and the bond between them (Bianco et al. 2010). The former approach is herein designated as "Experimental-base model", while the latter is named "Physical-mechanical-base model".

Experimental-Base Model

According to this model the force resulting from the tensile stress in the FRP laminates crossing the shear failure crack (F_f) is defined as,

$$F_f = n_f \cdot A_{fv} \cdot f_{fe} \tag{8.25}$$

where f_{fe} is the effective stress in the laminates, which is obtained multiplying the elastic modulus of the FRP, E_f, by the effective strain, ε_{fe}. In Eq. (8.25) A_{fv} is the cross-sectional area of a FRP shear reinforcement that is formed by two lateral elements:

$$A_{fv} = 2 \cdot a_f \cdot b_f \tag{8.26}$$

where a_f and b_f are the dimensions of the reinforcement cross-section. The number of reinforcements crossed by the shear failure crack (n_f) is obtained by the equation:

$$n_f = \frac{h_w \cdot \left(\text{cotg}\,\alpha + \text{cotg}\,\theta_f\right)}{s_f} \tag{8.27}$$

where (Fig. 8.25) h_w is the web depth of the beam (equal to the length of vertical reinforcements), α is the orientation of the shear failure crack, θ_f is the inclination of the FRP reinforcement with respect to the beam axis, and s_f is the spacing of laminates.

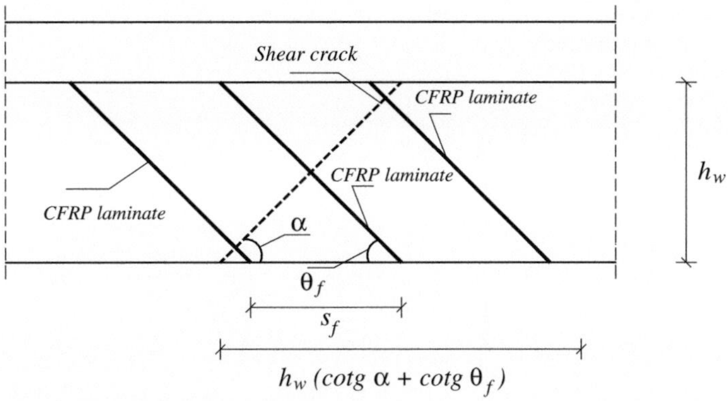

Fig. 8.25 Data for the analytical definition of the effective strain of the FRP

The vertical projection of the force F_f is the contribution of the FRP to the shear resistance of the beam (V_f):

$$V_f = F_f \cdot \sin\theta_f \tag{8.28}$$

Considering Eqs. (8.25)–(8.28) the value of V_f can be obtained from:

$$V_f = h_w \cdot \frac{A_{fv}}{s_f} \cdot \varepsilon_{fe} \cdot E_f \cdot \left(\text{cotg }\ \alpha + \text{cotg }\ \theta_f\right) \cdot \sin\theta_f \tag{8.29}$$

and, consequently:

$$\varepsilon_{fe} = V_f \Big/ \left(h_w \cdot \frac{A_{fv}}{s_f} \cdot E_f \cdot \left(\text{cotg }\ \alpha + \text{cotg }\ \theta_f\right) \cdot \sin\theta_f\right) \tag{8.30}$$

Based on the data derived from the experimental programs described in Dias (2008) and Dias and Barros (2013) the equation to obtain the value of the parameter ε_{fe} for the possible distinct NSM shear strengthening configurations is the following one:

$$\varepsilon_{fe} = 3.76888 \cdot e^{\left(-0.1160261\,\theta_f + 0.0010437\,\theta_f^2\right)} \cdot \left[\left(E_f\rho_f + E_s\rho_{sw}\right)/\left(f_{cm}^{2/3}\right)\right]^{-0.460679 \cdot e^{\left(0.0351199\,\theta_f - 0.0003431\,\theta_f^2\right)}} \tag{8.31}$$

where f_{cm} is the average concrete compressive strength, $\rho_f = \left(\left(2 \cdot a_f \cdot b_f\right)/\left(b_w \cdot s_f \cdot \sin\theta_f\right)\right)$ is the percentage of FRP shear reinforcement, and E_s and $\rho_{sw} = \left(A_{sw}/\left(b_w \cdot s_w\right)\right)$ are the elastic modulus and the percentage of existing steel stirrups, respectively.

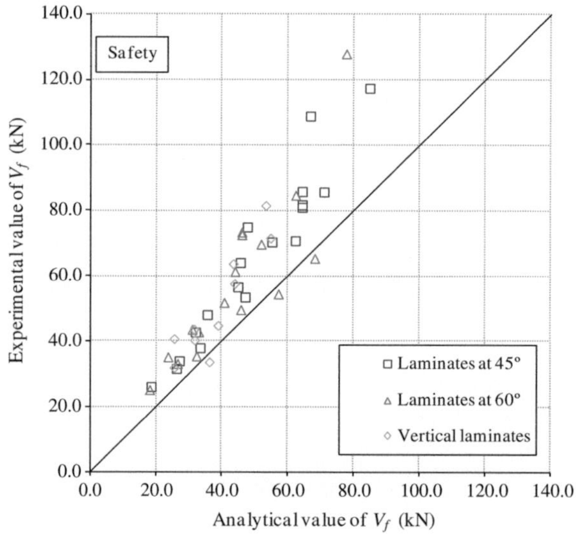

Fig. 8.26 Comparison between the experimental and analytical values of the CFRP contribution for the shear resistance (V_f)

Figure 8.26 compares the experimental and the analytical values of V_f (V_f^{exp} and V_f^{ana}, respectively) for the beams considered in the development of the analytical formulation (Dias 2008; Dias and Barros 2013). According to this figures 95 % of the considered beams are in the safety zone (at the left side of diagonal line). Furthermore, the average value of the k parameter ($k = V_f^{exp}/V_f^{ana}$) considering the safety factor γ_f equal to 1.3 was 1.31, and the corresponding standard deviation value was 0.18.

Physical-Mechanical-Base Model

The physical-mechanical-base model assumes that the possible failure modes that can affect the ultimate behaviour of a NSM FRP reinforcement comprise: loss of bond (debonding); concrete semi-pyramidal tensile fracture; mixed shallow-semi-pyramid-plus-debonding and FRP tensile rupture (Fig. 8.27). This follows from the successive simplifications of the original model (Bianco et al. 2010) in order to result a closed form approach suitable for design purposes (Bianco et al. 2013).

The input parameters of this model include (Figs. 8.28 and 8.29) (Bianco et al. 2013): beam cross-section web's depth h_w and width b_w;inclination angle of both CDC and FRP with respect to the beam longitudinal axis, θ and β, respectively; FRP spacing measured along the beam axis, s_f; angle α between axis and principal generatrices of the semi-pyramidal fracture surface (Fig. 8.28c–d); concrete average compressive strength, f_{cm}; FRP tensile strength, f_{fu}, and Young's modulus E_f; thickness, a_f, and width, b_f, of the FRP cross-section, and values of bond stress, τ_0, and slip, δ_1, defining the adopted local bond stress-slip relationship (Fig. 8.28b).

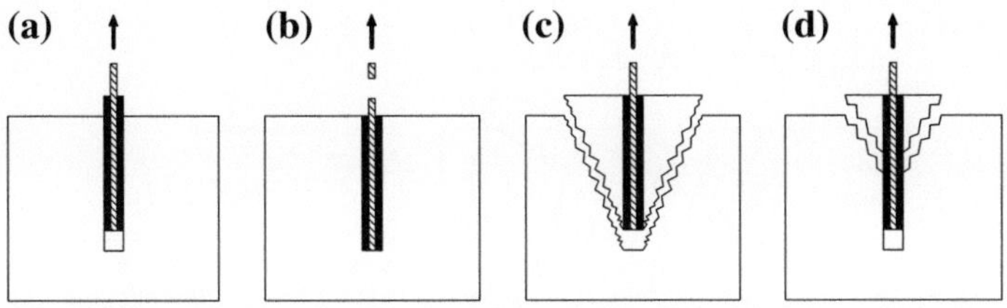

Fig. 8.27 Possible failure modes of an NSM FRP strip: **a** Debonding, **b** FRP tensile rupture, **c** concrete semi pyramidal fracture, **d** mixed shallow semi pyramid plus debonding

The implementation of the proposed calculation procedure comprehends the following steps (Fig. 8.29) (Bianco et al. 2013): (1) evaluation of the average value of the available (resisting) bond length $\bar{L}_{Rfi}$ (Eq. 8.32) and of the minimum integer number $N^{l}_{f,\text{int}}$ of NSM reinforcements that can effectively cross the CDC (Eq. 8.33); (2) evaluation of various constants, both integration and geometric ones (Eqs. 8.34–8.39); (3) evaluation of the bond-modeling constants (Eqs. 8.40 and 8.41); (4) evaluation of the reduction factor η of the average value of the available (resisting) bond length (Eqs. 8.42–8.44) and of the equivalent value of the average resisting bond length $\bar{L}^{eq}_{Rfi}$ (Eq. 8.45); (5) evaluation of the value of the imposed end slip δ_{Lu} in correspondence of which the peak value of the force transmissible through bond by the equivalent value of the resisting bond length $\bar{L}^{eq}_{Rfi}$ can be attained (Eqs. 8.46–8.48); (6) evaluation of the maximum effective capacity that a NSM of bond length $\bar{L}^{eq}_{Rfi}$ can attain during the beam loading process $(V^{\max}_{fi,\,eff})$ (Eqs. 8.49 and 8.50); (7) evaluation of the FRPs shear strength contribution V_f (Eq. 8.51).

$$\bar{L}_{Rfi} = \frac{h_w \cdot \sin\theta \cdot (\cot\theta + \cot\beta)}{4 \cdot \sin(\theta + \beta)} \tag{8.32}$$

$$N^{l}_{f,\text{int}} = \text{round off}\left[h_w \cdot \frac{(\cot\theta + \cot\beta)}{s_f}\right] \tag{8.33}$$

$$L_p = 2 \cdot b_f + a_f \tag{8.34}$$

$$A_c = s_f \cdot \frac{b_w}{2} \tag{8.35}$$

$$L_d = \frac{h_w}{\sin\theta} \tag{8.36}$$

$$V^{tr}_f = a_f \cdot b_f \cdot f_{fu} \tag{8.37}$$

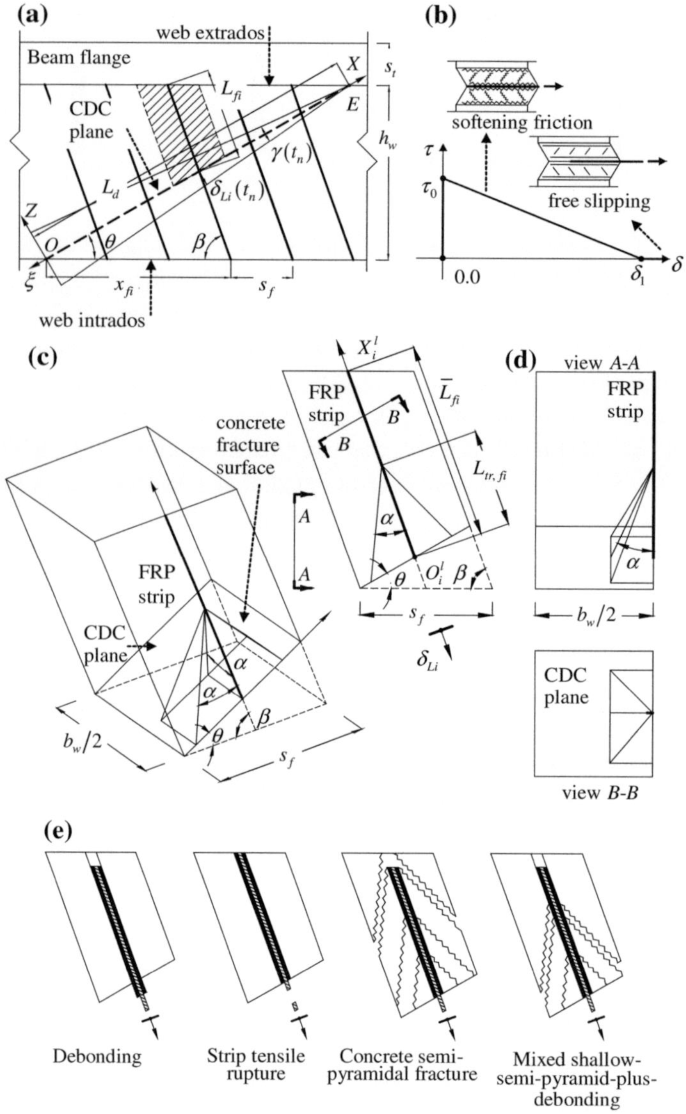

Fig. 8.28 Main physical mechanical features of the theoretical model and calculation procedure: **a** Average available bond length of the NSM reinforcement and concrete prism of influence, **b** adopted local bond stress slip relationship, **c** NSM reinforcement confined to the corresponding concrete prism of influence and semi pyramidal fracture surface, **d** sections of the concrete prism (Bianco et al. 2013)

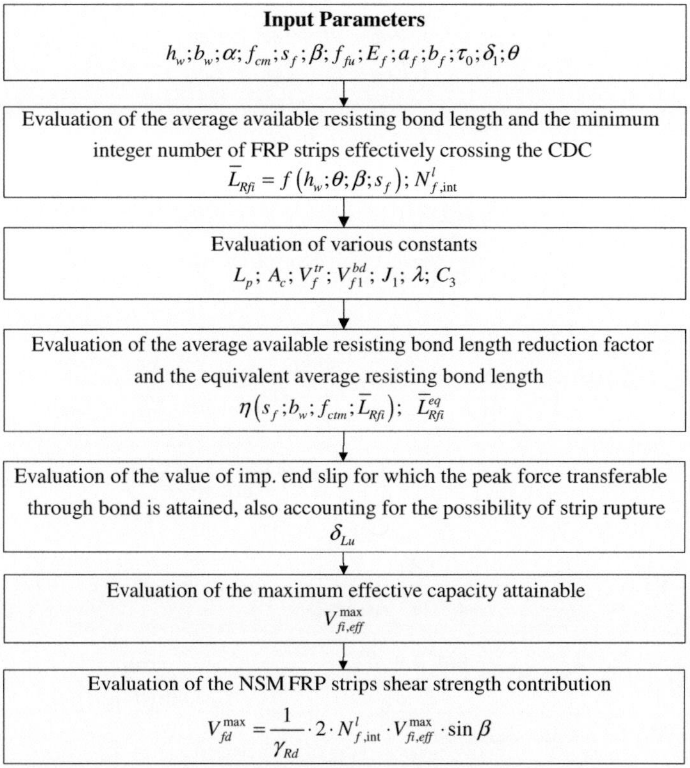

Fig. 8.29 Main algorithm of the calculation procedure (Bianco et al. 2013)

$$f_{ctm} = 1.4 \cdot ((f_{cm} - 8)/10)^{\frac{2}{3}} \tag{8.38}$$

$$E_c = 2.15 \cdot 10000 \cdot (f_{cm}/10)^{\frac{1}{3}} \tag{8.39}$$

$$J_1 = \frac{L_p}{A_f} \cdot \left[\frac{1}{E_f} + \frac{A_f}{A_c \cdot E_c}\right] \quad \frac{1}{\lambda^2} = \frac{\delta_1}{\tau_0 \cdot J_1} \quad C_3 = \frac{V_f^{tr} \cdot J_1}{L_p \cdot \lambda} \tag{8.40}$$

$$L_{Rfe} = \frac{\pi}{2 \cdot \lambda}; \quad V_{f1}^{bd} = \frac{L_p \cdot \lambda \cdot \delta_1}{J_1} \tag{8.41}$$

$$\eta(s_f; b_w; f_{cm}; \bar{L}_{Rfi}) = \begin{cases} f_{ctm}/f^*_{ctm} & se\ f_{ctm} < f^*_{ctm} \\ 1 & se\ f_{ctm} \geq f^*_{ctm} \end{cases} \tag{8.42}$$

$$f^*_{ctm} = \frac{L_p \cdot \lambda \cdot \delta_1 \cdot \sin(\lambda \cdot L_{Rfi})}{J_1 \cdot \min\left(L_{Rfi} \cdot \tan\alpha; \frac{b_w}{2}\right) \cdot \min(s_f \cdot \sin\beta; 2 \cdot L_{Rfi} \cdot \tan\alpha)} \tag{8.43}$$

$$L_{Rfi} = \begin{cases} \bar{L}_{Rfi} & if \quad \bar{L}_{Rfi} \leq L_{Rfe} \\ L_{Rfe} & if \quad \bar{L}_{Rfi} > L_{Rfe} \end{cases} \tag{8.44}$$

$$\bar{L}_{Rfi}^{eq} = \bar{L}_{Rfi} \cdot \eta\left(s_f; b_w; f_{cm}; \bar{L}_{Rfi}\right) \tag{8.45}$$

$$\delta_{Lu} = \begin{cases} \delta_{L1}\left(\bar{L}_{Rfi}^{eq}\right) & if \quad V_{f1}^{bd} < V_f^{tr} \\ \min\left[\delta_{L1}\left(\bar{L}_{Rfi}^{eq}\right); \delta_{Li}\left(V_f^{tr}\right)\right] & if \quad V_{f1}^{bd} \geq V_f^{tr} \end{cases} \tag{8.46}$$

$$\delta_{L1}\left(\bar{L}_{Rfi}^{eq}\right) = \begin{cases} \delta_1 \cdot \left[1 - \cos\left(\lambda \cdot \bar{L}_{Rfi}^{eq}\right)\right] & \text{for } \bar{L}_{Rfi}^{eq} \leq L_{Rfe} \\ \delta_1 & \text{for } \bar{L}_{Rfi}^{eq} > L_{Rfe} \end{cases} \tag{8.47}$$

$$\delta_{Li}\left(V_f^{tr}\right) = \delta_1 \cdot \left\{1 - \cos\left[-\arcsin\frac{C_3}{\delta_1}\right]\right\} \tag{8.48}$$

$$V_{fi,eff}^{\max} = V_{fi,eff}(\gamma_{\max}) = \frac{\delta_1 \cdot A_2}{2 \cdot L_d \cdot A_3 \cdot \gamma_{\max}} \cdot \left[\frac{\pi}{2} - \arcsin\psi - \psi \cdot \sqrt{1 - \psi^2}\right] \tag{8.49}$$

$$A_2 = \frac{L_p \cdot \lambda}{J_1}; \quad A_3 = \frac{\sin(\theta + \beta)}{2 \cdot \delta_1}; \quad \gamma_{\max} = \frac{2 \cdot \delta_{Lu}}{L_d \cdot \sin(\theta + \beta)}; \\ \psi = 1 - A_3 \cdot \gamma_{\max} \cdot L_d \tag{8.50}$$

$$V_{fd} = \frac{1}{\gamma_{Rd}} \cdot V_f = \frac{1}{\gamma_{Rd}} \cdot \left(2 \cdot N_{f,\text{int}}^l \cdot V_{fi,eff}^{\max} \cdot \sin\beta\right) \tag{8.51}$$

The predictive performance of this model was assessed by Bianco et al. (2013) by using available experimental results. Assuming for the angle α the value of 28.5° for all the experimental programs, considering for the f_{ctm} the values determined from the formulae of the CEB-FIP Model Code 90 (1993), and adopting for the local bond stress-slip relationship the values $\tau_0 = 20.1$ MPa and $\delta_1 = 7.12$ mm, this formulation provided very satisfactory estimates of the experimental recordings, resulting the ratio of the prediction versus the experimental value characterized by a mean value and a standard deviation of 0.69 and 0.29, respectively.

Acknowledgments The first author would like to thank the support by FEDER funds through the Operational Program for Competitiveness Factors—COMPETE and National Funds through FCT—Portuguese Foundation for Science and Technology under the project CutInDur PTDC/ECM/112396/2009.

References

ACI (2008). Guide for the design and construction of externally bonded FRP systems for strengthening concrete structures. *Report by ACI Committee 440.2R-08*, American Concrete Institute, Farmington Hills, MI, USA, 76 pp.

Alkhrdaji, T., Barker, M., Chen, G., Mu, H., Nanni, A., & Yang, X. (2001). Destructive and Non-Destructive Testing of Bridge J857, Phelps County, Missouri. Vol. III- Strengthening and Testing to Failure of Bridge Piers, Centre for Infrastructure Engineering Studies, University of Missouri at Rolla, Report No.RDT01–002C, 90 pp.

Al-Mahmoud, F., Castel, A., François, R., & Tourneur, C. (2009). Strengthening of RC members with near-surface mounted CFRP rods. *Composite Structures, 91*(2), 138–147.

Asplund, S. Q. (1949). Strengthening bridge slabs with grouted reinforcement. *ACI Journal, 52*(6), 397–406.

Barros, J. A. O., & Dias, S. J. E. (2006). Near surface mounted CFRP laminates for shear strength-ening of concrete beams. *Journal Cement and Concrete Composites, 28*(3), 276–292.

Barros, J. A. O., Dias, S. J. E., & Lima, J. L. T. (2007). Efficacy of CFRP-based techniques for the flexural and shear strengthening of concrete beams. *Cement & Concrete Composites, 29*(3), 203–217.

Barros, J. A. O., & Fortes, A. S. (2005). Flexural strengthening of concrete beams with CFRP laminates bonded into slits. *Journal of Cement & Concrete Composites, 27*, 471–480.

Barros, J. A. O., Varma, R. K., Sena-Cruz, J. M., & Azevedo, A. F. M. (2008). Near surface mounted CFRP strips for the flexural strengthening of RC columns: Experimental and numerical research. *Engineering Structures, 30*(12), 3412–3425.

Bianco, V., Barros, J. A. O., & Monti, G. (2010). New approach for modeling the contribution of NSM FRP strips for shear strengthening of RC beams. *ASCE Composites for Construction Journal, 14*(1), 36–48.

Bianco, V., Monti, G., & Barros, J.A.O. (2013). Design formula to evaluate the NSM FRP strips shear strength contribution to a RC beam. *Composites Part B: Engineering.*

Bilotta, A., Ceroni, F., Di Ludovico, M., Nigro, E., Pecce, M., & Manfredi, G. (2011). Bond efficiency of EBR and NSM FRP systems for strengthening concrete members. *Journal of Composites for Construction, 15*(5), 757–772.

Bilotta, A., Ceroni, F., Di Ludovico, M., Nigro, E., Pecce, M., & Manfredi, G. (2012). Experimental bond test on concrete members strengthened with NSM FRP systems: influence of groove dimensions and surface treatment. In *Proceedings of the CICE 2012, Rome, Italy* (p. 8).

Blaschko, M. (2001). *Zum tragverhalten von betonbauteilen mit in schlitze eingeklebten CFK-lamellen. Bericht 8/2001 aus dem Konstruktiven Ingenieurbau (in German).* Germany: TU München. 147 pp.

Blaschko, M. (2003). Bond behaviour of CFRP strips glued into slits. In *Proceedings of the FRPRCS-6. Singapore: World Scientific* (pp. 205–214).

Bonaldo, E. (2008). *Composite materials and discrete steel fibres for the strengthening of thin concrete structures.* Civil Engineering Departament: University of Minho.

Bonaldo, E., Barros, J.A.O., & Lourenço, P.B. (2008). Efficient Strengthening Technique to Increase the Flexural Resistance of Existing RC Slabs. *Journal of Composites for Construction*, 149–159.

Bournas, D., Lontou, P., Papanicolaou, C. G., & Triantafillou, T. C. (2007). Textile-reinforced mortar (TRM) versus FRP confinement in reinforced concrete columns. *ACI Structural Journal, 104*(6), 740–748.

Bournas, D. A., & Triantafillou, T. C. (2009). Flexural strengthening of RC columns with NSM FRP or stainless steel. *ACI Structural Journal, 106*(5), 495–505.

Bournas, D. A., & Triantafillou, T. C. (2013). Biaxial bending of RC columns strength-ened with externally applied reinforcement combined with confinement. *ACI Structural Journal, 110*(2), 193–204.

Burke, P.J. (2008). Low and high temperature performance of near surface mounted FRP strengthened concrete slabs. *MSc Thesis*, Department of Civil Engineering, Queen's University, 193 pp.

Castro, E.K., Melo, G.S., & Nagato, Y. (2007). Flexural strengthening of RC "T" beams with near surface mounted (NSM) FRP reinforcements. In *Proceedings of the eight international conference on fibre-reinforced plastics for reinforced concrete structures (FRPRSCS-8)*, Patras, Greece.

CEB-FIP Model Code 90 (1993). Bulletin d'Information N° 213/214. *Final version printed by Th. Telford*, London, 460 pp.

Ceroni, F., Pecce, M., Bilotta, A., & Nigro, E. (2012). Bond behavior of FRP NSM systems in concrete elements. *Composites Part B Engineering, 43*(2), 99–109.

Cholostiakow, S., Kotynia, R., & Przygocka, M. (2013). Flexural strengthening of reinforced concrete structures with near surface mounted FRP composites. In *Proceedings of the 9th Central European Congress on Concrete Engineering*, Wroclaw.

Coelho, M., Sena-Cruz, J.M., & Neves, L.A.C. (2013). State of the art of bond tests with FRP NSM systems in concrete. Report no. 13-DEC/E-12, University of Minho, Guimarães, Portugal, 104 pp.

Coelho, M., Sena-Cruz, J.M., & Neves, L.A.C. (2014). A review on the bond behavior of FRP NSM systems in concrete. *Submited to Construction and Building Materials.*

Costa, I. G., & Barros, J. A. O. (2010). Flexural and shear strengthening of RC beams with composite materials – The influence of cutting steel stirrups to install CFRP strips. *Cement & Concrete Composites, 32*, 544–553.

Costa, I., & Barros, J. (2011). Assessment of the bond behaviour of NSM FRP materials by pullout tests. In *Proceedings of the First Middle East Conference on Smart Monitoring, Assessment and Rehabilitation of Civil Structures, Dubai, UAE* (p. 8).

Dalfré, G. M. (2013). *Flexural and shear strengthening of RC elements*. Civil Engineering Departament: University of Minho.

Dalfré, G., & Barros, J. (2011). Assessing the effectiveness of a NSM-CFRP flexural strengthening technique for continuous RC slabs by experimental research. In *Proceedings of the First Middle East Conference on Smart Monitoring Assessment and Rehabilitation of Civil Structures (SMAR 2011)*, Dubai.

De Lorenzis, L., Lundgren, K., & Rizzo, A. (2004). Anchorage length of near-surface mounted fiber-reinforced polymer bars for concrete strengthening — Experimental investigation and numerical modeling. *ACI Structural Journal, 101*(2), 269–278.

De Lorenzis, L., & Nanni, A. (2002). Bond between near-surface mounted fiber-reinforced polymer rods and concrete in structural strengthening. *ACI Structural Journal, 99*(2), 123–132.

De Lorenzis, L., Nanni, A., & La Tegola, A. (2000). Flexural and shear strengthening of reinforced concrete structures with near surface mounted FRP rods. In J. Humar & AG Razaqpur (Eds.), *Proceedings of the third international conference on advanced composite materials in bridges and structures, Ottawa, Canada* (pp. 521–528).

De Lorenzis, L., Rizzo, A., & La Tegola, A. (2002). A modified pull-out test for bond of near-surface mounted FRP rods in concrete. *Composites Part B Engineering, 33*(8), 589–603.

De Lorenzis, L., & Teng, J. G. (2007). Near-surface mounted FRP reinforcement: An emerging technique for strengthening structures. *Composites Part B Engineering, 38*(2), 119–143.

Dias, S. J. E. (2008). *Experimental and analytical research on the shear strengthening of RC beams according to the NSM CFRP technique. Ph.D.* Departamento de Engenharia Civil: Universidade do Minho (in Portuguese). 391 pp.

Dias, S. J. E., & Barros, J. A. O. (2012). NSM shear strengthening technique with CFRP laminates applied in high-strength concrete beams with or without pre-cracking. *Composites Part B: Engineering Journal, 43*(2), 290–301.

Dias, S. J. E., & Barros, J. A. O. (2013). Shear strengthening of RC beams with NSM CFRP laminates: experimental research and analytical formulation. *Composite Structures Journal, 99*, 477–490.

El-Hacha, R., & Rizkalla, S. (2004). Near-Surface-Mounted fiber-reinforced polymer reinforcements for flexural strengthening of concrete structures. *ACI Structural Journal*, 101(5), 717–726.

El-Hacha, R., & Wagner, M. (2009). Shear strengthening of reinforced concrete beams using near-surface mounted CFRP strips. In *Proceedings of the 9th International Symposium on Fiber Reinforced Polymers Reinforcement for Concrete Structures (FRPRCS-9), Sydney, Australia* (p. 10).

En-Core and fib TG 9.3 (2010). Internal reports: Round robin testing exercise – universities of gent, minho and budapest. Retrieved from http://cigroup.shef.ac.uk/encore/rrt/index.php.

fib (2001). Externally bonded FRP reinforcement for RC structures. In *Bulletin No. 14, International Federation for Structural Concrete, Lausanne, Switzerland* (138 pp).

Galati, D., & De Lorenzis, L. (2009). Effect of construction details on the bond performance of NSM FRP bars in concrete. *Advances in Structural Engineering, 12*(5), 683–700.

Hassan, T., & Rizkalla, S. (2003). Investigation of bond in concrete structures strengthened with near surface mounted carbon fiber reinforced polymer strips. *Journal of Composites for Construction, 7*(3), 248–257.

Kalayci, A.S. (2008). *Development of surface flaw thresholds for pre-cured fiber reinforced polymer and groove size tolerance for near surface mounted fiber reinforced polymer retrofit systems*. Ph.D. Thesis, Department of Civil and Environmental Engineering, Florida International University, 160 pp.

Kotynia, R. (2007a). Analysis of the flexural response of NSM FRP-strengthened concrete beams. In *Proceedings of the eight international conference on fibre-reinforced plastics for reinforced concrete structures (FRPRSCS-8)*. Patras, Greece.

Kotynia, R. (2007b). Shear strengthening of RC beams with NSM CFRP laminates. In *Proceedings of the 8th Inter-national Symposium on Fiber Reinforced Polymer Reinforcement for Concrete Structures (FRPRCS-8)*. Patras, Greece.

Liu, I. S. T., Oehlers, D. J., & Seracino, R. (2006). Tests on the ductility of reinforced con-crete beams retrofitted with FRP and steel near-surface mounted plates. *Journal of Composites for Construction, 10*(2), 106–114.

Maaddawy, T. E., & Dieb, S. E. (2011). Near-Surface-Mounted composite system for repair and strengthening of reinforced concrete columns subjected to axial load and biaxial bending. *ASCE Journal of Composites for Construction, 15*(4), 602–614.

Novidis, D.G., & Pantazopoulou, S.J. (2007). Beam tests of NSM-FRP laminates in concrete. In *Proceedings of the eight international conference on fibre-reinforced plastics for reinforced concrete structures (FRPRSCS-8)*. Patras, Greece.

Novidis, D. G., & Pantazopoulou, S. J. (2008). Bond tests of short NSM-FRP and steel bar anchorages. *Journal of Composites for Construction, 12*(3), 323–333.

Palmieri, A., Matthys, S., Barros, J., Costa, I., Bilotta, A., Nigro, E., Ceroni, F., szabó, Z.K., & Balázs, G.L. (2012). Bond of NSM FRP strengthened concrete: round robin test initiative. In *Proceedings of the CICE 2012, Rome, Italy* (p. 8).

Perrone, M., Barros, J. A. O., & Aprile, A. (2009). A CFRP-based strengthening technique to increase the flexural and energy dissipation capacities of RC column. *ASCE Journal of Composites for Construction, 13*(5), 372–383.

Prota, A., Manfredi, G., Balsamo, A., Nanni, A., & Cosenza, E. (2005). Innovative technique for seismic upgrade of RC square columns. In *Proceedings of the FRPRCS-7 International Symposium, Kansas, USA* (pp. 1289–1304). ACI SP-230-73.

Realfonzo, R., & Napoli, A. (2009). Cyclic behavior of RC columns strengthened by FRP and steel devices. *Journal of Structural Engineering, 135*(10), 1164–1176.

Sena Cruz, J. M., & Barros, J. A. O. (2004). Bond between near-surface mounted carbon-fiber-reinforced polymer laminate strips and concrete. *Journal of Composites for Construction, 8*(6), 519–527.

Sena Cruz, J., Barros, J., Gettu, R., & Azevedo, Á. (2006). Bond behavior of near-surface mounted CFRP laminate strips under monotonic and cyclic loading. *Journal of Composites for Construction, 10*(4), 295–303.

Seracino, R., Jones, N. M., Ali, M. S. M., Page, M. W., & Oehlers, D. J. (2007). Bond strength of near-surface mounted FRP strip-to-concrete joints. *Journal of Composites for Construction, 11* (4), 401–409.

Täljsten, B., Carolin, A., & Nordin, H. (2003). Concrete structures strengthened with near surface mounted reinforcement of CFRP. *AN International Journal, 6*(3), 201–213.

Teng, J. G., De Lorenzis, L., Wang, B., Li, R., Wong, T. N., & Lam, L. (2006). Debonding failures of RC beams strengthened with near surface mounted CFRP strips. *Journal of Composites for Construction, 10*(2), 92–105.

Triantafillou, T., & Antonopoulos, C. (2000). Design of concrete flexural members strengthened in shear with FRP. *Journal of Composites for Construction, 4*(4), 198–205.

Vrettos, I., Kefala, E., & Triantafillou, T. C. (2013). Innovative flexural strengthening of RC columns using carbon spike anchors. *ACI Structural Journal, 110*(1), 63–70.

Dias, S., Coelho, M., & Sena Cruz, J.M. (2011). Presentation on the 6th meeting of RILEM TC 234-DUC: NSM Systems.

Yost, J. R., Gross, S. P., Dinehart, D. W., & Mildenberg, J. J. (2007). Flexural behavior of concrete beams strengthened with near-surface-mounted CFRP strips. *ACI Structural Journal, 104*(4), 430–437.

Chapter 9
Fiber Reinforced Composites with Cementitious (Inorganic) Matrix

Christian Carloni, Dionysios A. Bournas, Francesca G. Carozzi, Tommaso D'Antino, Giulia Fava, Francesco Focacci, Giorgio Giacomin, Giovanni Mantegazza, Carlo Pellegrino, Carlo Perinelli and Carlo Poggi

Abstract Fibre reinforced composite systems are increasingly used in civil engineering infrastructure applications for strengthening and rehabilitation of reinforced concrete (RC) structures. Composite materials represent a sustainable alternative to new construction because they allow for an extension of the original service life and therefore prevent demolition of existing structures. Promising newly-developed types of matrix that potentially represent a valid, sustainable, and durable alternative to epoxy, employed in fibre-reinforced polymer (FRP) composites, are the so-called inorganic matrices. Within the broad category of inorganic matrices, cement-based mortars have raised some interest in recent years. This chapter intends to highlight the potentials of this new category of fibre-reinforced composites as a viable alternative to traditional FRP systems. The latest advancements in this field and the new challenges that researchers will face in the future are presented and discussed.

Keywords FRCM · Strengthening · Cementitious matrix · Reinforced concrete

---

C. Carloni (✉)
University of Bologna, Bologna, Italy
e-mail: christian.carloni@unibo.it

D.A. Bournas · G. Giacomin
University of Nottingham, Nottingham, UK

F.G. Carozzi · G. Fava · C. Poggi
Politecnico of Milan, Milan, Italy

T. D'Antino · C. Pellegrino
Univeristy of Padua, Padua, Italy

F. Focacci
Univeristy eCampus, Novedrate, Italy

C. Perinelli
G&P Intech Srl, Altavilla Vicentina, Italy

G. Mantegazza
Ruredil Spa, San Donato Milanese, Italy

C. Pellegrino and J. Sena-Cruz (eds.), *Design Procedures for the Use of Composites in Strengthening of Reinforced Concrete Structures*, RILEM State-of-the-Art Reports 19, DOI 10.1007/978-94-017-7336-2_9

Introduction

Strengthening and rehabilitation of reinforced concrete (RC) structures with externally-bonded composite materials represent a sustainable alternative to new construction because they allow for an extension of the original service life and therefore prevent demolition of existing structures. In the last two decades fibre-reinforced polymer (FRP) composites have been the most common type of composite used for structural applications. FRP comprises of continuous fibres (usually carbon, glass, or aramid) and a thermosetting (organic) resin, typically epoxy, as the matrix. Promising newly-developed types of matrix that potentially represent a valid, sustainable, and durable alternative to epoxy are the so-called inorganic matrices. Within the broad category of inorganic matrices, polymer-modified cement-based mortars have raised some interest in recent years. Composite materials that employ modified cement-based mortars are usually referred to as fibre-reinforced cementitious matrix (FRCM) composites. Alternative names have been proposed in the literature and refer to different types of matrix or application. Among the others, the most common names are TRM (Textile Reinforced Mortar) (e.g. Bisby et al. 2009; Triantafillou 2010), TRC (Textile Reinforced Concrete) (e.g. Banholzer et al. 2006; Brückner et al. 2006; Hartig et al. 2008; Hegger et al. 2006; Peled et al. 2008; Wiberg 2003; Zastrau et al. 2008), MBC (Mineral Based Composites) (Täljsten and Blanksvärd 2007) or FRC (Fiber Reinforced Cement) (e.g. Wu and Sun 2005). Similar materials used for masonry structures strengthening applications are identified in the technical literature with the acronyms CMG (Cementitious Matrix-Grid system) (Prota et al. 2006; Lignola et al. 2009), IMG (Inorganic Matrix Grid system) (Parisi et al. 2011), CFCM (Carbon Fiber Cement Matrix) (Kolsch 1998). FRCM is used as the acronym of fabric reinforced cementitious matrix in the ACI guideline (ACI 2013). In the following sections the acronyms FRCM as well as TRM will be used as synonyms to refer to composites that employ inorganic matrices.

In FRCM composites, fibres are typically bundled (rovings), and the fibre pattern can be modified from unidirectional to bidirectional textile weaves or fabrics in an attempt to improve the bond characteristics. The still-limited available literature reports that FRCM composites can be used effectively for strengthening and rehabilitation of RC structures. In FRP-concrete joints, it is well-understood that interfacial crack propagation typically occurs within a thin layer of the substrate close to the FRP composite, and therefore the concrete mechanical and fracture properties and the surface treatment play a fundamental role in the evaluation of the strengthening performance. In FRCM-concrete joints the interfacial debonding might occur within the composite for some commercially available FRCM systems; hence the substrate, on which the composite is applied, may not play a key role in the design of the strengthening system, which is an interesting aspect of this composite. Large slips at the interface between fibres and matrix have been observed during the debonding of FRCM composites from a concrete substrate (D'Ambrisi et al. 2012, 2013a; D'Antino et al. 2014; Sneed et al. 2014; Carloni et al. 2014). When the debonding occurs at the matrix-fibre interface, the phenomenon itself is complicated by the telescopic behaviour observed among the fibre filaments of a fibre bundle where the core filaments have a different mechanism of stress transfer with respect to the outer filaments, mainly due to the different impregnation of the fibres by the matrix. Although real applications of FRCM composites on structure subjected to fatigue loading were performed (D'Ambrisi et al. 2015), studies regarding the behavior of FRCM composites subjected to fatigue loading are very limited (D'Antino et al. 2015).

In the next section a brief description of some of the FRCM systems commercially available is presented to show the great variability of the matrix-fibre combinations, which in turn determines also a variability of the response of the FRCM-concrete interface. After a brief presentation of the available FRCM systems, the tensile behaviour, the bond characteristics, and finally the use of FRCM systems for flexural and confinement applications are presented.

FRCM Materials

In this section a brief description of some of the commercially available FRCM systems is reported.

Materials by G&P

G&P introduced in the international market different composite systems such as FRCM (carbon, AR glass and basalt grids) and SRG (UHTSS high resistance steel fabric) used with inorganic matrix (like cementitious or structural lime) for structural reinforcement of buildings. The main advantages of these systems are fire resistance, easy application on rough and moist surfaces, and ductility. In addition, FRCM and SRG systems are also compatible with historical buildings and monuments.

The main applications of FRCM and SRG systems are in the following areas:

- Restoration of historical buildings and particularly the reinforcement of masonry structures such as walls, arches, vaults, and domes. In this case it is important to highlight that these systems are highly compatible with the historical masonries also when there are frescoes on the opposite surface of application.
- Improvement of structures in seismic areas with increase of strength and ductility of masonry and reinforced concrete elements.
- Reinforcement of tunnels and structures which need high fire resistance.

The fibres employed are obtained with carbon, glass, basalt grids, and steel fabric with the following mechanical properties.

- Carbon grid uni-bi directional C-NET (filament)

– Weight	g/m^2	100–170–200–220
– Elastic modulus	GPa	240
– Tensile strength	MPa	>4500
– Strain at failure	%	>1.5

- AR Glass grid bidirectional G-NET (filament)

– Weight	g/m^2	120–250–320
– Elastic modulus	GPa	65–74
– Tensile strength	MPa	>3000
– Strain at failure	%	>3

- Basalt grid bidirectional B-NET (filament)

– Weight	g/m^2	250–350
– Elastic modulus	GPa	90
– Tensile strength	MPa	>3200
– Strain at failure	%	>3

- Steel fabric UHTSS unidirectional STEEL NET

TYPE		150	190
– Weight	g/m^2	1528	1910
– Elastic modulus	GPa	190	190
– Tensile strength	MPa	3345	3345
– Strain at failure	%	>2.2	>2.2

The following inorganic matrices are technically approved and certified to be used with carbon, glass, and basalt grids and steel fabrics.

- CONCRETE ROCK W cementitious mortar with reactive nano compound additives specific for low thickness, no-shrinking, sulphates resistant.
- CONCRETE ROCK V-V2 cementitious mortar, one or two components, no-shrinking, with high resistance for concrete repair.
- LIMECRETE lime hydraulic mortar, with high resistance and adhesion to the support for masonry and historical buildings.
- LIMECRETE FR lime hydraulic mortar for low thickness, with high resistance and adhesion to the support for masonry and historical buildings.

FRCM Materials by Ruredil

The FRCM system patented by Ruredil worldwide features the following advantages:

- high heat resistance: once the matrix has hardened, the system is not affected by the outdoor temperature.

- excellent reaction to fire: the system reacts in the same way as the substrate, because the inorganic matrix maintains its properties up to a temperature of 550 °C, is not combustible, emits very little smoke and does release off incandescent particles;
- high durability even under damp conditions.
- effective even if applied over a damp substrate: humidity promotes adhesion to the hydraulic matrix, whereas it reduces the adhesion of inorganic resins to the substrate;
- easy handling: inorganic matrix is prepared in the same way as any hydraulic product;
- applicable even on rough, irregular surfaces: does not require preliminary smoothing of the surface;
- applicable under a great variety of environmental conditions;
- non-toxicity of the matrices employed for workers and the environment: they may in fact be considered similar to a traditional inorganic mortar.
- easy cleaning of utensils: water is sufficient, with no need for the solvents required to clean off resins, which are harmful to human health and the environment.

This breakthrough technology has resulted in a product line that has obtained certification from ICC Evaluation Services (ICC-ES) according to AC434-13 (2013). AC434-13 establishes guidelines for the necessary tests and calculations required to receive a product research report from ICC-ES. Thus, this product can now be accepted by code officials under Section 104.11.1 of the International Building Code, which allows research reports to be used as a source of information to show building code compliance of alternative materials. Example of applications with PBO and Carbon FRCM materials by Ruredil are shown in Figs. 9.1 and 9.2, respectively.

(a) (b) (c)

Fig. 9.1. Structural retrofitting of cooling towers with PBO-FRCM (by Ruredil): **a** Power plant cooling tower. **b** and **c** application of the FRCM composite

Fig. 9.2 Reconstruction and seismic retrofitting of the Cathedral Church of Noto (Italy) using C-FRCM (by Ruredil): **a** The Cathedral (today) Noto, Siracusa Italy; **b** the collapse of the structure due to earthquake; **c** and **d** reconstruction and seimic retrofitting of the arches of the main nave

Mechanical Characterization of FRCM Systems: Tensile Test

This section reports a summary of the experimental work, carried out at the Politecnico di Milano and University of Padova, to characterize FRCM composites in tension.

Finding a reliable and shared experimental procedure to test FRCM composites in tension is of particular importance due to the great number of materials and fibre-matrix combination available. The scientific literature reports several difficulties in characterizing the performance of FRCM composite. Different test setups have been used and, although some authors already proposed some recommendations, a shared test procedure has not been published yet (Hartig et al. 2012; Häußler-Combe and Hartig 2007; Jesse et al. 2005, 2009). The response of FRCM composites in tension may be influenced by: specimen production, dimensions, and shape, and load application. Hartig et al. (2012) classified two types of load application: "rigid load application" in which the main transfer mechanism between specimen and clamping is adhesive tension and shear; and "soft clamping" with friction load transfer allowing for gradual load application. There are different procedures to analyse the deformations (Contamine et al. 2011): (a) displacements of the clamps testing machine; (b) strain gauges, that could be inadequate in the

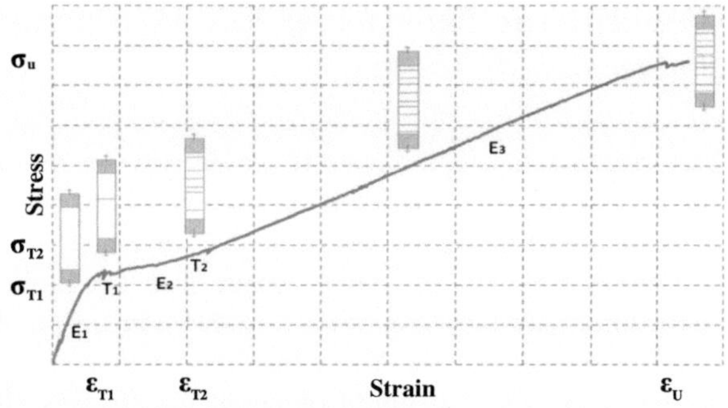

Fig. 9.3 Stress–strain behavior of a FRCM composite subject to tensile test

case of multi-cracking behaviour; (c) LVDT displacement transducers. The deformations are measured using an extensometer which analyses a length equal to the 30 % of the total specimen length, and compared with the ratio between the displacement of the tensile machine clamps and the length of the specimens.

The characteristic behaviour of FRCM materials under tension can be considered tri-linear. The first branch represents the uncracked state, where the slope of the stress–strain curve reflects the elastic modulus of the matrix. The second branch corresponds to the crack-formation: in this state several cracks gradually form due to the increase in the tensile stress applied. The length and smoothness of this portion of the curve depend on the quality of the bond between the fibres and matrix and the volume proportion of fibres in the composite activated by load transfer. The third branch represents the crack-widening. In this region only few new cracks appear and the existing cracks become wider. The specimen fails when the tensile strength of the fibres is reached. In the third region the slope reflects the elastic modulus of the dry fibres. Figure 9.3 shows the different phases of the stress-strain curve, the cracking of the specimens, and the failure mode.

The main parameters that could be analysed are:

- tensile stress and strain in the transition point between two phases (point T_1 and T_2), σ_{T1}, σ_{T2}, ε_{T1}, ε_{T2}
- elastic modulus of the three phases, E_1, E_2, E_3
- ultimate tensile stress and strain, ε_u, σ_u

In the first phase, the tensile stress is calculated by dividing the applied load by the area of the FRCM coupon in order to compare the elastic modulus with the one of the mortar, and the stress (σ_{T1}) of the uncracked mortar with the maximum tensile stress of the mortar. In the third phase, the elastic modulus and the maximum stress are computed with respect to the area of the longitudinal fibres.

Characteristic Behavior of FRCM Materials Under Tension (Politecnico di Milano)

Test results of tensile tests herein presented were defined in accordance with AC434 (2013) and the acceptance criteria for FRCM composites outlined in the draft of the Italian guidelines, which are yet to be published.

Tensile coupons are usually made in a flat mold by applying a first layer of cementitious mortar (approximately 4–5 mm), a layer of the fibre mesh which is evenly wetted with the fresh material, and a second layer of the cementitious mortar. The coupons should be cured for 28 days. The dimensions of the tensile coupons were 400 mm (length) × 40 mm (width) × 10 mm (thickness) (Fig. 9.4). During the curing phase attention should be paid to the possibility that some micro-cracks develop due to non-homogeneous shrinkage. Micro-cracking, a non-perfect planarity in the sample, and a non-constant thickness could have great influence on the test results.

At the ends of the samples FRP tabs of glass or carbon fibres were bonded in order to have a good stress distribution during the test and to avoid damages in the sample. Three different FRCM materials are presented as an example of material characterization via tensile test: FRCM with a mesh in poliparaphenylene benzobisoxazole fibres (PBO-FRCM), a mesh in carbon fibres (C-FRCM) and a mesh in glass fibres (G-FRCM).

Tensile tests were carried out with a testing machine with capacity of 100 kN under displacement control at an initial rate equal to 0.3 mm/min. The rate was increased to 0.5 mm/min after the first cracking phase. Axial strain was measured using an extensometer with a gauge length (between 100 and 200 mm) greater than 30 % of the total specimen length. The strain measured was then compared with the one obtained by dividing the stroke of the testing machine by the length of the specimens.

The tensile tests on PBO-FRCM coupons were characterized by a tri-linear behaviour; the results presented a large variability, in particular in the values corresponding to the point between the first and the second branch (T_1) and between the second and the third branch (T_2). This phenomenon is due to the non-constant dimension of the specimen section and to the first crack location with respect to the extensometer. The results showed a similarity between the stress in the mortar at point T_1, when the first crack appeared (3.5 MPa), and the tensile strength of the mortar defined by an indirect tensile test (4.7 MPa). The elastic modulus of the first branch corresponded to the one of the mortar. The elastic modulus of the third phase (218 GPa) was comparable with the elastic modulus of the dry PBO mesh (216 GPa) obtained from a tensile test on dry PBO rovings.

Tensile tests on G-FRCM coupons were characterized by a tri-linear behaviour in which it could be difficult to distinguish the second and third phase. The elastic modulus of the last branch (57 GPa) could be compared with the elastic modulus of the glass fibre grid (55 GPa).

In Fig. 9.5 a comparison between the results on the different materials is shown (Carozzi and Poggi 2015).

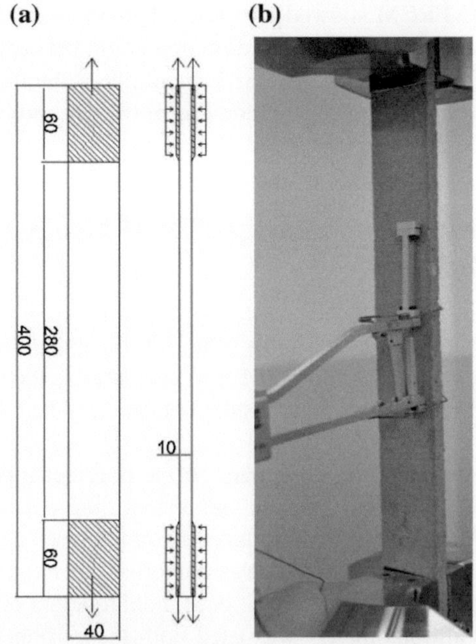

Fig. 9.4 **a** Specimens size. **b** Tensile test set-up

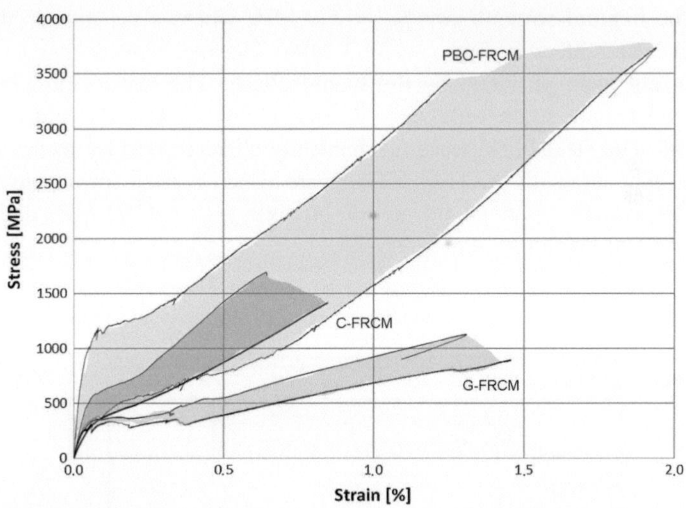

Fig. 9.5 Tensile tests results

Tensile tests on C-FRCM showed a tri-linear behaviour. Slippage was observed in many tests due to issues with bond and impregnation between the mortar and the dry carbon fibres. For this reason the elastic modulus of the third phase (191 GPa) could not be compared with the elastic modulus of the dry carbon textile (200 GPa).

Experimental Test to Characterize FRCM Composites (University of Padova)

The test procedure suggested by Hartig et al. (2007) and Jesse et al. (2009) was adopted by Pellegrino and D'Antino (2013) to characterize the overall behaviour of two different FRCM composites. The results obtained are here briefly revisited and commented on.

The specimens tested were comprised of a polymer-modified cementitious matrix used to embed a carbon fibre net and a steel fibre net. The carbon fibre net was comprised of longitudinal and transversal bundles with a width of approximately 5 mm each. The steel fibre net was comprised of longitudinal steel strands with diameter of approximately 0.48 mm^2 each, held together through a plastic fibre net. The same cementitious matrix was used both the carbon and steel fibre nets. The matrix was characterized in compression and bending according to the European Standard UNI EN 1015-11 (2007). Nine matrix prisms with dimensions 40 mm × 40 mm × 160 mm were cast from the same batch used to cast the FRCM tensile test specimens and were tested under bending and compression. The mean value of the flexural strength was $f_{flex} = 5.0$ MPa whereas the mean value of the compressive strength was $f_{c,matrix} = 39.3$ MPa. The mechanical characteristics of the carbon net were provided by the manufacturer. The elastic modulus, tensile strength, and tensile strain were $E_f = 240$ GPa, $f_f = 3800$ MPa, and $\varepsilon_f = 0.015$, respectively. The steel fibres were mechanically characterized by means of tensile tests. The indications of ASTM (1996) were used as base for the tests, which provided an average value of the tensile strength $f_{f,sf} = 3350$ MPa and a corresponding ultimate strain of $\varepsilon_{f,sf} = 0.0225$.

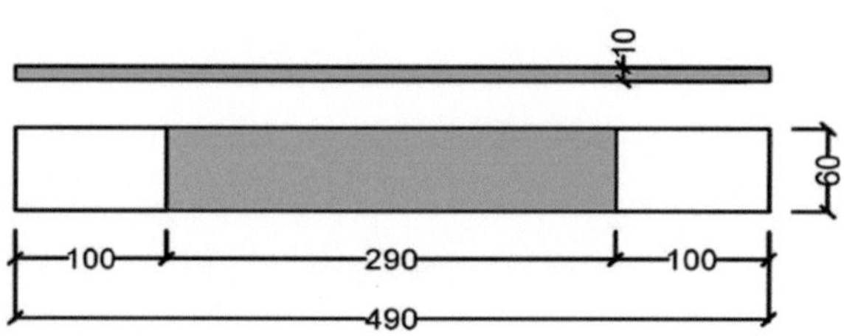

Fig. 9.6 FRCM specimen, dimension in mm

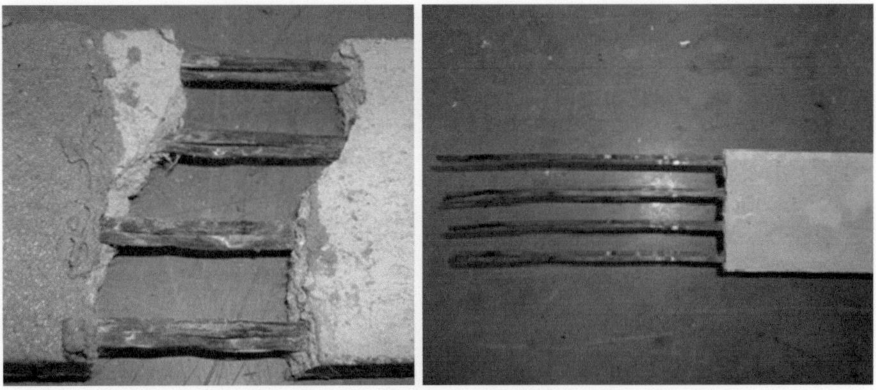

Fig. 9.7 Particulars of carbon FRCM specimens after failure

Twenty FRCM coupons, 10 with carbon fibres and 10 with steel fibres, were cast and tested in tension. The specimens were comprised of a fibre net ply embedded between two matrix layers that form a matrix prism (coupon) 490 mm long, 60 mm wide and 10 mm thick (Fig. 9.6).

In order to reduce the matrix roughness and avoid possible stress concentration, a thin layer of gypsum was applied at the ends of each specimen for a length approximately equal to 100 mm. Two steel plates, bolted together to assure uniform pressure on the clamped area of the specimen, were used to grip both ends. According to Hartig et al. (2012) a rubber layer was placed between the gypsum and the steel plates to avoid any possible local stress concentrations at the ends of the specimen. A DD1 strain transducer (gauge length = 100 mm) was applied at the centre of the specimen, and a load cell was used to record the load history.

The tests performed on carbon FRCM specimens did not provide significant results. Once the applied load reached the matrix tensile strength in one or more sections, the matrix cracked and the embedded fibres slid within the matrix without any increase of the tensile force (Fig. 9.7).

Although setup modifications were attempted in order to obtain a more effective clamping without inducing stress concentration, better results were not obtained. The failure mode observed may suggest that the length of clamping has to be properly designed but also that the surface treatment of the carbon fibres was not appropriate for this applications. Carbon fibres appeared smooth and clean after failure (Fig. 9.7), and this circumstance could indicate low bond properties between the fibres and the matrix. The results obtained from these tests indicate that the clamping system has to be properly designed and the surface treatment of the fibres has to be taken into account.

Steel fibres showed better adhesion to the cement-based matrix leading to good results, though complete failure of the composite due to steel rupture was not always reached. The opening of several transversal cracks in the cementitious matrix followed by the sudden failure of most of the steel strands was observed. During the tests, longitudinal splitting phenomena were also observed (Fig. 9.8).

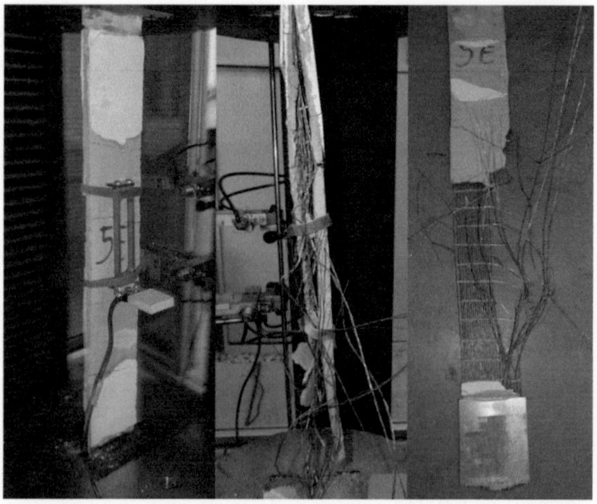

Fig. 9.8 Cracking, splitting and failure of the steel FRCM specimens

This phenomenon can be due to the fact that, although the cementitious matrix and the steel net have been properly designed for FRCM applications, steel strands were very close to each other, and the matrix penetration through the fibres was probably limited. This suggests that the spacing between strands has to be increased to improve the tensile behaviour of the FRCM system. The average ultimate stress obtained from these tests was 3290 MPa, which is very close to that obtained with the tensile test on the bare fibres reported above. Although the results of the tests on steel FRCM specimens were acceptable results in terms of ultimate load, the strain transducer did not provide useful measurements since it was strongly disturbed by matrix crack growth and propagation at different cross sections.

The test setup adopted does not seem to be suitable to characterize the FRCM composites presented in this study since it did not provide any information on their behaviour but was only able to find the ultimate load in case of steel FRCM specimens. Further investigations are needed to find a reliable, repeatable, and effective experimental test setup to characterized FRCM composites.

Analysis of the FRCM-Concrete Bond Behaviour

This section report the main results of an extensive experimental campaign carried out on FRCM-concrete joints to study the stress-transfer mechanism between the concrete support and the strengthening composite. The FRCM employed was comprised of a polyparaphenylene benzobisoxazole (PBO) fibre net embedded within a polymer-modified cementitious matrix. The parameters varied were the composite bonded length and bonded width. In addition, strain gauges were applied to the fibre net to study the stress-transfer mechanism. The PBO FRCM-concrete joints showed in this section were tested using a single-lap direct-shear test set-up.

Results on PBO FRCM-concrete joints tested using a double-lap direct-shear test set-up can be found in Sneed et al. (2015).

Test Set-up and Material Characteristics

Eighty-two FRCM-concrete joint specimens, herein presented, were tested using the single-lap direct-shear test set-up. The classical push-pull configuration was adopted where the fibres were pulled while the concrete prism was restrained (Smith and Teng 2002; Subramaniam et al. 2007, 2011; Carloni and Subramaniam 2012; Carloni et al. 2013; D'Antino et al. 2013, 2014). Two different concrete prisms were used, both had the same cross Section (125 mm width × 125 mm depth), but different lengths ($L = 375$ mm or $L = 510$ mm). The faces of the concrete blocks were sandblasted prior to applying the first (internal) layer of matrix. The nominal width b^* and average thickness t^* of one longitudinal fibre bundle were 5 and 0.092 mm, respectively. The matrix was applied only in the bonded area to embed the fibres and bond the composite to the concrete substrate (Fig. 9.9). Fibres were bare outside the bonded area. A single layer of PBO fibre net was applied onto the internal matrix layer and the transversal fibre bundles, which were all on one side of the longitudinal fibre bundles, were placed against the internal layer of matrix for some specimens. A second (external) 4 mm layer of matrix was applied over the PBO fibre net. The matrix layers had a thickness of 4 mm each as per the manufacturer's recommendations. The bonded width (b_1) and length (ℓ) of the composite were varied. Two aluminium plates (Fig. 9.9) were attached with a thermosetting epoxy to the end of the fibre strip to improve gripping during testing.

The concrete prism was restrained against movement by a steel frame bolted to the testing machine base. The direct-shear tests were conducted under displacement control using a close-loop servo-hydraulic universal testing machine. Two linear variable displacement transducers (LVDTs) were mounted on the concrete surface

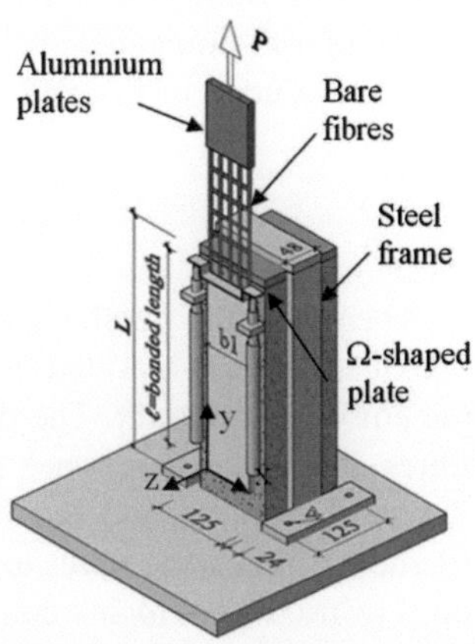

Fig. 9.9 Single-lap direct-shear test set-up

close to the edge of the composite bonded region. The LVDTs reacted off of a thin aluminium Ω shaped bent plate that was attached to the PBO transversal fibre bundle surface adjacent to the beginning of the bonded area (Fig. 9.9). The average of the two LVDT measurements, defined as the global slip g, was used to control the test with a constant rate of 0.00084 mm/s. The readings of the two LVDTs can be also used to study the distribution of the applied load among the width of the composite (Carloni et al. 2014). The applied load is termed P in this section.

From the same batch used to cast the concrete prisms, twelve (6 + 6) 100 mm × 200 mm cylinders were cast. Their average compressive strength (ASTM 2011a) and splitting tensile strength (ASTM 2011b) were 42.5 MPa (CoV = 0.013) and 3.4 MPa (CoV = 0.113) for the shorter blocks (L = 375 mm), and 33.5 MPa (CoV = 0.085) and 3.0 MPa (CoV = 0.042) for the longer blocks (L = 510 mm). At least two 50 mm × 100 mm cylinders were cast from each batch of matrix used to cast the FRCM composite. The average compressive (ASTM 2011a) and splitting tensile strengths (ASTM 2011b) of the matrix were 28.4 MPa (CoV = 0.092) and 3.5 MPa (CoV = 0.231), respectively. The bare PBO fibres were tested in tension as prescribed in ASTM (1996). The average measured tensile strength, ultimate strain, and elastic modulus were 3014 MPa (CoV = 0.068), 0.0145 (CoV = 0.104), and 206 GPa (CoV = 0.065), respectively.

Experimental Tests

The direct-shear test specimens were named following the notation DS_X_Y_S_D_Z^{T}, where X = bonded length (ℓ) in mm, Y = bonded width (b_1) in mm, S (if present) indicates that strain gauges were mounted on the specimen, D (if present) denotes that the specimen was tested until a constant load at the end of the test was measured, and Z = specimen number (Table 9.1). A superscript T after Z indicates that the fibre net was oriented with the transversal fibre bundles directly against the matrix internal layer. The peak load P^* of each specimens tested is reported in Table 9.1.

In order to compare specimens with different bonded widths, the ultimate stress σ^* is introduced in Eq. (9.1):

$$\sigma^* = \frac{P^*}{nt^*b^*} \tag{9.1}$$

where n is the number of longitudinal bundles.

All specimens reported in Table 9.1 failed due to debonding of the fibres within the embedding matrix. The debonding was associated with considerable slip of the fibres. For this reason, once the debonding initiates, the applied load is due to both the residual bond and the friction between fibres and matrix and between fibre filaments. When the fibres are completely debonded the applied load, which is due only to friction, remains constant ad the global slip increases. As an example, the

Table 9.1 Specimen tested in D'Antino et al. (2014)

Name	P^*(kN)	Name	P^*(kN)	Name	P^*(kN)
DS_100_34_1^T	1.92	DS_330_34_1^T	3.00	DS_330_80_2	8.84
DS_100_34_2^T	0.97	DS_330_34_2^T	3.51	DS_330_80_3	8.28
DS_100_34_3^T	1.62	DS_330_34_7	4.07	DS_330_80_D_1	8.90
DS_100_60_1	3.69	DS_330_34_8	4.02	DS_330_80_D_2	8.68
DS_100_60_2	3.83	DS_330_34_9	3.44	DS_330_80_D_3	8.90
DS_100_60_3	3.77	DS_330_43_1^T	4.43	DS_330_80_D_4	8.42
DS_150_34_1^T	2.22	DS_330_43_2^T	5.25	DS_330_80_D_5	8.58
DS_150_34_2^T	1.55	DS_330_43_3	5.27	DS_450_34_1	3.77
DS_150_34_3^T	2.87	DS_330_43_5	4.79	DS_450_34_2	3.85
DS_150_34_4^T	2.34	DS_330_43_6	5.09	DS_450_34_3	3.97
DS_150_60_1	5.25	DS_330_43_S_1^T	4.48	DS_450_60_1	6.40
DS_150_60_2	5.04	DS_330_43_S_2^T	5.12	DS_450_60_2	6.34
DS_150_60_3	3.05	DS_330_43_S_3^T	3.03	DS_450_60_3	6.44
DS_200_34_1	3.05	DS_330_43_S_5	4.03	DS_450_60_4	5.77
DS_200_34_2	2.52	DS_330_60_1^T	7.05	DS_450_60_5	6.51
DS_200_34_3	3.44	DS_330_60_2^T	6.56	DS_450_60_6	6.79
DS_200_60_2	5.66	DS_330_60_3^T	6.06	DS_450_60_7	6.65
DS_200_60_3	5.44	DS_330_60_4^T	6.50	DS_450_60_D_1	7.01
DS_200_60_4	6.58	DS_330_60_5^T	6.28	DS_450_60_D_2	6.67
DS_250_34_1^T	2.61	DS_330_60_6	7.01	DS_450_60_D_3	7.33
DS_250_34_2^T	2.11	DS_330_60_D_1	8.29	DS_450_60_S_1	6.63
DS_250_34_3^T	2.82	DS_330_60_D_2	7.12	DS_450_80_1	8.62
DS_250_34_4	3.21	DS_330_60_D_3	6.56	DS_450_80_2	9.07
DS_250_34_5	2.89	DS_330_60_D_4	5.24	DS_450_80_3	9.32
DS_250_34_6	3.61	DS_330_60_D_5	6.69	DS_450_80_4	8.86
DS_250_60_1	6.68	DS_330_60_S_1	6.30	DS_450_80_5	10.04
DS_250_60_2	6.17	DS_330_60_S_2	7.31		
DS_250_60_3	5.70	DS_330_80_1	8.47		

load responses of specimen DS_330_60_D_5, DS_330_80_D_1, and DS_450_60_D_1 are reported in Fig. 9.10.

It should be noted that some specimens, not reported in Table 9.1, failed due to fibre rupture outside the bonded area caused by the non-uniform distribution of the load among the different longitudinal bundles (D'Antino et al. 2014).

Effect of the Bonded Width

The effect of the bonded width on the behaviour of FRCM-concrete joints was investigated by comparing the ultimate stress σ^* with different composite bonded

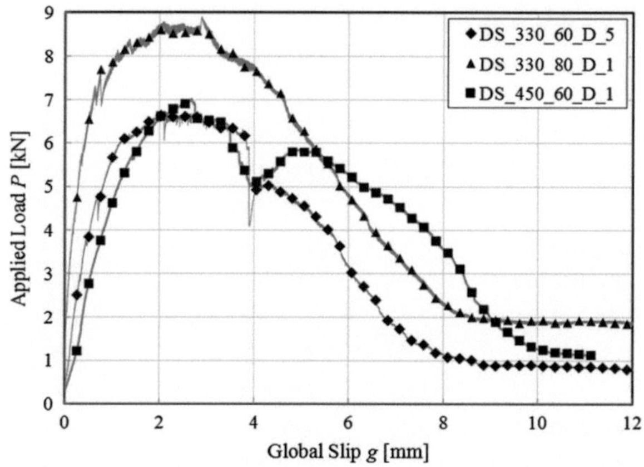

Fig. 9.10 Applied load P^* versus global slip g plot for specimens DS_330_60_D_5, DS_330_80_D_1, and DS_450_60_D_1

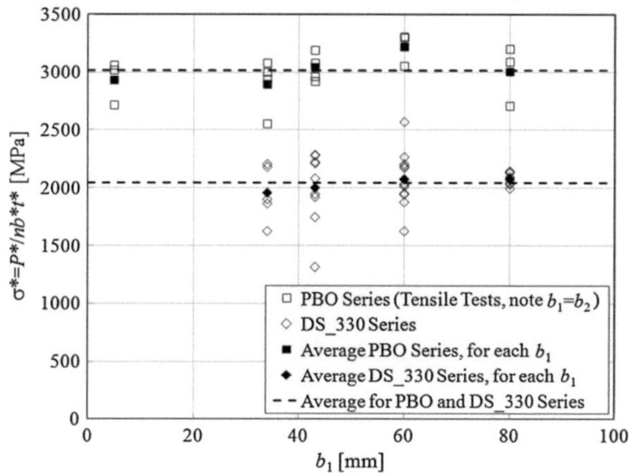

Fig. 9.11 Comparison of the ultimate stress σ^* versus bonded width b_1 for single-lap direct-shear tests (DS_330 Series) and tensile tests (PBO Series)

widths (b_1) for the test specimens with a bonded length $\ell = 330$ mm. Figure 9.11 reports the variation of the σ^* for different widths in case of the direct-shear tests (DS_330 Series) and of the bare fibre tensile test (PBO Series). For each width the average value of the ultimate stress is reported with a black filled marker. The average values of the ultimate stress for all tensile tests and all single-lap shear tests plotted in Fig. 9.11 are reported with dashed lines. Although it is possible that a width effect exists considering a single bundle of fibres (Banholzer 2004), Fig. 9.11 suggests that a *global* width effect does not exist among multiple bundles and therefore for the entire width of the composite.

Strain Measurements

Nine specimens were instrumented with strain gauges applied directly to the central fiber bundle along the bonded length and to the central and edge fiber bundles outside the bonded area.

The axial strain profiles corresponding to different stages of the load response of specimen DS_330_43_S_5 are plotted in Fig. 9.12. The reference system is shown in Fig. 9.9. The load response of DS_330_43_S_5 is plotted in Fig. 9.13 to show the points corresponding to the strain profiles.

The strain profiles of Fig. 9.12 are similar to those obtained from direct-shear tests of the FRP-concrete interface (Carloni and Subramaniam 2012; Pellegrino et al. 2008), which suggests that a cohesive material interfacial law can be obtained for FRCM-concrete joints. In this study, the strain profiles obtained experimentally will be approximated by Eq. (9.2) (Carloni and Subramaniam 2012):

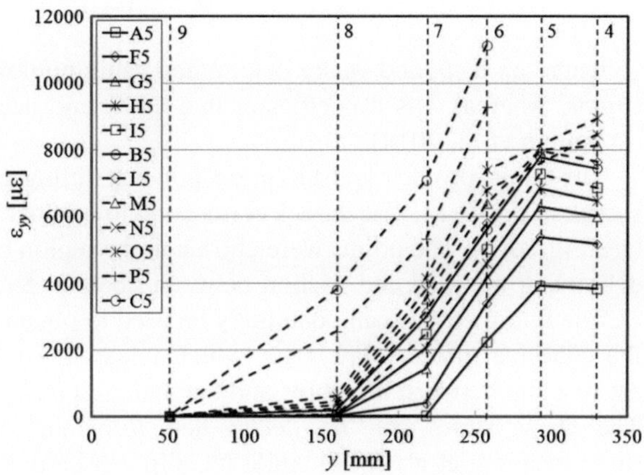

Fig. 9.12 Axial strain profile of specimen DS_330_43_S_5

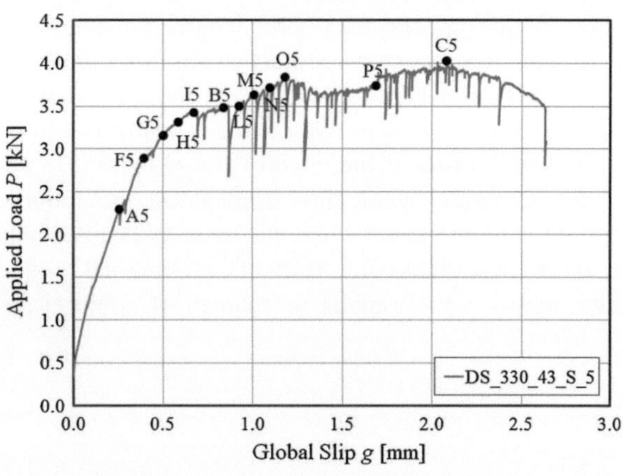

Fig. 9.13 Load response of specimen DS_330_43_S_5

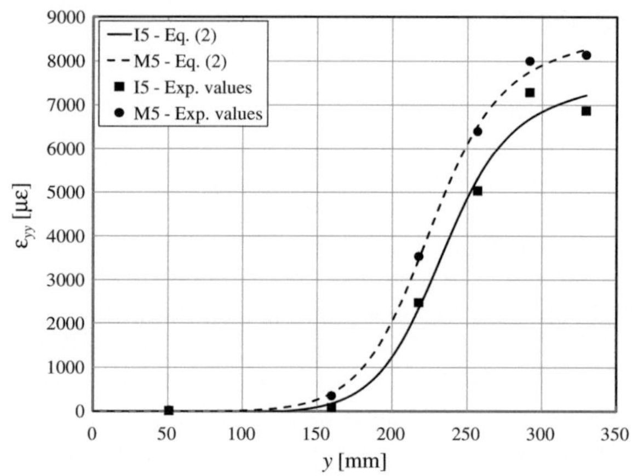

Fig. 9.14 Strain profiles and fitting curves corresponding to points I5 and M5 in the load response of DS_330_43_S_5 (Fig. 9.5)

$$\varepsilon_{yy} = \varepsilon_0 + \frac{\alpha + ky}{1 + e^{-\frac{y-y_0}{\beta}}} \tag{9.2}$$

where ε_0, α, β, and y_0 are determined using nonlinear regression analysis of the strains, whereas k is a coefficient that takes into account the presence of friction (D'Antino et al. 2014).

The strain profiles and the corresponding fitting curves based on Eq. (9.2) are shown in Fig. 9.14. The curves correspond to points I5 and M5 in the load response (see Fig. 9.13). The points were chosen in the region of the load response where the debonding initiated and friction between fibers and matrix is not present yet.

The fitting curves can potentially be used to obtain the effective bond length and the cohesive material law τ_{zy}-s, which relates the interfacial shear stress and the relative slip between the fibres and the matrix at any point along the bonded length. The shear stress can be determined from the gradient of the deformation (Subramaniam et al. 2007) while the slip can be determined by integration of ε_{yy}. The strain analysis carried out on the specimens equipped with strain gauges showed that an effective bond length, i.e. the minimum length needed to fully establish the stress-transfer mechanism (Subramaniam et al. 2011), exists. Provided that the friction between fibres and matrix and between fibre filaments is clearly identified, the strain profiles corresponding to the debonding load were analysed providing a value of the effective bond length l_{eff} of approximately 260 mm.

Strain gauges were also applied on the longitudinal fiber bundles of PBO FRCM-concrete joints where the external layer of matrix was omitted. Results were used to investigate the role of the internal and external matrix layer in the fiber-matrix stress-transfer mechanism (Carloni et al. 2014).

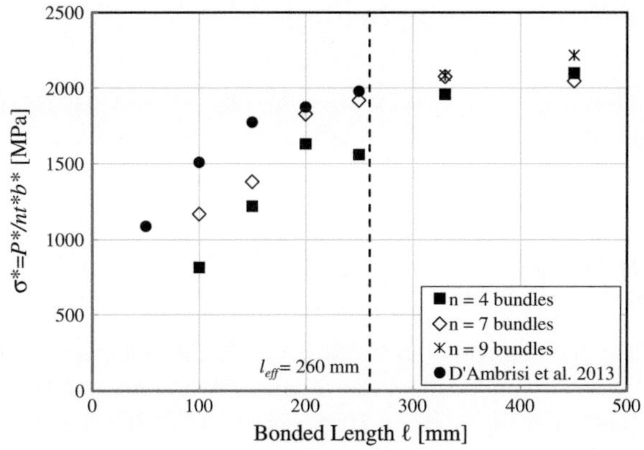

Fig. 9.15 Variation of the ultimate stress σ^* as a function of the bonded length ℓ

Effect of the Bonded Length

Figure 9.15 shows the average of the ultimate stress σ^* for each bonded width tested as a function of the bonded length ℓ for the tests herein presented. Three different bonded widths, namely 34 mm (n = 4), 60 mm (n = 7), and 80 mm (n = 9) are considered. The results of the double-lap shear tests published by D'Ambrisi et al. (2012, 2013b, c) are included for comparison in terms of the average of the ultimate stress σ^* for each bonded width. The results of Fig. 9.15 show that the ultimate stress σ^* increases as the bonded length ℓ increases up to a value equal to 450 mm. However, the ultimate stress appears to increase linearly when the bonded length is greater than 250 mm, confirming the existence of the effective bond length l_{eff}. The increase of the applied load when $\ell > l_{eff}$ is due to the friction between fibres and matrix and between fibre filaments.

Final Remarks on the Bond Behaviour for PBO-FRCM Composites

The results of single-lap direct-shear tests conducted on PBO FRCM-concrete indicate that the failure is characterized by debonding of the fibre within the embedding matrix. After the onset of debonding, the applied load is due both to the residual bond and the friction between matrix and fibres and between fibre filaments. The analysis of the peak stress for specimens with the same bonded length but different bonded width showed that, although a width effect within the single bundle can be recognized, a global width effect does not exist among multiple bundles and therefore for the entire width of the composite. Finally, the strain profiles along the bonded length in the load direction allowed for determining the value of the effective bond length.

Flexural Strengthening

This section provides an overview on the use of FRCM materials for flexural strengthening of RC structures. Experimental works are presented and discussed.

Introduction

The mechanical effectiveness of FRCM materials is strongly influenced by the bond between single fibers and matrix (Badanoiu and Holmgren 2003; Banholzer 2004; Banholzer et al. 2006; Hartig et al. 2008; Häuβler-Combe and Hartig 2007; Hegger et al. 2006; Zastrau et al. 2008), which in turn is related to the matrix capacity of wetting single filaments (Badanoiu and Holmgren 2003; Banholzer 2004; Banholzer et al. 2006; Hartig et al. 2008; Häuβler-Combe and Hartig 2007; Hegger et al. 2006; Zastrau et al. 2008), the bond between external fibers, directly in contact with the matrix, and internal fibers (Badanoiu and Holmgren 2003; Hartig et al. 2008; Häuβler-Combe and Hartig 2007), the contribution of joints between longitudinal and transverse fibers (Soranakom and Mobasher 2009; Peled et al. 2008), and the cracking of the cement based matrix (Curbach et al. 2006; Ortlepp et al. 2004, 2006). Moreover, the matrix moderate capacity to penetrate into the free spaces among the fibers and the poor shear transfer among the single filaments can cause non uniformity of tensile stress in fibers of a roving and consequent fibers telescopic failure (Banholzer 2004; Banholzer et al. 2006; Hartig et al. 2008; Häuβler-Combe and Hartig 2007; Hegger et al. 2006; Zastrau et al. 2008). In addition to these peculiarities, which are related to the characteristics of the FRCM system itself, in the case of FRCM materials used for the external strengthening of RC elements the strengthening effectiveness is also affected by bond between the cement-based matrix and the concrete substrate (Curbach et al. 2006; Ortlepp et al. 2004, 2006) and by the strength of the RC concrete substrate (Brückner et al. 2006; Curbach et al. 2006; Ortlepp et al. 2004, 2006). All these phenomena depend on the type, the surface treatment, and the geometrical arrangement of fibres, on the composition and grain fineness of the matrix, on the quality and the surface treatment of the concrete substrate. The shape of the fabric is of great importance (Curbach et al. 2006; Ortlepp et al. 2004, 2006) as the embedding of the rovings is ensured by the free spaces present in between the bundles because of the moderate cement based matrix capacity to impregnate single fibres due to its granularity. For this reason the surface of the matrix/roving interface should be maximized, e.g. by reducing the rovings dimension and simultaneously increasing the number of rovings, (Badanoiu and Holmgren 2003). Moreover, the presence of the fabric itself reduces the capacity of the matrix to transfer shear and tensile stresses through the thickness of the composite due to the surface reduction (Ortlepp et al. 2004, 2006).

Literature Review

Experimental and theoretical research works show that FRCM materials can be used effectively for the flexural strengthening of RC structures. Nevertheless, the comparison of results published by different researchers is difficult because of the differences in materials considered and test methods employed. FRCM systems comprised of different types of fibers, namely carbon, PBO, AR glass, steel, and basalt fibers, and different matrices have been considered by researchers for flexural strengthening applications of RC beams and slabs. In particular, carbon fibers were considered by Hashemi and Al-Mahaidi (2012a, b), Wiberg (2003), Pareek et al. (2007), Triantafillou (2010) and Täljsten and Blanksvärd (2007), PBO fibers were considered by Ombres (2009, 2011, 2012), and Bisby et al. (2009), basalt fibers were considered by Elsanadedy et al. (2013), and AR glass fibers were considered by Brückner et al. (2006). A comparison of FRCM systems with different types of fibers are reported in the works of Weiland et al. (2006) (AR glass vs. carbon), D'Ambrisi and Focacci (2011) (PBO vs. carbon) and Pellegrino and D'Antino (2013) (carbon vs. steel). In almost all of the aforementioned papers (Täljsten and Blanksvärd 2007; Pareek et al. 2007; Hashemi and Al-Mahaidi 2012a, b; Ombres et al. 2009; Bisby et al. 2009; Elsanadedy et al. 2013; D'Ambrisi and Focacci 2011; Pellegrino and D'Antino 2013; Triantafillou 2010) the flexural behavior of FRCM strengthened is compared with similar beams strengthened with FRP composites. The comparison indicated a similar strengthening effectiveness for FRCM and FRP composites, although in the most cases the FRP composites appear to be more effective than the FRCM composites when the same fibre cross section is considered. This fact is due to the better fibre impregnation by the epoxy resin, when compared to the cement-based mortar, which induces a uniform tensile stress among the fibres of the bundle.

FRCM materials have been used in several strengthening applications. Some applications are described in ACI (2013); the design criteria adopted for the FRCM-strengthening of a railway bridge are described in D'Ambrisi et al. (2013a, 2015).

Mechanical Effectiveness

The effectiveness of FRCM composites for flexural strengthening of RC beams is usually evaluated performing four points bending tests of unstrengthened and FRCM-strengthened RC beams (Fig. 9.16). Load deflection curves of unstrengthened and FRCM strengthened beams are compared in terms of maximum load and stiffness.

Referring to Fig. 9.16, experimental results available in the technical literature are relative to beams with ratio H/B ranging between 0.4 and 1.8 (prestressed beams and slabs are excluded) and ratio L_1/L ranging between 0.3 and 0.42. A very wide range can be observed in the ratio $\rho_f = A_f/A_c$ where A_f is the fiber cross-sectional area and A_c is the gross concrete cross-sectional area: this parameter ranged

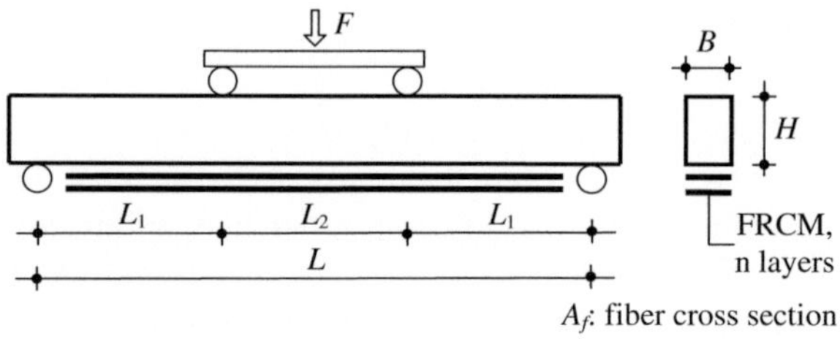

Fig. 9.16 Typical test setup for FRCM strengthened beams

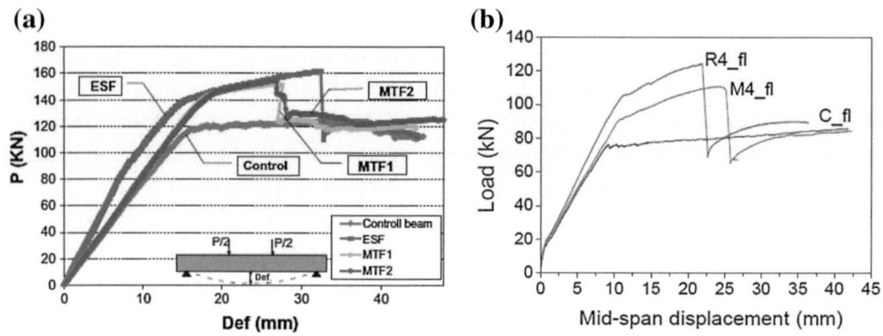

$B = 140$ mm, $H = 260$ mm, $A_s = 339$ mm^2, $A_f = 28.2$ mm^2, $L = 2300$ mm, $L_1 = 700$ mm. ESF: epoxy resin, MTF: cementitious mortar.

$B = 150$ mm, $H = 250$ mm, $A_s = 226$ mm^2, $A_f = 22.6$ mm^2, n = 4, $L = 2000$ mm, $L_1 = 750$ mm. R4_fl: epoxy resin, M4_fl: cementitious mortar.

Fig. 9.17 Experimental results of carbon FRCM strengthened RC beams obtained by **a** Hashemi and Al-Mahaidi (2012a, b) and **b** Triantafillou (2010); A_s is the cross-sectional area of steel rebar in tension

between 0.018 and 0.587 %. The number of fibre layers ranged between 1 and 10 but 2, 3, and 4 layers are the most frequently adopted numbers of layers.

The percentage flexural capacity increase depends on the specimens shape and the ratio between the area of steel and the area of fibers. The flexural capacity for carbon-FRCM composites increased from 10 to 100 %. It should be observed that the highest values correspond to the adoption of polymer-coated carbon fibre grid bonded to the concrete substrate with a cement based mortar (Täljsten and Blanksvärd 2007; Pareek et al. 2007). In the cases of dry fibres the flexural capacity increase typically ranged between 15 and 45 % with the application of 1–4 layers of fabric (Fig. 9.17).

In the case of PBO fibres the flexural capacity increase ranged between 15 and 150 %; typical values are around 20–50 % with the adoption of 1–4 layers of fabric (Fig. 9.18). In the case of steel fibres an increase of 24 % was found by Pellegrino and D'Antino (2013) in their experimental work on precast prestressed beams (Fig. 9.19).

Finally, in the case of basalt fibres a flexural capacity increase between 40 and 100 % was found in the experimental work of Elsanadedy et al. (2013) with the adoption of 5–10 fabric layers (Fig. 9.20).

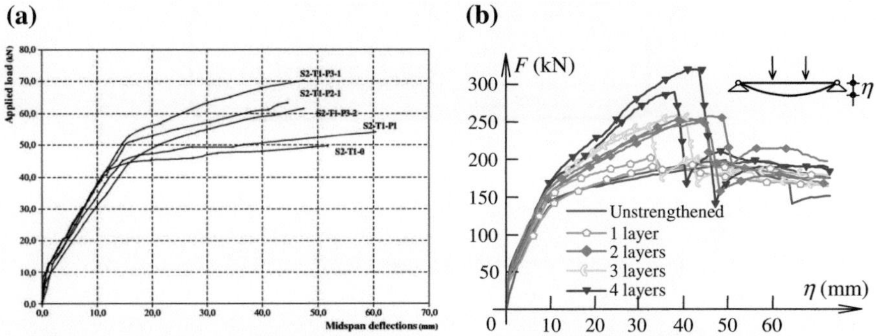

$B = 150$ mm, $H = 240$ mm, $A_s = 157$ mm^2, $A_f = 6.75$ mm^2/layer, $L = 2700$ mm, $L_1 = 900$ mm. S2-T1-P1: 1 layer; S2-T1-P2: 2 layers; S2-T1-P3: 3 layers.

$B = 400$ mm, $H = 250$ mm, $A_s = 462$ mm^2, $A_f = 18.0$ mm^2/layer, $L = 2200$ mm, $L_1 = 750$ mm.

Fig. 9.18 Experimental results of PBO FRCM strengthened RC beams obtained by **a** Ombres (2011) and **b** D'Ambrisi and Focacci (2011); A_s is the cross sectional area of steel rebar in tension

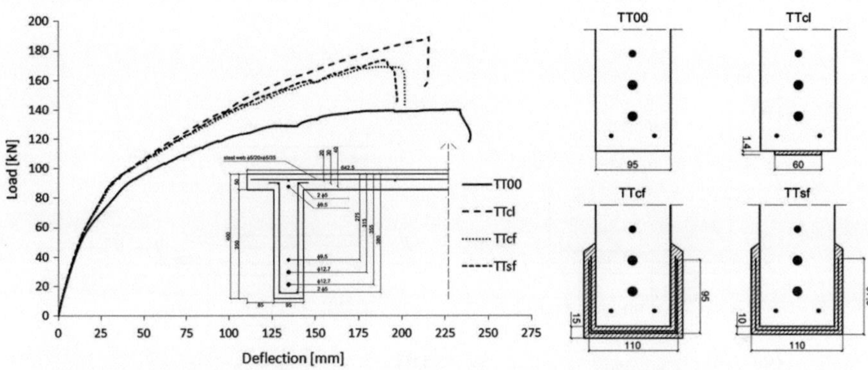

$L = 10$ m; $L_1 = 3588$ mm;

TT00: unstrengthened; TT cl: FRP strengthened with pultruded laminate ($A_f = 168$ mm$_2$, $n = 1$);

TTcf: carbon FRCM strengthened ($A_f = 70.2$ mm^2, $n = 1$);

TTsf: steel FRCM strengthened ($A_f = 134.2$ mm^2, $n = 1$).

Fig. 9.19 Experimental results of prestressed concrete beams strengthened with carbon and steel FRCM composites (Pellegrino and D'Antino 2013)

Failure Modes

The majority of the FRCM strengthened beams tested by the aforementioned authors failed due to the loss of composite action related to the debonding of the strengthening material from the supporting concrete. In FRC-strengthened beams the debonding can occur: at the matrix-fibre interface, at the matrix-concrete interface, and within the concrete. Depending on the type of fibres and matrix,

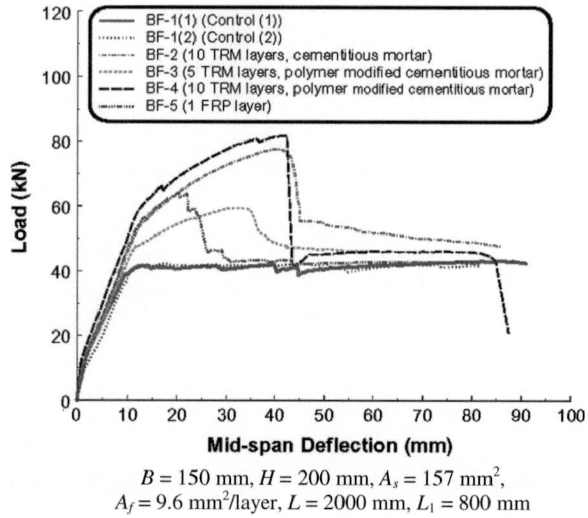

Fig. 9.20 Experimental results RC beams strengthened with basalt FRCM (Elsanadedy et al. 2013)

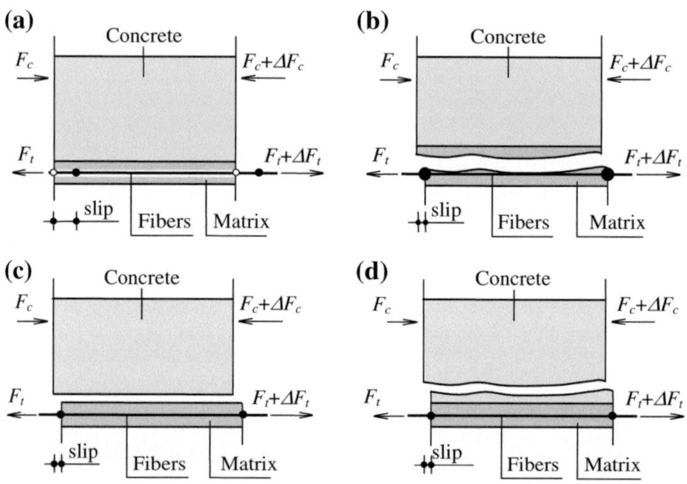

Fig. 9.21 Different possible debonding surfaces

different flexural debonding failure modes can be identified (D'Ambrisi and Focacci 2011), as schematically shown in Fig. 9.21.

Referring to Fig. 9.21, the failure modes reported in the literature are:

(a) Debonding of fibres from the matrix with high matrix-fibre slips (up to 1–2 mm) in the maximum bending moment region (Fig. 9.21a); this mode is typical for FRCM composites with dry carbon fibres embedded in a cement-based matrix; although it has been also observed in a few cases with dry PBO fibres embedded in a cement-based matrix. During the matrix-fibres slip, the total force carried by fibres gradually decreases due to the gradual rupture of the fibres, eventually damaged by friction (Badanoiu and Holmgren 2003; Banholzer 2004; Banholzer et al. 2006; Häuβler-Combe and Hartig 2007; Zastrau et al. 2008). This type of debonding essentially involves the matrix-fibre and the fibre–fibre bond surfaces and it is not affected by the mechanical properties of the concrete substrate.

(b) Delamination with fracture surface within the matrix, preceded by considerable matrix-fibre slips (Fig. 9.21b); this failure mode has been observed in several beams strengthened with dry PBO fibres embedded in a cement-based matrix. It is essentially related by the matrix-fibre bond and the matrix tensile and shear strengths, while it is not affected by the mechanical properties of the supporting concrete.

(c) Deboning at the matrix-concrete interface, (Fig. 9.21c); this very brittle mode has been rarely observed and can generally be avoided with a proper preparation of the concrete surface before the application of the strengthening material. It involves the matrix-concrete bond.

(d) Debonding of the strengthening material with fracture surface within the concrete (a thin layer of concrete remains attached to the debonded strengthening material) (Fig. 9.21d). This very brittle mode has been rarely observed for FRCM strengthening systems. It is typical for FRP strengthened beams and essentially involves the mechanical properties of the supporting concrete.

The aforementioned debonding mechanisms are consistent with those described in (Ortlepp et al. 2004, 2006), D'Ambrisi et al. (2012, 2013a, b) and Carloni et al. (2013) for the case of simple or double shear tests.

Figure 9.22 shows the schematic shape of load-deflection diagrams of FRCM flexurally strengthened RC beams. After the steel bars yield (point A), the slope of the strengthened beams diagrams is related to the axial stiffness of the intrados strengthening material. In this phase small matrix-fibre slips occur locally at the

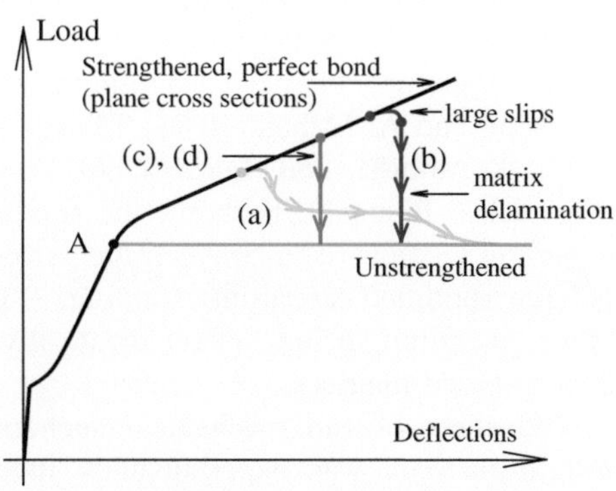

Fig. 9.22 Typical load-deflection curves of FRCM strengthened RC beams

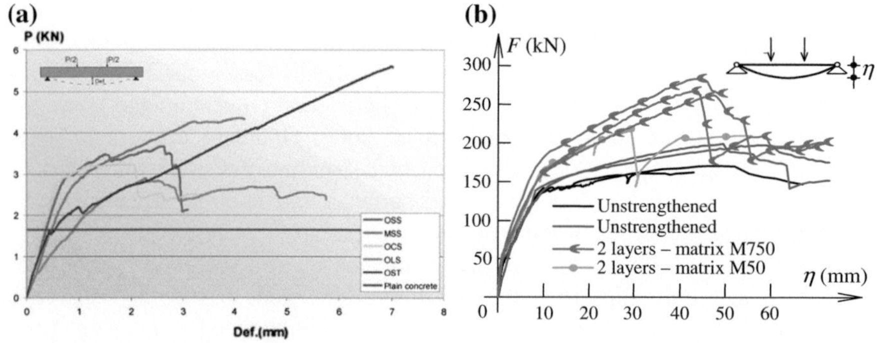

Carbon fibers. $A_f = 0$, $L = 750$ mm.
OCS: ordinary Portland cement - sheet.
OSS: silica fume incorporated – sheet.
OLS: polymer-modified mortar – sheet.
MSS: micro cement added to OC mortar – sheet.
OST: silica fume incorporated – textile.

PBO fibers. $B = 400$ mm, $H = 250$ mm,
$A_s = 462$ mm^2, $A_f = 36$ mm^2,
$L = 2200$ mm, $L_1 = 750$ mm.

Fig. 9.23 Effect of the matrix composition: experimental results of FRCM strengthened RC beams obtained by **a** Hashemi and Al-Mahaidi (2008) and **b** D'Ambrisi and Focacci (2011)

main flexural crack locations; nevertheless, the assumption of plane cross sections, including the strengthening material, allows for a good prediction of the experimental results. The post-yielding branch of the load-deflection diagram continues up to the loss of the strengthening action.

The loss of strengthening action can be sudden or more gradual, depending on the failure mode (a), (b), (c) or (d) (Fig. 9.21), corresponding to the schematic descending branches (a), (b), (c) or (d) of the load-deflection diagrams reported in Fig. 9.22.

The analytical prevision of the failure load requires the preliminary determination of the expected debonding failure mode among the four described above. This is not a simple task because it requires the determination of the hierarchy among the strengths related to the different bond failure surfaces. The hierarchy, in turn, depends on the type of fibres, type of matrix, fibres arrangement in the fabric, mechanical properties of matrix, and supporting concrete. Moreover, the strength hierarchy could also depend on the adopted number of fabric layers.

The important role played by the matrix-fibre bond induced several researchers (Hashemi and Al-Mahaidi 2008; Täljsten and Blanksvärd 2007; Elsanadedy et al. 2013; D'Ambrisi and Focacci 2011) to test different matrix compositions (Fig. 9.23). Figure 9.20 shows the results published by Elsanadedy et al. 2013 regarding a comparison between the performances of a cementitious mortar and a polymer modified cementitious mortar. These results confirm that, for a given fibre fabric, the strengthening effect drastically change with the adoption of different cement-based matrices.

The effect of end mechanical anchors applied to the flexural strengthening FRCM material was experimentally investigated by Hashemi and Al-Mahaidi

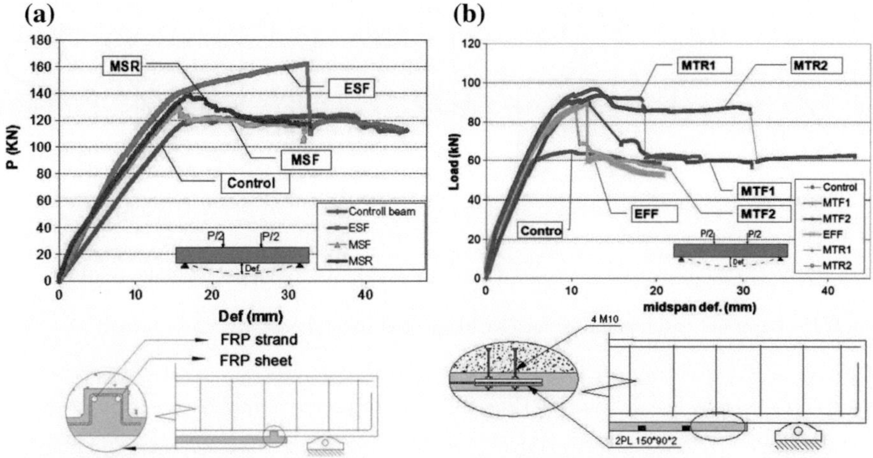

Carbon fibers. $B = 140$ mm, $H = 260$ mm, $A_s = 339\ \text{mm}^2$, $A_f = 28.2\ \text{mm}^2$, $n = 2$; $L = 2300$ mm, $L_1 = 700$ mm. ESF: epoxy resin, MTF: cementitious mortar without connectors, MSR: cementations mortar with end connectors.

Carbon fibers. $B = 120$ mm, $H = 180$ mm, $A_s = 226\ \text{mm}^2$, $A_f = 30\ \text{mm}^2$, $L = 1300$ mm, $L_1 = 550$ mm. EFF: epoxy resin, MTF: cementitious mortar without connectors, MTR: cementations mortar with end connectors.

Fig. 9.24 Effect of the end anchorage of FRCM strengthening materials: results obtained by **a** Hashemi and Al-Mahaidi (2012a) and **b** Hashemi and Al-Mahaidi (2012b)

(2012a, b). Their results are sumarized in Fig. 9.24, where it can be observed that this type of devices can effectively increase the FRCM debonding load.

Design Considerations

As described in the previous section, in the case of FRCM external strengthening of RC elements the bond stress transfer is a very complex phenomenon depending on type, surface treatment, geometrical arrangement of fibres, composition and grain fineness of the matrix, and on quality and surface treatment of the concrete. Despite this complexity, all the proposed design approaches (ACI 2013; Wiberg 2003; Curbach et al. 2006; Brückner et al. 2006; Ombres et al. 2009; Ombres 2011) are based on the assumption (Bernoulli) that cross sections remain plane during the deformation up to the flexural capacity (Fig. 9.25). The presence of the FRCM layer(s) is accounted for by introducing an additional term to the equations usually employed for un-strengthened RC elements. In a real application the FRCM material is applied when a preexisting tensile strain ε_0 exists (Fig. 9.25) in the concrete surface where the strengthening material is applied on, due to the dead loads. This approach requires the definition of the maximum tensile strain ε_{fe} attained by the fibres, called effective strain. Since the experimental works show that the failure of FRCM strengthened

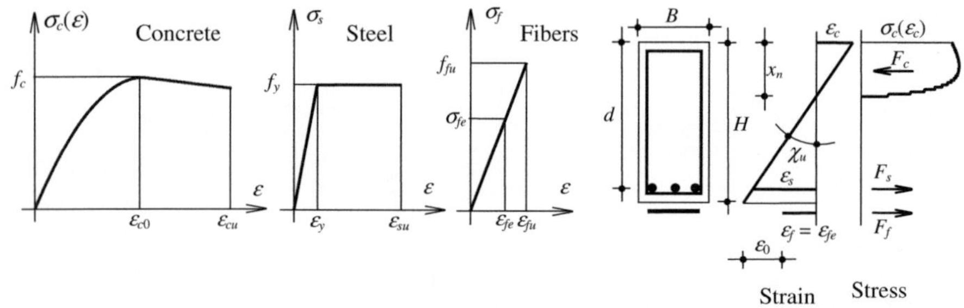

Fig. 9.25 Bernoulli assumption for an FRCM flexural strengthened RC cross section

beams is caused by the debonding (Fig. 9.21), the effective strain has to account for the loss of bond.

Following the guidelines ACI (2013), the effective strain is experimentally evaluated according to ICC (2013). Assuming that the member failure is reached when the effective tensile strain ε_{fe} in the FRCM reinforcement is attained and that the steel is yielded at failure, the flexural bending capacity is

$$M_R = B \cdot \int_0^{x_n} \sigma_c(\chi_u \cdot \xi) \cdot \xi \, d\xi + A_s \cdot f_{yd} \cdot (d - x_n) + A_f E_f \cdot \varepsilon_{fe} \cdot (H - x_n) \quad (9.3)$$

where $\sigma_c(\varepsilon)$ is the concrete compressive constitutive law, A_s is the cross-sectional area of the steel longitudinal reinforcement, A_f is the fibre cross-sectional area, E_f is the fibre elastic modulus. χ_u is the curvature at failure obtained as:

$$\chi_u = \frac{\varepsilon_{fe} + \varepsilon_0}{H - x_n} \quad (9.4)$$

The depth of neutral axis x_n is given by the equation

$$B \cdot \int_0^{x_n} \sigma_c(\chi_u \cdot \xi) \cdot d\xi = A_s \cdot f_y + A_f E_f \cdot \varepsilon_{fe} \quad (9.5)$$

In Eq. (9.4) ε_0 is the tensile strain of the concrete surface where the FRCM material is applied at the time of the FRCM application; the remaining symbols adopted in Eqs. (9.1)–(9.5) are defined in Fig. 9.25. In the most cases the flexural capacity can be approximated by

$$M_R = 0.9 \cdot A_s \cdot f_y \cdot d + 0.9 \cdot H \cdot A_f E_f \cdot \varepsilon_{fe} \quad (9.6)$$

The application of this approach requires only the definition of the effective strain of the fibres, which represents a strain limit globally accounting for the loss of bond. Similar approaches are suggested in CNR (2004), FIB (2001) and ACI (2008) for the case of flexural strengthening of RC structures with FRP composites. An important difference between the cases of FRP and FRCM materials is related to the type of debonding failure. In the case of flexural strengthening with FRP composites, the debonding usually occurs within the concrete. Therefore the effective strain ε_{fe} strongly depends on the concrete mechanical properties. In the case of flexural strengthening with FRCM composites debonding can occur at different bond surfaces, depending on the fibre type and arrangement, the type of matrix, and the mechanical properties of the concrete substrate. Therefore, the effective strain has to be evaluated for the particular FRCM material adopted. Since the most frequently observed failure modes are cases (a) and (b) represented in Fig. 9.21, the debonding strain should depend on the matrix-fiber interface properties and on the matrix mechanical properties rather than on the concrete properties, at least for the concrete substrates considered in the available literature.

Some of the cited experimental works (Hashemi and Al-Mahaidi 2012a, b; Pareek et al. 2007; Pellegrino and D'Antino 2013; Ombres 2011; Elsanadedy et al. 2013) involved the direct measure of the fibres strain during the beams flexural tests. In other cases this strain has been evaluated by means of finite element (Hashemi and Al-Mahaidi 2012a, b; Elsanadedy et al. 2013) or analytical (D'Ambrisi and Focacci 2011) investigations. The strain measured at the beams failure are summarized in Fig. 9.26, versus the equivalent fibres thickness adopted in the FRCM materials; only the results of those specimens that failed due to debonding of the strengthening material are included in Fig. 9.26.

In order to include also an estimate of the maximum fibres strain reached during the experimental work where a direct strain measure is not included, a simple estimation is performed applying the equation

$$\varepsilon_{fe} = \frac{\Delta M_R}{0.9 \cdot H \cdot A_f E_f} \tag{9.7}$$

where ΔM_R is the FRCM contribution to the failure bending moment, evaluated as

$$\Delta M_R = \frac{\Delta F_R}{2} \cdot L_1 \tag{9.8}$$

ΔF_R is evaluated as shown in Fig. 9.27. It should be observed that Eq. (9.7) gives a rough estimate of the maximum fibres strain because it considers the assumptions: (i) the internal moment arm at failure is $0.9H$ and it does not depend on the geometrical parameters of the cross section and on the mechanical properties of the materials; (ii) the contribution of the steel reinforcement to the flexural capacity is

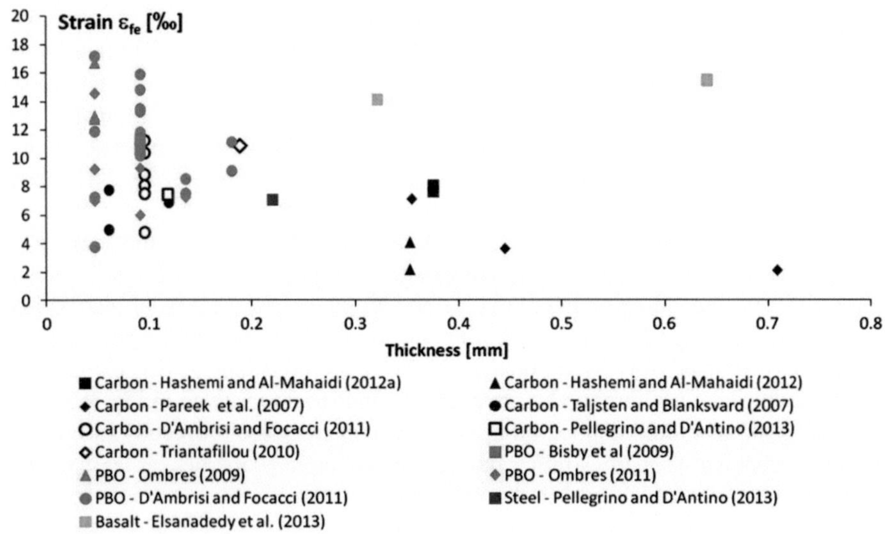

Fig. 9.26 Maximum fibres strain reached during the flexural tests of FRCM strengthened RC beams

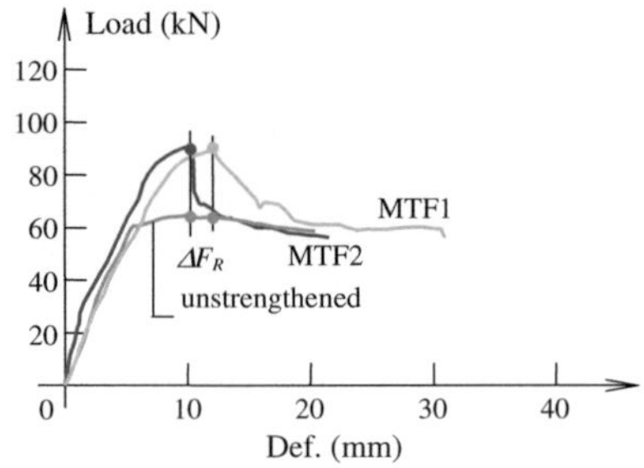

Fig. 9.27 Evaluation of ε_{fe} based on the experimental results of Hashemi and Al-Mahaidi (2012b)

the bending moment acting on the companion unstrengthened specimens at the same deflection. Nevertheless this approach allows for simply estimating the maximum fibres strain reached during the experimental tests. Moreover this estimated strain value has to be regarded as an average value on the fibre cross section; the phenomenon of the so-called telescopic failure, i.e. the successive fibre failure layer by layer from the outer to the inner fibres, (Badanoiu and Holmgren 2003; Banholzer et al. 2006; Häußler-Combe and Hartig 2007; Hegger et al. 2006; Zastrau et al. 2008) is not considered. However, the strain value is meaningful under the assumption that the debonding phenomenon is only related to the fibre local strain value. The experimental results reported in papers where a direct measure of ε_{fe} is performed were used to compare the measured and the analytically determined values of ε_{fe}. On average a difference of 17 % was found. As an example, Fig. 9.27 shows the determination of ε_{fe} based on the results described in Hashemi and Al-Mahaidi (2012a). In this case for the specimen MTF1 it results

$$\Delta F_R = 26.95\,\text{kN} \tag{9.9a}$$

$$\Delta M_R = \frac{\Delta F_R}{2} \cdot L_1 = 7.41\,\text{kN} \tag{9.9b}$$

Consequently

$$\varepsilon_{fe} = \frac{\Delta M_R}{0.9 \cdot H \cdot A_f E_f} = 7.62\,‰ \tag{9.10}$$

where $H = 180$ mm is the cross-section height, and $E_f = 200$ GPa and $A_f = 30$ mm^2 are the elastic modulus and cross section of the fibres, respectively. Similarly, considering the specimen MTF2 the calculated value is $\varepsilon_{fe} = 7.14$. The measured value reported by the authors of ε_{fe} is 7.65 ‰ for the specimens MTF.

The results represented in Fig. 9.26 are quite scattered both in the case of carbon (black symbols) and in the case of PBO (red symbols) fibers. Moreover, a clear dependence of the strain ε_{fe} on the fibres thickness cannot be observed. In the case of carbon fibre the average value of ε_{fe} reported in Fig. 9.26 is 7.00 ‰ with a standard deviation of 2.7 ‰ (coefficient of variation 0.37), while in the case of PBO fibre the average value of ε_{fe} reported in Fig. 9.26 is 10.85 ‰ with a standard deviation of 3.24 ‰ (coefficient of variation 0.30). In the cases of basalt and steel fibres a smaller number of experimental results is currently available. The average ε_{fe} for the basalt fibre is 15.05 ‰ while the only available value for the steel fibres is 7.1 ‰. Figure 9.28 shows the average fibres tensile stress at debonding, obtained multiplying the strains of Fig. 9.26 by the elastic modulus of the fibres. In this case the PBO fibres seems to be more effective due to the fibres elastic modulus (270 GPa), which is higher than the carbon (200–240 GPa), the steel (205 GPa), and the basalt (32 GPa) elastic moduli of the fibres.

The experimental results considered show that the debonding strain covers a wide range depending on the specific FRCM material (type of fibres, type of matrix, fibres/matrix coupling, and arrangement of fibres). It is therefore necessary that the manufacturer of the strengthening material furnishes the proper parameters for the evaluation of the debonding strain. Moreover further research is needed to assess mechanical models able to estabilish the correct bond strength hierarchy among the bond strengths related to the different possible debonding surfaces.

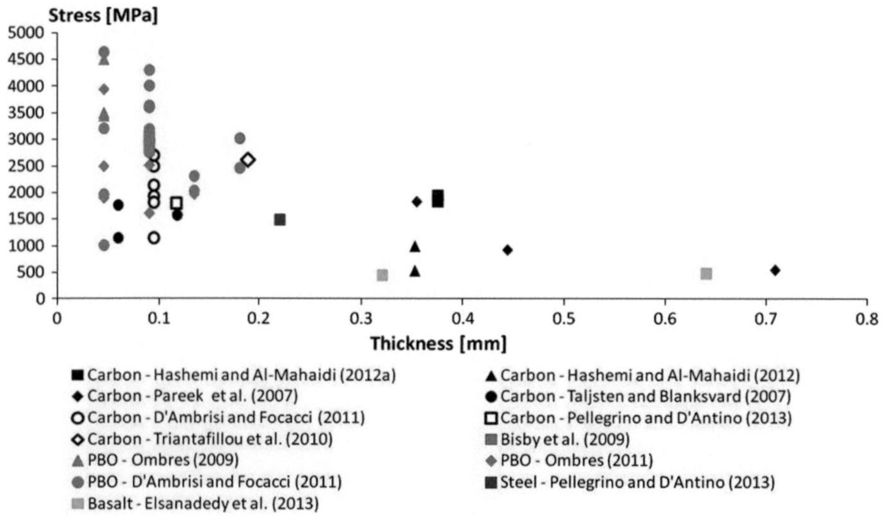

Fig. 9.28 Maximum fibres stress reached during the flexural tests of FRCM strengthened RC beams

Confinement

Confinement is generally applied to members in compression to enhance their axial load-carrying capacity, or to increase the deformation capacity of members subjected to seismic loading. Fibre-reinforced cementitious matrix composites, similarly to FRP, have an elastic behaviour and therefore exert an increasing confining action up to failure. This results in: (a) increasing concrete compressive strength and ultimate strain and (b) increasing the deformation capacity (ductility) of columns under seismic loading.

Jacketing of RC columns in existing structures is an increasingly attractive retrofit option. Among all jacketing techniques, the use of FRP has gained increasing popularity, due to the favourable properties possessed by these materials. However, certain problems associated with epoxy resins are still to be addressed. A solution of great potential, involving the combination of textiles with cement-based mortars for concrete confinement has been explored in very recent studies. TRM jackets have been investigated experimentally for confinement of plain or reinforced concrete and as a means of confining poorly detailed RC columns with limited deformation capacity under seismic loads. Comparison with FRP-retrofitted counterpart specimens allows for the evaluation of the effectiveness of TRM versus FRP jackets.

Behaviour of TRM-Confined Concrete in Axial Compression

Triantafillou et al. (2006) investigated experimentally for the first time the confinement effectiveness of various TRM jacketing schemes for plain concrete. That investigation was carried out on: (1) cylindrical specimens with a diameter of 150 mm and a height of 300 mm; (2) short column-type specimens with a rectangular cross section of 250 × 250 mm and a height of 700 mm. The four corners of all rectangular prisms were rounded at a radius equal to 15 mm. All specimens were unreinforced, as the jacket-reinforcement interactions were outside the scope of that study. Based on the response of confined cylinders, it was concluded that: (1) textile-mortar confining jackets provide substantial gain in compressive strength and deformability. This gain is higher as the number of confining layers increases and depends on the tensile strength of the mortar, which determines whether failure of the jacket will occur due to fibre fracture or debonding; (2) compared with their resin-impregnated counterparts, mortar-impregnated textiles may result in reduced effectiveness, in the order of approximately 80 % for strength and 50 % for ultimate strain, for the specific mortar used in that study; and (3) failure of mortar-impregnated textile jackets is less abrupt compared with that of their resin-impregnated counterparts, due to the slowly progressing fracture of individual fibre bundles. From the response of rectangular columns, it was concluded that mortar-impregnated textile jackets are quite effective in confining columns of rectangular cross sections for strength and axial deformability. In comparison with their epoxy-based counterparts, mortar-impregnated textile jackets provide approximately the same effectiveness in terms of strength and a slightly inferior one in terms of ultimate strain.

Bournas et al. (2007) went one step further by investigating experimentally the use of TRM jackets as a means of confining reinforced concrete. The experimental program aimed to compare the effectiveness of TRM versus FRP jackets as a measure of confining RC members. To examine this, 15 short RC prisms were tested under concentric compression. Specimens had a 200 mm ξ 200 mm cross section and a height of 380 mm. The prisms were divided in three series, with five specimens each. The first series comprised specimens with no internal steel reinforcement (Series U). The prisms in the second and third series were reinforced with four longitudinal 12 mm diameter bars placed at the corners of the cross section and with 8 mm diameter stirrups (Fig. 9.29).

The main interest in that study as far as the steel reinforcement is concerned was the spacing of stirrups. Hence, the second series comprised stirrups at a relatively large spacing of 200 mm (Series s200), to emulate old detailing practices. And in the last series (s100) the spacing was much smaller, equal to 100 mm, to represent current detailing practices. Each of the three series was comprised five different specimens: the control specimen (without wrapping), specimens wrapped with two or three layers of FRP and specimens wrapped with four or six layers of TRM. Note that the layers in the TRM-jacketed prisms were twice as many compared with their FRP counterparts, resulting in two "equivalent" confining systems that is with equal

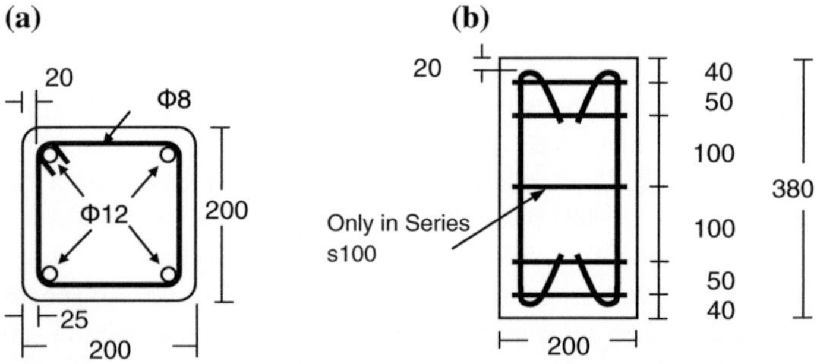

Fig. 9.29 **a** Cross section of prisms. **b** Configuration of reinforcement. (Dimensions in mm). (Bournas et al. 2007)

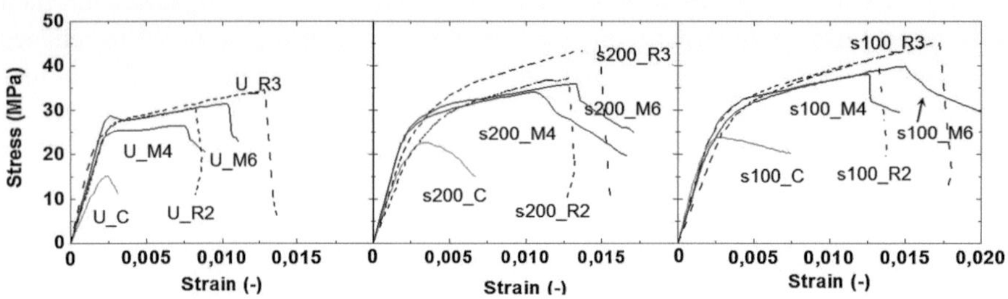

Fig. 9.30 Stress–strain curves for specimens: **a** without reinforcement; **b** with stirrups at 200 mm; **c** with stirrups at 100 mm (Bournas et al. 2007)

stiffness and strength in the circumferential direction. The notation of specimens is X_YN, where X refers to the internal steel reinforcement (U, s200, s100), Y denotes the type of jacket (C for the unjacketed—control—prisms, R for resin-based jackets and M for mortar-based jackets) and N denotes the number of layers.

The stress–strain plots recorded for all specimens are given in Fig. 9.30. All plots of the confined specimens are characterized by an ascending branch, followed by a second one, close to linear, that drops at a point where the jacket fractured due to hoop stresses (Fig. 9.31a); it is this point where peak stress and ultimate strain is defined, except for the control (unjacketed) specimens, where ultimate strain is defined conventionally at 15 % peak stress reduction. In some of the TRM-jacketed prisms (s200_M4, s100_M6) fracture of the fibres was accompanied by debonding at the end of the lap (Fig. 9.31b). In several occasions jacket rupture occurred simultaneously with bar buckling. Hence, failure of the jackets was due to stretching both by concrete dilation and by the outwards bending of the longitudinal bars in the middle of the specimens (Fig. 9.31c). Another important aspect of the response is that, contrary to FRP jackets, TRM jackets, did not fail abruptly since TRM jackets fracture initiates from a limited number of fibre bundles (when the

(a) (b) (c)

Fig. 9.31 **a** Fracture of jacket. **b** Debonding at the end of the lap. **c** Buckling of bar at corner (Bournas et al. 2007)

hoop stresses reach their tensile capacity) and then propagates rather slowly in the neighbouring bundles, resulting in a more ductile failure mechanism compared to FRP.

Overall, the concentric compression tests performed by Bournas et al. (2007) on RC prisms show that TRM confining jackets provide substantial gain in compressive strength and ultimate strain; this gain increases with the volumetric ratio of the TRM wrap. Compared with equal stiffness and strength FRP jackets, the TRM jackets used in Bournas et al. (2007) are slightly less effective in terms of increasing strength and deformation capacity by about 10 %. This reduction in effectiveness did not seem to depend on the volumetric ratio of the embedded stirrup reinforcement.

Behaviour of TRM-Confined Columns Under Cyclic Loading

More recently Bournas et al. (2009) investigated experimentally the use of TRM jackets as a means of confining poorly detailed old-type RC columns, which suffer from limited deformation capacity under seismic loads due to either buckling of the longitudinal bars or bond failure at lap splice regions. A total of 10 large-scale old-type RC columns were tested under cyclic uniaxial flexure with constant axial load (Fig. 9.32a). In Bournas et al. (2009) four specimens were constructed with continuous longitudinal reinforcement (Series L0). One specimen was tested without retrofitting, as control (L0_C), the second one was retrofitted with a double-layered CFRP jacket (specimen L0_R2), the third one was retrofitted with an equal (to its FRP counterpart) stiffness and strength carbon fibre TRM jacket comprising four layers (specimen L0_M4), and the last specimen was retrofitted

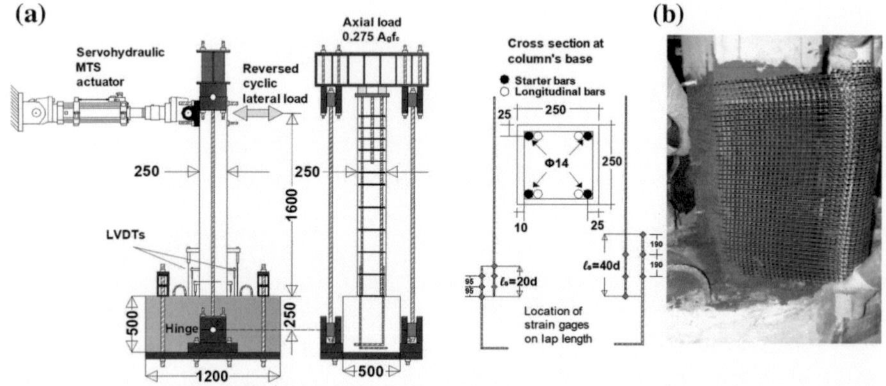

Fig. 9.32 **a** Schematic of test setup and cross section of the columns. **b** Application of a TRM jacket

with a lower stiffness and strength four-layered glass fibre TRM jacket (specimen L0_M4G), which represents a rehabilitation solution of lower cost in comparison with specimen L0_R2 and L0_M4.

The effectiveness of TRM versus FRP jackets, applied at the ends of old-type RC columns for specimens constructed with lap splicing of longitudinal reinforcement above the column base, was evaluated for two different lap lengths, which were selected equal to 20 and 40 bar diameters, as shown in Fig. 9.32a. In summary, the notation of specimens is LX_YN, where X defines the lap splice length above the column base (0 for continuous reinforcement, 20d for a lap splice length of 20 rebar diameters, 40d for a lap splice length of 40 rebar diameters), Y denotes the type of jacket (C for the unjacketed—control—columns, R for resin-based jackets, and M for mortar-based jackets) and N denotes the number of layers. For the specimen strengthened with a glass fibre TRM jacket the letter G was added after letter N. The jackets extended from the base of each column to a height of 430 mm except for the two columns with longer lap splices (L40d_R2 and L40d_M4) where the jackets were extended to a height of 600 mm. The overlapping length of the jacket was equal to 150 mm. Prior to jacketing, the four corners of the columns which received jacketing were rounded at a radius equal to 25 mm. A photograph of the application method of textile combined with mortar binder to provide jacketing in one of the specimens used in Bournas et al. (2009) is shown in Fig. 9.32b.

The response of the columns tested in Bournas et al. (2009) is given in Fig. 9.33 in the form of load-drift ratio loops. Key results are also presented in Table 9.2, which includes: (a) The peak resistance in the two directions of loading. (b) The drift ratio corresponding to peak resistance in the two directions of loading. (c) The drift ratio at conventional "failure" of the column, defined as reduction of peak resistance in a cycle below 80 % of the maximum recorded resistance in that direction of loading. (d) The observed failure mode.

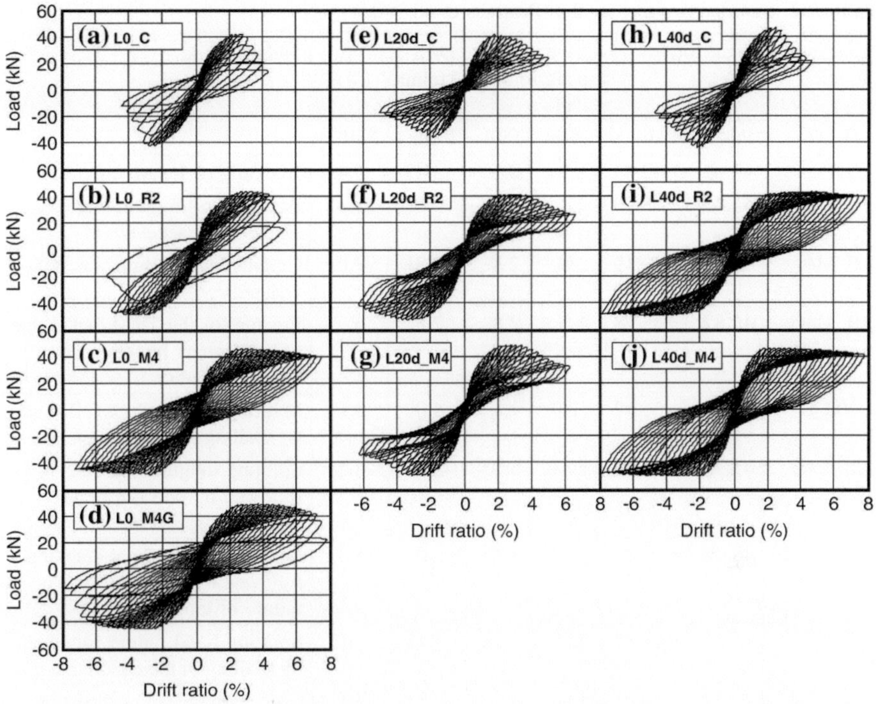

Fig. 9.33 Load versus drift ratio curves for all specimens tested in Bournas et al. (2009)

The failure mode of the unretrofitted specimen with continuous bars was controlled by bar buckling (Fig. 9.34a) with a sustained drift ratio at failure of 3.43 %.

The confinement provided by the FRP jacket to specimen L0_R2 restrained the outward bending of longitudinal bars inside the FRP jacket region, but bar buckling finally occurred above the FRP jacket (Fig. 9.34b) at a drift equal to 5.15 %. Contrary to specimen L0_R2, rebar buckling in columns L0_M4 and L0_M4G developed gradually inside the TRM-jacketed area, as the compressive force released from early buckled bars was carried by the surrounded confined concrete inside the jackets. This is possible to occur in this confining system, as TRM jackets are able to deform outwards without early fibre rupture, due to the low composite action between fibres and mortar, which allows for higher local deformations (e.g. slip of fibres within rovings). In specimen L0_M4 the carbon fibre TRM jacket remained intact until the test was terminated at drift ratio equal to 7.81 % (Fig. 9.34c), while in specimen L0_M4G fracture of the glass fibre TRM jacket (at a drift ratio equal to 7.2 %) led to failure (Fig. 9.34d).

The failure mode of the control specimens with lap splices was characterized by splitting cracks, developed along the splice length of lapped bars, for both unretrofitted specimens L20d_C and L40d_C, respectively. The drift ratio at failure sustained by unretrofitted columns L20d_C and L40d_C was 3.59 and 3.28 %,

Table 9.2 Summary of test results (Bournas et al. 2009)

Specimen notation	Peak force (kN)		Drift at peak force (%)		Drift at "failure" (%)		Failure mode
	Push	Pull	Push	Pull	Push	Pull	
L0_C	41.63	−42.48	2.5	2.5	3.43	3.43	Buckling of longitudinal bars
L0_R2	43.46	−48.70	2.8	3.1	5.0	−5.31	Buckling of longitudinal bars above FRP jacket
L0_M4	45.77	−49.19	2.8	2.8	>7.81	>7.81	Conventional failure was not reached
L0_M4G	48.82	−45.28	4.0	2.8	7.5	6.9	Fracture of the TRM jacket due to both rebars buckling and concrete dilation
L20d_C	41.50	−36.62	1.87	1.87	4.06	3.12	Splitting bond failure followed by spalling of the concrete cover
L20d_R2	41.26	−52.86	2.81	3.12	5.31	6.25	Splitting longitudinal cracking followed by pull out bond failure of lapped bars
L20d_M4	48.46	−49.80	3.12	2.18	5.0	5.0	Splitting longitudinal cracking followed by pull out bond failure of lapped bars
L40d_C	46.26	−43.82	2.5	2.18	3.43	3.12	Splitting bond failure followed by spalling of the concrete cover
L40d_R2	42.97	−49.93	4.68	5.0	>7.81	>7.81	Conventional failure was not reached
L40d_M4	45.90	−50.48	1.87	3.75	>7.81	>7.81	Conventional failure was not reached

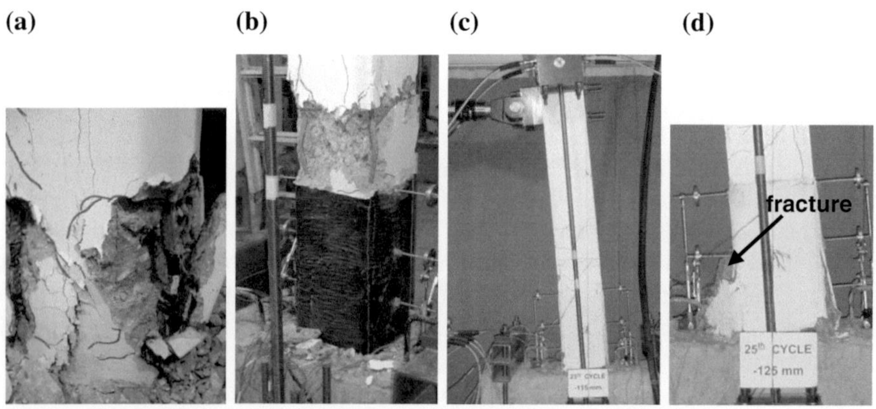

Fig. 9.34 **a** Disintegration of concrete and bar buckling. **b** Buckling of longitudinal bars above the FRP jacket. **c** Undamaged carbon TRM jacket at end of test. **d** Fracture of glass TRM jacket due to bar buckling

respectively. TRM and FRP jacketed columns, with either short or long lap length, responded far better than their unretrofitted counterparts. Confinement provided sufficient resistance against splitting cracks and lateral expansion of concrete. Specimens L20d_R2 and L20d_M4 (with short lap lengths) sustained reversed deformation cycles up to 6.3 % drift at failure. The mean strength increase for columns L20d_R2 and L20d_M4 was 20.3 and 25.6 %, respectively, in comparison with the control specimen (L20d_C), while the corresponding increase in deformation capacity was 64.7 and 38.8 %, respectively. Columns with longer lap splices (L40d_R2 and L40d_M4) behaved in an identical manner until the end of the test at a drift ratio of 7.81 % (maximum stroke of piston was reached), resulting in an increase of the member deformation capacity by a factor of more than 2.5. Peak resistance was practically the same as in the unretrofitted column (L40d_C).

The tests on columns with continuous longitudinal reinforcement (Series L0) show that TRM jackets are quite effective as a means of increasing the cyclic deformation capacity and the energy dissipation of old-type RC columns with poor detailing, by delaying bar buckling. Compared with equal stiffness and strength FRP, TRM jacketing has a higher effectiveness by about 50 %. From the tests on columns with lap-spliced longitudinal reinforcement (Series L20d and L40d), it may be concluded that TRM confining jackets provide substantial gain in lateral strength and deformation capacity of cyclically loaded reinforced concrete columns with lap splices at the base of the column. Compared with equal stiffness and strength FRP jackets, they are characterized by a slightly reduced effectiveness in terms of deformation capacity for columns with short lap splices and with the same effectiveness for columns with longer lap lengths. Detailed results on the bond strength of lap-spliced bars and bar buckling in TRM confined concrete are presented in Bournas et al. (2011a, b) and Bournas and Triantafillou (2011).

Deformation Capacity of TRM-Confined RC Members

The cyclic deformation capacity of RC columns, a key property in displacement-based design used in seismic rehabilitation applications, is typically expressed through the attained drift ratio of the members at failure. This important parameter for all specimens tested is compared in this section with predictions given by Eurocode 8 (EN 1998-3 2005). The drift ratio, which is defined as chord rotation capacity at ultimate in EN 1998-3 (2005), is given by the following empirical expression:

$$\theta_u = k0.016(0.3^{v})\left[\frac{\max(0.01,\omega')}{\max(0.01,\omega)}f_c\right]^{0.225}\left(\frac{L_V}{h}\right)^{0.35}25^{c}\left(1.25^{100\rho_d}\right) \qquad (9.11)$$

where f_c = compressive strength of concrete (MPa); ω and ω' = mechanical reinforcement ratio of tension and compression longitudinal reinforcement, respectively; $v = N/bhf_c$ = normalized axial force (compression taken as positive); b = width of compression zone; h = cross section side parallel to the loading direction; $L_V = M/V$ = moment-to-shear ratio at the end section; $c = \alpha\rho_{sx}f_{yw}/f_c$; $\rho_{sx} = A_{sw}/bs_h$ = transverse steel ratio parallel to the direction x of loading; f_{yw} = yield stress of stirrups; s_h = spacing of stirrups; A_{sw} = area of transverse steel reinforcement parallel to the direction x within s_h; $k = 0.825$ for columns with deformed bars, without detailing for earthquake resistance; ρ_d = geometric ratio of diagonal reinforcement, if any; and α = effectiveness coefficient for confinement with stirrups.

If a column is retrofitted with an FRP or TRM jacket in the plastic hinge region, it is logical: (a) to take k equal to 1 instead of 0.825, as the lack of detailing for earthquake resistance has been compensated by the external confinement; and (b) to adopt the expression in Eq. (9.11) with c given by the sum of two terms: one to account for the contribution of stirrups and a second one to account for the contribution of the jacket, as follows (Bournas et al. 2007):

$$c = a\rho_{sx}\frac{f_{yw}}{f_c} + a_f\rho_{fx}\frac{f_{fe}}{f_c} \tag{9.12}$$

where $\rho_{fx} = 2nt_f/b$; n = number of layers of the fibre sheet or textile; t_f = thickness of one fibre sheet or textile layer; f_{fe} = effective stress of jacket at conventional failure of the column; and α_f = effectiveness coefficient for confinement with fibres (TRM or FRP jackets), equal to:

$$\alpha_f = \beta\left[1 - \frac{(b-2R)^2+(h-2R)^2}{3bh}\right] \tag{9.13}$$

where R = radius at corners of the cross section. The coefficient β in Eq. (9.13) accounts for the reduced or enhanced effectiveness of TRM versus FRP jackets in terms of ultimate strain. On the basis of concentric compression tests on reinforced concrete prisms presented in Triantafillou et al. (2006) and summarized above, this value is about 0.9. But if jacket failure has not been reached at conventional failure of the column, no reduction (nor enhancement) should be made and β should be taken equal to 1 (i.e. specimen L0_M4). It should be noted that in view of the relatively limited experimental database on TRM jacket failures, this value of β should be carefully selected. Other materials (e.g. different mortars) may result in different values for the effectiveness of TRM versus FRP. Therefore for such novel materials much more experimental work is needed to propose design values of β.

References

American Concrete Institute (ACI). (2008). Guide for the design and construction of externally bonded FRP systems for strengthening concrete structures. ACI 440.2R-08, Farmington Hills, MI.

American Concrete Institute (ACI). (2013). Guide to Design and Construction of Externally Bonded Fabric-Reinforced Cementitious Matrix (FRCM) Systems for Repair and Strengthening Concrete and Masonry Structures. ACI 549.4R-13, Farmington Hills, MI.

ASTM. (1996) Standard test method for tensile properties of polymer matrix composite materials. D3039/D3039 M, ASTM International, 6 pages.

ASTM. (2011a) Standard test method for compressive strength for cylindrical concrete specimens. C39/C39 M-12, ASTM International, 7 pages.

ASTM. (2011b) Standard test method for splitting tensile strength of cylindrical concrete specimens. C496/C496 M, ASTM International, 5 pages.

Badanoiu, A., & Holmgren, J. (2003). Cementitious composites reinforced with continuous carbon fibers for strengthening of concrete structures. *Cement & Concrete Composites, 25*, 387–394.

Banholzer, B. (2004). *Bond behaviour of a multi-filament yarn embedded in a cementitious matrix*. Ph.D. Thesis, RWTH Aachen University.

Banholzer, B., Brockmann, T., & Brameshuber, W. (2006). Material and bonding characteristics for dimensioning and modeling of textile reinforced concrete (TRC) elements. *Materials and Structures, 39*, 749–763.

Bisby, L. A., Roy, E. C., Ward, M., & Stratford, T. J. (2009). Fibre reinforced cementitious matrix systems for fire-safe flexural strengthening of concrete: Pilot testing at ambient temperature. In *Proceedings Advanced Composites in Construction, Edinburgh, UK.*

Bournas, D., Lontou, P., Papanicolaou, C. G., & Triantafillou, T. C. (2007). Textile-Reinforced Mortar (TRM) versus FRP Confinement in Reinforced Concrete Columns. *ACI Structural Journal, 104*(6), 740–748.

Bournas, D. A., & Triantafillou, T. C. (2011a). Bar Buckling in RC Columns Confined with Composite Materials. *Journal of Composites for Construction, 15*(3), 393–403.

Bournas, D. A., & Triantafillou, T. C. (2011b). Bond Strength of Lap Spliced Bars in Concrete Confined with Composite Materials. *Journal of Composites for Construction, 15*(2), 156–167.

Bournas, D. A., Triantafillou, T. C., Zygouris, K., & Stavropoulos, F. (2009). Textile-reinforced mortar (TRM) versus FRP jacketing in seismic retrofitting of RC columns with continuous or lap-spliced deformed bars. *Journal of Compososites for Construction, 13*(5), 360–371.

Brückner, A., Ortlepp, R., & Curbach, M. (2006). Textile reinforced concrete for strengthening in bending and shear. *Materials and Structures, 39*, 741–748.

Carloni, C., D'Antino, T., Sneed, L. H., & Pellegrino, C. (2014). Role of the matrix layers in the stress-transfer mechanism of FRCM composites bonded to a concrete substrate. *Journal of Engineering Mechanics, ASCE.* doi:10.1061/(ASCE)EM.1943-7889.0000883.

Carloni, C., Sneed, L. H., & D'Antino, T. (2013). Interfacial bond characteristics of fiber reinforced cementitious matrix for external strengthening of reinforced concrete members. In *Proceedings of the 8th international Conference on Fracture Mechanics of Concrete and Concrete Structures (FraMCoS-8), Toledo (Spain)* (pp. 129–137).

Carloni, C., & Subramaniam, K. V. (2012). Application of fracture mechanics to debonding of FRP from RC members. ACI SP-286: 10-1-10-14.

Carozzi, F. G., & Poggi, C. (2015). Mechanical properties and debonding strength of Fabric Reinforced Cementitious Matrix (FRCM) systems for masonry strengthening. *Composites: Part B, 70*, 215–230.

Contamine, R., Si LArbi, A., & Hamelin, P. (2011). Contribution to direct tensile testing of textile reinforced concrete (TRC) composites, *Material Science and Engineering A, 528,* 8589–8598.

Curbach, M., Ortlepp, R., & Triantafillou, T. C. (2006). TRC for rehabilitation. In W. Brameshuber (Ed.), *Textile Reinforced Concrete* (pp. 221–236), RILEM Report 36.

D'Ambrisi, A., & Focacci, F. (2011). Flexural strengthening of RC beams with cement based composites. *Journal of Composites for Construction, 15*(2), 707–720.

D'Ambrisi, A., Feo, L., & Focacci, F. (2012). Bond-slip relations for PBO-FRCM materials externally bonded to concrete. *Compos Part B*, *43*(8), 2938–2949.

D'Ambrisi, L., Focacci, F., & Caporale, A. (2013a). Strengthening of masonry–unreinforced concrete railway bridges with PBO-FRCM materials. *Composite Structures*, *102*, 193–204.

D'Ambrisi, A., Feo, L., & Focacci, F. (2013b). Experimental analysis on bond between PBO-FRCM strengthening materials and concrete. *Composites—Part B, 44*(1), 524–532.

D'Ambrisi, A., Feo, L., & Focacci, F. (2013c). Experimental and analytical investigation on bond between carbon-FRCM materials and masonry. *Compos Part B*, *46*, 15–20.

D'Ambrisi, A., Focacci, F., Luciano, R., Alecci, V., & De Stefano, M. (2015). Carbon-FRCM materials for structural upgrade of masonry arch road bridges. *Compos Part B*, *75*, 355–366.

D'Antino, T., Carloni, C., Sneed, L. H., & Pellegrino, C. (2014). Matrix-fiber bond behavior in PBO FRCM composites: a fracture mechanics approach. *Engineering Fracture Mechanics, 117*, 94–111.

D'Antino, T., Carloni, C., Sneed, L. H., & Pellegrino, C. (2015). Fatigue and Post-fatigue Behavior of PBO FRCM-Concrete Joints. *International Journal of Fatigue*. doi:10.1016/j.ijfatigue.2015.06.008.

D'Antino, T., Sneed, L.H., Carloni, C., & Pellegrino, C. (2013). Bond behavior of the FRCM-concrete interface. In *Proceedings of 11th international symposium on fiber reinforced polymer reinforcement for concrete structures (FRPRCS-11), Guimaraes, Portugal.*

Elsanadedy, H. M., Almusallam, T. H., Alsayed, S. H., & Al-Salloum, Y. A. (2013). Flexural strengthening of RC beams using textile reinforced mortar—Experimental and numerical study. *Composite Structures, 97*, 40–55.

Féderation Internationale du Béton (FIB) (2001). Externally bonded FRP reinforcement for RC structures. Bulletin 14, Lausanne, Switzerland.

Hartig, J., Häuβler-Combe, U., & Schicktanz, K. (2008). Influence of bond properties on the tensile behaviour of Textile Reinforced Concrete. *Cement & Concrete Composites, 30*, 898–906.

Hartig, F., Jesse, F., Schicktanz, K., & Häußler-Combe, U. (2012). Influence of experimental setups on the apparent uniaxial tensile load-bearing capacity of textile reinforced concrete specimens. *Materials and Structures, 45*, 433–446.

Hashemi, S., & Al-Mahaidi, R. (2008). Cement based bonding material for FRP strengthening of RC structures. In *Proceedings of 11th International Inorganic-Bonded Fiber Composites Conference (IIBC), Toledo, Spain.*

Hashemi, S., & Al-Mahaidi, R. (2012a). Experimental and finite element analysis of flexural behavior of FRP-strengthened RC beams using cement-based adhesives. *Construction and Building Materials, 26*, 268–273.

Hashemi, S., & Al-Mahaidi, R. (2012b). Investigation of flexural performance of RC beams strengthened with CFRP textile and cement based adhesives. In *Proceedings of the 3rd Asia-Pacific Conference on FRP in Structures (APFIS 2012)*, Hokkaido, Japan.

Häuβler-Combe, U., & Hartig, J. (2007). Bond and failure mechanisms of textile reinforced concrete (TRC) under uniaxial tensile loading. *Cement and Concrete Composites, 29*, 297–289.

Hegger, J., Will, N., Bruckermann, O., & Voss, S. (2006). Load-bearing behavior and simulation of textile reinforced concrete. *Materials and Structures, 39*, 765–776.

ICC Evaluation Service. (2013). Acceptance Criteria for Masonry and Concrete Strengthening Using Fiber-reinforced Cementitious Matrix (FRCM) Composite Systems. AC434-2013.

Italian National Research Council (CNR). (2004). Guide for the design and construction of externally bonded FRP systems for strengthening existing structures. CNR-DT200/2004, Rome.

Jesse, F., Schicktanz, K., & Curbach, M. (2009). Obtaining Characteristic Material Strength of Textile Reinforced Concrete (TRC) from Laboratory Tests. In *Proceedings of the 9th International Symposium on Ferrocement (Ferro9), Bali* (pp. 305–318.

Jesse, F., Weiland, S., Curbach, M. (2005). Flexural strengthening of rc-structures with textile reinforced concrete, Textile Reinforced Concrete (TRC)—German/International Experience. ACI Special Publication, SP-250CD-4.

Kolsch, H. (1998). Carbon fiber cement matrix overlay system for masonry strengthening. *Journal of Composites for Construction, 2*(2), 105–109.

Lignola, G. P., Prota, A., & Manfredi, G. (2009). Nonlinear Analyses of tuff masonry walls strengthened with cementitious matrix grid composites. *Journal of Composites for Construction, 14*(4), 243–251.

Ombres, L. (2009). Failure modes in reinforced concrete beams strengthened with PBO fiber reinforced mortars. In *Proceedings of Fiber-Reinforced Polymer Reinforcement for Concrete Structures, FRPRCS-9, Sydney, Australia.*

Ombres, L. (2011). Flexural analysis of reinforced concrete beams strengthened with a cement based high strength composite material. *Composite Structures, 94*(1), 143–155.

Ombres, L. (2012). Debonding analysis of reinforced concrete beams strengthened with fibre reinforced cementitious mortar. *Engineering Fracture Mechanics, 81*, 94–109.

Ortlepp, R., Hampel, U., & Curbach, M. (2006). A new approach for evaluating bond capacity or TRC strengthening. *Cement & Concrete Composites, 28*, 589–597.

Ortlepp, R., Ortlepp, S., & Curbach, M. (2004). Stress transfer in the bond joint of subsequently applied Textile Reinforced Concrete strengthening. In *Proceedings of Fibre-Reinforced Concretes (FRC), RILEM, PRO 39, Varenna, Italy.*

Pareek, S., Suzuki, Y., & Kobayashi, A. (2007). Flexural and shear strengthening of RC beams using newly developed CFRP and polymer-cement pastes as bonding agents. In *Proceedings of Fiber-Reinforced Polymer Reinforcement for Concrete Structures, FRPRCS-8, Patras, Greece.*

Parisi, F., Lignola, G. P., Augenti, N., Prota, A., & Manfredi, G. (2011). Nonlinear Behavior of a Masonry Subassemblage Before and After Strengthening with Inorganic Matrix-Grid Composites. *Journal of Composites for Construction, 15*(5), 821–832.

Peled, A., Zaguri, E., & Marom, G. (2008). Bonding characteristics of multifilament polymer yarns and cement matrices. *Composites: Part A, 39*, 930–939.

Pellegrino, C., & D'Antino, T. (2013). Experimental behavior of existing precast prestressed reinforced concrete elements strengthened with cementitious composites. *Composites: Part B, 55*, 31–40.

Pellegrino, C., Tinazzi, D., & Modena, C. (2008). An experimental study on bond behavior between concrete and FRP reinforcement. *Journal of Compososites for Construction, 12*(2), 180–189.

Prota, A., Marcari, G., Fabbrocino, G., Manfredi, G., & Aldea, C. (2006). Experimental in-plane behavior of tuff masonry strengthened with cementitious matrix grid composites. *Journal of Composites for Construction, 10*(3), 223–233.

Smith, S. T., & Teng, J. G. (2002). FRP-strengthened RC beams-II: assessment of debonding strength models. *Engineering Structures, 24*(4), 397–417.

Sneed, L. H., D'Antino, T., & Carloni, C. (2014). Investigation of bond behavior of PBO fiber-reinforced cementitious matrix composite—concrete interface. *ACI Materials Journal, 111*(1–6), 1–12.

Sneed, L. H., D'Antino, T., Carloni, C., & Pellegrino, C. (2015). A comparison of the bond behavior of PBO-FRCM composites determined by single-lap and double-lap shear tests. *Cement and Concrete Composites*. doi:10.1016/j.cemconcomp.2015.07.007.

Soranakom, C., & Mobasher, B. (2009). Geometrical and mechanical aspects of fabric bonding and pullout in cement composites. *Materials and Structures, 42*, 765–777.

Subramaniam, K.V., Carloni, C., & Nobile, L. (2007). Width effect in the interface fracture during debonding of FRP from concrete. *Engineering Fracture Mechanics, 74*, 578–594.

Subramaniam, K.V., Carloni, C., & Nobile, L. (2011). An understanding of the width effect in FRP-concrete debonding. *Strain, 47*, 127–137.

Täljsten, B., & Blanksvärd, T. (2007). Mineral-Based bonding of Carbon FRP to strengthen concrete structures. *Journal of Composites for Construction, 11*(2), 120–128.

Triantafillou, T. C. (2010). Innovative Textile-based Composites for Strengthening and Seismic Retrofitting of Concrete and Masonry Structures. In *Proceedings of the 5th International Conference on FRP Composites in Civil Engineering, CICE-2010, Beijing, China.*

Triantafillou, T.C., Papanicolaou, C.G., Zissimopoulos, P., & Laourdekis, T. (2006). Concrete confinement with textile-reinforced mortar jackets. *ACI Structural Journal, 103*(1), 28–37.

UNI EN 1015-11. (2007). Methods of test for mortar for masonry—Part 11: Determination of flexural and compressive strength of hardened mortar, Comité Européen de Normalisation, Brussels, Belgium.

UNI EN 1998-3. (2005). Eurocode 8: Design of Structures for Earthquake Resistance—Part 3: Assessment and Retrofitting of Buildings, European Committee for Standardization, Brussels, Belgium.

Weiland, S., Ortlepp, R., & Curbach, M. (2006). Strengthening of preformed slabs with textile reinforced concrete. In *Proceedings of the 2nd International fib Congress, Naples, Italy.*

Wiberg, A. (2003). *Strengthening of Concrete Beams Using Cementitious Carbon Fibre Composites*. Doctoral Thesis, Royal Institute of Technology, Stockholm, Sweden.

Wu, H. C., & Sun, P. (2005). Fiber reinforced cement based composite sheets for structural retrofit. In *Proceedings of the International Symposium on Bond Behaviour of FRP in Structures, BBFS 2005, Hong Kong, China.*

Zastrau, B., Lepenies, I., & Richter. M. (2008). On the multi scale modeling of textile reinforced concrete. *Technische Mechanik, 28*, 53–63.

MIX
Papier aus verantwortungsvollen Quellen
Paper from responsible sources
FSC® C105338

If you have any concerns about our products,
you can contact us on
ProductSafety@springernature.com

In case Publisher is established outside the EU,
the EU authorized representative is:
Springer Nature Customer Service Center GmbH
Europaplatz 3, 69115 Heidelberg, Germany

Printed by Libri Plureos GmbH
in Hamburg, Germany